Metagenomic Analysis for Unveiling Agricultural Microbiome

Editors

Yong-Xin Liu
Peng Yu

MDPI

Basel • Beijing • Wuhan • Barcelona • Belgrade • Novi Sad • Cluj • Manchester

Editors

Yong-Xin Liu
Food Science Center
Agricultural Genomics
Institute at Shenzhen
Shenzhen
China

Peng Yu
Institute of Crop Science and
Resource Conservation
University of Bonn
Bonn
Germany

Editorial Office
MDPI
St. Alban-Anlage 66
4052 Basel, Switzerland

This is a reprint of articles from the Special Issue published online in the open access journal *Agronomy* (ISSN 2073-4395) (available at: www.mdpi.com/journal/agronomy/special_issues/S70BHD8E5H).

For citation purposes, cite each article independently as indicated on the article page online and as indicated below:

Lastname, A.A.; Lastname, B.B. Article Title. *Journal Name* **Year**, *Volume Number*, Page Range.

ISBN 978-3-7258-1322-3 (Hbk)
ISBN 978-3-7258-1321-6 (PDF)
doi.org/10.3390/books978-3-7258-1321-6

Contents

About the Editors . . . . **vii**

Yao Wang, Peng Yu and Yong-Xin Liu
Metagenomic Analysis for Unveiling Agricultural Microbiome
Reprinted from: *Agronomy* **2024**, *14*, 981, doi:10.3390/agronomy14050981 . . . . **1**

Jiao Xi, Zanbo Ding, Tengqi Xu, Wenxing Qu, Yanzhi Xu and Yongqing Ma et al.
Maize Rotation Combined with *Streptomyces rochei* D74 to Eliminate *Orobanche cumana* Seed Bank in the Farmland
Reprinted from: *Agronomy* **2022**, *12*, 3129, doi:10.3390/agronomy12123129 . . . . **5**

Wen Luo, Yuanli Li, Ruiping Luo, Gehong Wei, Yongxin Liu and Weimin Chen
Dodder Parasitism Leads to the Enrichment of Pathogen *Alternaria* and Flavonoid Metabolites in Soybean Root
Reprinted from: *Agronomy* **2023**, *13*, 1571, doi:10.3390/agronomy13061571 . . . . **21**

Mao Li, Xuejuan Zi, Rong Sun, Wenjun Ou, Songbi Chen and Guanyu Hou et al.
Co-Ensiling Whole-Plant Cassava with Corn Stalk for Excellent Silage Production: Fermentation Characteristics, Bacterial Community, Function Profile, and Microbial Ecological Network Features
Reprinted from: *Agronomy* **2024**, *14*, 501, doi:10.3390/agronomy14030501 . . . . **38**

Zilia Y. Muñoz-Ramírez, Román González-Escobedo, Graciela D. Avila-Quezada, Obed Ramírez-Sánchez, Victor M. Higareda-Alvear and Emiliano Zapata-Chávez et al.
Exploring Microbial Rhizosphere Communities in Asymptomatic and Symptomatic Apple Trees Using Amplicon Sequencing and Shotgun Metagenomics
Reprinted from: *Agronomy* **2024**, *14*, 357, doi:10.3390/agronomy14020357 . . . . **55**

Li Wen, Fengqiu Huang, Zhongxiu Rao, Kaikai Cheng, Yong Guo and Haiming Tang
Paddy-*Lilium* Crop Rotation Improves Potential Beneficial Soil Fungi and Alleviates Soil Acidification in *Lilium* Cropping Soil
Reprinted from: *Agronomy* **2024**, *14*, 161, doi:10.3390/agronomy14010161 . . . . **72**

Lidong Ji, Xing Xu, Fengju Zhang, Haili Si, Lei Li and Guilian Mao
The Preliminary Research on Shifts in Maize Rhizosphere Soil Microbial Communities and Symbiotic Networks under Different Fertilizer Sources
Reprinted from: *Agronomy* **2023**, *13*, 2111, doi:10.3390/agronomy13082111 . . . . **87**

Junbo Zhou, Zhenjiang Jin, Wu Yuan, Weijian Chen, Xuesong Li and Liyuan Xiong et al.
Microbial Communities and Soil Respiration during Rice Growth in Paddy Fields from Karst and Non-Karst Areas
Reprinted from: *Agronomy* **2023**, *13*, 2001, doi:10.3390/agronomy13082001 . . . . **103**

Fan Chang, Feng-An Jia, Min Guan, Qing-An Jia, Yan Sun and Zhi Li
Effects of Transplantation and Microhabitat on Rhizosphere Microbial Communities during the Growth of American Ginseng
Reprinted from: *Agronomy* **2023**, *13*, 1876, doi:10.3390/agronomy13071876 . . . . **122**

Qiuling Pang, Mohammad Murtaza Alami, Weilong Yu, Zhen Ouyang, Shaohua Shu and Daiqun Tu et al.
A Meta-Analysis in Nine Different Continuous Cropping Fields to Find the Relationship between Plant Species and Rhizosphere Fungal Community
Reprinted from: *Agronomy* **2023**, *13*, 1827, doi:10.3390/agronomy13071827 . . . . **141**

Lijun Wu, Yan Ren, Xiangsong Zhang, Guanghui Chen, Chuantang Wang and Qi Wu et al.
Effect of Root-Knot Nematode Disease on Bacterial Community Structure and Diversity in Peanut Fields
Reprinted from: *Agronomy* **2023**, *13*, 1803, doi:10.3390/agronomy13071803 . . . . **158**

Rui Liu, Yang Liu, Yuan Gao, Fazhu Zhao and Jun Wang
The Nitrogen Cycling Key Functional Genes and Related Microbial Bacterial Community $\alpha-$Diversity Is Determined by Crop Rotation Plans in the Loess Plateau
Reprinted from: *Agronomy* **2023**, *13*, 1769, doi:10.3390/agronomy13071769 . . . . **177**

Fan Yang, Huayan Jiang, Shen Liang, Gaozheng Chang, Kai Ma and Lili Niu et al.
High-Throughput Sequencing Reveals the Effect of the South Root-Knot Nematode on Cucumber Rhizosphere Soil Microbial Community
Reprinted from: *Agronomy* **2023**, *13*, 1726, doi:10.3390/agronomy13071726 . . . . **192**

Jian Xiao, Jianglin Zhang, Yajie Gao, Yanhong Lu, Xue Xie and Changyu Fang et al.
Long-Term Chemical Fertilization Drove Beneficial Bacteria for Rice Soil to Move from Bulk Soil to the Rhizosphere
Reprinted from: *Agronomy* **2023**, *13*, 1645, doi:10.3390/agronomy13061645 . . . . **206**

Beibei Zhou, Ruirui Chen, Shuang Peng, Jianwei Zhang, Xiangui Lin and Yiming Wang
Variations in Methanogenic and Methanotrophic Communities Resulted in Different Methane Emissions from Paddy Soil Applied with Two Types of Manure
Reprinted from: *Agronomy* **2023**, *13*, 1268, doi:10.3390/agronomy13051268 . . . . **221**

Zeyu Cheng, Lingbo Meng, Tengjiao Yin, Ying Li, Yuhang Zhang and Shumin Li
Changes in Soil Rhizobia Diversity and Their Effects on the Symbiotic Efficiency of Soybean Intercropped with Maize
Reprinted from: *Agronomy* **2023**, *13*, 997, doi:10.3390/agronomy13040997 . . . . **237**

Anna Muratova, Svetlana Gorelova, Sergey Golubev, Dilyara Kamaldinova and Murat Gins
Rhizosphere Microbiomes of *Amaranthus* spp. Grown in Soils with Anthropogenic Polyelemental Anomalies
Reprinted from: *Agronomy* **2023**, *13*, 759, doi:10.3390/agronomy13030759 . . . . **254**

Jian Xiao, Tian Liang, Shangdong Yang and Hongwei Tan
Can Sugarcane Yield and Health Be Altered with Fully Mechanized Management?
Reprinted from: *Agronomy* **2023**, *13*, 153, doi:10.3390/agronomy13010153 . . . . **270**

About the Editors

Yong-Xin Liu

Yong-Xin (Young) Liu has a major in bioinformatics and microbiome. He is a Professor at AGIS, CAAS, an Executive Editor of iMeta, and a participant in two QIIME projects. He has published >50 papers in *Nature Biotechnology, Nature Microbiology, Nature Protocols, Science China Life Sciences*, etc. These papers have been cited >18000 times. He is ranked among the world's top 2% scientists, and has reviewed 200+ times for 70+ journals, including *Nature Communications, Microbiome, ISME*, and *NAR*.

Peng Yu

Dr. Peng Yu's major study focus is root functional biology, aiming to understand the genetic control and adaptive strategy of root growth and development interacting with soil microbes under environmental stresses in crops. He works at the Institute of Crop Science and Resource Conservation (INRES), with research topics in nutrition and health, and development and change.

agronomy

Editorial

Metagenomic Analysis for Unveiling Agricultural Microbiome

Yao Wang [1], Peng Yu [2,*] and Yong-Xin Liu [1,*]

1 Shenzhen Branch, Guangdong Laboratory of Lingnan Modern Agriculture, Genome Analysis Laboratory of the Ministry of Agriculture and Rural Affairs, Agricultural Genomics Institute at Shenzhen, Chinese Academy of Agricultural Sciences, Shenzhen 518120, China; wangyao01@caas.cn

2 Emmy Noether Group Root Functional Biology, Institute of Crop Science and Resource Conservation, University of Bonn, Bonn 53113, Germany

* Correspondence: yupeng@uni-bonn.de (P.Y.); liuyongxin@caas.cn (Y.-X.L.)

Microbial communities play crucial roles in sustaining agricultural ecosystems, influencing both crop health and productivity [1] (Figure 1). To foster sustainable agricultural practices and food production, it is vital to grasp and utilize the capabilities of the agricultural microbiome. Microorganisms, particularly bacteria and fungi in the soil, are pivotal in the cycling of essential nutrients like nitrogen, phosphorus, and potassium. They convert these elements into forms that plants can readily absorb, thereby directly impacting soil fertility and plant nutrition [2,3]. Soil microorganisms are also known to produce hormones and other biochemicals that promote plant growth. For example, specific taxa of rhizobacteria have been found to enhance root growth, thereby increasing a plant's ability to absorb nutrients and water [4,5]. Beyond growth, the microbiome serves as a protective agent against pathogens through various mechanisms including the production of antimicrobial compounds, competing for resources, and triggering plant immune response [6,7]. This biological control is instrumental in managing crop diseases and reducing dependency on chemical pesticides [8,9]. Moreover, microorganisms play a critical role in helping plants cope with abiotic stresses, such as drought, salinity, and extreme temperatures [10]. Diversity within microbial communities also contributes to the resilience of agricultural ecosystems, enabling them to maintain function under changing environmental conditions [11]. Such biodiversity is essential for robust agricultural systems that need to adapt to and mitigate the impacts of climate change.

Grasping the complex interactions and functionalities of microbial communities in agricultural environments is pivotal for advancing sustainable farming practices. Metagenomics, a cutting-edge technology that delves into the structure and composition of these communities, is radically enhancing our comprehension of the agricultural microbiome [12]. This approach facilitates a detailed characterization of microbial diversity and functionality, setting the stage for precise interventions aimed at maximizing the benefits of microbes in agriculture. This Special Issue, titled "Metagenomic Analysis for Unveiling Agricultural Microbiome", encompasses a series of 17 papers that explore the profound impact of microorganisms on various agricultural ecosystems. The research presented here highlights the essential role of the agricultural microbiome in several key areas:

Crop Management Innovations: The study by Wen et al. [13] showcases how paddy–*Lilium* crop rotation enhances beneficial fungal communities and alleviates soil acidification, demonstrating a sustainable practice to mitigate the adverse effects of continuous cropping on soil health. Li and colleagues [14] illustrate how co-ensiling cassava with corn stalk optimizes silage quality by modifying microbial community functions, which could significantly impact livestock feed efficiency.

Disease Dynamics and Plant Health: In the research by Muñoz-Ramírez et al. [15], the dynamics of microbial communities in the rhizosphere of both symptomatic and asymptomatic apple trees are explored, highlighting how shifts in microbial populations can influence plant health and disease outcomes. The interplay between fungal communities

Citation: Wang, Y.; Yu, P.; Liu, Y.-X. Metagenomic Analysis for Unveiling Agricultural Microbiome. *Agronomy* **2024**, *14*, 981. https://doi.org/10.3390/agronomy14050981

Received: 23 April 2024
Accepted: 30 April 2024
Published: 7 May 2024

and pathogen suppression is presented in the study on soybean and maize intercropping by Cheng et al. [16], where increased nitrogen-fixing bacterial diversity correlates with improved plant health. Luo et al. [17] demonstrate that dodder parasitism leads to the enrichment of pathogen Alternaria and flavonoid metabolites in soybean root.

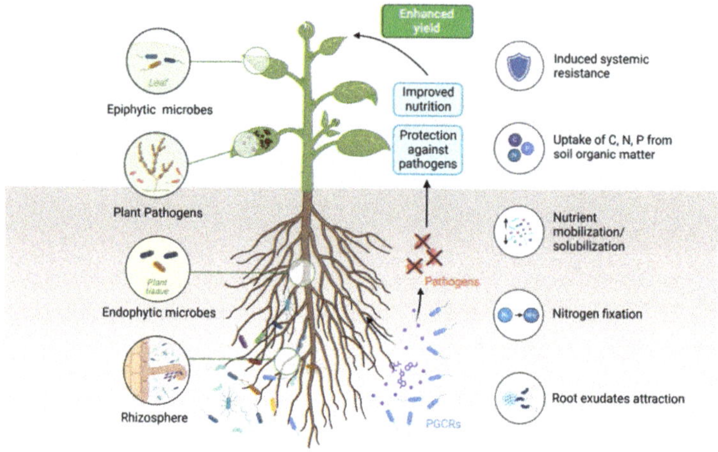

Figure 1. Relationships between plants and their surrounding microbial communities are pivotal in sustaining agricultural ecosystems. Microorganisms occupy various niches in relation to the plant, such as epiphytic microbes on leaf surfaces, endophytic microbes within plant tissues, and microbes in the rhizosphere, contributing to plant health and growth. These microbes facilitate nutrient absorption and utilization, and disease and pest resistance, thereby enhancing crop yield. The diversity of these microbial communities is crucial for supporting the overall health and productivity of crops. Created with BioRender.com.

Nutrient Cycling and Soil Health: Microorganisms are instrumental in transforming essential nutrients such as nitrogen, phosphorus, and potassium into bioavailable forms. This process is crucial for enhancing soil fertility and plant nutrition. A study by Ji et al. [18] on different fertilizer sources in maize cultivation shows how organic and inorganic fertilizers distinctly influence soil microbial communities, affecting nutrient cycling and availability. Zhou et al. [19] provide insights into how microbial community diversity under different manure treatments influences methane emissions in paddy fields, suggesting strategies to manage greenhouse gas emissions while maintaining soil fertility.

In conclusion, the contributions in this Special Issue underscore the transformative potential of metagenomic technologies in understanding and leveraging the agricultural microbiome for sustainable practices. These insights not only advance our scientific knowledge but also pave the way for practical applications that enhance crop production, protect plant health, and promote ecological sustainability. It is our hope that this issue inspires continued research and innovation, driving forward the integration of metagenomic insights into effective, sustainable agricultural practices. This Special Issue (https://www.mdpi.com/journal/agronomy/special_issues/S70BHD8E5H) (accessed on 1 December 2022) was closed on 31 December 2023, having been viewed 24,413 times. Due to the sustained interest from numerous authors in this topic and the continuous evolution of research, we have since launched "Metagenomic Analysis for Unveiling Agricultural Microbiome—2nd Edition" (https://www.mdpi.com/journal/agronomy/special_issues/YOQL5J6Y9F) on 1 April 2024. Contributions that advance the development of the field are warmly welcomed. We invite not only the submission of research on agricultural animals and plants [20] but also relevant systematic reviews [21] and analytical methods [22].

Author Contributions: Conceptualization, Y.-X.L.; writing—original draft preparation, Y.W.; writing—review and editing, Y.W., P.Y. and Y.-X.L. All authors have read and agreed to the published version of the manuscript.

Funding: This study was financially supported by the Agricultural Science and Technology Innovation Program (CAAS-ZDRW202308), the Natural Science Foundation of China (U23A20148).

Data Availability Statement: No new data were created or analyzed in this study.

Conflicts of Interest: The author declares no conflicts of interest.

References

1. Trivedi, P.; Leach, J.E.; Tringe, S.G.; Sa, T.; Singh, B.K. Plant-microbiome interactions: From community assembly to plant health. *Nat. Rev. Microbiol.* **2020**, *18*, 607–621. [CrossRef] [PubMed]
2. Oldroyd, G.E.D.; Leyser, O. A plant's diet, surviving in a variable nutrient environment. *Science* **2020**, *368*, eaba0196. [CrossRef]
3. Richardson, A.E.; Simpson, R.J. Soil microorganisms mediating phosphorus availability update on microbial phosphorus. *Plant Physiol.* **2011**, *156*, 989–996. [CrossRef] [PubMed]
4. Gouda, S.; Kerry, R.G.; Das, G.; Paramithiotis, S.; Shin, H.-S.; Patra, J.K. Revitalization of plant growth promoting rhizobacteria for sustainable development in agriculture. *Microbiol. Res.* **2018**, *206*, 131–140. [CrossRef] [PubMed]
5. Zhang, J.; Liu, Y.-X.; Zhang, N.; Hu, B.; Jin, T.; Xu, H.; Qin, Y.; Yan, P.; Zhang, X.; Guo, X.; et al. NRT1.1B is associated with root microbiota composition and nitrogen use in field-grown rice. *Nat. Biotechnol.* **2019**, *37*, 676–684. [CrossRef]
6. Pieterse, C.M.J.; Zamioudis, C.; Berendsen, R.L.; Weller, D.M.; van Wees, S.C.M.; Bakker, P.A.H.M. Induced systemic resistance by beneficial microbes. *Annu. Rev. Phytopathol.* **2014**, *52*, 347–375. [CrossRef] [PubMed]
7. Mendes, R.; Kruijt, M.; de Bruijn, I.; Dekkers, E.; van der Voort, M.; Schneider, J.H.M.; Piceno, Y.M.; DeSantis, T.Z.; Andersen, G.L.; Bakker, P.A.H.M.; et al. Deciphering the rhizosphere microbiome for disease-suppressive bacteria. *Science* **2011**, *332*, 1097–1100. [CrossRef] [PubMed]
8. van Lenteren, J.C.; Bolckmans, K.; Köhl, J.; Ravensberg, W.J.; Urbaneja, A. Biological control using invertebrates and microorganisms: Plenty of new opportunities. *BioControl* **2018**, *63*, 39–59. [CrossRef]
9. Collinge, D.B.; Jensen, D.F.; Rabiey, M.; Sarrocco, S.; Shaw, M.W.; Shaw, R.H. Biological control of plant diseases—What has been achieved and what is the direction? *Plant Pathol.* **2022**, *71*, 1024–1047. [CrossRef]
10. Timm, C.M.; Carter, K.R.; Carrell, A.A.; Jun, S.-R.; Jawdy, S.S.; Vélez, J.M.; Gunter, L.E.; Yang, Z.; Nookaew, I.; Engle, N.L.; et al. Abiotic stresses shift belowground *Populus*-associated bacteria toward a core stress microbiome. *mSystems* **2018**, *3*. [CrossRef] [PubMed]
11. Prudent, M.; Dequiedt, S.; Sorin, C.; Girodet, S.; Nowak, V.; Duc, G.; Salon, C.; Maron, P.-A. The diversity of soil microbial communities matters when legumes face drought. *Plant Cell Environ.* **2020**, *43*, 1023–1035. [CrossRef]
12. Gao, Y.; Li, D.; Liu, Y.-X. Microbiome research outlook: Past, present, and future. *Protein Cell* **2023**, *14*, 709–712. [CrossRef] [PubMed]
13. Wen, L.; Huang, F.; Rao, Z.; Cheng, K.; Guo, Y.; Tang, H. Paddy-*Lilium* crop rotation improves potential beneficial soil fungi and alleviates soil acidification in *Lilium* cropping soil. *Agronomy* **2024**, *14*, 161. [CrossRef]
14. Li, M.; Zi, X.; Sun, R.; Ou, W.; Chen, S.; Hou, G.; Zhou, H. Co-ensiling whole-plant cassava with corn stalk for excellent silage production: Fermentation characteristics, bacterial community, function profile, and microbial ecological network features. *Agronomy* **2024**, *14*, 501. [CrossRef]
15. Muñoz-Ramírez, Z.Y.; González-Escobedo, R.; Avila-Quezada, G.D.; Ramírez-Sánchez, O.; Higareda-Alvear, V.M.; Zapata-Chávez, E.; Borrego-Loya, A.; Muñoz-Castellanos, L.N. Exploring microbial rhizosphere communities in asymptomatic and symptomatic apple trees using amplicon sequencing and shotgun metagenomics. *Agronomy* **2024**, *14*, 357. [CrossRef]
16. Cheng, Z.; Meng, L.; Yin, T.; Li, Y.; Zhang, Y.; Li, S. Changes in soil rhizobia diversity and their effects on the symbiotic efficiency of soybean intercropped with maize. *Agronomy* **2023**, *13*, 997. [CrossRef]
17. Luo, W.; Li, Y.; Luo, R.; Wei, G.; Liu, Y.; Chen, W. Dodder parasitism leads to the enrichment of pathogen *Alternaria* and flavonoid metabolites in soybean root. *Agronomy* **2023**, *13*, 1571. [CrossRef]
18. Ji, L.; Xu, X.; Zhang, F.; Si, H.; Li, L.; Mao, G. The preliminary research on shifts in maize rhizosphere soil microbial communities and symbiotic networks under different fertilizer sources. *Agronomy* **2023**, *13*, 2111. [CrossRef]
19. Zhou, B.; Chen, R.; Peng, S.; Zhang, J.; Lin, X.; Wang, Y. Variations in methanogenic and methanotrophic communities resulted in different methane emissions from paddy soil applied with two types of manure. *Agronomy* **2023**, *13*, 1268. [CrossRef]
20. Yu, P.; He, X.; Baer, M.; Beirinckx, S.; Tian, T.; Moya, Y.A.T.; Zhang, X.; Deichmann, M.; Frey, F.P.; Bresgen, V.; et al. Plant flavones enrich rhizosphere *Oxalobacteraceae* to improve maize performance under nitrogen deprivation. *Nat. Plants* **2021**, *7*, 481–499. [CrossRef] [PubMed]

21. Wang, Z.; Song, Y. Toward understanding the genetic bases underlying plant-mediated "cry for help" to the microbiota. *iMeta* **2022**, *1*, e8. [CrossRef]
22. Wen, T.; Niu, G.; Chen, T.; Shen, Q.; Yuan, J.; Liu, Y.-X. The best practice for microbiome analysis using R. *Protein Cell* **2023**, *14*, 713–725. [CrossRef] [PubMed]

agronomy

Article

Maize Rotation Combined with *Streptomyces rochei* D74 to Eliminate *Orobanche cumana* Seed Bank in the Farmland

Jiao Xi [1], Zanbo Ding [1], Tengqi Xu [1], Wenxing Qu [1], Yanzhi Xu [1], Yongqing Ma [2], Quanhong Xue [3], Yongxin Liu [4,*] and Yanbing Lin [1,*]

1 College of Life Sciences, Northwest A&F University, Yangling 712100, China
2 State Key Laboratory of Soil Erosion and Dry Land Farming, Institute of Soil and Water Conservation, Chinese Academy of Sciences and Ministry of Water Resources, Yangling 712100, China
3 College of Natural Resources and Environment, Northwest A&F University, Yangling 712100, China
4 Shenzhen Branch, Guangdong Laboratory of Lingnan Modern Agriculture, Genome Analysis Laboratory of the Ministry of Agriculture and Rural Affairs, Agricultural Genomics Institute at Shenzhen, Chinese Academy of Agricultural Sciences, Shenzhen 518120, China
* Correspondence: liuyongxin@caas.cn (Y.L.); linyb2004@nwsuaf.edu.cn (Y.L.)

Abstract: *Orobanche cumama* wallr. is the sunflower root parasitic weed with special life stage in which seed germination and parasitism take place in the soil. In practice, applying microbial agents and trapping crop rotation are utilized separately, or just one of them is selected to control *O. cumana*. The development of the sunflower industry is severely constrained on the farmland, where there is high density of *O. cumana*'s seed banks. In this study, two biological control methods were combined to solve the problem of *O. cumana* parasitism. The bioassay experiment showed that the high concentration fermentation filtrates of *Streptomyces rochei* D74 could effectively inhibit the germination and growth of the germ tube of *O. cumana* seeds. As the concentration was increased to 3.1 mg/mL, *O. cumana* was almost unable to sprout. A two-year pot experiment revealed that the use of D74 agents and sunflower–maize–sunflower rotation together promoted sunflower growth, as shown by the biomass accumulation, plant height, and denser root systems. The combined method resulted in a significant decrease in the number of *O. cumana* parasitism, compared to one method alone. Additionally, it affected the bacterial community composition of sunflower rhizosphere, mostly leading to an increase in *Streptomyces* and *Brevibacterium* and a decrease in *Arthrobacter*. This experiment, combined with multiple biological control, means significantly reducing the parasitism of *O. cumana*, which provides an effective foundation for practical application.

Keywords: *Orobanche cumama*; *Streptomyces rochei*; maize rotation; sunflower; rhizosphere microbial

Citation: Xi, J.; Ding, Z.; Xu, T.; Qu, W.; Xu, Y.; Ma, Y.; Xue, Q.; Liu, Y.; Lin, Y. Maize Rotation Combined with *Streptomyces rochei* D74 to Eliminate *Orobanche cumana* Seed Bank in the Farmland. *Agronomy* **2022**, *12*, 3129. https://doi.org/10.3390/agronomy12123129

Academic Editor: Ivo Toševski

Received: 8 November 2022
Accepted: 8 December 2022
Published: 9 December 2022

1. Introduction

Orobanche cumana Wallr is a fully parasitic weed that lacks chlorophyll and has no photosynthetic capacity [1]. Its seeds can detect strigolactone (SLs), a germination signal substance secreted by the root of the host plant, and initiate the germinating process [2,3]. The end of the germ tube of the germinating seeds can form haustorium, a special nodular protruding tissue that can be adsorbed on the host root surface, penetrate the root epidermis, grow to the xylem and phloem, and form a continuous xylem between the host and the *Orobanche* [4]. *Orobanche* feeds itself by absorbing carbohydrates, water, and mineral elements from the host [5], and then grows to the surface, blooms, and produces tons of tiny seeds eventually [6,7]. *O. cumana* has specific host requirements and can only parasitize sunflowers. *Orobanche* has affected many countries. In China, the area affected by *Orobanche* reached 20,000 hectares, with yield losses ranging from 20% to 50% [8]. In France, the economic losses caused by *Orobanche* are incalculable, and the Ministry of Agriculture claimed that, if no effective measures are taken, yield loss could reach 90% (https://doi.org/10.1079/pwkb.species.37741 (accessed on 1 November 2022)). The serious parasitism of *O. cumana* reduces sunflower yield and quality, affecting farmers' income and stifling agricultural development. Manual uprooting, herbicide spraying, soil fumigation, irrigation, trapping crop rotation, and allelopathy are currently the most commonly used methods [9–12].

Maize was used as a trap crop in this study because its root exudates can stimulate the germination of *Orobanche* seeds in the surrounding soil, but cannot establish an effective connection with the crop, and it will eventually die due to nutrient deficiency [13]. Planting trap crops repeatedly, making the active parasitic weed seeds germinate in vain without producing new seed replenishment. Additionally, this has been shown to effectively reduce soil seed bank content [14]. In recent years, more and more studies have been conducted on the influence of microorganisms on *Orobanche* parasitism [15,16]. *Pseudomonas fluorescens* and *Myrothecium verrucaria* isolated from rhizosphere soils reduced the germination rate of *O. foetida* and *O. ramosa*, respectively [17,18]. Studies have shown that *Streptomyces enissocaesilis* can inhibit the germination of *Orobanche* seeds by 47% [19]. *Streptomyces rochei* D74, an actinomycete with germination inhibitory potential discovered by Chen in the early stages [19], was used in this experiment. *S. rochei* D74 has been shown in studies to promote the growth of a variety of crops, including *Amorphophallus konjac*, *Aconitum carmichaelii*, tomato, pepper, sunflower, and jujube [20–22].

Because of the nature of *Orobanche* seed germination and parasitism occurring under the ground, it has already caused crop damage and displayed infection symptoms before reaching the surface. More studies have used one method alone to prevent the parasitism of *Orobanche* in the past, and there was a lack of research on combining multiple methods to reduce parasitism. Therefore, in order to achieve the purpose of long-term and effective control of parasitism, this study combined biological control and crop rotation, for the first time, to study the effects of *Streptomyces* and maize rotation on parasitism.

2. Materials and Methods

2.1. Preparation of *S. rochei D74* Fermentation Filtrates

S. roche D74 was incubated into Gauze's synthetic medium [23] and was shaken at 28 °C and 100 rpm for 14 days. Afterwards, the fermentation filtrates of D74 above were centrifuged at 4 °C and 10,000 rpm to separate supernatant and precipitate. Filter the supernatant with a 0.45 μm microporous membrane to remove excess impurities. The collected precipitate was dried in a drying oven at 35 °C and weighed afterwards. The concentration of D74 fermentation filtrates was 3.1 mg/mL, which was calculated from the volume of the supernatant and the mass of the dried precipitate. The obtained D74 fermentation filtrates was diluted in a gradient of 10–10^5 times of the original filtrates. The diluted fermentation filtrates concentration was in the range of 3.1×10^{-1} mg/mL–3.1×10^{-5} mg/mL. Finally, we dispensed the diluted solution and kept it in −80 °C refrigerator.

2.2. Petri Dish Bioassay Experiments

The pretreatment of the *O. cumana*'s seed: sterilize the seeds' surfaces by dipping them in 75% ethanol for 3 min, cleaning them from impurities, and then drying them naturally. Lay a round layer of plain filter paper with a diameter of about 9 cm flat on the bottom of the Petri dish. Place the 5 mm diameter round glass fiber filter paper neatly on the round plain filter paper in order. Then, soak the filter paper sheet with sterilized water and then spread the seeds evenly on it (about 40–60 *O. cumana*'s seeds on each piece of filter paper). Put the Petri dishes into the artificial climate incubator at 28 °C for 5 days.

Prepare some new Petri dishes. Place the sterilized glass fiber filter paper with a diameter of 5 mm to the bottom of the Petri dish in an arc shape. Pipette 25 μL of the prepared D74 fermentation filtrates at different concentrations into Petri dishes, while the blank control group was added with an equal amount of sterilized water. Each treatment was replicated 5 times. A pre-cultured seed piece of *O. cumana* was placed on the top layer of each filter paper sheet, and 25 μL of the artificial hormone GR24 at 1 ppm was added in turn. For the control group, sterile water was used instead of GR24 to verify whether the seeds could germinate in the absence of hormonal stimulation. Place a moistened triangular filter paper sheet in the center of each Petri dish to avoid the death of seeds due to water loss. The dishes were then sealed with Parafilm and incubated at 28 °C for 7–10 days in the dark. After the culture was completed, the seeds were examined microscopically, and the number of germinations and the number of seeds of which the lengths of germ tubes were between 0–1 mm (short) and 1–3 mm (long) were counted [19]. The germination rate of sunflower seeds and the percentage of seeds of which the lengths of germ tubes were between different ranges were calculated according to the following formula (1):

$$\text{Percentage of 0–1 mm} = \frac{(\text{Number of 0–1 mm seeds})}{(\text{Total number of germinated}) \times 100\%} \quad (1)$$

2.3. Co-Culture Experiments

Pretreatment of sunflower seeds: prepare 100-hole cavity trays. Then, mix them according to the ratio of substrate:vermiculite = 2:1 and put them into the cavity trays. Sterilize the seeds in 75% ethanol for 3 min, and then wash off impurities on the surface with sterilized water and press them into the cavity tray (2 seeds in each hole). Water with 20 mL water and incubate in an artificial climate chamber (28 °C; light incubation: dark incubation = 12 h: 12 h) for 5–8 days. When sunflower plants grow to about 5 cm in height, take them out with roots. Rinse off excess soil and impurities from the roots with sterile water, and the seedlings for testing are ready. The pretreatment method of seed is the same as in Section 2.2. Simply sprinkle the seeds of *O. cumana* on a 9 cm glass fiber filter paper, incubate under the same conditions for 5 days, and they are ready for use.

A 9 cm diameter piece of sterilized glass fiber filter paper was laid flat on the bottom of the Petri dish in a clean bench. Then, we added 2 mL of D74 fermentation filtrates at a concentration of 3.1 mg/mL, while the blank control group (CK) was added with an equal amount of sterilized water. After soaking for 1.5 h, the filter paper sheets were removed to dry in Petri dishes. We placed the dried filter paper sheet in the middle of the PE bag with tweezers, and then overlaid a glass fiber filter paper sheet that has been evenly sprinkled with pre-cultivated *O. cumana* seeds. We inserted the cleaned sunflower seedling plants through the gap above the cut PE bag to make the roots of the sunflower plants lay evenly and flatly on the filter paper sheet and make full and close contact with the *O. cumana* seeds. The prepared Hoagland nutrient solution [24] was then added in the amount of 15 mL/bag and incubated in an artificial climate chamber (28 °C; light incubation: dark incubation = 12 h: 12 h) for 30 days [16]. During this period, the PE bags were supplemented with 5 mL of Hoagland nutrient solution daily, and we examined the germination and parasitism of *O. cumana* seeds by microscopy after 30 days.

2.4. Design of Pot Experiments

The design of experimental treatment group: the experimental sample site of this study is located at the Institute of Soil and Water Conservation, Northwest A&F University. The maize variety used was Zheng Dan 958, and the sunflower variety was 363. To investigate the effects of maize rotation and addition of D74 agents on the growth of sunflower and the emergence of *O. cumana*, we set up a crossover test with three groups of factors. With (L) and without (N) the addition of *O. cumana* seeds; with (J) and without (N) the addition of D74 agents; with (Y) and without (N) the rotation of maize. A total of 10 treatment groups were set up in this study, with 7 pot replicates for each treatment. Added *O. cumana* groups: CKL (blank control soil); LXJY (crop rotation + D74); LXJN (addition of D74); LXNY (maize rotation); LXNN (sunflower planting only). Non-addition of *O. cumana* group: CKN (blank control soil); NXJY (rotation + D74); NXJN (addition of D74); NXNY (rotation of maize); NXNN (sunflower planting only). With three sowings, the pot planting trial was divided into three crop rotation stages, namely sunflower–maize–sunflower. To ensure consistent planting conditions, the work was performed in the same pot (20 cm in diameter, 25 cm in height) for each crop rotation stage. Stage 1, plant sunflower with 4 seeds per pot in

April 2019, according to treatment group. When the sunflower seedlings had grown for about 15 days, we kept two plants in each pot with the least difference in growth. The planters in different treatment groups were moved every two weeks to ensure that the environmental conditions, such as light and water, were consistent between treatments, and they were watered as needed throughout the period. After 2 months of plant growth, we harvested the aboveground and underground parts of the sunflower and measured their physiological parameters. We harvested and recorded the number of *O. cumana* above the ground, as well as the total number of parasites. Collect the rhizosphere soil for future research. Stage 2, following sunflower harvest, fertilizer, and D74 were reapplied to the original soil in June 2019, in accordance with the treatment group settings. After 1 week of resting, the maize hybrid Zhengdan 958 was planted according to the treatment group setting for crop rotation. Each pot was planted with 10 maize seeds and picked to two plants per pot according to growth after the maize seeds sprouted out of the soil. The aboveground portion of maize was harvested in August 2019 after 45 days of planting to evaluate its physiological parameters, while the underground root portion was left in the soil to facilitate its continued production of root secretions for subsequent research on the effect of maize root secretions on *O. cumana*'s seeds. Stage 3, in April 2020, we opened the pots, removed the underground portion of the maize root system, and re-mixed the soil with fertilizer and D74. We resowed sunflower seeds in the original soil in May 2020, 10 seeds per pot. According to the growth situation, we kept 2 plants in each pot to continue to grow. We harvested the sunflower when it was fully parasitized. In June 2020, fresh samples of above- and underground sunflower plants were harvested, and their physiological parameters were measured, as well as the number of unearthed *O. cumana* and total parasites. Fresh samples of rhizosphere soils were collected and promptly transported back to the laboratory at low temperatures before being divided into two parts. One part of the rhizosphere soil was stored at 4 °C and isolated culturable microorganisms by serial dilution plating. We compared the number and species of culturable microorganisms in the soil of different treatments by counting. We screened and identified the dominant strains. Another part of sunflower rhizosphere soil was stored at −80 °C for extracting total soil DNA extraction. Using 16S rRNA gene amplicon sequencing, we analyzed the structure of the soil microorganism community.

Treatment of sunflower seeds: the full-grained sunflower seeds were selected and divided into two parts based on the needs of the treatment groups. We took out a part of sunflower seeds, weighed D74 at 4% of the total mass, and mixed them with D74 using sodium carboxymethyl cellulose (CMC-Na) at a concentration of 0.6% as a binder, so that the D74 was fully coated with the shell of sunflower seeds. Another portion of sunflower seeds were coated with only an equivalent concentration of 0.6% CMC-Na as a control treatment. Maize seeds were soaked in D74 fermentation filtrates for 10 h and then coated with 0.6% CMC-Na at 15% of their total weight.

Treatment of soil: the test soil was a small mound collected from the cultivated land around Yangling. The soil was obtained from farmland and passed through 5 mm sieve to remove debris, and each pot was filled with 8 kg of soil mixture. In each pot, 25 g of organic fertilizer, 3.44 g of urea, and 1.2 g of calcium superphosphate were included. In the treatment group, D74 and *O. cumana* seeds were evenly mixed with the above soil at 1.5 g/kg and 3.4 mg/kg, respectively. The packed soil was left to stand for 1 week, and then the plants were planted.

2.5. Measurement of Plant Physiological Parameters

We measured the plant height, stem diameter, and root length and weighed the dry stem and root of sunflowers at maturity. *O. cumana* was collected and counted at the same time and then divided into the part aboveground and the part underground.

2.6. Sunflower Rhizosphere Soil Collection, Isolation and Identification of Bacteria

Silwet L-77 was added at 0.02% of the volume of phosphate buffer (pH 7.0), and it was mixed as the rhizosphere soil eluted. In the laboratory, we kneaded the fresh sunflower root samples with sterile gloves at low temperature to allow the soil in the root zone to fall naturally. We put the plant roots in a 100 mL sterile centrifuge tube and added an appropriate amount of phosphate buffer. We vortexed for 15 s at the maximum speed to disperse the rhizosphere soil. We removed impurities such as plant debris and large sediments with forceps, then transferred the roots to a new 100 mL sterile centrifuge tube, added appropriate amount of phosphate buffer, centrifuged at 3200 rpm for 15 min, and collected the sediment in a 10 mL centrifuged tube, then centrifuged at 8000 rpm for 10 min and removed the supernatant. One part of the precipitate was stored at 4 °C for isolation and identification of soil rhizosphere bacteria, and one part of the precipitate was stored in a refrigerator at −80 °C for extraction of total DNA [25]. Isolation and identification of sunflower rhizosphere bacteria were carried out on TSB and R2A media to analyze the differences in the composition, and the detailed methods are described in [16].

2.7. Extraction and Sequencing of Total DNA from Rhizosphere Soil

Microbial DNA from soil was extracted using FastDNA® SPIN Kit for Soil (MP Biomedicals, Solon, OH, USA), and the extraction procedure was based on the operation manual. Sequencing libraries were constructed by TruSeq® DNA PCR-Free Sample Preparation Kit (Illumina, Inc., San Diego, CA), and the libraries were quality assessed by Qubit 2.0 fluorometer. Finally, libraries were sequenced on the NovaSeq PE250 platform (Thermo Fisher Scientific, Waltham, MA, USA) to generate 250 bp sequences of paired-end reads. Sequences with ≥97% similarity in clean reads were clustered into identical OTUs (operational taxonomic units) for species annotation.

2.8. Statistical Analyses

SPSS IBM Statistics software (Version 21) [26] was used for two-way ANOVA test at 0.05 level. The plotting was performed using Origin software (Version 2018) [27]. Amplicon analysis was performed through the Novomagic Cloud Platform (https://magic.novogene.com). PCoA plots were obtained using the ImageGP web server (http://www.ehbio.com) [28,29].

3. Results

3.1. Effects of D74 Fermentation Filtrates on the Germination of O. cumana Seeds

The results of the effect of D74 fermentation filtrates on the germination of *O. cumana* seeds are shown in Figure 1a. The germination rate of *O. cumana* seeds in the control group (GR24) without the addition of D74 fermentation filtrates was 83.31%, and the germination rate of *O. cumana* seeds in the treated group with the addition of D74 fermentation filtrates was lower than control group ($p < 0.05$). The germination rate of *O. cumana* seeds was 0% when D74 fermentation filtrates on concentration of 3.1 mg/mL was added, indicating that D74 fermentation filtrates had a significant inhibition effect on the germination of *O. cumana* seeds ($p < 0.05$), and the inhibition rate could reach 100%. When D74 fermentation filtrates was diluted 10–10^5 times, the germination rate of *O. cumana* seeds was reduced by 19.82%, 17.31%, 17.51%, 12.26%, and 8.00%, respectively, compared with the control. The results indicated that the high concentration of D74 fermentation filtrates had a better effect on the germination inhibition of *O. cumana* seeds.

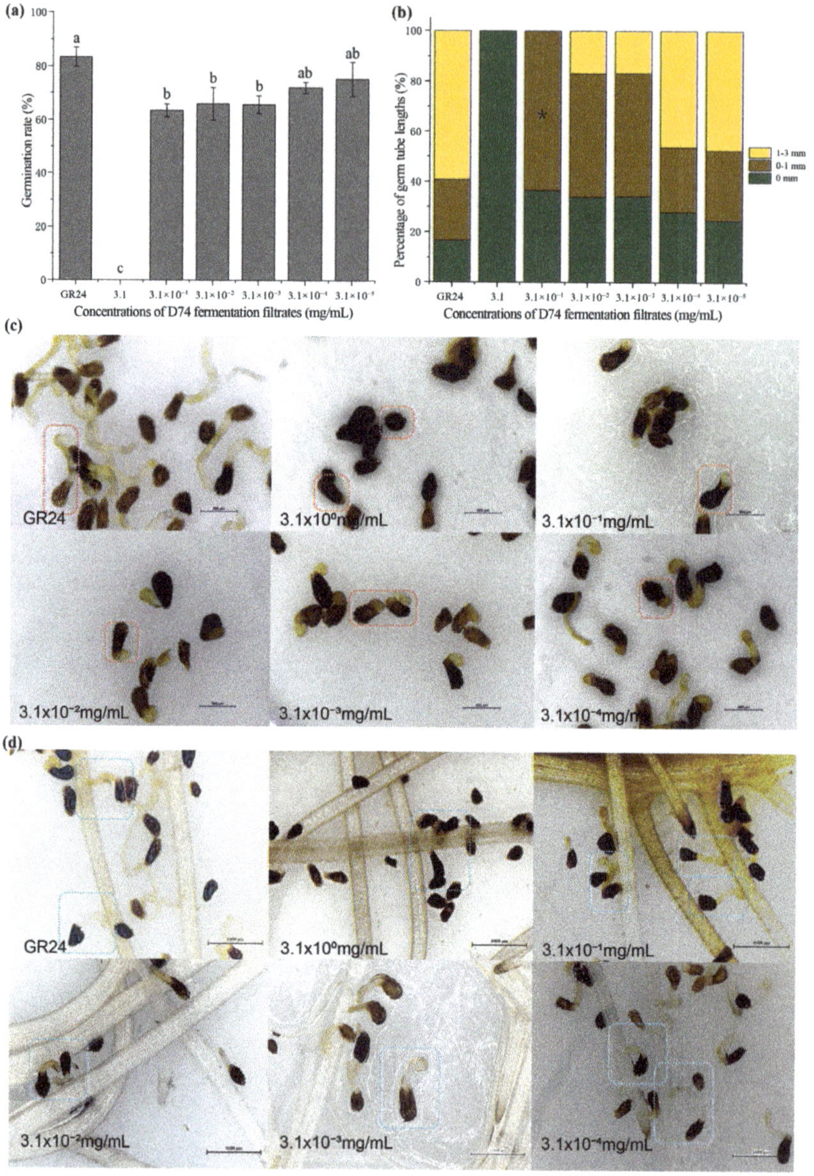

Figure 1. Effects of different concentrations of D74 fermentation filtrates on the germination rate (**a**,**c**), germ tube length (**b**), and parasitism (**d**) of *O. cumana* seeds. Different letters indicate significant differences between the groups (Tukey's test, $p < 0.05$), bar: 1000 µm. * Critical level of significance of constructs at 5%.

The germination rate of *O. cumana* seeds increased with increasing dilution of D74 fermentation filtrates. Additionally, more typical haustorial papillae, probably due to the presence of haustorial-inducing substances in the D74 fermentation filtrates, which induced haustorial formation on *O. cumana* tubes. Tubes of germinating seeds were measured (Figure 1b), and 59.33%·of which in the control group ranging from 1–3 mm. It was significantly higher than the 23.98% with 0–1 mm of tubes length. After the fermentation filtrates was diluted 10 times, the seeds of *O. cumana* started to germinate, and the length of the shoot tubes were all concentrated in the 0–1 mm range (63.49%), which was significantly higher than the control group ($p < 0.05$). When the fermentation filtrate was diluted 10^2–10^5 times, the percentages of *O. cumana* with shoot tube lengths in the 1–3 mm range were 16.9%, 16.78%, 46.29%, and 47.25%, respectively. As observed from the germination phenotype of *O. cumana*, the seeds of *O. cumana* treated with D74 fermentation filtrates were brown, expanded, and spherical at the end of shoot tube, showing a non-healthy state (Figure 1c).

The results of the sunflower–*O. cumana* co-culture experiment (Figure 1d) were consistent with the results of the germination experiment, and high concentrations of D74 fermentation filtrates had a significant inhibitory effect on the germination of *O. cumana* seeds. With the increase of the dilution of D74 fermentation filtrates, the inhibitory effect on the germination of *O. cumana* seeds gradually diminished and began to parasitize.

3.2. *Effects of Maize Rotation and S. rochei D74 Agents on Sunflower Growth*

We harvested sunflowers at maturity and observed their growth. The sunflowers in the group with *O. cumana* seeds (L) had weaker growth, while the sunflowers in the group without *O. cumana* seeds (N) had normal growth. The short plant, thinner stalks, and shriveled leaves indicated that *O. cumana* can severely restrict the growth of the host sunflower (Figure S1). In the first year, we took samples primarily to study the effect of D74 alone on sunflower growth. After adding *O. cumana*, the dry matter accumulation, plant height, and root length of the sunflower above and below the ground were significantly lower than those without *O. cumana*. (Figure 2a, Table 1). The dry weights of plants (LXJY, LXJN, NXJY, NXJY) after the addition of D74 were higher than those in the control groups (LXNY, LXNN, NXNY, NXNN), regardless of the presence or absence of *O. cumana*, indicating that D74 can promote sunflower biomass accumulation to varying degrees.

Table 1. The growth index of sunflower in two consecutive years under different treatments.

Sampling Time	Group	Plant Height/cm	Stalk Diameter /mm	Root Length/cm
First year	LXJY	62.00 ± 1.59 b	9.13 ± 0.17 bc	13.71 ± 0.32 bc
	LXJN	60.50 ± 1.4 b	9.28 ± 0.19 bc	10.64 ± 0.92 c
	LXNY	60.75 ± 1.28 b	8.99 ± 0.21 d	12.29 ± 0.21 c
	LXNN	61.00 ± 2.25 b	8.75 ± 0.12 d	11.06 ± 1.05 c
	NXJY	126.76 ± 5.65 a	9.63 ± 0.36 bc	17.38 ± 2.38 b
	NXJN	140.81 ± 5.23 a	10.57 ± 0.64 a	22.31 ± 2.68 a
	NXNY	135.34 ± 7.65 a	9.58 ± 0.21 bc	23.24 ± 1.97 a
	NXNN	137.10 ± 7.91 a	10.05 ± 0.21 ab	24.50 ± 1.93 a
Second year	LXJY	103.06 ± 0.10 c	8.23 ± 0.22 ab	14.70 ± 0.51 c
	LXJN	79.84 ± 0.60 e	7.85 ± 0.18 b	10.27 ± 0.36 e
	LXNY	92.67 ± 2.38 d	8.28 ± 0.19 ab	13.14 ± 0.56 cd
	LXNN	87.26 ± 1.53 d	7.05 ± 0.17 c	10.87 ± 0.68 de
	NXJY	116.29 ± 2.14 a	7.81 ± 0.08 b	28.90 ± 1.09 a
	NXJN	108.94 ± 1.58 b	7.79 ± 0.24 b	21.71 ± 1.33 b
	NXNY	114.16 ± 2.55 ab	8.42 ± 0.18 a	23.79 ± 1.01 b
	NXNN	117.97 ± 2.86 a	8.03 ± 0.08 ab	22.36 ± 1.11 b

Different letters indicate significant differences at $p < 0.05$, as determined by ANOVA, followed by Tukey's test. Groups with *O. cumana* added: LXJY (maize rotation + D74); LXJN (D74 only); LXNY (maize rotation only); LXNN (sunflower planting only). Groups without *O. cumana* added: CKN (blank control); NXJY (maize rotation + D74); NXJN (D74 only); NXNY (maize rotation only); NXNN (sunflower planting only).

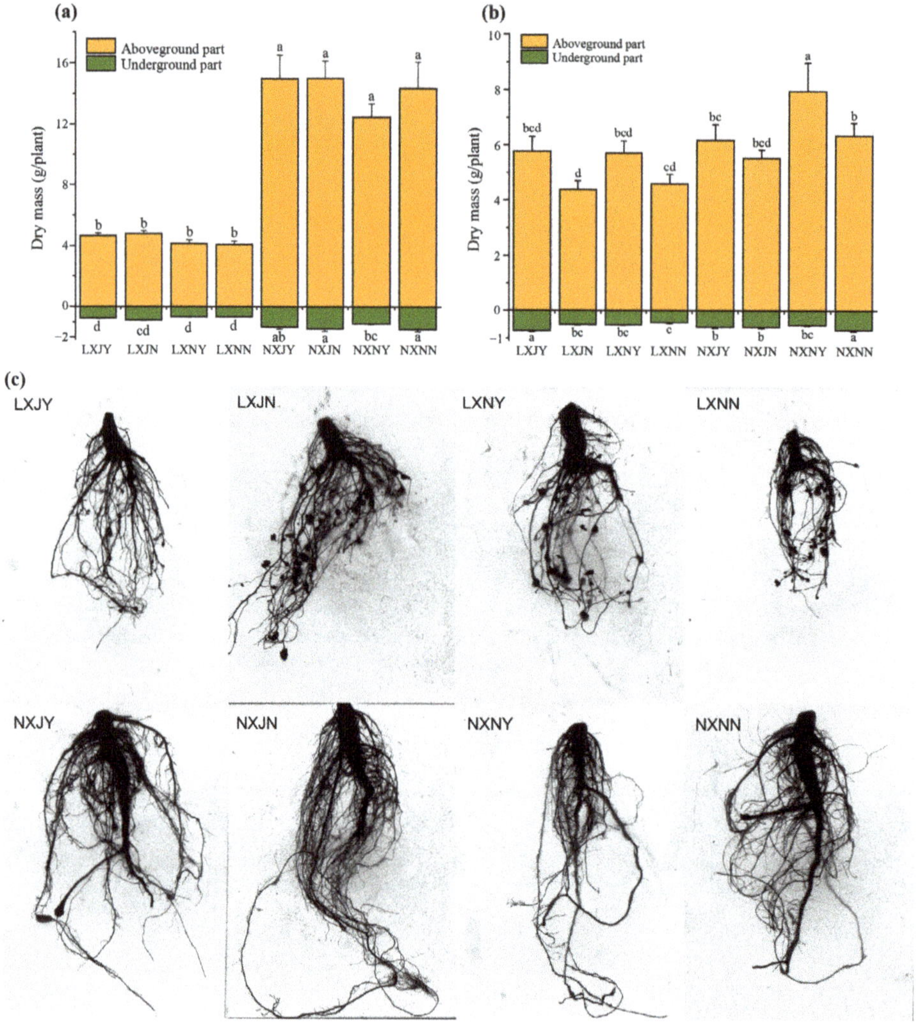

Figure 2. The D74 agent and maize rotation promote sunflower growth. Aboveground and underground dry weights of sunflowers at maturity in the first (**a**) and second (**b**) year. Scanning image (**c**) of sunflower roots at mature stage in second year. Different letters indicate significant differences between the groups (Tukey's test, $p < 0.05$). Groups with *O. cumana* added: LXJY (maize rotation + D74); LXJN (D74 only); LXNY (maize rotation only); LXNN (sunflower planting only). Groups without *O. cumana* added: CKN (blank control); NXJY (maize rotation + D74); NXJN (D74 only); NXNY (maize rotation only); NXNN (sunflower planting only).

The second year, we took samples to study the effects of D74 and maize rotation on the growth of sunflowers. When *O. cumana* was present, dry matter accumulation was higher in the maize rotation group (LXJY, LXNY) than in the non-maize rotation group (LXJN, LXNN), and it was highest in the group that had both maize rotation and the addition of D74 (LXJY) (Figure 2b). The sunflower biomass (including plant height, stem diameter, and root length) was highest in the LXJY group, with plant height increasing by 29.08%, 11.21%, and 18.11% ($p < 0.05$), when compared to the other three groups (LXJN, LXNY, and LXNN). Additionally, the root length increased by 43.14%, 11.87%, and 35.23%, respectively (Figure 2c, Table 1, $p < 0.05$), while the difference between stem diameter and other treatment groups was minor. When only the effect of maize rotation was considered, the plant height, stem diameter, and root length of LXJN and LXNN increased by 29.08% and 6.2%, 4.84% and 17.45%, and 43.14% and 20.88%, respectively, compared to the non-maize rotation groups (LXJY and LXNY, Table 1). In the absence of *O. cumana*, the root length of sunflower in the maize rotation and D74 treatment group (NXJY) increased by 33.12%, 21.48%, and 29.25% ($p < 0.05$), compared to the other treatment groups (NXJN, NXNY, NXNN). The length of the main root and the number of fibrous roots and lateral roots increased significantly with the addition of D74 groups (NXJY and NXJN), compared to the control group (NXNY and NXNN). The roots, particularly in the NXJY group, were well-developed and formed a root network. This aids in the absorption of nutrients and moisture from the soil.

Based on the results of two years, the plant height, stem diameter, dry weight of above-ground and underground parts, and root length of sunflower added *O. cumana* seeds were significantly lower than those without *O. cumana* seeds ($p < 0.05$), indicating that *O. cumana* can cause a severe depletion of sunflower plant biomass. However, under the parasitic conditions of *O. cumana*, the combination of maize rotation and D74 has a significant effect on sunflower growth, which is higher than that of D74 or maize rotation alone.

3.3. Effect of Maize Rotation and S. rochei D74 Agents on O. cumana Parasitism

In the first year, sunflowers were planted to study the effect of D74 alone on their growth (Figure 3a). The emergence rate of *O. cumana* in the group with the addition of D74 (LXJY) was lower than that of LXNY and LXNN in the group without the addition of D74. However, using D74 alone reduced the *O. cumana* unearthed, as well as parasitism, but the effect was not statistically significant. The next year, sunflowers were planted to study the effects of D74 and maize rotation alone or in combination on its growth (Figure 3b). The number of *O. cumana* unearthed and parasites in the maize rotation group (LXNY) were significantly inhibited. Compared with the control group, LXNN was significantly reduced by 65.22% and 64.36% ($p < 0.05$). In comparison to LXJN, the group of maize rotation combined with D74 application (LXJY) had the best inhibition effect, with the 78.26% inhibition rate on the unearthed number of *O. cumana*, and the inhibition rate against the total number of parasites was 48.51%.

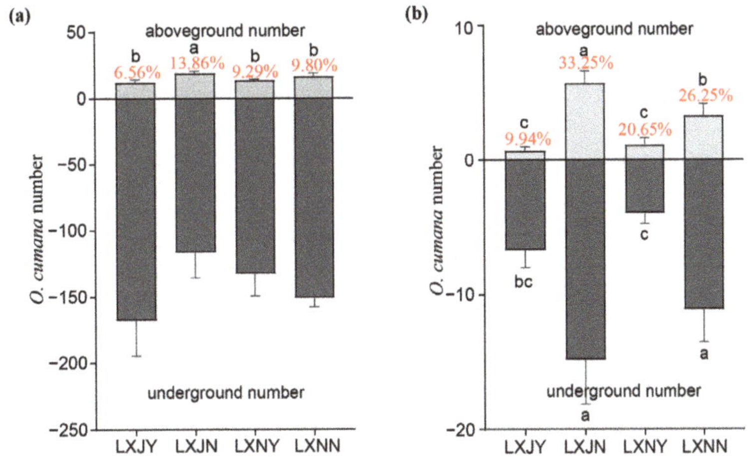

Figure 3. Effects of D74 agents and maize rotation on *O. cumana* parasitism in the first (**a**) and second (**b**) year. The red numbers represent the unearthed rate, different letters indicate significant differences between the groups (uncorrected Fisher's LSD, $p < 0.05$).

In conclusion, the unearthed number and total amount of parasitism of *O. cumana* in the different treatment groups at harvest in the second year were significantly lower than in the first year. This indicates that the continuous planting of crops for induction can effectively reduce the number of *O. cumana* seed banks in the soil. *O. cumana* parasitism can be reduced by either maize rotation alone or D74 application alone. When the two measures were used in combination, the inhibitory effect on *O. cumana* was the best ($p < 0.05$).

3.4. Effect of Maize Rotation and *S. rochei* D74 Agents on Rhizosphere Microbiota

3.4.1. Comparison between Culturable Microbial Dominant Strains in Sunflower Rhizosphere Soil

We obtained similar results of culturable bacterial species and relative abundance in sunflower rhizosphere soil from different treatment groups by using TSB and R2A media. The culturable bacteria we obtained were all distributed in three phyla: Actinobacteria, Firmicutes, and Proteobacteria. On TSB medium, 14 genera of culturable dominant bacteria were isolated. *Streptomyces*, *Paenibacillus*, *Microbacterium*, *Brevibacterium*, and *Ensifer* were the top 5 genera, accounting for 44.46%, 14.01%, 12.65%, 7.03%, and 4.35%, respectively (Figure 4a, Table S1 and Figure S2). On R2A medium, 21 genera of culturable dominant bacteria were isolated. *Streptomyces*, *Microbacterium*, *Ensifer*, *Cellulosimicrobium*, and *Brevibacterium* were the top 5 genera, accounting for 48.51%, 10.64%, 8.77%, 4.23%, and 3.26%, respectively (Figure 4b and Figure S3 and Table S2).

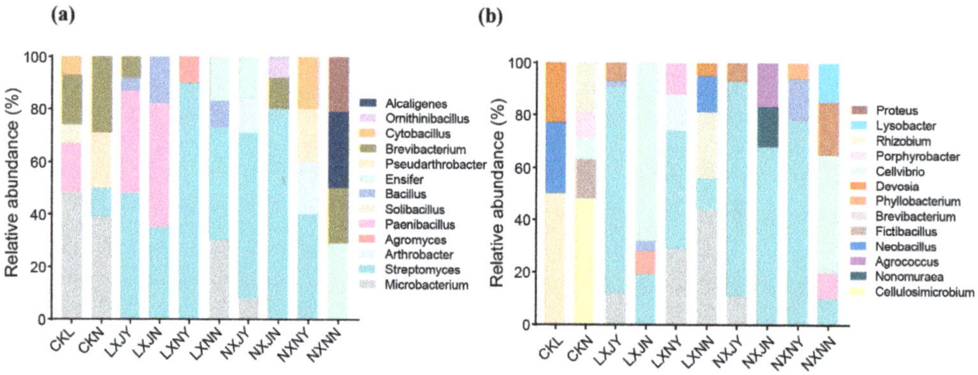

Figure 4. Relative abundance of bacterial genus levels in TSB (**a**) and R2A (**b**) media. CKL (blank control with *O. cumana* added), CKN (blank control without *O. cumana* added).

The number of culturable microorganisms in sunflower rhizosphere soil can be affected by the addition of D74 and maize rotation. When *O. cumana* present, the highest number of bacteria was isolated from the treatment group that was applied with D74 and maize rotation (LXJY) by TSB and R2A medium (TSB: 1.46×10^7 CFU/g; R2A: 2.05×10^7 CFU/g), while 1.28×10^7 CFU/g and 9.71×10^6 CFU/g were obtained from LXNY (*O. cumana* with maize rotation). Regardless of the presence or absence of *O. cumana*, the number of bacteria isolated after D74 application, maize rotation, or co-treatment with both was higher than the corresponding control group. *Streptomyces* numbers increased significantly in the sunflower rhizosphere soil in the D74 application group, possibly due to the D74 multiplied on the sunflower roots (Tables S1 and S2). In the unplanted control group (CKL), *Solibacillus* (TSB) and *Pseudomonas* (R2A) were isolated from the two media, respectively. In conclusion, different treatments have different effects on the types and numbers of culturable microorganisms in the sunflower rhizosphere soil. The isolation results in the R2A medium were similar to those in the TSB medium, and the results showed that the total number and proportion of *Streptomyces* were the largest in the culturable bacteria in the rhizosphere of sunflowers after D74 treatment.

3.4.2. Effects of *S. rochei* D74 Agents and Maize Rotation on Sunflower Rhizosphere Bacterial Community

The top five phyla of sunflower rhizosphere bacterial community in different treatment groups were Proteobacteria, Acidobacteria, Actinobacteria, Gemmatimonadetes, and Bacteroidetes, with relative abundance of 27.71%, 18.77%, 12.78%, 9.38%, and 8.12%, respectively (Figure 5a,b and Figure S4). Compared with the CKL group, the relative abundance of Bacteroidetes increased by 7.59%, 12.51%, 11.37%, and 7.77%; Actinobacteria increased by 6.48%, 6.32%, 5.98%, and 6.83%; Acidobacteria decreased by 3.84%, 2.69%, 4.69%, and 3.11%; and Gemmatimonadetes decreased by 3.13%, 5.66%, 3.49%, and 3.62%, respectively, in the treatment groups (LXJY, LXJN, LXNY, and LXNN groups) that added *O. cumana*. Compared with the CKN group, the relative abundance of Bacteroidetes increased by 6.54%, 13.82%, 5.6%, and 29.39%; Actinobacteria increased by 10%, 6.72%, 6.08%, and 2.77%; Acidobacteria decreased by 5.47%, 5.75%, 2.43%, and 5.81%; Gemmatimonadetes decreased by 1.57%, 5.45%, 0.85%, and 5.84% in the no *O. cumana* treatment group (NXJY, NXJN, NXNY, and NXNN), respectively. The results showed that, regardless of the presence or absence of *O. cumana*, compared with the control group (CKL and CKN), the maize rotation or the addition of D74 agent significantly increased the relative abundance of Bacteroidetes and Actinobacteria and decreased the relative abundance of Acidobacteria and Gemmatimonadetes.

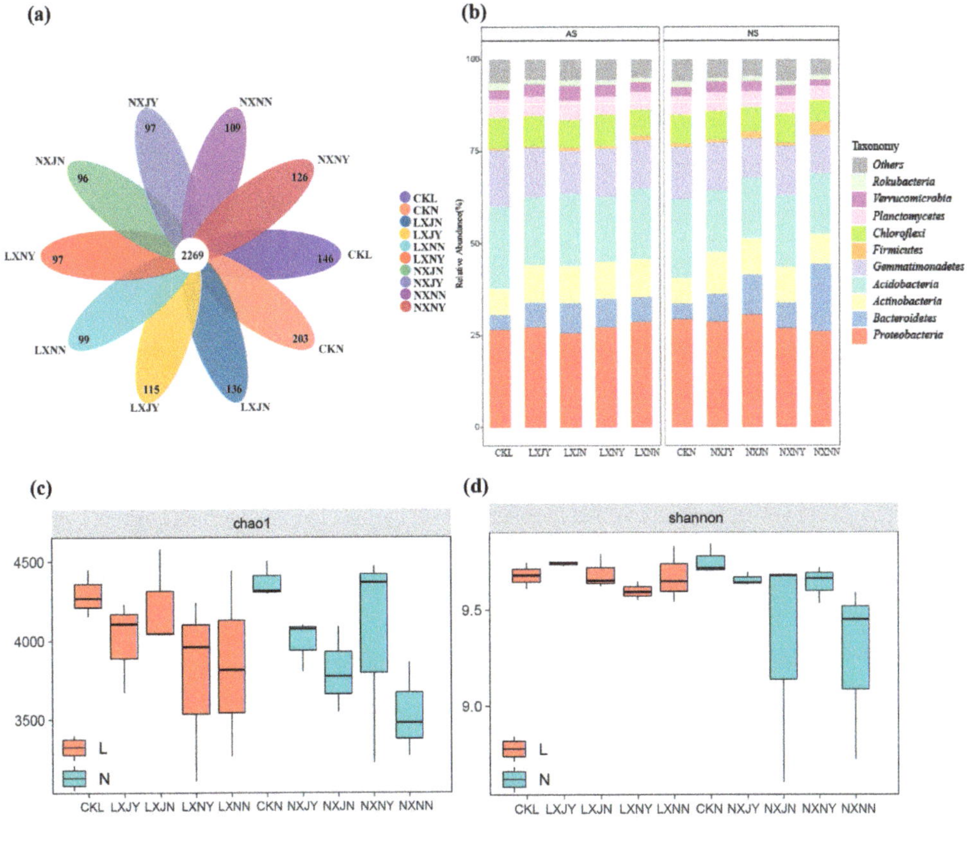

Figure 5. Effects of D74 agents and maize rotation on the α-diversity of sunflower rhizosphere bacterial community. Include OTUs petal map (**a**), relative abundance of Top 10 in phylum level (**b**), AS: added *O. cumana*, NS: unadded *O. cumana*), chao1 index (**c**), and shannon index (**d**).

For the α-diversity of the sunflower rhizosphere bacterial community, the chao1 index and shannon index of the no-plant control group (CKN) were the highest among all treatments, followed by CKL, while the α-diversity index of the sunflower-only group (NXNN) was the lowest (Figure 5c,d and Table S3). The difference of α-diversity index among the groups was small after adding *O. cumana* weeds to the soil, and the chao1 index and Shannon index of the D74 agents added groups (LXJY and LXJN) were higher than those of the corresponding control groups LXNY and LXNN. When no *O. cumana* was present in the soil, the chao1 index and Shannon index of the rotated maize groups (NXJY and NXNY) were higher than those of the non-rotated maize groups (NXJN and NXNN). The α-diversity of soil microbes in either the D74 alone (NXJN) or rotation maize alone (NXNY) groups were significantly higher than that in the control group (NXNN). This indicated that the addition of D74 or maize rotation increased the bacterial community diversity in the rhizosphere of sunflower, and the effect of rotated maize was greater than that of D74. The effect of *O. cumana* seeds on the α-diversity index of microorganisms was small. The results of PCoA for treatment groups (Figure 6a,b) showed that the composition of rhizosphere bacterial community was significantly different ($p < 0.05$), regardless of whether the *O. cumana* was added or not. Based on Bray–Curtis distance, the first two axes explained 20.00% and 26.30% of the total variance. In the *O. cumana*-free soil, the microbial communities were mainly separated along the PCoA1 axis, and the D74 agent and maize rotation had the greatest effect on the community difference, followed by the application of D74 agents alone, and the rotation of maize had the least effect. T-test analysis indicated that maize rotation significantly increased ($p < 0.05$) the relative abundances of *Sphingomonas*, *Altererythrobacter*, *Lechevalieria*, *Aeromicrobium*, and *Acidibacter* in bacterial communities. The D74-added group (LXJY) significantly increased the abundance of *Aquicella* and unidentified Alphaproteobacteria, compared to the LXNY group (Figure 6c,d).

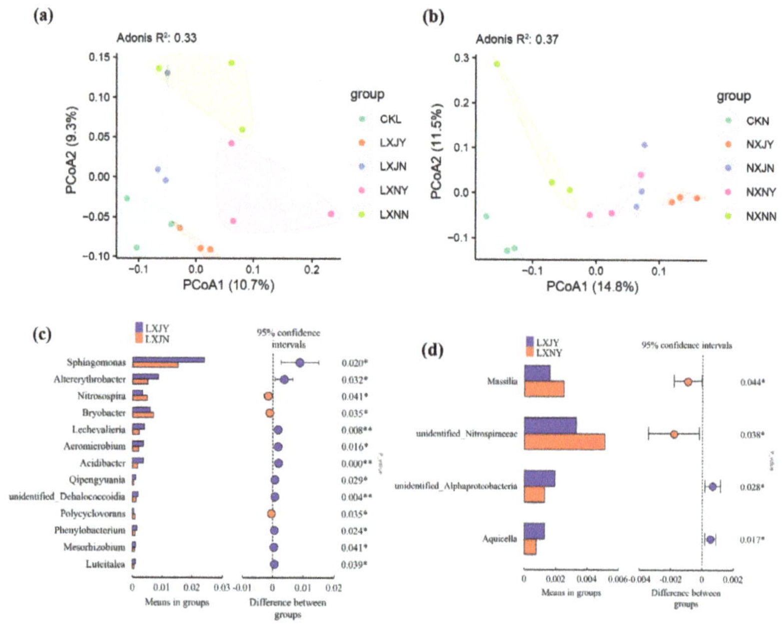

Figure 6. Effects of D74 agents and maize rotation on bacterial community structure. The PCoA based on Bray–Curtis distance with Adonis test ((**a**) added *O. cumana*, (**b**) unadded *O. cumana*, $p < 0.05$). Analysis of species with significant difference in genus level ((**c**) maize rotation, (**d**) added D74 agents, T-test, $p < 0.05$). Critical level of significance of constructs at * 5%, ** 1%.

4. Discussion

4.1. Effects of D74 Fermentation Filtrates on the Germination of O. cumana Seeds

The early growth stages of parasitic plants, such as seed germination, attachment host, and nodule development, are critical and ideal periods for controlling *O. cumana* parasitism. Therefore, the use of soil microorganisms to interfere with the development of parasitic weeds during the above period is an effective management strategy. Many fungi and bacteria can infect *Orobanche* and prevent its infestation to improve crop growth. In 2007, Zermane discovered natural soils that inhibited *Orobanche* parasitism [17]. Studies have shown that in the rhizosphere of plants, intensive and important interactions occur between parasitic plants, hosts, and microorganisms, such as biochemical reactions and the exchange of signaling molecules [30–32]. The use of microorganisms to control parasites is an environmentally friendly and easy to implement measure. Existing studies have shown that microorganisms not only reduce the number of parasites, but also significantly promote hosts growth. For example, *Fusarium oxysporum* f. sp. *orthoceras* (FOO) inhibited the germination of *O. cumana* and increased sunflower yield [33]. The use of *F. camptoceras* and *F. chlamydosporum* significantly reduced the biomass of *O. cumana* [34,35]. *Bacillus atrophaeus* inhibited *O. aegyptiaca* seeds germination and germ tubes growth, resulting in reduced parasitism number [36]. The *S. rochei* D74 used in this study not only has a good inhibitory effect on a variety of soil-borne diseases of crops, but also has a significant promotion effect on crop growth [20]. The D74 fermentation filtrates significantly inhibited the parasitism ($p < 0.05$), the effect was the best at high concentration, and the length of the seed germ tube was in the range of 0–1mm. In this case, the seeds show enlarged tops and darker colors, resulting in poor growth and reduced germination vigor.

4.2. Effects of Maize Rotation and S. rochei D74 Agents on Sunflower Growth and O. cumana Parasitism

The addition of D74 alone can promote the growth and development of sunflower roots, mainly in promoting the growth of main roots, increasing the number of lateral roots and fibrous roots, and providing more nutrients for the growth of the aerial parts. Additionally, the sunflower stem diameter, stem dry weight, and root dry weight increased after using D74 agents, which promoted the accumulation of sunflower dry matter. Both D74 inoculum or maize rotation could reduce the number of parasites, but the synergistic use of the two (LXJY) had the best inhibitory effect on *O. cumana*. Studies have shown that maize, as a trap crop, and its root exudates can induce *O. cumana* "suicide germination". However, after germination, *O. cumana* died quickly because there was no host to supply nutrients, thus reducing the number of *O. cumana* seed banks in the soil significantly [14,37]. At the same time, D74 inhibited the germination of *O. cumana*, and the parasitic number of *O. cumana* was significantly reduced in the second year. Additionally, after those two are used together, the sunflower biomass is increased due to the reduction of nutrient loss, which is of great significance for improving economic benefits.

4.3. Effects of Maize Rotation and S. rochei D74 Agents on Rhizosphere Microbial Community

The interaction between the root exudates and rhizosphere microorganisms is a very important process. Plant roots affect the species, quantity, and distribution of rhizosphere microorganisms through the secretion of various secondary metabolites and have a selective effect on the rhizosphere microbial community structure [38,39]. Research shows that changes in plant–microbe interactions mediated by root exudates are extremely important for soil fertility, health, and plant growth and development [40–42]. This study investigated the effects of D74 agents and maize rotation on the number and diversity of culturable and non-culturable bacterial communities in the rhizosphere of sunflower. Through the isolation and identification of rhizosphere microorganisms in two mediums, the bacteria isolated on both TSB and R2A media were distributed in three phyla, which were Actinobacteria, Firmicutes, and Proteobacteria. A total of 14 genera and 21 genera were isolated on TSB and R2A medium, respectively. The common genera were *Streptomyces*, *Microbacterium*,

Ensifer, and *Brevibacterium*. The addition of D74 agent significantly increased the number of *Streptomyces* and *Bacillus*. The *S. rochei* has been isolated from tomato, pepper, and other plant tissues. This strain suppressed the growth of the Chinese cabbage seedlings' pathogen *Pythium aphanidermatum* and *Sclerotium rolfsii* [43,44]. In summary, the addition of D74 agents increased the number of beneficial bacteria in the plant rhizosphere soil, thereby increasing soil fertility and promoting plant growth [45,46]. Both addition of D74 and maize rotation can reduce the total number of soil OTUs and the number of unique OTUs, compared with control soil. After analyzing the α-diversity of sunflower rhizosphere bacterial community, the results revealed that the richness and diversity of bacterial community in the addition of D74 increased, to a certain extent, compared with the control group. The bacterial community diversity in only the rotation maize treatment group decreased. In the absence of *O. cumana*, the D74 and maize rotation significantly increased the richness and diversity of bacterial communities, either alone or in combination. The results showed that the addition of D74 and the rotation of maize increased the types of soil microorganisms. The combined approach increased the abundance of Proteobacteria and Actinomycetes and decreased the abundance of Bacteroidetes, compared to the no-plant control (NXNN).

5. Conclusions

In order to reduce the parasitism of *O. cumana* in sunflower, we used a combination of *S. rochei* D74 and maize rotation for control. The combined prevention has proved to be quite effective. It reduced the parasitism of *O. cumana* well, with a reduction rate of 48.51%, and at the same time, could increase the accumulation of sunflower material and promote the growth of sunflowers. This method breaks the traditional idea of using a single method to control *O. cumana*. For practical purposes, we used microbial agents, instead of chemical control, which improves soil microbial diversity, has low environmental impact, adheres to the concept of green environment, and is conducive to the promotion of agriculture.

Supplementary Materials: The following supporting information can be downloaded at: https://www.mdpi.com/article/10.3390/agronomy12123129/s1, Figure S1: Growth of sunflower with different treatments (mature stage); Figure S2: Effects of different treatments in TSB medium on sunflower rhizosphere soil-cultivated bacteria; Figure S3: Effects of different treatments in R2A medium on sunflower rhizosphere soil-cultivated bacteria; Figure S4: Comparison of differences in OTUs of sunflower rhizosphere bacterial communities with (A) and without (B) *O. cumana*; Table S1: Effects of different treatments in TSB medium on the number of dominant strains in sunflower rhizosphere soil ($\times 10^5$ CFU/g); Table S2: Effects of different treatments in R2A medium on the number of dominant strains in sunflower rhizosphere soil ($\times 10^5$ CFU/g); Table S3: α-diversity index analysis of sunflower rhizosphere bacterial community.

Author Contributions: Conceptualization, Y.M. and Y.L. (Yanbing Lin); methodology, J.X. and Q.X.; investigation, W.Q., Z.D. and Y.X.; data curation, J.X.; writing—original draft preparation, J.X. and W.Q.; writing—review and editing, Y.L. (Yanbing Lin) and Y.L. (Yongxin Liu); visualization, J.X. and T.X. All authors have read and agreed to the published version of the manuscript.

Funding: This research was funded by the Fundamental Research Funds for the Central Universities, grant numbers 2452019183 and 2452021163.

Data Availability Statement: Raw sequence data of soil microbe is available in the National Center for Biotechnology Information (NCBI) database, with the accession number PRJNA898994.

Acknowledgments: We warmly thank Yongqing Ma for providing the guidance and yard for this study.

Conflicts of Interest: The authors declare no conflict of interest.

References

1. Yoshida, S.; Cui, S.; Ichihashi, Y.; Shirasu, K. The haustorium, a specialized invasive organ in parasitic plants. *Annu. Rev. Plant Biol.* **2016**, *67*, 643–667. [CrossRef] [PubMed]
2. Westwood, J.H.; Yoder, J.I.; Timko, M.P.; Depamphilis, C.W. The evolution of parasitism in plants. *Trends Plant Sci.* **2010**, *15*, 227–235. [CrossRef]
3. Uraguchi, D.; Kuwata, K.; Hijikata, Y.; Yamaguchi, R.; Imaizumi, H.; Am, S.; Rakers, C.; Mori, N.; Akiyama, K.; Irle, S.; et al. A femtomolar–range suicide germination stimulant for the parasitic plant *Striga hermonthica*. *Science* **2018**, *362*, 1301–1305. [CrossRef] [PubMed]
4. Plakhine, D.; Joel, D.M. Ecophysiological consideration of *Orobanche cumana* germination. *Helia* **2010**, *33*, 13–18. [CrossRef]
5. Mutuku, J.M.; Cui, S.; Yoshida, S.; Shirasu, K. Orobanchaceae parasite–host interactions. *New Phytol.* **2021**, *230*, 46–59. [CrossRef]
6. Goldwasser, Y.; Rodenburg, J. Integrated agronomic management of parasitic weed seed banks. In *Parasitic Orobanchaceae*; Joel, D.M., Ed.; Springer: Berlin, Germany, 2013; pp. 393–413.
7. Krupp, A.; Rücker, E.; Heller, A.; Spring, O. Seed structure characteristics of *Orobanche cumana* populations. *Helia* **2015**, *38*, 1–14. [CrossRef]
8. Parker, C. Observations on the current status of *Orobanche* and *Striga* problems worldwide. *Pest Manag. Sci.* **2009**, *65*, 453–459. [CrossRef]
9. Cimmino, A.; Masi, M.; Rubiales, D.; Evidente, A.; Fernández–Aparicio, M. Allelopathy for parasitic plant management. *Nat. Prod. Commun.* **2018**, *13*, 289–294. [CrossRef]
10. Goldwasser, Y.; Kleifeld, Y. Recent approaches to *Orobanche* management. *Weed Biol. Manag.* **2004**, *1*, 439–466.
11. Duriez, P.; Vautrin, S.; Auriac, M.-C.; Bazerque, J.; Boniface, M.-C.; Callot, C.; Carrère, S.; Cauet, S.; Chabaud, M.; Gentou, F. A receptor–like kinase enhances sunflower resistance to *Orobanche cumana*. *Nat. Plants* **2019**, *5*, 1211–1215. [CrossRef]
12. Van Bruggen, A.H.; He, M.; Shin, K.; Mai, V.; Jeong, K.; Finckh, M.; Morris Jr, J. Environmental and health effects of the herbicide glyphosate. *Sci. Total Environ.* **2018**, *616*, 255–268. [CrossRef] [PubMed]
13. Parker, C. A personal history in parasitic weeds and their control. *Plants* **2021**, *10*, 2249. [CrossRef] [PubMed]
14. Ma, Y.; Zhang, M.; Li, Y.; Shui, J.; Zhou, Y. Allelopathy of rice (*Oryza sativa* L.) root exudates and its relations with *Orobanche cumana* Wallr. and *Orobanche minor* Sm. germination. *J. Plant Interact.* **2014**, *9*, 722–730. [CrossRef]
15. Fernández-Aparicio, M.; García-Garrido, J.M.; Ocampo, J.A.; Rubiales, D. Colonisation of field pea roots by arbuscular mycorrhizal fungi reduces *Orobanche* and *Phelipanche* species seed germination. *Weed Res.* **2010**, *50*, 262–268. [CrossRef]
16. Xi, J.; Lei, B.; Liu, Y.X.; Ding, Z.; Liu, J.; Xu, T.; Hou, L.; Han, S.; Qian, X.; Ma, Y. Microbial community roles and chemical mechanisms in the parasitic development of *Orobanche cumana*. *iMeta* **2022**, *1*, e31. [CrossRef]
17. Zermane, N.; Souissi, T.; Kroschel, J.; Sikora, R. Biocontrol of broomrape (*Orobanche crenata* Forsk. and *Orobanche foetida* Poir.) by *Pseudomonas fluorescens* isolate Bf7–9 from the faba bean rhizosphere. *Biocontrol Sci. Technol.* **2007**, *17*, 483–497. [CrossRef]
18. Gonsior, G.; Buschmann, H.; Szinicz, G.; Spring, O.; Sauerborn, J. Induced resistance–an innovative approach to manage branched broomrape (*Orobanche ramosa*) in hemp and tobacco. *Weed Sci.* **2004**, *52*, 1050–1053. [CrossRef]
19. Chen, J.; Xue, Q.; McErlean, C.; Zhi, J.; Ma, Y.; Jia, X.; Zhang, M.; Ye, X. Biocontrol potential of the antagonistic microorganism *Streptomyces enissocaesilis* against *Orobanche cumana*. *BioControl* **2016**, *61*, 781–791. [CrossRef]
20. Li, Y.; Li, H.; Han, X.; Han, G.; Xi, J.; Liu, Y.; Zhang, Y.; Xue, Q.; Guo, Q.; Lai, H. Actinobacterial biofertilizer improves the yields of different plants and alters the assembly processes of rhizosphere microbial communities. *Appl. Soil Ecol.* **2022**, *171*, 104345. [CrossRef]
21. Li, Y.; Guo, Q.; He, F.; Li, Y.; Xue, Q.; Lai, H. Biocontrol of root diseases and growth promotion of the tuberous plant *Aconitum carmichaelii* induced by Actinomycetes are related to shifts in the rhizosphere microbiota. *Microb. Ecol.* **2020**, *79*, 134–147. [CrossRef]
22. Fei, H.; Zhong–liang, Z.; Ming, C.; Quan–hong, X.; Dong–sheng, W. Disease prevention and growth promotion effects of actinomycete strain D74 on *Amorphophallus konjac*. *Acta Hortic. Sin.* **2015**, *42*, 367.
23. Shi, L.; Nwet, T.T.; Ge, B.; Zhao, W.; Liu, B.; Cui, H.; Zhang, K. Antifungal and plant growth–promoting activities of *Streptomyces* roseoflavus strain NKZ–259. *Biol. Control* **2018**, *125*, 57–64. [CrossRef]
24. Hothem, S.D.; Marley, K.A.; Larson, R.A. Photochemistry in Hoagland's nutrient solution. *J. Plant Nutr.* **2003**, *26*, 845–854. [CrossRef]
25. Zhang, J.; Liu, Y.-X.; Guo, X.; Qin, Y.; Garrido-Oter, R.; Schulze-Lefert, P.; Bai, Y. High–throughput cultivation and identification of bacteria from the plant root microbiota. *Nat. Protoc.* **2021**, *16*, 988–1012. [CrossRef] [PubMed]
26. Perry, Z.H.; Barak, A.-T.; Neumann, L.; Levy, A. Computer–based learning: The use of SPSS statistical program for improving biostatistical competence of medical students. *J. Biomed. Educ.* **2014**, *2014*, 298140. [CrossRef]
27. Moberly, J.G.; Bernards, M.T.; Waynant, K.V. Key features and updates for origin 2018. *J. Cheminf.* **2018**, *10*, 5. [CrossRef]
28. Chen, T.; Liu, Y.X.; Huang, L. ImageGP: An easy-to-use data visualization web server for scientific researchers. *iMeta* **2022**, *1*, e5. [CrossRef]
29. Chen, T.; Zhang, H.; Liu, Y.; Liu, Y.-X.; Huang, L. EVenn: Easy to create repeatable and editable Venn diagrams and Venn networks online. *J. Genet. Genom.* **2021**, *48*, 863–866. [CrossRef]
30. Bouwmeester, H.J.; Roux, C.; Lopez–Raez, J.A.; Becard, G. Rhizosphere communication of plants, parasitic plants and AM fungi. *Trends Plant Sci.* **2007**, *12*, 224–230. [CrossRef]

31. Cardoso, C.; Ruyter–Spira, C.; Bouwmeester, H.J. Strigolactones and root infestation by plant–parasitic *Striga*, *Orobanche* and *Phelipanche* spp. *Plant Sci.* **2011**, *180*, 414–420. [CrossRef]
32. Albert, M.; Axtell, M.J.; Timko, M.P. Mechanisms of resistance and virulence in parasitic plant–host interactions. *Plant Physiol.* **2021**, *185*, 1282–1291. [CrossRef] [PubMed]
33. Müller–Stöver, D.; Batchvarova, R.; Kohlschmid, E.; Sauerborn, J. Mycoherbicidal management of *Orobanche cumana*: Observations from three years of field experiments. In Proceedings of the 10th International World Congress of Parasitic Plants, Kusadasi, Turkey, 8–12 June 2009; p. 86.
34. Boari, A.; Vurro, M. Evaluation of *Fusarium* spp. and other fungi as biological control agents of broomrape (*Orobanche ramosa*). *Biol. Control* **2004**, *30*, 212–219. [CrossRef]
35. Anteyi, W.O.; Klaiber, I.; Rasche, F. Diacetoxyscirpenol, a *Fusarium* exometabolite, prevents efficiently the incidence of the parasitic weed *Striga hermonthica*. *BMC Plant Biol.* **2022**, *22*, 84. [CrossRef] [PubMed]
36. He, W.; Li, Y.; Luo, W.; Zhou, J.; Zhao, S.; Xu, J. Herbicidal secondary metabolites from *Bacillus velezensis* JTB8–2 against *Orobanche aegyptiaca*. *AMB Express* **2022**, *12*, 52. [CrossRef]
37. Ma, Y.; Jia, J.; An, Y.; Wang, Z.; Mao, J. Potential of some hybrid maize lines to induce germination of sunflower broomrape. *Crop. Sci.* **2013**, *53*, 260–270. [CrossRef]
38. Guorong, N.; Guoquan, T.; Saijin, W.; Jianfu, W.; Qinghua, S.; Chunhuo, Z.; Xiaohua, P. Effects of combination of straw returning and a microbial agent on microorganisms and enzyme activity in rhizosphere soil and yield of late rice. *Meteorol. Environ. Res.* **2017**, *8*, 78–82.
39. Singh, G.; Mukerji, K.G. Root exudates as determinant of rhizospheric microbial biodiversity. In *Microbial Activity in the Rhizoshere;* Springer: Berlin, Germany, 2006; pp. 39–53.
40. Zhalnina, K.; Louie, K.B.; Hao, Z.; Mansoori, N.; da Rocha, U.N.; Shi, S.; Cho, H.; Karaoz, U.; Loqué, D.; Bowen, B.P.; et al. Dynamic root exudate chemistry and microbial substrate preferences drive patterns in rhizosphere microbial community assembly. *Nat. Microbiol.* **2018**, *3*, 470–480. [CrossRef]
41. Jiang, N.; Wu, M.; Li, G.; Li, P.; Liu, M.; Li, Z. Comparative effects of two humic substances on microbial dysbiosis in the rhizosphere soil where cucumber (*Cucumis sativus* L.) is grown. *Land Degrad. Dev.* **2022**, *33*, 1944–1953. [CrossRef]
42. Mendes, R.; Kruijt, M.; De Bruijn, I.; Dekkers, E.; Van Der Voort, M.; Schneider, J.H.; Piceno, Y.M.; DeSantis, T.Z.; Andersen, G.L.; Bakker, P.A. Deciphering the rhizosphere microbiome for disease-suppressive bacteria. *Science* **2011**, *332*, 1097–1100. [CrossRef]
43. Ezziyyani, M.; Requena, M.; Egea-Gilabert, C.; Candela, M. Biological control of Phytophthora root rot of pepper using *Trichoderma harzianum* and *Streptomyces rochei* in combination. *J. Phytopathol.* **2007**, *155*, 342–349. [CrossRef]
44. Suwitchayanon, P.; Chaipon, S.; Chaichom, S.; Kunasakdakul, K. Potentials of *Streptomyces rochei* ERY1 as an endophytic actinobacterium inhibiting damping–off pathogenic fungi and growth promoting of cabbage seedling. *Chiang Mai J. Sci* **2018**, *45*, 692–700.
45. Meena, R.K.; Singh, R.K.; Singh, N.P.; Meena, S.K.; Meena, V.S. Isolation of low temperature surviving plant growth–promoting rhizobacteria (PGPR) from pea (*Pisum sativum* L.) and documentation of their plant growth promoting traits. *Biocatal. Agric. Biotechnol.* **2015**, *4*, 806–811. [CrossRef]
46. Hua, L.Q.; Yang, S.Q.; Xia, Z.F.; Zeng, H. Application of Sophora alopecuroides organic fertilizer changes the rhizosphere microbial community structure of melon plants and increases the fruit sugar content. *J. Sci. Food Agric.* **2022**, *103*, 164–175. [CrossRef] [PubMed]

agronomy

Article

Dodder Parasitism Leads to the Enrichment of Pathogen *Alternaria* and Flavonoid Metabolites in Soybean Root

Wen Luo [1], Yuanli Li [1], Ruiping Luo [2], Gehong Wei [1], Yongxin Liu [3,*] and Weimin Chen [1,*]

[1] State Key Laboratory of Crop Stress Biology in Arid Areas, Shaanxi Key Laboratory of Agricultural and Environmental Microbiology, College of Life Sciences, Northwest A&F University, Xianyang 712100, China
[2] Crop Research Institute of Ningxia Academy of Agriculture and Forestry Sciences, Yinchuan Comprehensive Experimental Station of National Soybean Industry Technology System, Yinchuan 750105, China
[3] Shenzhen Branch, Guangdong Laboratory of Lingnan Modern Agriculture, Genome Analysis Laboratory of the Ministry of Agriculture and Rural Affairs, Agricultural Genomics Institute at Shenzhen, Chinese Academy of Agricultural Sciences, Shenzhen 518120, China
* Correspondence: liuyongxin@caas.cn (Y.L.); chenwm029@nwsuaf.edu.cn (W.C.)

Abstract: Dodders (*Cuscuta chinensis*) are rootless and holoparasitic herbs that can infect a variety of host plants, including the vitally important economic and bioenergy crop soybean (*Glycine max*). Although dodder parasitism severely affects the physiology of host plants, little is known about its effects on fungal communities and root secondary metabolites in hosts. In this study, variations in root-associated fungal communities and root metabolites of soybean under different parasitism conditions were investigated using ITS rRNA gene sequencing and UPLC–MS/MS metabolome detection technologies. The results showed that dodder parasitism significantly altered the composition and diversity of the fungal communities in the rhizosphere and endosphere of soybean. The relative abundance of the potential pathogenic fungus *Alternaria* significantly increased in the root endosphere of dodder-parasitized soybean. Furthermore, correlation analysis indicated that the fungal community in the root endosphere was susceptible to soil factors under dodder parasitism. Meanwhile, the content of soil total nitrogen was significantly and positively correlated with the relative abundance of *Alternaria* in the rhizosphere and endosphere of soybean. Metabolomic analysis indicated that dodder parasitism altered the accumulation of flavonoids in soybean roots, with significant upregulation of the contents of kaempferol and its downstream derivatives under different parasitism conditions. Taken together, this study highlighted the important role of dodder parasitism in shaping the fungal communities and secondary metabolites associated with soybean roots, providing new insights into the mechanisms of multiple interactions among dodder, soybean, microbial communities and the soil environment.

Keywords: *Cuscuta*; flavonoid metabolite; legume plant; parasitism status; rhizosphere; root-associated microbiome

Citation: Luo, W.; Li, Y.; Luo, R.; Wei, G.; Liu, Y.; Chen, W. Dodder Parasitism Leads to the Enrichment of Pathogen *Alternaria* and Flavonoid Metabolites in Soybean Root. *Agronomy* **2023**, *13*, 1571. https://doi.org/10.3390/agronomy13061571

Academic Editor: Federico Vita

Received: 13 May 2023
Revised: 4 June 2023
Accepted: 8 June 2023
Published: 9 June 2023

1. Introduction

Parasitic plants are heterotrophs that obtain resources for growth and reproduction from host plants [1]. They attach to the host root system or stem through a specialized organ called the haustorium. The life history of parasitic plants severely affects the growth of host plants [2]. In particular, the holoparasitic dodder is completely dependent on the host plant to complete its life history [3] and is considered a common noxious weed that can reduce crop and forage production in agroecosystems [4]. In contrast to parasitic plants, beneficial root-associated microbes have significantly positive effects on their host plants. Numerous studies demonstrated the important role of beneficial microbes in promoting plant growth or stress resistance [5–7] and even in alleviating the damage caused by parasitism [8]. Therefore, exploring the host microbial community under plant parasitism is essential to the management of parasitic plants and their hosts.

The rhizosphere is a critical link in soil nutrient cycling and crop nutrient acquisition in agricultural fields [9]. In the narrow rhizosphere environment, close interactions between roots and microbes can strongly promote the nutrient acquisition processes of plants [10,11]. In particular, the rhizosphere harbors various functional guilds of fungi with varying trophic patterns, such as endophytes, saprophytes or pathogens [12]. Environmental factors, soil characteristics and cultivation practices can affect the diversity and composition of the rhizosphere fungal community [13–15]. Consequently, these factors also alter plant growth, thereby jointly regulating the rhizosphere microbial community [9]. These studies also demonstrate that the growth status of plants plays an important role in influencing rhizosphere microbial communities [14,16]. Dodder parasitism, as an important biotic stress, has serious negative impacts on host growth by also acquiring fixed carbon, water and mineral resources [17]. To tolerate detrimental parasitism, host plants need to modify their physiological or metabolic activities and attract beneficial microbes from their surrounding environment into the rhizosphere and even the root endosphere [18–20]. Some studies investigated the effects of parasitic plants on root-associated bacterial communities in host plants [21–23]. However, little is known about the effect of parasitic plant, such as dodder, on host root-associated fungal communities.

Root metabolites of plants have essential associations with the surrounding fungal communities. Metabolomic analyses indicate that plant root tissues and exudates contain hundreds of secondary metabolites [24]. Some of these secondary metabolites can act as signaling molecules, nutrient sources or toxins toward the fungal community in the rhizosphere and root endosphere [25]. For example, studies demonstrated that flavonoids produced by plant roots can enhance mycorrhizal symbioses by promoting the germination of fungal spores, increasing fungal colonization of the roots and stimulating hyphal growth [26,27]. Benzoxazinoids and herbicolin A have a significant effect on pathogenic fungal taxa in the maize [28] and wheat [29]. Biotic stress imposed by parasitic plants may change the metabonomic characteristics of the host. Mistletoe (*Phoradendron perrottetii*) infection had negative effects on the soluble carbohydrates in branches of *Tapirira guianensis* [30]. Parasitism by dodder increased the levels of phenolic acids and flavonols in the leaves of cranberry cultivars [31]. Biotic stress not only induces defense response metabolites in aboveground plant parts but also may influence chemical substances in the roots [32]. Nonetheless, there is limited knowledge on how the accumulation of these secondary metabolites within host roots is affected by dodder parasitism.

Soybean (*Glycine max*) is an important crop for addressing global food insecurity, providing a valuable source of plant protein, oil and biofuel [33]. However, increasing challenges, including rapid human population growth, soil degradation and climate change, threaten global food production [34]. The holoparasitic plant dodder (*Cuscuta chinensis*), a parasitic plant that usually infests soybean crops, can severely disrupt soybean growth and yield [35,36]. Dodder and soybean are regarded as an excellent model system for studying the interactions between parasitic plants and hosts [37,38]. We conducted this research in a major soybean-growing region in northwestern China, which is persistently infected with dodder. This research could be beneficial for understanding the impact of the interaction between dodder and soybean on the composition of root-associated fungal communities and the root metabolite profiling in soybean. We identified the composition of the root-associated fungal communities and root metabolites from dodder-parasitized and nonparasitized soybeans. There were three objectives: (1) to investigate the impact of dodder parasitism on the composition and recruitment of root-associated fungal communities in soybean, (2) to assess how key soil factors influence the composition and specific taxa of root-associated fungal communities in soybean under dodder parasitism and (3) to determine the effect of dodder parasitism on the metabolite profiles and the accumulation of specific metabolites in soybean roots.

2. Materials and Methods

2.1. Sample Collection

The study area was located in Yinchuan, Ningxia Hui Autonomous Region, China. The dominant soil type was Anthrosol. The area was in an semi-arid region, which belongs to the temperate continental climate. Soybean (*Glycine max*) was one of the major economic crops in the region, and suffered heavy infestation by the holoparasitic plant dodder (*Cuscuta chinensis*). Soybeans were parasitized by dodder for more than 10 years in some localized areas, as dodder is also considered an economically important herbal medicine. Samples were collected from three different sites, including S1 (38°6′55″ N, 106°10′45″ E), S2 (38°12′56″ N, 106°11′54″ E) and S3 (38°9′44″ N, 106°12′25″ E) [39]. Based on the growth status (Figure S1), two types of soybeans with dodder parasitism were selected: soybeans with slightly inhibited growth (P1) and soybeans with significantly inhibited growth (P2). Soybeans that were not subjected to dodder parasitism were used as controls. At each sampling site, five random plots (3×3 m) were utilized as replicates. Fifteen plants of each soybean type were randomly collected and mixed as a replicate sample in each plot. Bulk soil (BS) samples were collected from 0 to 20 cm depth and 20 cm away from the main stems of the plants. The root compartment samples were collected according to a previous method [40]: the entire soybean plant was carefully removed from the soil, and its root system was gently shaken to collect loose soil adhering to the roots. The loose soil was used as the root zone soil (ZS) sample. Then, the tightly adhered soil (<1 mm) was washed from the roots with 1 M phosphate-buffered solution and used as the rhizosphere soil (RS) sample. Finally, ultrasonic treatment was performed on the roots two or more times to remove root surface microbes, and the remaining roots were preserved as root endosphere (RE) samples. The BS and ZS samples were separated into two subsamples. One subsample was kept for the subsequent analysis of soil physicochemical properties, while the other subsample and the RS and RE samples were stored at −80 °C for microbial analysis.

2.2. DNA Extraction, Sequencing and Bioinformatics Analysis

According to the manufacturers' instructions, the genomic DNA was extracted from soil and plant samples using the FastDNA SPIN Kit for Soil (MP, Santa Ana, CA, USA) and the Power Plant Pro DNA kit (MB, Santa Ana, CA, USA). The primers 1737F (5′-GGAAGTAAAAGTCGTAACAAGG-3′) and 2043R (5′-GCTGCGTTCTTCATCGATGC-3′) were used to amplify the sequence of the fungal ITS1 region [41]. The polymerase chain reaction (PCR) amplifications were performed in triplicate for each sample, and the PCR products were mixed in equidensity ratios. The obtained PCR products were checked on a 2.0% agarose gel and purified using the GeneJET Gel Recovery Kit from Thermo Scientific. DNA libraries were constructed using the Ion Plus Fragment Library Kit from Thermo Fisher Scientific. The DNA library quality was assessed using the Qubit@ 2.0 Fluorometer (Thermo Fisher Scientific, Waltham, MA, USA). Finally, the library was sequenced on Illumina MiSeq platform (Illumina Inc., San Diego, CA, USA) for high-throughput double-end sequencing at Novogene Bioinformatics Technology Co., Ltd. (Beijing, China).

Raw sequences were analyzed using DADA2 (version 1.8.0) [42], and unique microbial taxa were inferred based on amplicon sequence variants (ASVs). The DADA2 pipeline mainly included: sequence trimming and filtering, removal of redundant sequences, removal of chimeric sequences, generation of ASV abundance tables and fungal taxonomic annotation (UNITE database) [43]. To minimize the effect of sequencing artifacts, singletons were removed. Considering the differences in sequencing depth per sample, the data were normalized with the minimal sequence method and were then subjected to downstream analysis. Fungal functional annotation library FUNGuild was used to predict the potential environmental functions of fungal communities [12].

2.3. Determination of Physical and Chemical Properties of Soils

Sixteen soil factors were measured using standard testing methods [44,45], including soil pH, soil organic matter (SOM), total nitrogen (TN), ammonium nitrogen (NH_4^+-N), nitrate nitrogen (NO_3^--N), total phosphorus (TP), available phosphorus (AP), total potassium (TK), available potassium (AK), soluble sodium (Na), soluble potassium (K), soluble calcium (Ca), soluble magnesium (Mg), percentage of clay, sand and silt.

2.4. Widely-Targeted Metabolomic Analysis

For the widely-targeted metabolomic analysis, a set of three biological samples were randomly chosen as replicates. The sample preparation and extraction process was carried out as described below: (1) plant root samples were freeze-dried using a vacuum freeze-dryer (Scientz-100F, Scientz, Ningbo, China); (2) the freeze-dried samples were ground into a fine powder using a mixer mill (MM 400, Retsch, Haan, Germany) at 30 Hz for 1.5 min; (3) 100 mg of the powder was weighed and dissolved in 1.2 mL of a 70% methanol solution; (4) the sample was vortexed for 30 s every 30 min, for a total of 6 times, and then stored in a refrigerator at 4 °C overnight; (5) the sample was subjected to centrifugation at 12,000 rpm for 10 min and then filtered through a microporous membrane (0.22 μm) before being stored for UPLC–MS/MS analysis.

UPLC–MS/MS (UPLC, SHIM-PACK UFLC Shimadzu CBM30A system; MS, Applied Biosystems 4500 QTRAP) was used to perform extensive targeted metabolomics on the prepared samples (Metware, Wuhan, China). The operation procedure was as follows: a Waters ACQUITY UPLC HSS T3 C18 column (1.8 μm, 2.1 mm × 100 mm) was used, and the mobile phase consisted of solvent A (water containing 0.04% acetic acid) and solvent B (acetonitrile containing 0.04% acetic acid). The linear gradient elution program was as follows: phase B increased from 5% to 95% during the first 9.0 min and was maintained at 95% for another 1.0 min, followed by 2.9 min of re-equilibrium (phase A/phase B: 95%/5%). The column temperature was set at 40 °C, and the injection volume was 4 μL.

In the present study, an electrospray ionization (ESI)-triple quadrupole-linear ion trap (QTRAP)-mass spectrometer (AB 4500 Q TRAP UPLC/MS/MS System) was utilized to perform linear ion trap (LIT) and triple quadrupole (QQQ) scans. An ESI turbo ion–spray interface was equipped and executed in the positive/negative ion mode. The scans were controlled by Analyst 1.6.3 software (AB Sciex). The ESI source operating parameters were as follows: ion source, turbo spray, source temperature 550 °C; ion spray voltage (IS): 5500 V (positive ion mode)/−4500 V (negative ion mode); ion source gas I (GSI), 50 psi; ion source gas II (GSII), 60 psi; curtain gas (CUR), 25 psi; the collision-induced ionization parameter, high. The QQQ scan was obtained by multiple reaction monitoring (MRM) experiments with the collision gas (nitrogen) set to medium. After further optimization of the collision energy (CE) and declustering potential (DP), a specific set of MRM pairs was monitored for each period according to the metabolites eluted within the period. Data acquisition and processing were performed as described previously [46]. Metabolites were annotated using the Metware in-house MS2 spectral tag (MS2T) library (Wuhan Metware Biotechnology Co., Ltd.; http://www.metware.cn, accessed on 8 May 2022, Wuhan, China).

2.5. Statistical Analysis

All statistical analyses were performed with R software (3.5.0, http://www.r-project.org, accessed on 1 April 2023). Unless stated, visualization of data relied on the ggplot2 package [47]. The significance was tested by one-way analysis of variance (ANOVA) and Tukey's HSD post hoc test in the stats and multcomp packages. Constrained principal coordinates analysis (CAP) based on Bray–Curtis distance was performed to visualize the relationships among samples using the vegan package [48]. Multivariate permutation analysis of variance (PERMANOVA/Adonis) was employed to determine differences in microbial communities between treatments by the vegan package. The correlations between the matrices of the microbial communities and soil properties were examined using the Mantel test in the ggClusterNet package [49]. Spearman correlation and ordinary least squares linear regression were used to analyze the relationship between dominant taxa and specific soil nutrients.

The metabolomic data were processed by several multivariate statistical analysis methods. Unsupervised principal component analysis (PCA) and hierarchical clustering analysis (HCA) were applied to determine the overall metabolite differences among different groups by the vegan package and ComplexHeatmap package [50], respectively. Supervised multiple regression orthogonal partial least squares discriminant analysis (OPLS-DA) was conducted to discriminate the differentially expressed metabolites based on the variable importance in projection (VIP) value in the test model by the MetaboAnalystR package [51]. The threshold VIP value ≥ 1 and fold change ≥ 2 (upregulated) or ≤ 0.5 (downregulated) were used for screening the differential metabolites. Differential metabolites identified in the OPLS-DA were used for further k-means clustering analysis to investigate variations in different clusters of the metabolites. Metabolites were mapped to the Kyoto Encyclopedia of Genes and Genomes (KEGG) pathway database (http://www.kegg.jp/kegg/pathway.html, accessed on 8 May 2022) to determine pathway associations. Pathway enrichment analysis was performed using Metabolite Sets Enrichment Analysis (MSEA; http://www.msea.ca, accessed on 8 May 2022). The significance was determined by Bonferroni-corrected *p* values.

3. Results

3.1. Alpha and Beta Diversity of the Fungal Community

PERMANOVA analysis revealed that the root-associated fungal community of soybean was significantly affected by sampling site ($R^2 = 0.16$, $p = 0.001$), root compartment ($R^2 = 0.12$, $p = 0.001$) and parasitism condition ($R^2 = 0.02$, $p = 0.032$) (Figures 1A and S2A). Additionally, both the sampling site and parasitism condition had a greater influence on the community in RS than in ZS and RE of soybean (Figure 1B). To elucidate the effect of parasitism on the root-associated fungal communities, further analysis was conducted to determine the variations in the fungal communities among different parasitism conditions at each sampling site. Parasitism significantly affected the alpha diversity in different root compartments ($p < 0.05$) and increased the richness index of the RE community at sampling site S3 (Figure S2B). Furthermore, paired PERMANOVA tests showed that different parasitism effects (P1 vs. control, P2 vs. control and P1 vs. P2) significantly affected the fungal community structure at each sampling site. Although different parasitism effects on fungal community composition varied with sampling site, the RE community composition was significantly affected by parasitism across all sampling sites ($p < 0.05$) (Figure S2C).

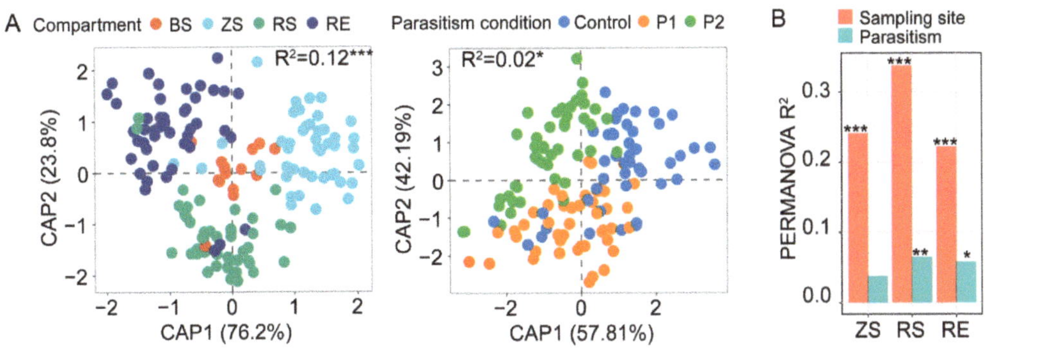

Figure 1. General response patterns of root-associated fungal community diversities in soybeans to parasitism. (**A**) Constrained analysis of principal coordinates (CAP) based on Bray–Curtis distances, showing the compositional variation explained by compartment and parasitism. (**B**) Bar chart showing the compositional variation of the root-associated fungal communities explained by parasitism condition and sampling site (PERMANOVA test, * $p < 0.05$, ** $p < 0.01$, and *** $p < 0.001$). BS, Bulk soil; ZS, root zone soil; RS, rhizosphere soil; RE, root endosphere.

3.2. Fungal Community Composition and the Enrichment of Dominant Taxa

According to the community abundance analysis, the dominant phyla in different root compartments were Ascomycota (BS: 57.25%, RZ: 60.33%, RS: 70.22%, RE: 63.60%) and Basidiomycota (BS: 10.81%, RZ: 8.19%, RS: 1.44%, RE: 8.09%), and their relative abundances varied with parasitism condition (Figure 2A). Specifically, the relative abundance of Ascomycota was higher under P1 and P2 conditions than under control condition in ZS and RE. The relative abundance of Ascomycota under P2 condition was higher than that under P1 condition in all root compartments. At the genus level, *Fusarium* (RZ, 15.83%; RS, 21.82%, RE, 22.29%) and *Alternaria* (RZ, 5.41%; RS, 17.05%, RE, 26.10%) were the dominant taxa in all root compartments. Compared with under the control condition, the relative abundance of *Alternaria* was significantly higher under P1 condition in RE ($p < 0.05$) and under P2 condition in RS ($p < 0.05$) (Figure 2B). Furthermore, functional prediction analysis of the fungal community showed that the relative abundance of plant pathogens and parasites significantly increased under P1 condition in RE ($p < 0.05$) and under P2 condition in RS ($p < 0.05$) compared to under control condition. These results suggest that dodder parasitism led to the enrichment of potentially pathogenic fungi in the host root system.

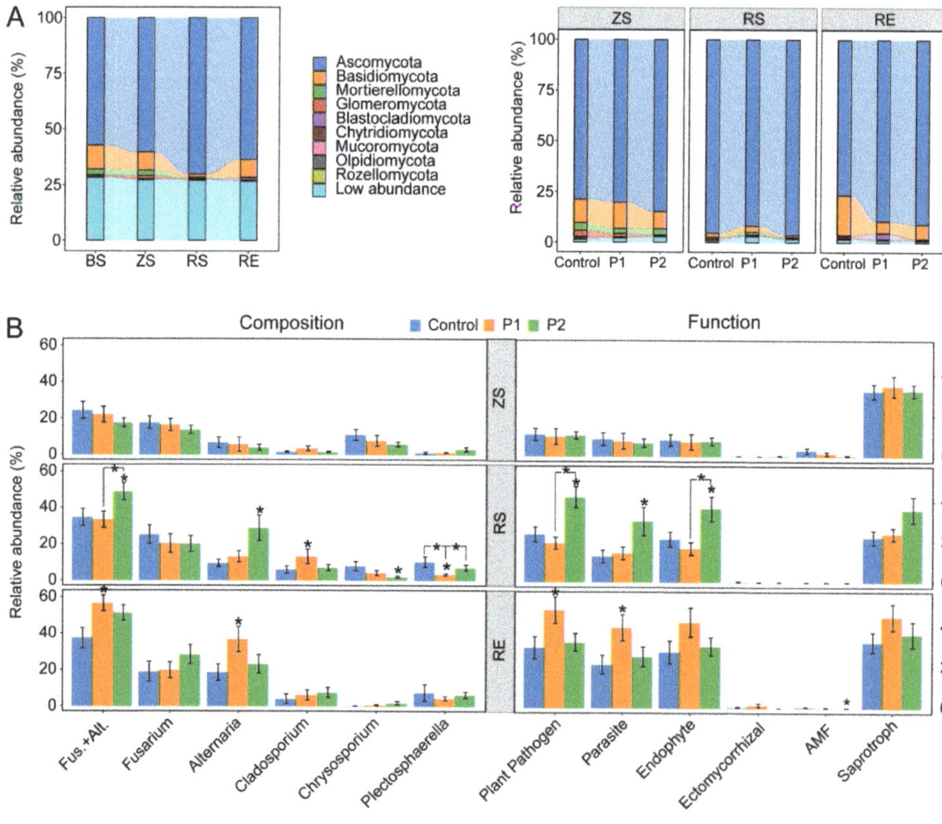

Figure 2. Composition and potential functions of fungal communities under different parasitism conditions. (**A**) The distribution of the fungal communities in different root compartments. (**B**) Variation in dominant taxa and their potential functions among different soybeans. (* $p < 0.05$). BS, bulk soil; ZS, root zone soil; RS, rhizosphere soil; RE, root endosphere.

3.3. Relationship between Fungal Communities and Soil Factors

Community dissimilarity analysis was used to investigate the effect of parasitism on the variation in the fungal community among different sampling sites. The results showed that fungal community dissimilarity in ZS and RS significantly increased under P1 and P2 conditions ($p < 0.05$) compared with under control condition (Figure S3). However, the RE community under the P1 and P2 conditions showed the opposite trend. The Mantel test further showed the correlations between fungal communities and different soil factors. More soil factors were significantly associated with the RE fungal community by parasitism (control: 6; P1: 10, P2: 13). In addition, the effect of important soil factors on the fungal communities was different among parasitism conditions (Figure 3A). Under control condition, fungal communities in ZS, RS and RE were mainly affected by silt (r = 0.47, $p = 0.001$), silt (r = 0.65, $p = 0.001$) and soluble K (r = 0.46, $p = 0.001$), respectively. Under P1 condition, these fungal communities were mainly influenced by soluble K (r = 0.60, $p = 0.001$), soluble Ca (r = 0.51, $p = 0.001$) and TN (r = 0.67, $p = 0.001$). Under P2 condition, these fungal communities were mainly affected by soluble Na (r = 0.60, $p = 0.001$), AP (r = 0.60, $p = 0.001$) and NH4+-N (r = 0.74, $p = 0.001$). In general, the RE fungal community under P1 and P2 conditions exhibited strong environmental sensitivity, especially to soil nitrogen.

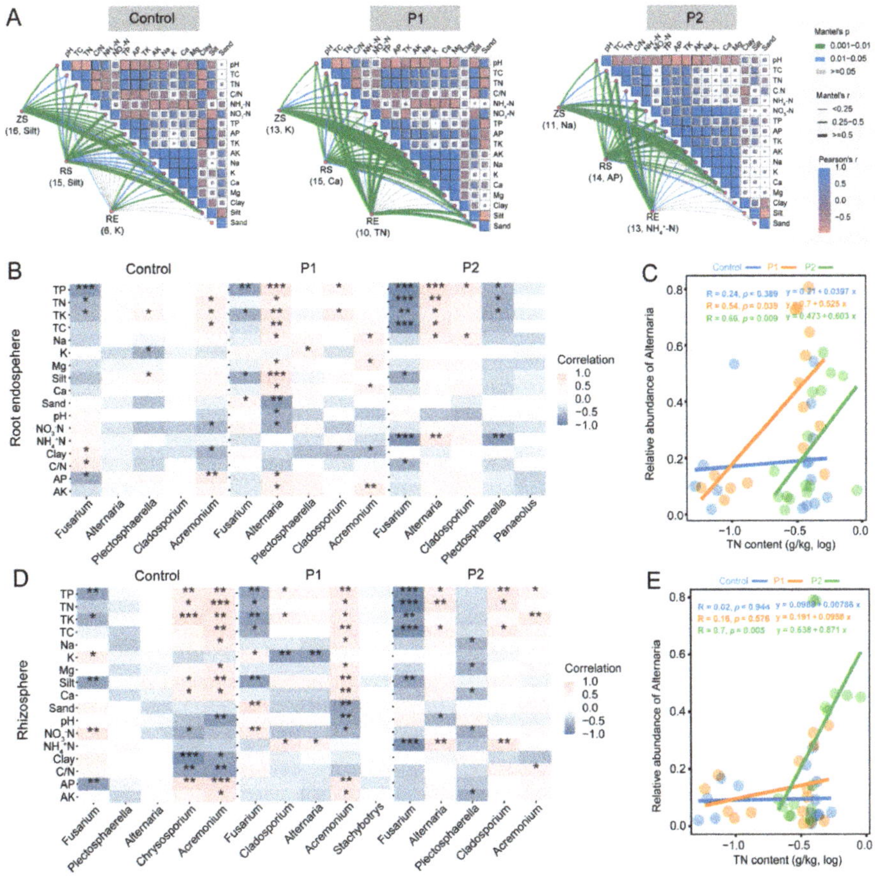

Figure 3. Effect of parasitism on the relationships between soil properties and the root-associated fungal communities of soybean. (**A**) Correlation analysis between soil properties and the root-associated

fungal communities (Bray–Curtis distances) based on the Mantel test. The color of the line represents the significance of the differences (p values). The thickness of the line represents correlation coefficients (Mantel's r). The number of significant soil factors and the strongest soil factors are displayed under each compartment. (**B**,**D**) Spearman correlation analysis between soil factors and dominant taxa. (* $p < 0.05$, ** $p < 0.01$, and *** $p < 0.001$). (**C**,**E**) The relationship between soil TN content and the relative abundance of *Alternaria*. The line represents the linear regression of the ordinary least squares model. BS, bulk soil; ZS, root zone soil; RS, rhizosphere soil; RE, root endosphere.

Spearman correlation analysis was performed to examine the relationship between soil factors and dominant fungal taxa. The results showed that dodder parasitism altered the correlations between soil factors and the top five fungal taxa in each root compartment. Notably, more soil nutrients were significantly correlated ($p < 0.05$) with the relative abundance of *Alternaria* in RS and RE under P1 and P2 conditions (Figure 3B). In the RS, the soil NH_4^+-N content was positively correlated with the relative abundance of *Alternaria* under the two parasitism conditions. In RE, the contents of TC, TN, TP, TK and soluble Na were positively correlated with the relative abundance of *Alternaria* under the two parasitism conditions. These results indicated close relationships between soil nitrogen (TN and NH_4^+-N) and *Alternaria* under dodder parasitism. The ordinary least square regression analysis further confirmed the strong and positive correlation between soil TN content and the relative abundance of *Alternaria* in RS and RE under P2 condition (Figures 3C,E and S4).

3.4. Accumulation of Metabolites in Soybean Roots

The microbial analysis showed that dodder parasitism had a significant effect on the fungal communities in both the rhizosphere and root endosphere at sampling site S1 (Figure S2C). Consequently, soybean root samples from this site were chosen for metabolomic analysis to gain a better understanding of how parasitism affects the metabolites. A total of 956 metabolites were detected in soybean roots, with flavonoids being the most abundant class (18.9%), followed by lipids (14.4%), phenolic acids (3.9%) and terpenoids (10.2%) (Figure 4A). Additionally, PCA showed that the metabolites exhibited a clear separation between different parasitism conditions based on the first principal component (PC1, explained 54.92% of the total variance) and the second principal component (PC2, explained 16.79% of the total variance) (Figure 4B). This result indicated that the total metabolites were greatly changed by dodder parasitism. HCA analysis classified the metabolites with the same characteristics into a group to identify the variation in the content of metabolites among different parasitism conditions. The heatmap showed that the metabolites were clearly divided into three distinct profiles by parasitism, further confirming the strong effect of dodder parasitism on the metabolites (Figure 4C). Differential metabolite analysis showed that a total of 474 metabolites significantly varied among different parasitism conditions, with flavonoids being the most abundant class (21.3%), followed by lipids (15.4%), phenolic acids (13.5%), and amino acids and their derivatives (9.7%) (Figure 4D).

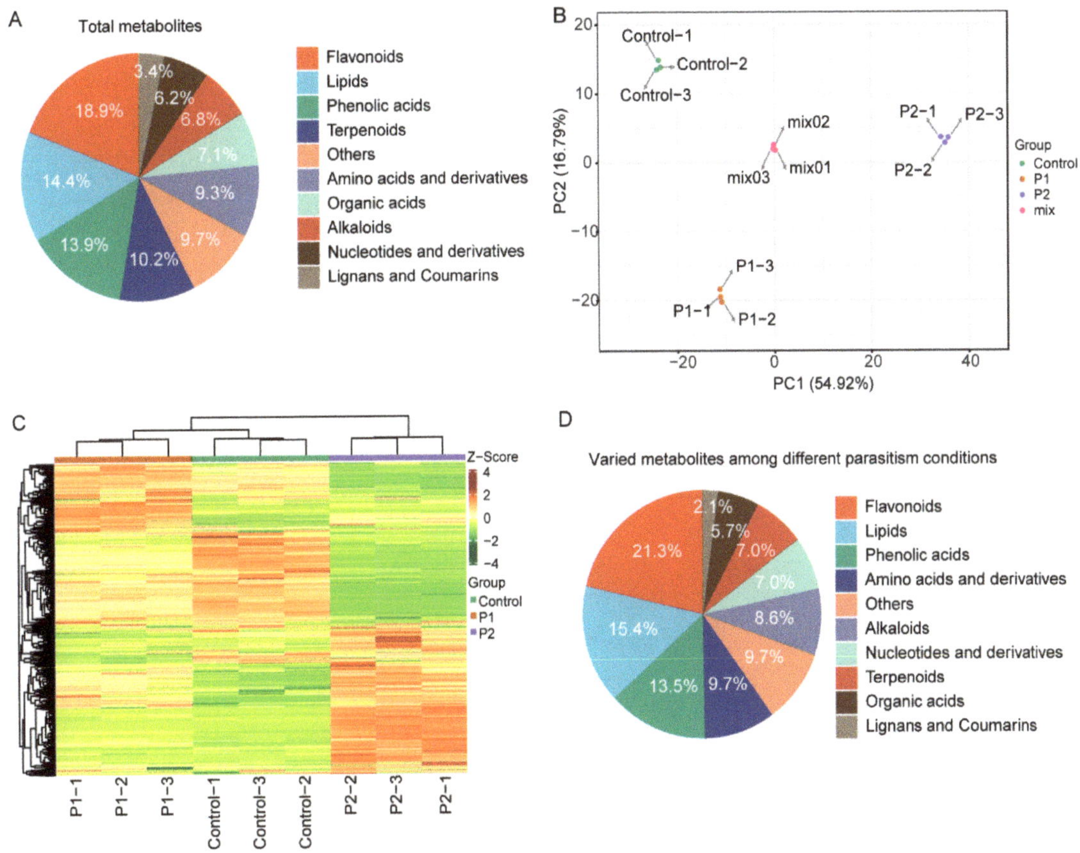

Figure 4. Qualitative and quantitative analysis of the metabolomics data of soybean roots under different parasitism conditions. (**A**) Bar chart showing the categories of identified metabolites. (**B**) PCA of metabolites identified from soybean roots. (**C**) HCA of the metabolites identified from soybean roots. (**D**) Bar chart showing the categories of the metabolites that significantly changed under different parasitism conditions.

OPLS-DA was used to compare the metabolic characteristics of soybeans under different parasitism conditions. The results showed a total of 150 metabolites that significantly differed between control condition and P1 condition, of which 120 metabolites were upregulated under P1 condition (Figure 5A). In contrast, a total of 411 metabolites were significantly different between the control condition and P2 condition, of which 201 were upregulated under the P2 condition (Figure 5B). To study the change trends of differential metabolites among different sample groups, k-means cluster analysis was then performed. These differential metabolites were divided into five subclasses (Figure 5C). Notably, the standardized relative content of 63 metabolites in subclass 3 was elevated in soybean roots under P1 and P2 conditions, with the highest levels under P2 condition. In contrast, the standardized relative content of 92 metabolites in subclass 5 sharply varied in the opposite way, showing a decreasing trend under both P1 and P2 conditions, with the lowest level under P2 condition.

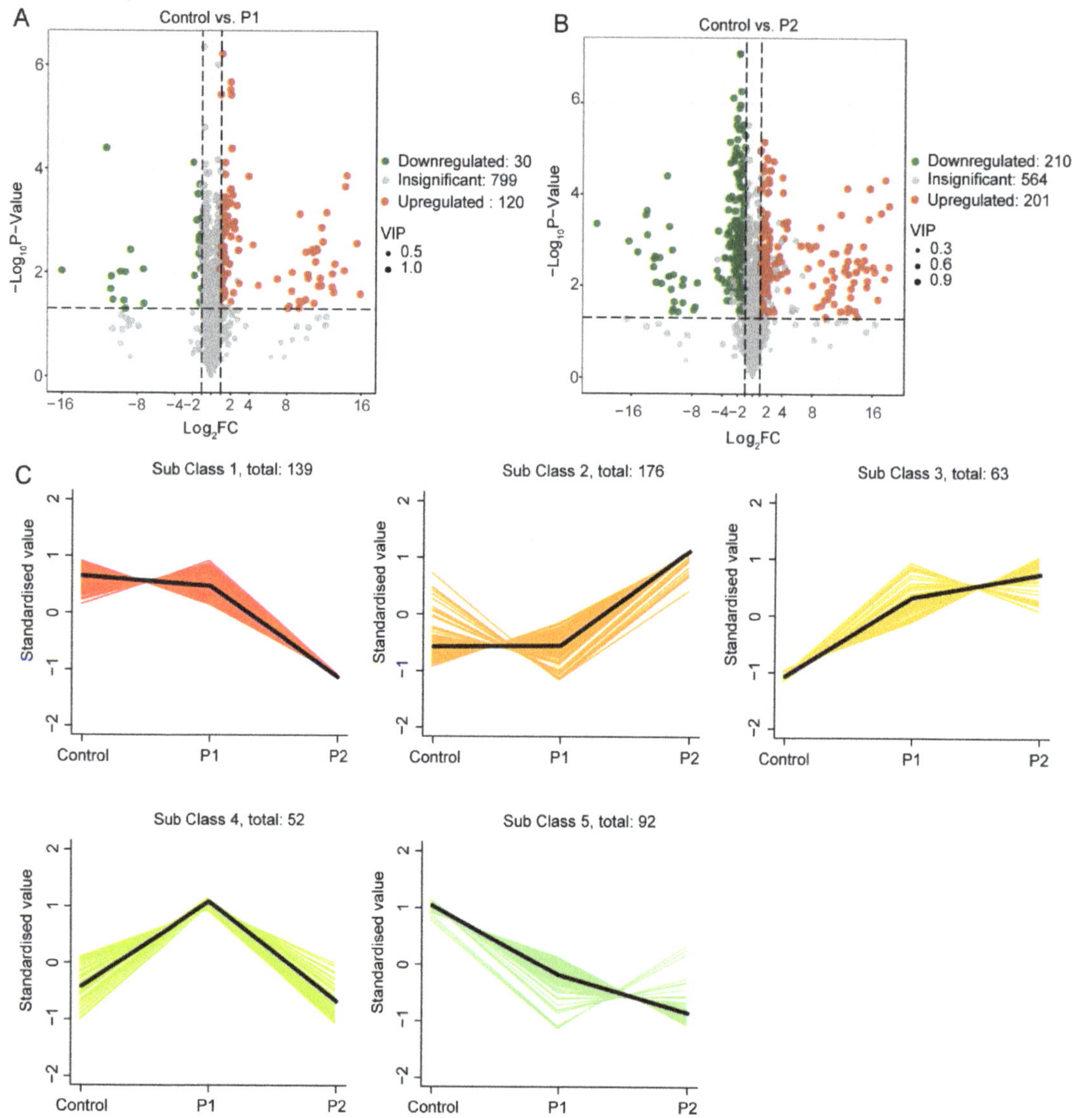

Figure 5. K-means analysis of all differential metabolites of soybean roots among different parasitism conditions. Volcano plots showing the expression levels of different metabolites in soybean roots under P1 (**A**) and P2 (**B**) conditions. (**C**) K-means clustering analysis of differential metabolites based on the fuzzy C-means algorithm (Mfuzz).

The differential enrichment analysis yielded 82 core metabolites coenriched in soybean roots under both P1 and P2 conditions, mainly composed of flavonoids (35.4%), followed by phenolic acids (12.2%), alkaloids (9.8%) and lipids (9.8%) (Figure 6A). Similarly, the metabolites in subclass 3 were mainly composed of flavonoids (41.3%) (Figure 6B). To identify core metabolites that are sensitive to parasitism, the coenriched metabolites and subclass 3 metabolites were compared. Twenty-nine core flavonoids were further identified and were mainly kaempferol, luteolin, quercetin and their derivatives (Table S1). Specifically, the contents of kaempferol (3,5,7,4′-tetrahydroxyflavone), hesperetin-5-*O*-glucoside, isorhamnetin-7-O-glucoside and neptin-7-O-alloside exhibited greater changes than other metabolites (Figure 6C).

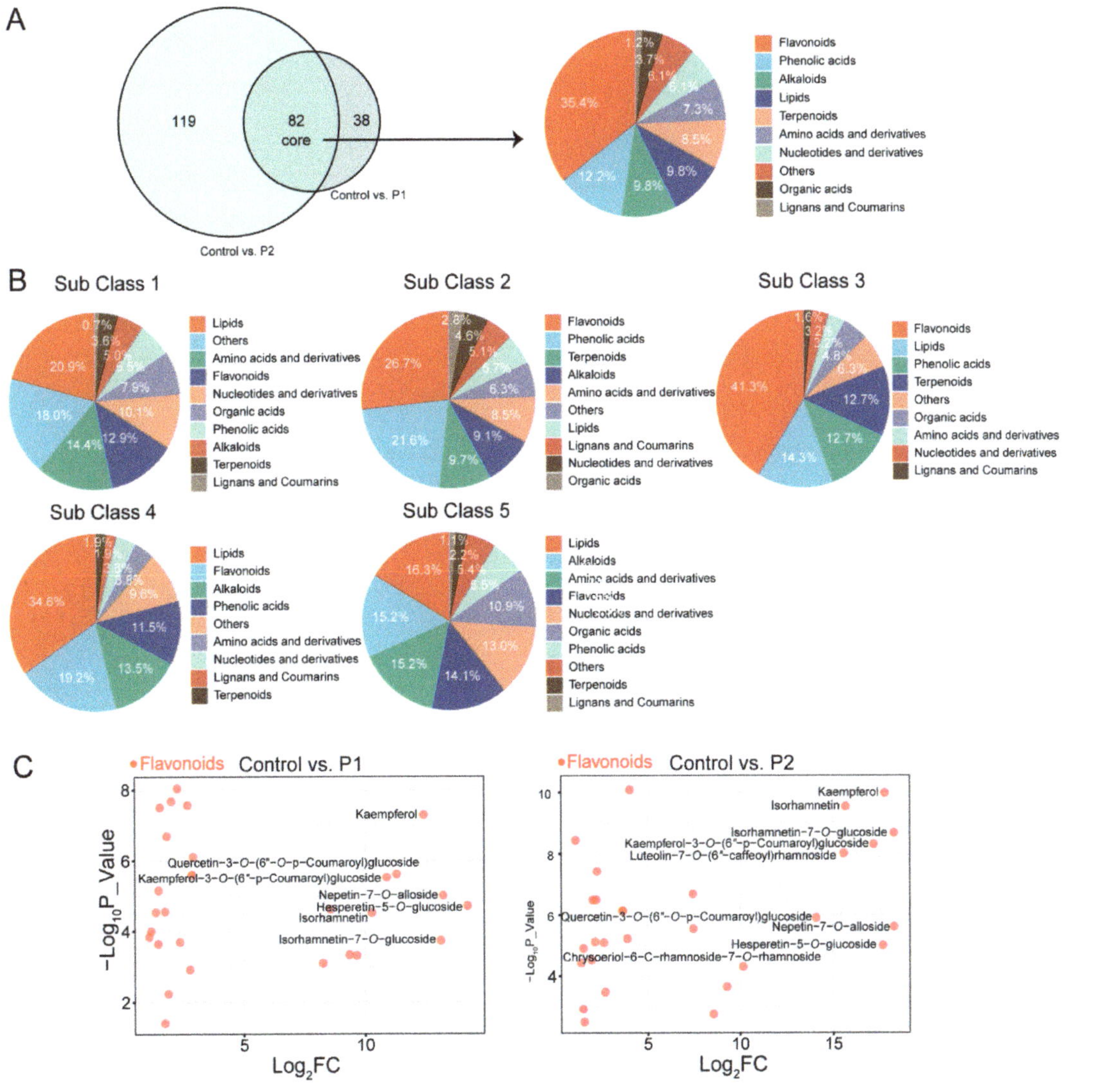

Figure 6. The distribution of coenriched metabolites (**A**) and K-means cluster metabolites (**B**). (**C**) Volcano plot of the core flavonoid metabolites that were significantly enriched under both P1 and P2 conditions and sensitive to parasitism. Metabolites with more than 10-fold change were labeled on the left or top of the dot in volcano plot.

3.5. Flavonoid Metabolic Reprogramming Induced by Dodder in Soybean

To deeply investigate the impact of dodder parasitism on the accumulation of flavonoid metabolites in soybean roots, the metabolic dataset was analyzed using the KEGG database in a point-by-point manner. The flavonoid metabolites that were significantly altered due to parasitism were assigned to a common metabolic pathway and subsequently interpreted in a flavonoid biosynthesis metabolite network (Figure 7). The results showed that many downstream metabolites of naringenin (a key intermediate and precursor in the flavonoid biosynthesis pathway) were significantly activated by dodder parasitism ($p < 0.05$), such as kaempferol, luteolin, quercetin and their metabolite derivatives. Under P1 and P2 conditions, neohesperidin, sakuranetin, scolymoside, kaempferol and astragalin metabolites were upregulated in soybean roots, while dihydrokaempferol was the opposite. Meanwhile, upstream (dihydrokaempferol) and downstream (quercetin) metabolites of kaempferol were downregulated, while its own and derived metabolites (astragalin) were upregulated under two parasitism conditions. Moreover, more metabolites, such as luteolin, isovitexin, isoquercitrin and rutin, were upregulated in soybean roots under P2 condition than under P1 condition.

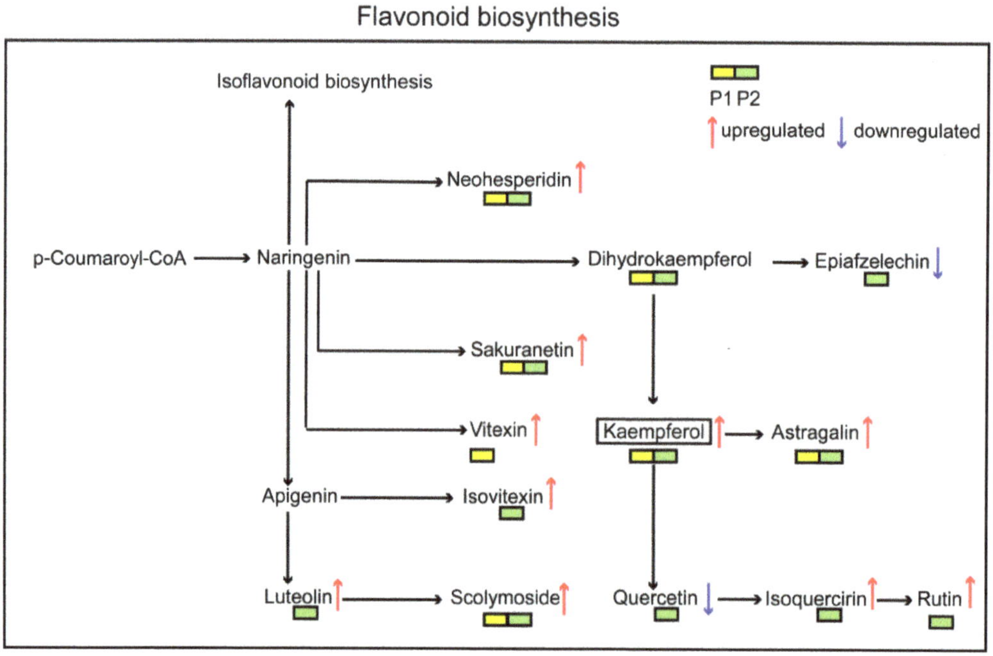

Figure 7. Regulation of the flavonoid metabolite biosynthesis network in soybean roots by parasitism. The yellow and green boxes represent metabolites that exhibit significant changes under P1 and P2 conditions, respectively. The red- and blue-colored arrows indicate the metabolites that were significantly upregulated and downregulated ($p < 0.05$), respectively.

4. Discussion

4.1. Dodder Parasitism Led to the Enrichment of Potential Pathogenic Fungus Alternaria in Soybean Root

Our results indicated that although the effects of soil factors, plant site and parasitism on the fungal communities of soybean were low ($R^2 < 0.4$), the effects were significant. The results were similar to some studies on plant root-associated microbial communities [21,52,53]. Thus, such findings can not be ignored. The possible reason might be that plants have a strong selective effect on root-associated microbes through metabolites such as root secretions [7] and plants can supply a stable environment to maintain the ecological niche of microbes [54]. Therefore, different factors may be difficult to strongly affect plant-associated microbes. Dodder parasitism had the least significant impact on the root-associated fungal community. This might be related to indirect regulation by dodder through the host plant. Previous studies demonstrated that the influence of the host plant genotype on plant-associated microbial communities was weaker than that of soil sources [53,55]. Compared with the rhizosphere and root zones, the fungal community in the root endosphere was more susceptible to different parasitism effects of dodder across different soil environments, indicating a close relationship between the cascading effect induced by parasitism and the root fungal community in the host. However, this potential cascading effect on the root fungal community might be negative for plants. Differential abundance analysis indicated that dodder parasitism resulted in the significant enrichment of potential pathogenic fungi of the genus *Alternaria* in soybean roots. There could be two potential explanations for this observation: on the one hand, it is possible that parasitic stress by dodder leads to disruption of the physiological metabolism and immune system of the host plant [4,56]; on the other hand, dodder parasitism may lead to the accumulation of nutrient resources (e.g., metabolites) in soybean roots, reducing the competition between microbes and favoring the persistence of a larger number of microbes. Interestingly, previous studies suggested the isolation of *Alternaria* from dodder for use as a dodder-resistant herbicide [57]. Whether the invasion of *Alternaria* in host plants is associated with dodder-resistance needs to be further investigated in future work.

4.2. Dodder Parasitism Strengthened the Relationship between Soil Nitrogen and the Genus *Alternaria* in Soybean Root

The distribution of fungal communities associated with plant roots is known to be closely linked to soil environmental factors [58]. The present study demonstrated that parasitism by dodder further strengthened this association in the host soybean. This is similar to previous research, indicating that external stress can alter the response of plant microbial communities to environmental changes [59]. External stress can stimulate the systemic immune response of the host, leading to the accumulation of metabolites in the host roots that can impact the recruitment and colonization of host root microbes [60]. In the case of soybean parasitized by dodder, this effect may be amplified, making the host root-associated microenvironment more unstable and vulnerable to environmental disturbances. Our results revealed that the fungal communities within the roots of soybean parasitized by dodder were affected by a greater number of soil factors, particularly total nitrogen. Nitrogen is known to have an inhibitory effect on the formation of soybean root nodules [61], which could significantly impact soybean growth. Moreover, we observed a positive correlation between total soil nitrogen content and the abundance of *Alternaria* in soybean, indicating that while biological nitrogen fixation in soybean may be compromised, pathogenic fungi may take advantage of the opportunity to invade. This raises serious concerns for the healthy growth of soybean and coping with dodder parasitism, underscoring the need for additional trade-offs in future management of soybean and dodder in agricultural systems.

4.3. Dodder Parasitism Promote the Accumulation of Flavonoid Metabolites in Soybean Root

Phenylpropanoids are known to play a crucial role in plant defense against biotic and abiotic stresses, including pathogen and pest attacks [62]. Flavonoids are a key product of the phenylpropanoid pathway. Our metabolomic analysis revealed that flavonoids were induced in soybean roots under dodder parasitism. The flavonoid synthesis pathway plays multiple roles in plant roots, including as a signaling molecule for symbiotic microorganisms and resistance to pathogenic microorganisms. Flavonoids can act as signaling molecules in the root zone to recruit rhizobia and promote the development of nodules, which help the host plant to grow [63]. It is noteworthy that the secondary metabolite kaempferol and its derivatives in the roots of soybean were highly sensitive to dodder parasitism, and their contents accumulate actively in response. A previous study indicated that the flavonol kaempferol acted as an auxin transport regulator and was critical to the nodulation of *Medicago truncatula* [64]. Furthermore, flavonoids are known to have fungicidal activity against potential pathogenic fungi [65,66]; for example, the parasitism-induced flavonoid rutin exhibited a potent inhibitory effect on the growth of *Alternaria* [67]. Hence, the accumulation of flavonoids in soybean roots under dodder parasitism could be beneficial in establishing systemic defense in soybean, potentially inhibiting pathogenic fungi and enhancing soybean resistance to the parasitic plant.

5. Conclusions

In this study, we investigated the impact of dodder parasitism on the root-associated fungal communities and root secondary metabolites in soybean, as well as the correlation between the microbial communities and soil properties. Our results demonstrated a significant effect of dodder parasitism on the fungal community in the soybean root endosphere across different soil environments. Additionally, the genus *Alternaria* was found to be enriched in soybean roots and was closely associated with soil nitrogen under dodder parasitism. These findings suggest that dodder parasitism increased the risk of potential pathogen invasion into soybean roots while also enhancing the environmental sensitivity of the root fungal community. Furthermore, our metabolomic analysis indicated that dodder parasitism significantly induced the accumulation of flavonoid metabolites in soybean roots. Overall, our study revealed the collaborative response of the root-associated fungal communities and root secondary metabolites in soybeans to dodder, potentially providing a new perspective for the management of dodder and soybean.

Supplementary Materials: The following supporting information can be downloaded at: https://www.mdpi.com/article/10.3390/agronomy13061571/s1, Figure S1: The sketch of soybeans with three parasitism status (Control, P1, P2) and three rhizocompartments (root zone soil, rhizosphere soil and root endosphere); Figure S2: The effect of dodder parasitism on the root-associated fungal community of host in each sampling site; Figure S3: The variation in the dissimilarity (Bray–Curtis distance) of the fungal community in each root compartment among different parasitism statuses; Figure S4: The relationship between soil NH_4^+-N content and the relative abundance of *Alternaria*; Table S1: The core flavonoid metabolites that significantly enriched under both P1 and P2 conditions based on the differential enrichment analysis and the K-means clustering analysis.

Author Contributions: Investigation, W.L. and R.L.; data curation, Y.L. (Yuanli Li); writing—original draft preparation, W.L.; conceptualization, G.W.; writing—review and editing, W.C. and Y.L. (Yongxin Liu) supervision and funding acquisition, W.C. All authors have read and agreed to the published version of the manuscript.

Funding: This study was funded by the National Natural Science Foundation of China, grant number 42177106 and 31870476.

Data Availability Statement: Raw sequence data are available in the GenBank database under the project accession number PRJNA967994.

Acknowledgments: We warmly thank Jiao Xi for providing the guidance and yard for this study.

Conflicts of Interest: The authors declare no conflict of interest.

References

1. Yoshida, S.; Cui, S.; Ichihashi, Y.; Shirasu, K. The haustorium, a specialized invasive organ in parasitic plants. *Annu. Rev. Plant Biol.* **2016**, *67*, 643–667. [CrossRef] [PubMed]
2. Cirocco, R.M.; Facelli, J.M.; Watling, J.R. A native parasitic plant affects the performance of an introduced host regardless of environmental variation across field sites. *Funct. Plant Biol.* **2018**, *45*, 1128–1137. [CrossRef] [PubMed]
3. Hartenstein, M.; Albert, M.; Krause, K. The plant vampire diaries: A historic perspective on *Cuscuta* research. *J. Exp. Bot.* **2023**, *74*, 2944–2955. [CrossRef]
4. Albert, M.; Axtell, M.J.; Timko, M.P. Mechanisms of resistance and virulence in parasitic plant–host interactions. *Plant Physiol.* **2021**, *185*, 1282–1291. [CrossRef] [PubMed]
5. Mendes, R.; Kruijt, M.; De Bruijn, I.; Dekkers, E.; Van Der Voort, M.; Schneider, J.H.; Piceno, Y.M.; Desantis, T.Z.; Andersen, G.L.; Bakker, P.A. Deciphering the rhizosphere microbiome for disease-suppressive bacteria. *Science* **2011**, *332*, 1097–1100. [CrossRef]
6. Panke-Buisse, K.; Poole, A.C.; Goodrich, J.K.; Ley, R.E.; Kao-Kniffin, J. Selection on soil microbiomes reveals reproducible impacts on plant function. *ISME J.* **2015**, *9*, 980–989. [CrossRef]
7. Santoyo, G. How plants recruit their microbiome? New insights into beneficial interactions. *J. Adv. Res.* **2022**, *40*, 45–58. [CrossRef]

8. Sui, X.; Zhang, T.; Tian, Y.; Xue, R.; Li, A. A neglected alliance in battles against parasitic plants: Arbuscular mycorrhizal and rhizobial symbioses alleviate damage to a legume host by root hemiparasitic *Pedicularis* species. *New Phytol.* **2019**, *221*, 470–481. [CrossRef]
9. Zhang, R.; Vivanco, J.M.; Shen, Q. The unseen rhizosphere root–soil–microbe interactions for crop production. *Curr. Opin. Microbiol.* **2017**, *37*, 8–14. [CrossRef]
10. Hashem, A.; Abd Allah, E.F.; Alqarawi, A.A.; Al-Huqail, A.A.; Wirth, S.; Egamberdieva, D. The interaction between arbuscular mycorrhizal fungi and endophytic bacteria enhances plant growth of *Acacia gerrardii* under salt stress. *Front. Microbiol.* **2016**, *7*, 1089. [CrossRef]
11. Chen, J.; Huang, X.; Tong, B.; Wang, D.; Liu, J.; Liao, X.; Sun, Q. Effects of rhizosphere fungi on the chemical composition of fruits of the medicinal plant *Cinnamomum migao* endemic to southwestern China. *BMC Microbiol.* **2021**, *21*, 206. [CrossRef]
12. Nguyen, N.H.; Song, Z.; Bates, S.T.; Branco, S.; Tedersoo, L.; Menke, J.; Schilling, J.S.; Kennedy, P.G. Funguild: An open annotation tool for parsing fungal community datasets by ecological guild. *Fungal Ecol.* **2016**, *20*, 241–248. [CrossRef]
13. Coller, E.; Cestaro, A.; Zanzotti, R.; Bertoldi, D.; Pindo, M.; Larger, S.; Albanese, D.; Mescalchin, E.; Donati, C. Microbiome of vineyard soils is shaped by geography and management. *Microbiome* **2019**, *7*, 140. [CrossRef]
14. Li, Y.; Wang, Z.; Li, T.; Zhao, D.; Han, J.; Liao, Y. Wheat rhizosphere fungal community is affected by tillage and plant growth. *Agric. Ecosyst. Environ.* **2021**, *317*, 107475. [CrossRef]
15. Suryanarayanan, T.S.; Shaanker, R.U. Can fungal endophytes fast-track plant adaptations to climate change? *Fungal Ecol.* **2021**, *50*, 101039. [CrossRef]
16. Qu, Q.; Zhang, Z.; Peijnenburg, W.J.G.M.; Liu, W.; Lu, T.; Hu, B.; Chen, J.; Chen, J.; Lin, Z.; Qian, H. Rhizosphere microbiome assembly and its impact on plant growth. *J. Agric. Food Chem.* **2020**, *68*, 5024–5038. [CrossRef]
17. Westwood, J.H.; Yoder, J.I.; Timko, M.P.; Depamphilis, C.W. The evolution of parasitism in plants. *Trends Plant Sci.* **2010**, *15*, 227–235. [CrossRef]
18. Press, M.C.; Phoenix, G.K. Impacts of parasitic plants on natural communities. *New Phytol.* **2005**, *166*, 737–751. [CrossRef] [PubMed]
19. Bouwmeester, H.J.; Roux, C.; Lopez-Raez, J.A.; Becard, G. Rhizosphere communication of plants, parasitic plants and am fungi. *Trends Plant Sci.* **2007**, *12*, 224–230. [CrossRef] [PubMed]
20. Li, W.; Wang, L.; Li, X.; Zheng, X.; Cohen, M.F.; Liu, Y. Sequence-based functional metagenomics reveals novel natural diversity of functioning CopA in environmental microbiomes. *Genom. Proteom. Bioinform.* 2022, *in press*. [CrossRef]
21. Brunel, C.; Beifen, Y.; Pouteau, R.; Li, J.; van Kleunen, M. Responses of rhizospheric microbial communities of native and alien plant species to *Cuscuta* parasitism. *Microb. Ecol.* **2020**, *79*, 617–630. [CrossRef] [PubMed]
22. Xi, J.; Ding, Z.; Xu, T.; Qu, W.; Xu, Y.; Ma, Y.; Xue, Q.; Liu, Y.; Lin, Y. Maize rotation combined with *Streptomyces rochei* D74 to eliminate *Orobanche cumana* seed bank in the farmland. *Agronomy* **2022**, *12*, 3129. [CrossRef]
23. Xi, J.; Lei, B.; Liu, Y.X.; Ding, Z.; Liu, J.; Xu, T.; Hou, L.; Han, S.; Qian, X.; Ma, Y. Microbial community roles and chemical mechanisms in the parasitic development of *Orobanche cumana*. *iMeta* **2022**, *1*, e31. [CrossRef]
24. Jacoby, R.P.; Koprivova, A.; Kopriva, S. Pinpointing secondary metabolites that shape the composition and function of the plant microbiome. *J. Exp. Bot.* **2021**, *72*, 57–69. [CrossRef]
25. Pascale, A.; Proietti, S.; Pantelides, I.S.; Stringlis, I.A. Modulation of the root microbiome by plant molecules: The basis for targeted disease suppression and plant growth promotion. *Front. Plant Sci.* **2020**, *10*, 1741. [CrossRef]
26. Shah, A.; Smith, D.L. Flavonoids in agriculture: Chemistry and roles in, biotic and abiotic stress responses, and microbial associations. *Agronomy* **2020**, *10*, 1209. [CrossRef]
27. Pei, Y.; Siemann, E.; Tian, B.; Ding, J. Root flavonoids are related to enhanced AMF colonization of an invasive tree. *AoB Plants* **2020**, *12*, plaa002. [CrossRef]
28. Cotton, T.A.; Pétriacq, P.; Cameron, D.D.; Meselmani, M.A.; Schwarzenbacher, R.; Rolfe, S.A.; Ton, J. Metabolic regulation of the maize rhizobiome by benzoxazinoids. *ISME J.* **2019**, *13*, 1647–1658. [CrossRef]
29. Xu, S.; Liu, Y.; Cernava, T.; Wang, H.; Zhou, Y.; Xia, T.; Cao, S.; Berg, G.; Shen, X.; Wen, Z. Fusarium fruiting body microbiome member *Pantoea agglomerans* inhibits fungal pathogenesis by targeting lipid rafts. *Nat. Microbiol.* **2022**, *7*, 831–843. [CrossRef]
30. Anselmo-Moreira, F.; Teixeira-Costa, L.; Ceccantini, G.; Furlan, C.M. Mistletoe effects on the host tree *Tapirira guianensis*: Insights from primary and secondary metabolites. *Chemoecology* **2019**, *29*, 11–24. [CrossRef]
31. Tjiurutue, M.C.; Sandler, H.A.; Kersch-Becker, M.F.; Theis, N.; Adler, L.A. Cranberry resistance to dodder parasitism: Induced chemical defenses and behavior of a parasitic plant. *J. Chem. Ecol.* **2016**, *42*, 95–106. [CrossRef]
32. Xiao, L.; Carrillo, J.; Siemann, E.; Ding, J. Herbivore-specific induction of indirect and direct defensive responses in leaves and roots. *AoB Plants* **2019**, *11*, plz003. [CrossRef] [PubMed]
33. Xia, L.; Robock, A.; Scherrer, K.; Harrison, C.S.; Bodirsky, B.L.; Weindl, I.; Jägermeyr, J.; Bardeen, C.G.; Toon, O.B.; Heneghan, R. Global food insecurity and famine from reduced crop, marine fishery and livestock production due to climate disruption from nuclear war soot injection. *Nat. Food* **2022**, *3*, 586–596. [CrossRef] [PubMed]
34. Baritz, R.; Wiese, L.; Verbeke, I.; Vargas, R. Voluntary guidelines for sustainable soil management: Global action for healthy soils. *Int. Yearb. Soil Law Policy* **2017**, *2018*, 17–36.
35. Mishra, J.S. Biology and management of *Cuscuta* species. *Indian J. Weed Sci.* **2009**, *41*, 1–11.

36. Song, S.; Lee, J.; Kang, J.; Ko, J.; Seo, M.; Woo, K.; Oh, B.; Nam, M. The growth and yield of soybean as affected by competitive density of *Cuscuta pentagona*. *Korean J. Weed Sci.* **2010**, *30*, 390–395. [CrossRef]
37. Zhang, J.; Xu, Y.; Xie, J.; Zhuang, H.; Liu, H.; Shen, G.; Wu, J. Parasite dodder enables transfer of bidirectional systemic nitrogen signals between host plants. *Plant Physiol.* **2021**, *185*, 1395–1410. [CrossRef]
38. Zhuang, H.; Li, J.; Song, J.; Hettenhausen, C.; Schuman, M.C.; Sun, G.; Zhang, C.; Li, J.; Song, D.; Wu, J. Aphid (*Myzus persicae*) feeding on the parasitic plant dodder (*Cuscuta australis*) activates defense responses in both the parasite and soybean host. *New Phytol.* **2018**, *218*, 1586–1596. [CrossRef]
39. Luo, W.; Li, Y.; Jia, Y.; Chen, Y.; Li, D.; Luo, R.; Wei, G.; Chou, M.; Chen, W. Positive response of host root-associated bacterial community and soil nutrients to inhibitory parasitism of dodder. *Plant Soil* **2023**. [CrossRef]
40. Xiao, X.; Chen, W.; Zong, L.; Yang, J.; Jiao, S.; Lin, Y.; Wang, E.; Wei, G. Two cultivated legume plants reveal the enrichment process of the microbiome in the rhizocompartments. *Mol. Ecol.* **2017**, *26*, 1641–1651. [CrossRef]
41. Jiao, S.; Chen, W.; Wang, J.; Du, N.; Li, Q.; Wei, G. Soil microbiomes with distinct assemblies through vertical soil profiles drive the cycling of multiple nutrients in reforested ecosystems. *Microbiome* **2018**, *6*, 146. [CrossRef]
42. Callahan, B.J.; Mcmurdie, P.J.; Rosen, M.J.; Han, A.W.; Johnson, A.J.A.; Holmes, S.P. DADA2: High-resolution sample inference from Illumina amplicon data. *Nat. Methods* **2016**, *13*, 581–583. [CrossRef]
43. Nilsson, R.H.; Larsson, K.; Taylor, A.F.S.; Bengtsson-Palme, J.; Jeppesen, T.S.; Schigel, D.; Kennedy, P.; Picard, K.; Glöckner, F.O.; Tedersoo, L. The unite database for molecular identification of fungi: Handling dark taxa and parallel taxonomic classifications. *Nucleic Acids Res.* **2019**, *47*, D259–D264. [CrossRef]
44. Fan, K.; Weisenhorn, P.; Gilbert, J.A.; Shi, Y.; Bai, Y.; Chu, H. Soil pH correlates with the co-occurrence and assemblage process of diazotrophic communities in rhizosphere and bulk soils of wheat fields. *Soil Biol. Biochem.* **2018**, *121*, 185–192. [CrossRef]
45. Bao, S. *Soil and Agricultural Chemistry Analysis*; Chinese Agricultural Publisher: Beijing, China, 2000.
46. Chen, W.; Gong, L.; Guo, Z.; Wang, W.; Zhang, H.; Liu, X.; Yu, S.; Xiong, L.; Luo, J. A novel integrated method for large-scale detection, identification, and quantification of widely targeted metabolites: Application in the study of rice metabolomics. *Mol. Plant* **2013**, *6*, 1769–1780. [CrossRef]
47. Ginestet, C. ggplot2: Elegant graphics for data analysis. *J. R. Stat. Soc.* **2011**, *174*, 245–246. [CrossRef]
48. Oksanen, J.; Kindt, R.; Legendre, P.; O'Hara, B.; Stevens, M.H.H.; Oksanen, M.J.; Suggests, M. The vegan package. *Community Ecol. Packag.* **2007**, *10*, 719.
49. Wen, T.; Xie, P.; Yang, S.; Niu, G.; Liu, X.; Ding, Z.; Xue, C.; Liu, Y.X.; Shen, Q.; Yuan, J. ggClusterNet: An R package for microbiome network analysis and modularity-based multiple network layouts. *iMeta* **2022**, *1*, e32. [CrossRef]
50. Gu, Z.; Eils, R.; Schlesner, M. Complex heatmaps reveal patterns and correlations in multidimensional genomic data. *Bioinformatics* **2016**, *32*, 2847–2849. [CrossRef]
51. Chong, J.; Xia, J. MetaboAnalystR: An R package for flexible and reproducible analysis of metabolomics data. *Bioinformatics* **2018**, *34*, 4313–4314. [CrossRef]
52. Xiong, C.; Zhu, Y.G.; Wang, J.T.; Singh, B.; Han, L.L.; Shen, J.P.; Li, P.P.; Wang, G.B.; Wu, C.F.; Ge, A.H.; et al. Host selection shapes crop microbiome assembly and network complexity. *New Phytol.* **2021**, *229*, 1091–1104. [CrossRef]
53. Brown, S.P.; Grillo, M.A.; Podowski, J.C.; Heath, K.D. Soil origin and plant genotype structure distinct microbiome compartments in the model legume *Medicago truncatula*. *Microbiome* **2020**, *8*, 139. [CrossRef] [PubMed]
54. Lareen, A.; Burton, F.; Schäfer, P. Plant root-microbe communication in shaping root microbiomes. *Plant Mol. Biol.* **2016**, *90*, 575–587. [CrossRef] [PubMed]
55. Liu, F.; Hewezi, T.; Lebeis, S.L.; Pantalone, V.; Grewal, P.S.; Staton, M.E. Soil indigenous microbiome and plant genotypes cooperatively modify soybean rhizosphere microbiome assembly. *BMC Microbiol.* **2019**, *19*, 201. [CrossRef] [PubMed]
56. Cook, J.C.; Charudattan, R.; Zimmerman, T.W.; Rosskopf, E.N.; Stall, W.M.; Macdonald, G.E. Effects of *Alternaria destruens*, glyphosate, and ammonium sulfate individually and integrated for control of dodder (*Cuscuta pentagona*). *Weed Technol.* **2009**, *23*, 550–555. [CrossRef]
57. Aneja, K.R.; Kumar, V.; Jiloha, P.; Kaur, M.; Sharma, C.; Surain, P.; Dhiman, R.; Aneja, A. Potential bioherbicides: Indian perspectives. In *Biotechnology: Prospects & Applications*; Springer: New Delhi, India, 2013; pp. 197–215. [CrossRef]
58. Chen, L.; Xin, X.; Zhang, J.; Redmile-Gordon, M.; Nie, G.; Wang, Q. Soil characteristics overwhelm cultivar effects on the structure and assembly of root-associated microbiomes of modern maize. *Pedosphere* **2019**, *29*, 360–373. [CrossRef]
59. Changey, F.; Bagard, M.; Souleymane, M.; Lerch, T.Z. Cascading effects of elevated ozone on wheat rhizosphere microbial communities depend on temperature and cultivar sensitivity. *Environ. Pollut.* **2018**, *242*, 113–125. [CrossRef]
60. Rolfe, S.A.; Griffiths, J.; Ton, J. Crying out for help with root exudates: Adaptive mechanisms by which stressed plants assemble health-promoting soil microbiomes. *Curr. Opin. Microbiol.* **2019**, *49*, 73–82. [CrossRef]
61. Xu, H.; Li, Y.; Zhang, K.; Li, M.; Fu, S.; Tian, Y.; Qin, T.; Li, X.; Zhong, Y.; Liao, H. *miR169c-NFYA-C-ENOD40* modulates nitrogen inhibitory effects in soybean nodulation. *New Phytol.* **2021**, *229*, 3377–3392. [CrossRef]
62. Ma, D.; Constabel, C.P. MYB repressors as regulators of phenylpropanoid metabolism in plants. *Trends Plant Sci.* **2019**, *24*, 275–289. [CrossRef]
63. Subramanian, S.; Stacey, G.; Yu, O. Distinct, crucial roles of flavonoids during legume nodulation. *Trends Plant Sci.* **2007**, *12*, 282–285. [CrossRef]

64. Zhang, J.; Subramanian, S.; Stacey, G.; Yu, O. Flavones and flavonols play distinct critical roles during nodulation of *Medicago truncatula* by *Sinorhizobium meliloti*. *Plant J.* **2009**, *57*, 171–183. [CrossRef]
65. Shemshura, O.N.; Seitbattalova, A.I.; Bekmahanova, N.E.; Ismailova, E.T.; Kaptagai, R.J. Components of the flavonoid nature of plants of the family Lamiaceae Lindl., having fungicidal activity against the phytopathogens of tomatoes and soybean. *Exp. Bio.* **2017**, *70*, 120–127.
66. Hendra, R.; Ahmad, S.; Sukari, A.; Shukor, M.Y.; Oskoueian, E. Flavonoid analyses and antimicrobial activity of various parts of *Phaleria macrocarpa* (scheff.) Boerl fruit. *Int. J. Mol. Sci.* **2011**, *12*, 3422–3431. [CrossRef]
67. Mohamed, A.A.; Salah, M.M.; El-Dein, M.M.Z.; El-Hefny, M.; Ali, H.M.; Farraj, D.A.A.; Hatamleh, A.A.; Salem, M.Z.M.; Ashmawy, N.A. Ecofriendly bioagents, *Parthenocissus quinquefolia*, and *Plectranthus neochilus* extracts to control the early blight pathogen (*Alternaria solani*) in tomato. *Agronomy* **2021**, *11*, 911. [CrossRef]

agronomy

MDPI

Article

Co-Ensiling Whole-Plant Cassava with Corn Stalk for Excellent Silage Production: Fermentation Characteristics, Bacterial Community, Function Profile, and Microbial Ecological Network Features

Mao Li [1,2,†], Xuejuan Zi [3,†], Rong Sun [3], Wenjun Ou [1], Songbi Chen [1], Guanyu Hou [1,2,*] and Hanlin Zhou [1,2,*]

1 Tropical Crops Genetic Resources Institute, Chinese Academy of Tropical Agricultural Sciences, Danzhou 571737, China; limao@catas.cn (M.L.); wenjunou@catas.cn (W.O.); songbichen@catas.cn (S.C.)
2 Zhanjiang Experimental Station, Chinese Academy of Tropical Agricultural Sciences, Zhanjiang 524000, China
3 Key Laboratory of Ministry of Education for Genetics and Germplasm Innovation of Tropical Special Trees and Ornamental Plants, Key Laboratory of Germplasm Resources of Tropical Special Ornamental Plants of Hainan Province, School of Tropical Agriculture and Forestry, Hainan University, Danzhou 571737, China; zixuejuan@hainanu.edu.cn (X.Z.); sunrong@hainanu.edu.cn (R.S.)
* Correspondence: guanyuhou@catas.cn (G.H.); zhouhanlin@catas.cn (H.Z.)
† These authors have contributed equally to this work.

Abstract: The objective of this study was to explore excellent silage production through co-ensiling whole-plant cassava and corn stalk, and different ratios of whole-plant cassava (0%, 10%, 20%, 30%, 40%, and 50%, fresh-matter basis) co-ensiled with corn stalk were analyzed based on the silage bacterial community, function profile, and microbial ecological network features. The results demonstrated that co-ensiling 30% whole-plant cassava with 70% corn stalk could be considered an efficient mode of production. The mixed silage showed great quality, as reflected by the reduced pH value and concentrations of acetic acid, butyric acid, and ammonia nitrogen and the enhanced lactic acid concentration, V-score, and nutritional value compared with corn stalk ensiled alone. Meanwhile, co-ensiling restricted the undesirable bacterial *Acetobacter fabarum* of corn stalk and *Pseudomonas aeruginosa* of whole-plant cassava and raised the abundance of lactic acid bacteria (LAB) such as *Levilactobacillus brevis*, *Lactiplantibacillus plantarum*, *Lactobacillus harbinensis*, etc. Besides that, the predicted functions of the bacterial community showed large differences in mixed silage compared with whole-plant cassava or corn stalk ensiled alone. Moreover, the analysis of co-occurrence networks showed that mixed silage affected microbial network features, module numbers, and bacterial relative abundances and weakened the complexity and stability of the networks compared with whole-plant cassava single silage. Furthermore, silage microbial community composition had a huge impact on the network properties, and undesirable *Pseudomonas aeruginosa* played a crucial role in the complexity and stability. Overall, this study revealed the characteristics of whole-plant cassava with corn stalk mixed-silage microbial communities and co-occurrence network modules, complexity, and stability and partly clarified the microbial mechanism of co-ensiling for producing high-quality silage. The findings of this study have important implications for deeply understanding the ensiling process and precisely regulating silage fermentation quality.

Keywords: co-ensiling; whole-plant cassava; corn stalk; fermentation quality; bacterial community; co-occurrence networks

Citation: Li, M.; Zi, X.; Sun, R.; Ou, W.; Chen, S.; Hou, G.; Zhou, H. Co-Ensiling Whole-Plant Cassava with Corn Stalk for Excellent Silage Production: Fermentation Characteristics, Bacterial Community, Function Profile, and Microbial Ecological Network Features. *Agronomy* **2024**, *14*, 501. https://doi.org/10.3390/agronomy14030501

Academic Editor: Caterina Morcia

Received: 7 November 2023
Revised: 30 December 2023
Accepted: 3 January 2024
Published: 28 February 2024

1. Introduction

The dietary structure of Chinese residents has undergone significant changes in recent years, with a considerable increase in the consumption of animal-based foods such as beef, mutton, and milk. This rising demand has led to the rapid development of "Grass-based Livestock Husbandry", which has greatly increased the total amount of ruminant livestock being raised. However, the shortage of high-quality forage has been accentuated due to the limited production of existing grasslands [1]. Consequently, large amounts of forage must be imported each year, and the import of feed has become a key constraint in the development of domestic animal husbandry [2,3]. The tropical regions of southern China are important areas for ruminant livestock production, but they lack large-scale natural grasslands. This mismatch causes a serious deficit in forage [4], necessitating the effective use of local plant resources to meet farming needs.

Cassava (*Manihot esculenta* Crantz) is an important crop in subtropical and tropical areas globally, and it is also widely planted in southern China to produce starch, biofuels, and feed [4,5]. The whole cassava plant includes the tubers and roots underground and the stems and leaves aboveground. It has high biomass and is rich in essential nutrients (including protein, fiber, vitamins, and minerals) for animal growth [6] and can be harvested conveniently. The cassava tuber is a vital source of calories for both humans and animals in tropical areas as it contains abundant carbohydrates [7–10]. Cassava leaves are suitable feed with good digestibility for ruminants, pigs, and poultry, as they are rich in biomass and protein with low fiber content [4,11,12]. The cassava stem is also a high-quality feed that has about 30% starch and only a moderate cellulose content [6,13–15]. Cassava leaves, stems, and tubers each have been used as feed resources [4,11,16,17], and whole-plant cassava represents an ideal feed source with the potential to aid the sustainable development of local animal husbandry in tropical regions. Due to the seasonal limitations of cassava cultivation, it is necessary to preserve whole-plant cassava through silage processing, but the fermentation characteristics of the silage and the techniques of silage manipulation remain to be further examined.

Corn (*Zea mays* L.) stalk is the most widely used forage in the world and is used in ruminant livestock production from cold to tropical regions [18,19]. Naturally fermented corn stalk can typically be well preserved due to its high content of water-soluble sugar [20], but Bernardes et al. [21] indicated that in tropical regions, due to the impact of high temperatures and the presence of undesirable bacteria, the corn stalk silage may have too much butyric acid or experience alcoholic fermentation and undergo aerobic deterioration. Furthermore, the single silage of corn stalk generally does not have sufficient protein, which is not helpful in balancing the carbon and nitrogen contents in the diet of ruminants [22]. Some modulation methods improve the storage of the corn stalk by enhancing silage quality and preserving nutrition [18].

Silage fermentation can be regarded as a dynamic microbial ecology process, and microbial co-occurrence networks, network modules, and network stability are the three most important characteristics of microbial ecological systems [18,23,24]. Due to the lack of epiphytic lactic acid bacteria (LAB), low sugar content, or high buffer capacity, additives are needed to produce silage from most forage [4,12]. Treating a single raw material with additives can significantly enhance the silage quality but typically does not improve the nutritional profile. The nutritional value of the silage is decreased when some nutrients are consumed during the anaerobic fermentation [21]. In addition, previous studies have shown that the silage quality and the microbial co-occurrence network interact with each other. The raw material epiphytic microbiota, LAB inoculants, and storage temperatures can affect the microbial co-occurrence networks and modules in the silage, thereby enhancing fermentation quality [3,18,25,26].

The co-ensiling of two or more raw materials to create mixed silage improves the fermentation quality by changing the physical and chemical properties of the materials and the composition of epiphytic microorganisms. Many studies have shown that the co-ensiling of lower-nutrition forage with rich-nutrition forage readily improves the fermentation quality and the nutritional value of the silage. We previously reported that the 1:1 co-ensiling of cassava leaves with king grass strongly improved the quality and the feeding value of the silage [12]. Numerous studies have demonstrated that co-ensiling corn stalk with high-protein-content forage (e.g., soybean, alfalfa, Neolamarckia cadamba) was beneficial to silage fermentation and nutrient preservation [22,27–29]. Since whole-plant cassava is rich in nutrients, we speculated that co-ensiling it with corn stalk may contribute to the fermentation. However, the underlying microbial principles, e.g., the correlation between the mixed silage and the complexity and stability of the bacterial networks, for the success of the co-ensiling practice is largely unknown. Analyzing the microbial ecological network in the mixed silage will help to understand the ensiling process and regulate the fermentation quality.

As stated in the above analyses, we speculate that the co-ensiling of whole-plant cassava and corn stalk may help to produce high-quality feed with good fermentation quality and excellent nutritional value. However, as far as we know, existing research has not explored the bacterial community, function, and microbial ecological network of the mixed silage of whole-plant cassava and corn stalk. The objective of this study is to optimize the production of animal feed by co-ensiling whole-plant cassava and corn stalk at different ratios, with a particular focus on silage bacterial community structure, function profile, microbial ecological network characteristics, and network complexity and stability.

2. Materials and Methods

2.1. Silage Preparation

The present study was conducted at the Chinese Academy of Tropical Agricultural Sciences (CATAS) in Danzhou, Hainan, China (109°30′ E, 19°30′ N, 149 m ASL). Two cultivars of materials were used. The whole-plant cassava (SC7) and corn (Huamei) were planted on 15 March 2021 and harvested on 15 September 2021 from 3 randomly selected plots. The corn was harvested at half milk-line.

The fresh raw materials were shredded to about 2 cm in size using a 9Z-2.5 grass chopper (Zhengzhou Jinhongxing Jixie Co. Ltd., Zhengzhou, China) and then mixed evenly according to the formulations given in Table 1 (fresh-matter basis). For each formulation, three vacuum-sealed plastic bags (30 cm × 10 cm × 4 cm) of the blended materials (500 g) were prepared, and the incubation was carried out at room temperature. The samples were analyzed on day 60 to determine the chemical composition, fermentation quality, and microbial community.

Table 1. Co-ensiling formulation settings.

Group	Corn Stalk (CS)	Whole-Plant Cassava (CF)
CS	100%	0%
CS10CF	90%	10%
CS20CF	80%	20%
CS30CF	70%	30%
CS40CF	60%	40%
CS50CF	50%	50%
CF	0%	100%

2.2. Chemical Composition and Fermentation Index

The weight of the dry matter (DM) was measured after heating the silage samples at 65 °C for 72 h. The dried materials were then ground and passed through a sieve (1 mm), and the water-soluble carbohydrates (WSC), crude protein (CP), neutral detergent fiber (NDF), and acid detergent fiber (ADF) were measured according to the protocols given by the Association of Official Analytical Chemists (AOAC, 1990), and NDF was assayed with a heat-stable amylase. The starch content was determined using a total starch determination kit based on the procedures described by Bai et al. [20]. The number of lactic acid bacteria, molds, and coliform bacteria was determined with MRS agar, potato dextrose agar, and violet red bile agar, respectively [29].

In addition, a mixture of the silage sample (50 g) in distilled water (200 mL) was incubated at 4 °C for 24 h and then filtered. Half of the filtrate was used for the analysis of the fermentation quality as follows [30]. First, the pH was measured with a glass electrode pH meter. The concentrations of lactic acid, acetic acid, propionic acid, and butyric acid were determined by high-performance liquid chromatography (HPLC, SHIMADZU-10A, Kyoto, Japan). The ammonia nitrogen (NH_3–N) was assayed by the phenol-sodium hypochlorite method according to the protocol described by Liu et al. [31]. The V-Score was calculated as the index of fermentation quality based on the ratio of NH_3–N to total N, the content of acetic acid and propionic acid, and the amount of higher volatile fatty acids [32]. All of the above indexes had three replicates. The other half of the filtrate was stored at −80 °C until use in the analysis of the microbial communities.

2.3. Microbial Communities and Functional Profile

The total bacteria DNA of the silage was extracted from the saved filtrate using the TGuide S96 Magnetic Soil /Stool DNA Kit from Tiangen Biotech (Beijing, China) Co., Ltd. The DNA concentration was determined using the Qubit dsDNA HS Assay Kit and a Qubit 4.0 Fluorometer (Invitrogen, Thermo Fisher Scientific, Waltham, MA, USA). The universal primer set (27F: AGRGTTTGATYNTGGCTCAG, 1492R: TASGGHTACCTTGTTASGACTT) was used to amplify the full-length 16S rRNA gene from the total genomic DNA extracted from each sample. The PCR protocols (amplification, purification, quantification) followed the work of Bai et al. [18]. The high-quality PCR products were sequenced on a PacBio Sequel platform (Pacific Biosciences, Menlo Park, CA, USA) according to the standard protocols of Biomarker Technologies (Beijing, China).

The bioinformatics analysis was performed at the BMK Cloud (Biomarker Technologies Co., Ltd., Beijing, China). The raw reads generated from sequencing were filtered and demultiplexed using the SMRT Link software (version 8.0) to obtain the circular consensus sequencing (CCS) reads, with the minPasses set at ≥5 and minPredictedAccuracy at ≥0.9. The CCS reads were assigned to the corresponding samples based on their barcodes using lima (version 1.7.0). The CCS reads containing no primers or outside the length range (1200–1650 bp) were discarded using Cutadapt (version 2.7). Chimera sequences were detected and removed using UCHIME (version 8.1) to obtain the clean reads. Sequences with ≥97% similarity were clustered into the same operational taxonomic unit (OTU) using USEARCH (version 10.0), and the OTUs were filtered if their redundancy was less than 0.005%.

The taxonomy annotation of the OTUs was performed based on the Naive Bayes classifier in QIIME2 using the SILVA database (release 138) with a similarity cut-off of 70% [33]. The alpha diversity indices (Shannon index and Simpson's index) were calculated using QIIME2 and visualized using the R software v4.2.3. The beta diversity (as determined by the principal coordinate analysis, PCoA) was determined using QIIME to evaluate the degree of similarity between microbial communities from different samples. The Linear Discriminant Analysis (LDA) effect size (LEfSe) was used to test the significant taxonomic difference among groups [34]. The bacterial functions were predicted using the Kyoto Encyclopedia of Genes and Genomes (KEGG) database and by applying the Phylogenetic Investigation of Communities by Reconstruction of Unobserved States 2 (PICRUSt2) [35]. The analyses of the silage microbial network (co-occurrence network, network modules, complexity, and stability) were performed using ggClusterNet (https://github.com/taowenmicro/ggClusterNet/, accessed on 15th May 2023) [24]. Network features include nodes, edges, network density, average degree, negative/positive edge ratios, etc. The basic elements in a network are nodes and edges, and the connections between nodes are called edges. Network density is the ratio of the actual number of edges present in a network to the upper limit of the number of edges it can accommodate. The average degree refers to the average number of edges connected to each node in a network graph. The correlation heatmaps were drawn and the canonical correlation analysis (CCA) between the network properties and microbial community was carried out to investigate the impact of the silage microbial communities on the microbial networks. Data were analyzed using the free online BMKCloud Platform (https://www.biocloud.net/). The sequencing data were deposited in the Sequence Read Archive (SRA, accession number PRJNA1011842).

2.4. Statistical Analysis

All statistical analyses were performed using SPSS 20 and GraphPad Prism 8.0. The results are reported as the mean of three replicates. For each treatment, the fermentation quality and the chemical composition were analyzed by one-way analysis of variance (ANOVA). The orthogonal polynomial contrasts (linear correlation) were used to evaluate the effects of the mixing ratio. The differences between the mean values were considered significant if Duncan's multiple-range test gave $p < 0.05$.

3. Results

3.1. Raw Material

Table 2 lists the chemical composition and epiphytic microbial number of the fresh whole-plant cassava and the corn stalk. For the whole-plant cassava, the DM content was 242.5 g/kg fresh matter (FM), and the CP, ADF, NDF, WSC, and starch contents were 158.6, 116.8, 158.9, 177.2, and 215.9 g/kg DM, respectively. For the corn stalk, the DM content was 305.1 g/kg FM, and the CP, ADF, NDF, WSC, and starch contents were 96.4, 226.3, 407.5, 128.4, and 104.6 g/kg DM, respectively. Both the whole-plant cassava and the corn stalk had a similar number of epiphytic lactic acid bacteria, but the whole-plant cassava had less mold and *Enterobacter*.

Table 2. Chemical composition of raw materials.

Items	Abbreviation	Whole-Plant Cassava	Corn Stalk
Dry matter (g/kg FM)	DM	242.5	305.1
Crude protein (g/kg DM)	CP	158.6	96.4
Acid detergent fiber (g/kg DM)	ADF	116.8	226.3
Neutral detergent fiber (g/kg DM)	NDF	158.9	407.5
Water-soluble carbohydrates (g/kg DM)	WSC	177.2	128.4
Starch (g/kg DM)		215.9	104.6
Lactic acid bacteria (Log cfu/g FM)	LAB	5.76	5.88
Mold (Log cfu/g FM)		2.43	4.52
Enterobacter (Log cfu/g FM)		2.68	4.73

Note: FM, fresh matter; DM, dry matter; CP, crude protein; ADF, acid detergent fiber; NDF, neutral detergent fiber; WSC, water-soluble carbohydrates; LAB, lactic acid bacteria.

3.2. Fermentation Quality

The mixed silage clearly had higher fermentation quality than the single silage (Table 3). Co-ensiling effectively reduced pH, acetic acid, propionic acid, butyric acid, and ammonia nitrogen content and significantly increased lactate content and improved the V-score. Among all treatment groups, CS30CF had the best fermentation quality, with the lowest pH value and acetic acid, propionic acid, butyric acid, and NH_3–N concentrations ($p < 0.05$), as well as the highest lactic acid content and V-score ($p < 0.05$).

Table 3. Fermentation quality of mixed silage.

Items	Treatments							SEM	*p*-Value	
	CS	CS10CF	CS20CF	CS30CF	CS40CF	CS50CF	CF		T	L
pH	4.95 a	4.56 b	4.23 c	3.97 d	4.18 c	4.23 c	4.27 c	0.12	<0.001	>0.05
Lactic acid (g/kg DM)	42.10 d	51.19 b	53.77 b	62.39 a	54.00 b	48.25 c	46.93 c	2.43	<0.001	>0.05
Acetic acid (g/kg DM)	25.73 a	17.92 b	14.87 b	9.29 c	13.96 b	16.64 b	19.28 a	1.92	<0.001	>0.05
Propionic acid (g/kg DM)	4.31	3.98	3.14	2.25	3.24	4.07	4.16	0.28	>0.05	>0.05
Butyric acid (g/kg DM)	1.08 a	0.69 a	0.44 b	N	0.31 b	0.55 b	0.88 a	0.11	N	N
NH_3–N (g/kg TN)	80.52 a	73.84 b	66.56 b	51.81 c	62.34 b	69.45 b	85.78 a	4.29	<0.001	>0.05
V-Score	75.26 c	79.71 b	83.17 b	92.30 a	85.05 b	81.71 b	75.80 c	2.21	<0.001	>0.05

Note: DM, dry matter, TN, total nitrogen. Within the same row, values with different letters denote statistically significant ($p < 0.05$) differences. SEM, standard error of the mean. N, not detected; T, treatment; L, linear.

Table 4 demonstrates the chemical compositions of the mixed silage. The contents of DM, ADF, and NDF were the highest for the single silage of corn stalk ($p < 0.05$) and gradually decreased with a rising amount of whole-plant cassava. In contrast, the contents of CP, WSC, and starch were the highest for the single silage of whole-plant cassava and gradually decreased with the addition of corn stalk. Significant differences existed in the chemical composition of mixed silage, and there was a notable linear correlation with the mixing ratio ($p < 0.01$). Both whole-plant cassava and corn stalk contained abundant nutrients, and the nutrients, especially the CP, WSC, and starch, were consumed to some extent after ensiling.

Table 4. Chemical composition of mixed silage.

Items	Treatments							SEM	*p*-Value	
	CS	CS10CF	CS20CF	CS30CF	CS40CF	CS50CF	CF		T	L
DM (g/kg FM)	282.5 a	276.2 b	269.9 b	263.5 bc	257.2 c	250.9 c	219.3 d	7.89	<0.001	<0.001
CP (g/kg DM)	83.4 d	89.3 cd	95.2 cd	101.1 c	107.0 b	113.0 b	142.5 a	7.38	<0.001	<0.001
ADF (g/kg DM)	198.6 a	188.2 b	177.7 c	167.3 d	156.8 de	146.4 e	94.1 f	13.05	<0.001	<0.001
NDF (g/kg DM)	379.7 a	355.0 b	330.2 c	305.5 cd	280.8 d	256.1 e	132.4 f	30.88	<0.001	<0.001
WSC (g/kg DM)	56.1 c	59.7 c	63.3 bc	66.9 bc	70.5 b	74.1 b	92.0 a	4.48	<0.001	<0.001
Starch (g/kg DM)	70.7 d	79.2 dc	87.6 c	96.1 bc	104.6 b	113.1 a	155.4	10.58	<0.001	<0.001

Note: DM, dry matter; CP, crude protein; ADF, acid detergent fiber; NDF, neutral detergent fiber; WSC, water-soluble carbohydrates. Within the same row, values with different letters denote statistically significant ($p < 0.05$) differences. SEM, standard error of the mean. T, treatment; L, linear.

3.3. *Bacterial Communities*

3.3.1. Diversity, Compositions, Structure, and Predicted Functions

Figure 1 plots the alpha and beta diversities of the bacterial communities in the silage. The Shannon diversity and Simpson's diversity indices were the lowest for the single silage of corn stalk (Group CS, $p < 0.05$) and increased by the co-ensiling of corn stalk and whole-plant cassava. The silage of all seven groups had 5928 OTUs in total and shared 25 OTUs in common (Figure 1C), and the unique OTU number was the lowest for CS (507) and the highest for CS10CF (1069). The structure of the microbial communities shifted in response to the mixing ratio, as the microbial communities in the silage varied significantly among the seven groups (Figure 1D).

The compositions of the bacterial communities in the silage were identified at the genus level and compared with each other (Figure 2A). The single silage of corn stalk (Group CS) had predominantly *Acetobacter* (36.67%) and *Levilactobacillus* (12.20%), and the abundance of *Acetobacter* decreased significantly after co-ensiling with whole-plant cassava ($p < 0.05$), reaching zero for CS50CF and CF. Meanwhile, the single silage of whole-plant cassava (Group CF) was dominated by *Lactiplantibacillus* (14.37%), *Levilactobacillus* (11.90%), and the undesirable *Pseudomonas* (5.79%). For all mixed silage, *Levilactobacillus*, *Schleiferilactobacillus*, and *Limosilactobacillus* were the top three genera, and their total abundance was the highest in CS30CF. At the species level, *Acetobacter fabarum* was the most abundant in CS, while *Levilactobacillus brevis*, *Lactiplantibacillus plantarum*, *Lactobacillus harbinensis*, and *Pseudomonas aeruginosa* were the top bacterial species in other groups (Figure 2B).

The differences in the bacterial communities among silage were analyzed by LEfSe to identify the microbial taxa specific to each group (Figure 2C, Supplementary Materials Table S1). The bacteria that were enriched significantly in a treatment group were deemed indicator bacteria. The indicator bacteria were *Clostridia*, *Oscillospirales*, *Lactiplantibacillus plantarum*, and *Companilactobacillus nuruki* for CF; *Acetobacter*, *Acetobacter fabarum*, *Lacticaseibacillus*, *Gluconobacter*, and *Gluconobacter japonicus* for CS; *Blautia hansenii* for CS20CF; *Schleiferilactobacillus*, *Levilactobacillus*, *Lentilactobacillus*, *Lactobacillus harbinensis*_DSM_16991, *Levilactobacillus brevis*, *Lentilactobacillus parabuchner*, *Lentilactobacillus buchneri*, and *Companilactobacillus paralimentarius* for CS30CF; and *Pseudomonas*, *Pseudomonas aeruginosa*, and *Lactobacillus* for CS50CF. While the indicator bacteria of CS and CS50CF were undesirable for the silage, CS30CF had many healthy lactic acid bacteria as its indicator bacteria.

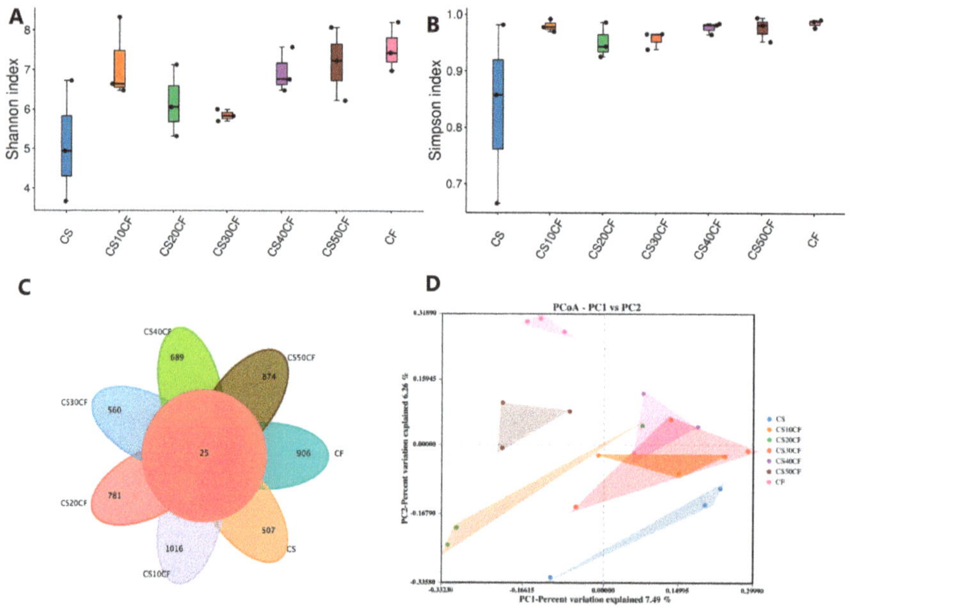

Figure 1. (**A**) The Shannon index and (**B**) the Simpson index of the bacterial communities in the silage. The number on the horizontal line represents the *p*-value. (**C**) Venn diagram of the bacterial OTUs. (**D**) Principal Coordinates Analysis (PCoA) and the beta diversity of the bacterial communities in the silage. Notations are as defined in Table 1.

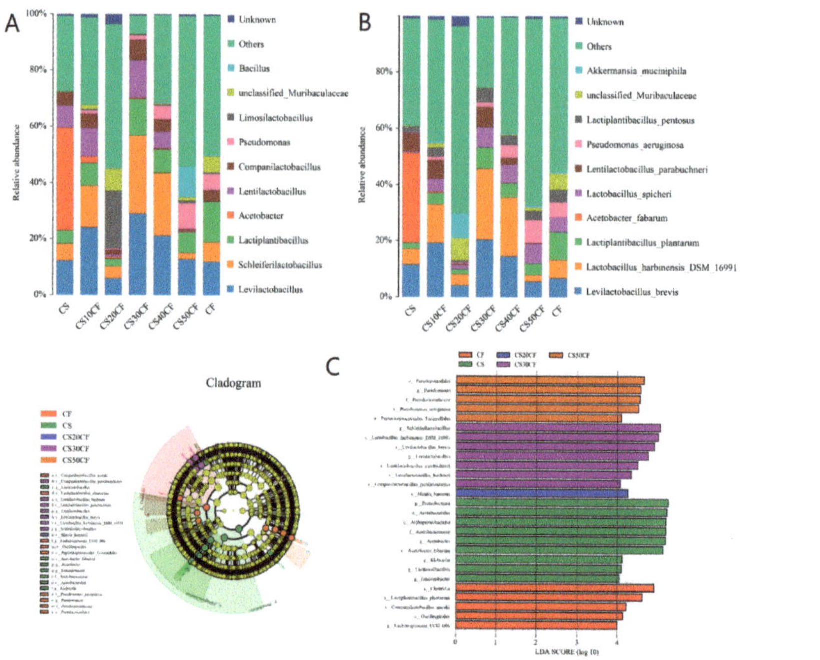

Figure 2. Relative abundance of bacteria at the (**A**) genus and (**B**) species levels in the silage. (**C**) Comparison of microbial variations and the identification of indicator bacteria using LEfSe.

In the silage of the seven treatment groups, the LAB included *Levilactobacillus*, *Schleiferilactobacillus*, *Lactiplantibacillus*, *Lentilactobacillus*, *Companilactobacillus*, and *Limosilactobacillus* and their corresponding species-level microorganisms (Figure 3A,C), and the undesirable bacteria included *Acetobacter*, *Pseudomonas*, and unclassified_*Muribaculaceae* and their corresponding species-level microorganisms (Figure 3B,D). We then calculated the total abundance of the LAB and the undesirable bacteria for each group (Figure 3). Co-ensiling clearly increased the lactic acid bacteria and reduced the undesirable bacteria compared to the single silage of whole-plant cassava or corn stalk, and CS30CF had the highest abundance of LAB and the lowest abundance of undesirable bacteria ($p < 0.05$), which is probably why the fermentation quality of CS30CF was the best.

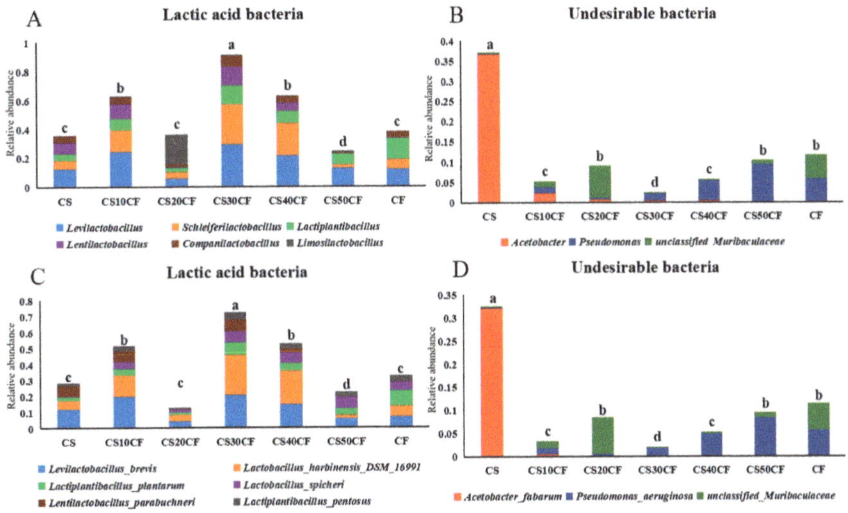

Figure 3. Abundance of lactic acid bacteria at the (**A**) genus and (**C**) species levels in the silage. Abundance of undesirable bacteria at the (**B**) genus and (**D**) species level in the silage. Different letters on bar chart denote statistically significant ($p < 0.05$) difference.

The potential functions of the bacterial communities were predicted using PICRUSt2 (Figure 4). The treatment groups CS, CF, and CS30CF were selected for functional prediction based on the specificity of raw materials, fermentation quality, and microbial composition. Figure 4A shows that there were 16 significantly different KEGG pathways between CF and CS30CF ($p < 0.05$). The four upregulated pathways in CF were (1) global and overview maps, (2) the metabolism of terpenoids and polyketides, (3) amino acid metabolism, and (4) the signal transduction and biosynthesis of other secondary metabolites. The twelve upregulated pathways in CS30CF were (1) carbohydrate metabolism, (2) nucleotide metabolism, (3) lipid metabolism, (4) the metabolism of other amino acids, (5) xenobiotics biodegradation and metabolism, (6) glycan biosynthesis and metabolism, (7) the metabolism of terpenoids and polyketides, (8) translation, (9) replication and repair, (10) folding, sorting, and degradation, (11) antimicrobial drug resistance, and (12) the endocrine system. Figure 4B shows that the two significantly ($p < 0.05$) downregulated KEGG pathways in Group CS compared to Group CS30CF were (1) replication and repair and (2) the metabolism of other amino acids. Moreover, the signal transduction pathways were enriched significantly ($p < 0.05$) in CF compared to CS (Figure 4C). Compared to single silage, co-ensiling significantly changed the microbial community structure, which rendered the obvious diversity in the predicted metabolic functions. The impact of co-ensiling on the predicted functions was mainly related to the metabolism of nutrients, e.g., amino acids, carbohydrates, and lipids.

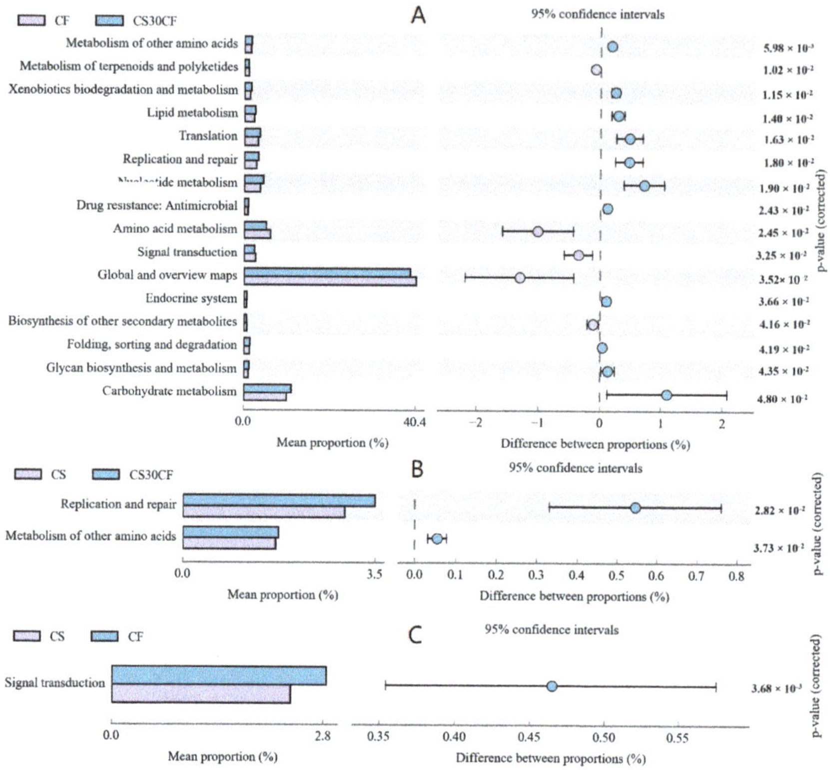

Figure 4. Comparison of microbial function pathways: detected pathway enrichment of (**A**) CF vs. CS30CF, (**B**) CS vs. CS30CF, and (**C**) CS vs. CF.

3.3.2. Co-Occurrence Network, Network Modules, and Stability

The co-occurrence networks of the bacterial communities varied with the silage. Figure 5 shows that the networks of Group CF had the greatest edges, network density, and average degree. The higher the number of edges, average degree, and network density, the tighter the network connection. That is, Group CF had more complicated microbial community structures and interactions than Group CS and other mixed silage groups. In addition, Group CF also had a significantly higher ($p < 0.05$) ratio of negative to positive interactions than all other groups. Because the microbial networks are more stable when the ratio of negative to positive edge interactions is higher, for the seven groups, the bacterial networks were less stable when the silage treatment used corn stalk.

Figure 6 shows the modules of the co-occurrence networks, and the details of the bacterial compositions of modules in different treatments are listed in Supplementary Materials Table S2–S8. The networks in Group CF had only one, but in the largest module (module_1), the dominant bacteria were all LAB, including *Lactiplantibacillus plantarum*, *Levilactobacillus brevis*, *Lactobacillus harbinensis*, *Lactobacillus spicheri*, *Lactiplantibacillus pentosus*, and *Companilactobacillus nuruki*. There were also some undesirable bacteria in module_1 with lower abundance, including unclassified_*Muribaculaceae*, *Pseudomonas aeruginosa*, and *Mycoplasma wenyonii*. In contrast, there were six modules in the bacterial networks of Group CS, and module_3 and module_5 were dominant. *Acetobacter fabarum* was the most dominant bacteria in module_3, but module_5 mainly contained three bacteria with high abundance, i.e., *Lentilactobacillus parabuchneri*, *Paucilactobacillus vaccinostercus*, and *Lactobacillus harbinensis*. The modules in the other five mixed silage groups had much more diverse compositions.

Interestingly, in the bacterial networks of Group CS50CF, *Pseudomonas aeruginosa* was the major bacteria in module_1, which was one of the dominant modules. Correspondingly, Group CS50CF had a lower abundance of lactic acid bacteria and a higher abundance of undesirable bacteria compared to other mixed silage groups, hence indicating that the module dominated by *Pseudomonas aeruginosa* affected the bacterial communities in the silage. Overall, co-ensiling significantly changed the numbers, pattern size, distributions, and relative abundances of the modules in the bacterial networks, and the diversity of the microbial network modules depended heavily on the mixing ratio of corn stalk to whole-plant cassava.

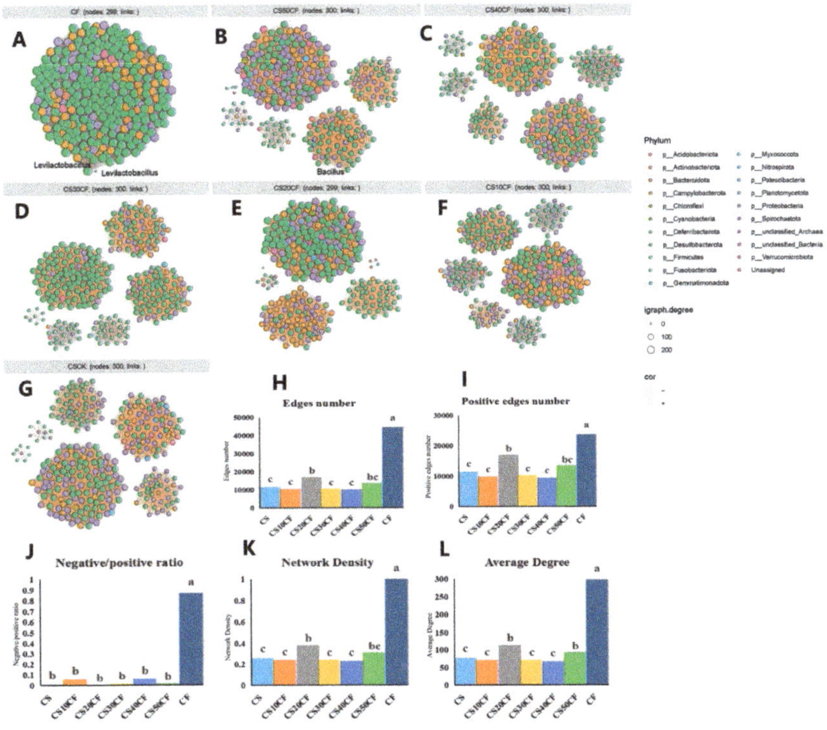

Figure 5. The bacterial community co-occurrence networks of (**A**) CF, (**B**) CS50CF, (**C**) CS40CF, (**D**) CS30CF, (**E**) CS20CF, (**F**) CS10CF, and (**G**) CS. (**H**) Numbers of edges, (**I**) numbers of positive edges, (**J**) negative/positive edge ratios, (**K**) network density, and (**L**) average degree of bacterial community co-occurrence networks.Different letters on bar chart denote statistically significant ($p < 0.05$) difference.

The diversity of the microbial network modules may impact the function and stability of the silage micro-ecosystem. We calculated multiple stability indices (robustness, natural connectivity, and negative correlation ratio) to assess how the microbial network stability in the silage was affected by the mixing ratio of corn stalk to whole-plant cassava (Figure 7). On the basis of either the random or targeted removal of module hubs, the silage of Group CF had noticeably higher robustness than that of all other groups (Figure 7A,B). The bacterial network appeared to have the greatest natural connectivity and thus the strongest resilience in Group CF (Figure 7C). Group CF also had a remarkably higher negative correlation ratio than all other groups (Figure 7D). In general, co-ensiling whole-plant cassava and corn stalk reduced the stability of the silage microbial network compared to ensiling whole-plant cassava alone, probably due to the decreased network complexity associated with the connectivity and relative modularity.

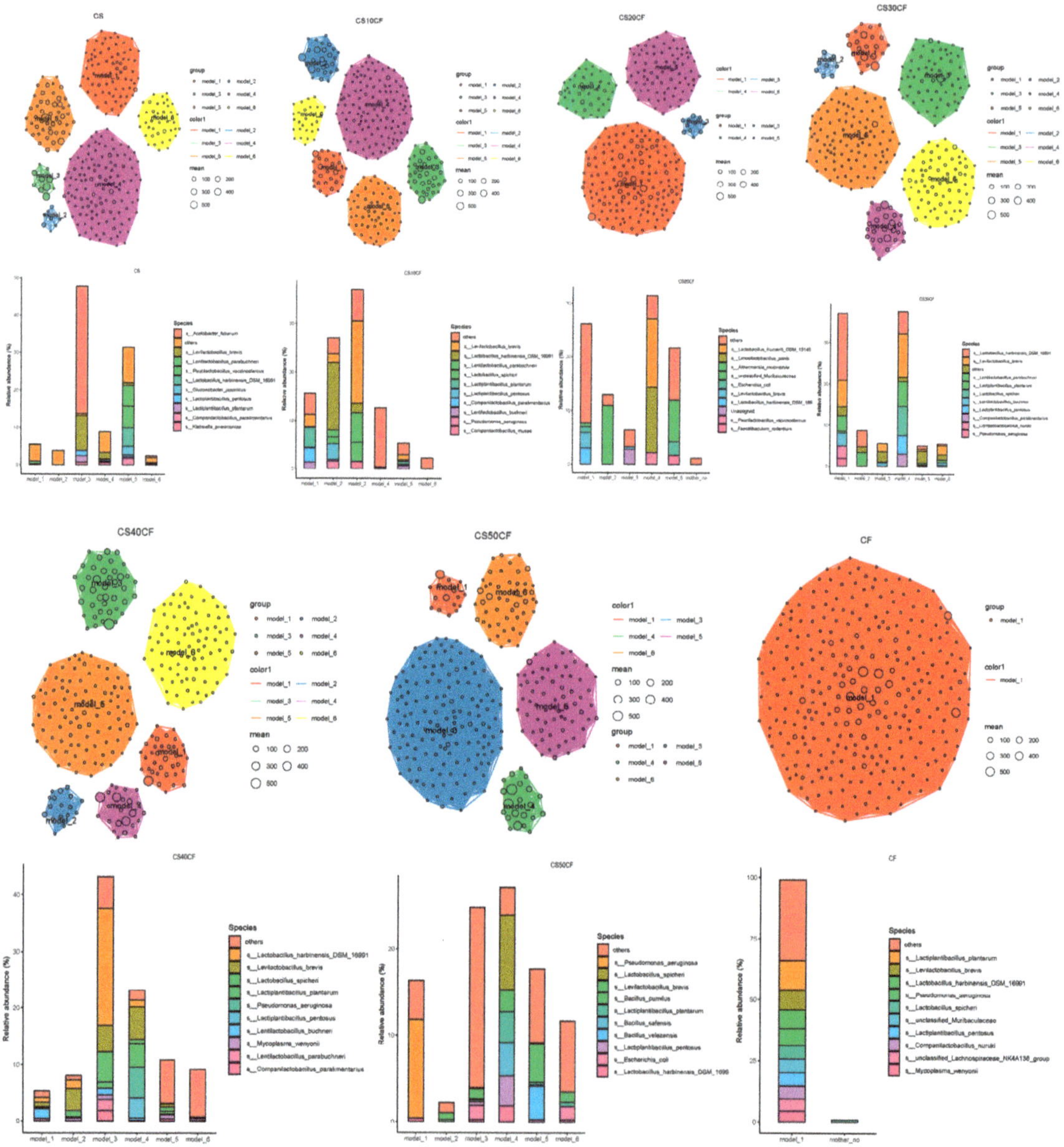

Figure 6. Modules in the bacterial community co-occurrence networks.

Figure 8 plots the correlation heatmap and the canonical correlation analysis (CCA) between the network properties and microbial communities to examine the impact of the microbial community composition on the microbial networks. In the correlation heatmap (Figure 8A), the edges, positive edges, edge density, and average degree had significant negative correlations with *Lactobacillus harbinensis*_DSM_16991, *Levilactobacillus brevis*, and *Lentilactobacillus parabuchneri* ($p < 0.05$). In addition, the negative edges and the negative/positive ratio had positive correlations with *Pseudomonas aeruginosa*, *Lactiplantibacillus plantarum*, *Lactiplantibacillus pentosus*, and *Lactobacillus spicheri* ($p < 0.05$). According to the canonical correlation analysis (Figure 8B), the positive edges, negative/positive ratio, density, and centralization degree were affected by the microbial communities in the silage. Positive edges and the negative/positive ratio had the strongest and weakest impact, respectively. The negative/positive ratio was closely related to *Pseudomonas aeruginosa* and *Lactobacillus spicheri*, and the centralization degree was closely related to *unclassified_Muribaculaceae*.

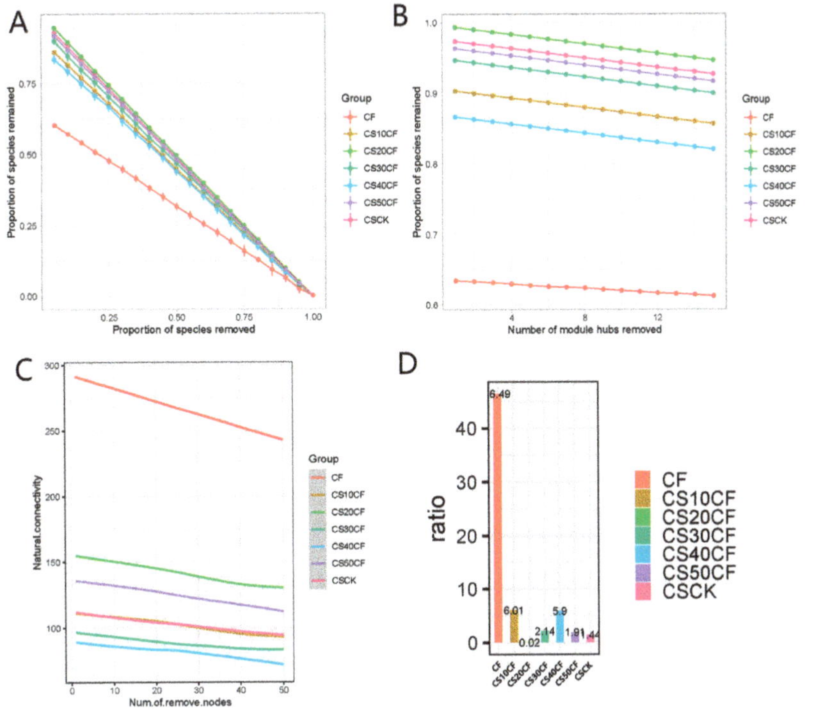

Figure 7. Robustness of microbial networks upon the (**A**) random and (**B**) targeted removal of module hubs. (**C**) Natural connectivity and (**D**) negative correlation ratio of the microbial networks.

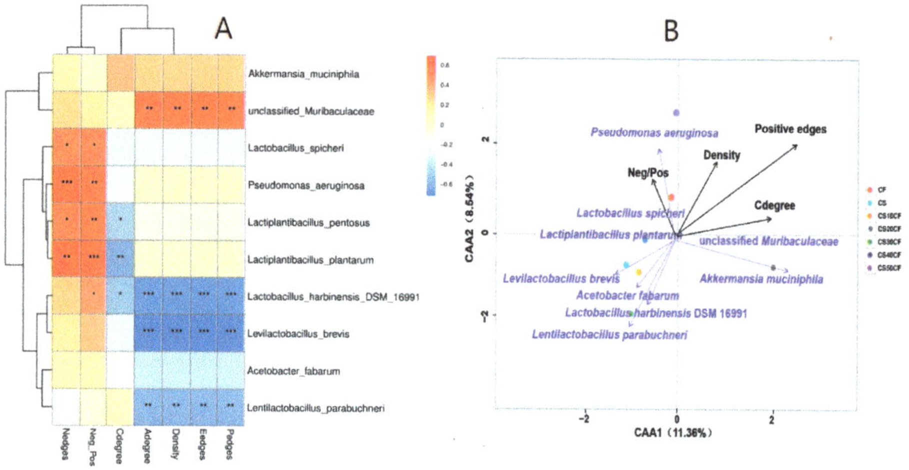

Figure 8. (**A**) Correlation heatmap of network properties and microbial communities. Positive and negative correlations are shown in red and blue, respectively. The color intensity is proportional to the correlation values. * $p < 0.05$, ** $p < 0.01$, *** $p < 0.001$. (**B**) Canonical correlation analysis of network properties and microbial communities.

4. Discussion

Whole-plant cassava has more balanced nutrition for animals than its leaves, stem, or tubers [10,11,15], but the feeding value of whole-plant cassava has not been examined. The data of the corn stalk agreed with those reported by Wang et al. [28] and Bai et al. [20], confirming that corn was an ideal feed for livestock with high DM, WSC, and starch content. We found that the whole-plant cassava had abundant CP, WSC, and starch with low fiber content. In addition, both whole-plant cassava and corn stalk had abundant epiphytic LAB and few undesirable bacteria, which is beneficial to creating high-quality silage through co-ensiling.

The pH value of the silage is a key parameter of the fermentation quality. The single silage of corn stalk gave the highest pH value. The single silage of whole-plant cassava gave an obviously lower pH compared to the single silage of cassava foliage, indicating that whole-plant cassava is a better silage material [4]. The addition of whole-plant cassava to corn stalk significantly reduced the pH value for the mixed silage, and all mixed silage except CS10CF had a pH value close to or lower than 4.2, which is the pH value threshold of well-preserved silage. Wang et al. [3] and Zeng et al. [29] reported similar pH values and trends for the mixed silages of corn with soybean or alfalfa. In addition, in our results, all silage groups had a relatively high content of lactic acid, likely because the low pH fermentation environment contributed to homofermentation to produce large amounts of lactic acid. Wang et al. [3] also noted an obvious increase in lactate in the mixed silage of *Sesbania cannabina* and sweet sorghum. Furthermore, the changes in butyric acid and ammonia nitrogen, both of which are harmful to the preservation of forage nutrients, were consistent with the variation in the pH value. Butyric acid and ammonia nitrogen are the product of protein decomposition by undesirable bacteria such as *Clostridium*, aerobic bacteria, and *Enterobacteriaceae*, all of which typically grow better at a higher pH. Both corn stalk and whole-plant cassava contain abundant nutrients that may also benefit the growth of undesirable bacteria, hence potentially increasing the contents of butyric acid and ammonia nitrogen. Some previous studies also reported that the mixed silage could effectively reduce undesirable microbial fermentation and inhibit the production of butyric acid and ammonia nitrogen [3,28,29]. In summary, co-ensiling whole-plant cassava and corn stalk improved the silage quality, which was characterized by low pH value, higher lactic acid content, and significantly reduced butyric acid and ammonia nitrogen content.

The ammonia nitrogen content can reflect the decomposition of proteins or amino acids [36]. Like the mixed silage system of corn stalk and forage soybean, the mixed silage of corn stalk and whole-plant cassava also had a significantly lower content of ammonia nitrogen, indicating the stronger preservation of CP [27]. On the other hand, the sharp reduction in WSC and starch contents revealed the efficient uptake of the fermentation substrates by the lactic acid bacteria to generate organic acids in high yields. Zeng et al. [29] and Wang et al. [3] also reported similar phenomena in the mixed silage of materials with different nutritional characteristics. In this study, the change in the nutritional compositions of the mixed silage was mainly due to the change in the mixing ratio of the two raw materials, and a more balanced nutrient composition was more conducive to silage fermentation.

Typically, the LAB occupy a dominant position as the fermentation proceeds during silage production, which simplifies the structure and reduces the alpha diversity of the bacterial communities in well-preserved silage [26,29]. Among all the mixed silage, Group CS30CF had the lowest alpha diversity and the best fermentation quality. Interestingly, lower alpha diversity was also found for the single silage of corn stalk (Group CS), likely due to the high abundance of the undesirable *Acetobacter* (36.67%). The mixed silage also had an obvious impact on the beta diversity of the bacterial communities, which were clearly separated based on their silage group. The results agreed well with the findings from the mixed silages of corn stalk and alfalfa or soybean [28,29].

Previous research suggested that corn stalk silage is generally well-preserved and dominated by LAB [18,20,28,29]. In our case, the corn stalk silage was dominated by *Acetobacter* (36.67%), which reduced the fermentation quality, possibly due to the secondary fermentation of the raw materials. In contrast, *Levilactobacillus brevis* and *Lactiplantibacillus plantarum* were dominant in the single silage of whole-plant cassava, and they contributed to the fermentation quality by promoting lactic acid fermentation. Some previous studies reported low fermentation quality for the silage of cassava foliage not treated with additives, in which *Pseudomonas*, *Bacillus*, *Paenibacillus*, and *Clostridium* were dominant [4,37]. We found that co-ensiling corn stalk and whole-plant cassava optimized the silage microbial community structure by increasing the abundance of lactic acid bacteria (up to 80%) and reducing undesirable bacteria (down to 0.05%). Similar benefits have been observed in other co-ensiling systems. For example, Wang et al. [3] found that in the mixed silage of *Sesbania cannabina* and sweet sorghum, the abundance of *Lentilactobacillus hilgardii* and *L. buchneri* increased to over 90% while the undesirable *Bacillus* and *Enterococcus* decreased to zero. Wang et al. [28] reported that the microbial communities in the mixed silage of alfalfa and corn were dominated by *L. buchneri* and *L. plantarum*, with a combined abundance of approximately 50%. Zeng et al. [29] reported that the abundance of *Lactobacillus* was approximately 70% in the mixed silage of forage soybean and corn. Interestingly, the well-preserved silage in the current study contained many more dominant bacteria, including *Levilactobacillus*, *Schleiferilactobacillus*, *Lactiplantibacillus*, *Lentilactobacillus*, *Companilactobacillus*, and *Limosilactobacillus*. Further research is needed to account for the flourishing of dominant bacteria in the mixed silage of corn stalk and whole-plant cassava.

Ensiling is a complex micro-ecological system process in which diverse microbial communities affect metabolic products to regulate the fermentation quality. The functional prediction of the microbial communities can to some extent reflect the role of the microbes in the fermentation systems [25]. In this study, the most abundant pathways were "global and overview maps", and they were upregulated in Group CF, which was consistent with the previous report by Du et al. [38] and indicated the critical role of these pathways in microbial metabolism. Amino acid metabolism was an important metabolic pathway related to the degradation of nitrogen and amino acids. Group CF had the highest CP and ammonia nitrogen content, along with remarkably upregulated amino acid metabolism pathways. Similar observations have been reported by Bai et al. [20] and Li et al. [39]. Carbohydrate metabolism is mainly affected by the abundance of lactic acid bacteria in the microbial communities. Group CS30CF had the most abundant lactic acid bacteria, which presumably led to the upregulation of the carbohydrate metabolism pathways, similar to the reports by Wang et al. [3] and Li et al. [39]. In addition, some pathways regarding genetic functions were also enriched in different treatment groups, possibly because the microbial communities were adapting to the acidic environments during ensiling [20]. The diverse microorganisms in the silage fermentation system enabled different metabolic pathways and functions, hence regulating the fermentation quality.

From an ecological perspective, each individual package of silage is a microbial ecosystem [3,25]. Numerous research articles suggest that the changes in the network structure of an ecosystem can affect the functionality and stability at a macroscopic scale, and the relationships between network complexity and stability have been clarified to some extent [40,41]. Yuan et al. [23] analyzed the complexity and stability of the molecular ecological networks of soil microbial communities. Recent analyses of the microbial community networks in silage revealed that environmental factors, modulation methods, the duration of fermentation, and silage materials all have a significant impact on the microbial networks. Dong et al. [26] demonstrated that the epiphytic microbiota of a sorghum–sudangrass hybrid harvested at various times of the day affected the complexity of the bacterial networks in the silage, and the AM silage was more stable. Bai et al. [18] studied the bacterial network properties of whole-plant corn silage and found that the storage temperature had a greater influence on the network complexity than treatment with added lactic acid bacteria. Wang et al. [3] showed that the bacterial network stability of the mixed silage increased as the duration of the fermentation was extended. However, the impact of co-ensiling on the features of bacterial network modules, complexity, and stability is still largely unknown. Our results revealed that compared to the single silage of whole-plant cassava, the mixed silage had microbial networks with less complexity, lower stability, higher module numbers, and different bacterial compositions. There were also strong interactions between the dominant bacterial species in the silage and the microbial network characteristics. As for the microbial networks, the complexity was influenced by the dominant lactic acid bacteria, and the stability was largely affected by the undesirable bacteria *Pseudomonas aeruginosa*. The current results suggested that the lower abundance of the undesirable bacteria in the silage played a crucial role in the complexity and stability of microbial ecological networks, and it appeared that the tested co-ensiling had a negative impact on the complexity and stability of the bacterial networks.

5. Conclusions

Co-ensiling whole-plant cassava with corn stalk is an efficient mode of silage production. The analysis of several kinds of ratio mixtures of corn stalk and whole-plant cassava revealed that the best silage was produced from 30% cassava and 70% corn, as reflected by excellent fermentation quality and nutritional value. The co-ensiling reduced the abundance of undesirable bacteria and increased the abundance of LAB. The analysis of the microbial co-occurrence networks revealed that co-ensiling affected microbial network features, module numbers, and bacterial relative abundances and weakened the complexity and stability of the networks. The composition of the microbial communities had a heavy impact on the network properties, and the undesirable *Pseudomonas aeruginosa* played a crucial role in the complexity and stability. The findings helped to understand the co-ensiling process by partly clarifying the microbial mechanisms of co-ensiling in producing high-quality silage.

Supplementary Materials: The following supporting information can be downloaded at: https://www.mdpi.com/article/10.3390/agronomy14030501/s1, Table S1: Species lefse LDA4 less strict, Table S2: CSCKplot tax module data, Table S3: CS10CFplot tax module data, Table S4: CS20CFplot tax module data, Table S5: CS30CFplot tax module data, Table S6: CS40CFplot tax module data, Table S7: CS50CFplot tax module data, Table S8: CFplot tax module data.

Author Contributions: M.L., X.Z., R.S., W.O., S.C., G.H. and H.Z. carried out the experimental design work. M.L., X.Z. and R.S. conducted the experiments. M.L., X.Z., R.S., W.O., S.C., G.H. and H.Z.

analyzed the data. M.L., X.Z. and R.S. wrote and revised the manuscript. All authors have read and agreed to the published version of the manuscript.

Funding: This study was funded by the key research and development projects of Hainan province (ZDYF2022XDNY153, HAIKOU2023-050), the Agriculture Research System of China (CARS-11), and the Central Public-interest Scientific Institution Basal Research Fund for Chinese Academy of Tropical Agricultural Sciences (No. 1630032022011).

Data Availability Statement: The data used to support the findings of this study are included within the article.

Conflicts of Interest: The authors declare no conflicts of interest. The funders had no role in the design of the study; in the collection, analyses, or interpretation of data; in the writing of the manuscript; or in the decision to publish the results.

References

1. Fang, J.; Jing, H.; Zhang, W.; Gao, S.; Duan, Z.; Wang, H.; Zhong, J.; Pan, Q.; Zhao, K.; Bai, W.; et al. The concept of "Grass-based Livestock Husbandry" and its practice in Hulun Buir, Inner Mongolia. *Chin. Sci. Bull.* **2018**, *63*, 1619–1631. [CrossRef]
2. Tang, S.X.; He, Y.; Zhang, P.H.; Jiao, J.Z.; Han, X.F.; Yan, Q.X.; Tan, Z.L.; Wang, H.R.; Wu, D.Q.; Yu, L.H.; et al. Nutrient digestion, rumen fermentation and performance as ramie (*Boehmeria nivea*) is increased in the diets of goats. *Anim. Feed. Sci. Technol.* **2019**, *247*, 15–22. [CrossRef]
3. Wang, T.; Zhang, J.; Shi, W.; Sun, J.; Xia, T.; Huang, F.; Liu, Y.; Li, H.; Teng, K.; Zhong, J. Dynamic Changes in Fermentation Quality and Structure and Function of the Microbiome during Mixed Silage of Sesbania cannabina and Sweet Sorghum Grown on Saline-Alkaline Land. *Microbiol. Spectr.* **2022**, *10*, e0248322. [CrossRef] [PubMed]
4. Li, M.; Zi, X.; Zhou, H.; Lv, R.; Tang, J.; Cai, Y. Effect of lactic acid bacteria, molasses, and their combination on the fermentation quality and bacterial community of cassava foliage silage. *Anim. Sci. J.* **2021**, *92*, e13635. [CrossRef]
5. Malik, A.I.; Kongsil, P.; Nguyễn, V.A.; Ou, W.; Srean, P.; Sheela, M.N.; Becerra López-Lavalle, L.A.; Utsumi, Y.; Lu, C.; Kittipadakul, P.; et al. Cassava breeding and agronomy in Asia: 50 years of history and future directions. *Breed. Sci.* **2020**, *70*, 145–166. [CrossRef] [PubMed]
6. Lyu, H.; Yang, S.; Zhang, J.; Feng, Y.; Geng, Z. Impacts of utilization patterns of cellulosic C5 sugar from cassava straw on bioethanol production through life cycle assessment. *Bioresour. Technol.* **2021**, *323*, 124586. [CrossRef] [PubMed]
7. Isamah, G.K.; Asagba, S.O.; Ekakitie, A.O. Lipid peroxidation, activities of superoxide dismutase and catalase during post-harvest deterioration of cassava (*Manihot esculenta* Crantz) root tubers. *Int. Biodeterior. Biodegrad.* **2003**, *52*, 129–133. [CrossRef]
8. Kayode, B.I.; Kayode, R.M.O.; Salami, K.O.; Obilana, A.O.; George, T.T.; Dudu, O.E.; Adebo, O.A.; Njobeh, P.B.; Diarra, S.S.; Oyeyinka, S.A. Morphology and physicochemical properties of starch isolated from frozen cassava root. *LWT* **2021**, *147*, 111546. [CrossRef]
9. Bogale, S.; Haile, A.; Berhanu, B.; Beshir, H.M. Cassava production practices in Ethiopia and its use as Ingredient for injera making. *Future Foods* **2022**, *6*, 100204. [CrossRef]
10. Ono, L.T.; Silva, J.J.; Soto, T.S.; Doná, S.; Iamanaka, B.T.; Fungaro, M.H.P.; Taniwaki, M.H. Fungal communities in Brazilian cassava tubers and food products. *Int. J. Food Microbiol.* **2023**, *384*, 109909. [CrossRef]
11. Anyanwu, C.N.; Ibeto, C.N.; Ezeoha, S.L.; Ogbuagu, N.J. Sustainability of cassava (*Manihot esculenta* Crantz) as industrial feedstock, energy and food crop in Nigeria. *Renew. Energ.* **2015**, *81*, 745–752. [CrossRef]
12. Li, M.; Zi, X.; Tang, J.; Zhou, H.; Cai, Y. Silage fermentation, chemical composition and ruminal degradation of king grass, cassava foliage and their mixture. *Grassl. Sci.* **2019**, *65*, 210–215. [CrossRef]
13. Zhu, W.; Lestander, T.A.; Örberg, H.; Wei, M.; Hedman, B.; Ren, J.; Xie, G.; Xiong, S. Cassava stems: A new resource to increase food and fuel production. *GCB Bioenergy* **2015**, *7*, 72–83. [CrossRef]
14. Ojo, I.; Apiamu, A.; Egbune, E.O.; Tonukari, N.J. Biochemical Characterization of Solid-State Fermented Cassava Stem (*Manihot esculenta* Crantz-MEC) and Its Application in Poultry Feed Formulation. *Appl. Biochem. Biotechnol.* **2022**, *194*, 2620–2631. [CrossRef]
15. Fanelli, N.S.; Torres-Mendoza, L.J.; Abelilla, J.J.; Stein, H.H. Chemical composition of cassava-based feed ingredients from South-East Asia. *Anim. Biosci.* **2023**, *36*, 908–919. [CrossRef]
16. Wanapat, M.; Kang, S. Cassava chip (*Manihot esculenta* Crantz) as an energy source for ruminant feeding. *Anim. Nutr.* **2015**, *1*, 266–270. [CrossRef] [PubMed]
17. Azad, M.A.K.; Jiang, H.; Ni, H.; Liu, Y.; Huang, P.; Fang, J.; Kong, X. Diets Partially Replaced With Cassava Residue Modulate Antioxidant Capacity, Lipid Metabolism, and Gut Barrier Function of Huanjiang Mini-Pigs. *Front. Vet. Sci.* **2022**, *9*, 902328. [CrossRef]
18. Bai, J.; Ding, Z.; Su, R.; Wang, M.; Cheng, M.; Xie, D.; Guo, X. Storage Temperature Is More Effective Than Lactic Acid Bacteria Inoculations in Manipulating Fermentation and Bacterial Community Diversity, Co-Occurrence and Functionality of the Whole-Plant Corn Silage. *Microbiol. Spectr.* **2022**, *10*, e0010122. [CrossRef]

19. Cui, Y.; Liu, H.; Gao, Z.; Xu, J.; Liu, B.; Guo, M.; Yang, X.; Niu, J.; Zhu, X.; Ma, S.; et al. Whole-plant corn silage improves rumen fermentation and growth performance of beef cattle by altering rumen microbiota. *Appl. Microbiol. Biotechnol.* **2022**, *106*, 4187–4198. [CrossRef]
20. Bai, J.; Franco, M.; Ding, Z.; Hao, L.; Ke, W.; Wang, M.; Xie, D.; Li, Z.; Zhang, Y.; Ai, L.; et al. Effect of Bacillus amyloliquefaciens and Bacillus subtilis on fermentation, dynamics of bacterial community and their functional shifts of whole-plant corn silage. *J. Anim. Sci. Biotechnol.* **2022**, *13*, 7. [CrossRef]
21. Bernardes, T.F.; Daniel, J.L.P.; Adesogan, A.T.; McAllister, T.A.; Drouin, P.; Nussio, L.G.; Huhtanen, P.; Tremblay, G.F.; Bélanger, G.; Cai, Y. Silage review: Unique challenges of silages made in hot and cold regions. *J. Dairy. Sci.* **2018**, *101*, 4001–4019. [CrossRef]
22. Zhou, W.; Pian, R.; Yang, F.; Chen, X.; Zhang, Q. The sustainable mitigation of ruminal methane and carbon dioxide emissions by co-ensiling corn stalk with Neolamarckia cadamba leaves for cleaner livestock production. *J. Clean. Prod.* **2021**, *311*, 127680. [CrossRef]
23. Yuan, M.M.; Guo, X.; Wu, L.; Zhang, Y.; Xiao, N.; Ning, D.; Shi, Z.; Zhou, X.; Wu, L.; Yang, Y.; et al. Climate warming enhances microbial network complexity and stability. *Nat. Clim. Chang.* **2021**, *11*, 343–348. [CrossRef]
24. Wen, T.; Xie, P.; Yang, S.; Niu, G.; Liu, X.; Ding, Z.; Xue, C.; Liu, Y.-X.; Shen, Q.; Yuan, J. ggClusterNet: An R package for microbiome network analysis and modularity-based multiple network layouts. *iMeta* **2022**, *1*, e32. [CrossRef]
25. Xu, D.; Wang, N.; Rinne, M.; Ke, W.; Weinberg, Z.G.; Da, M.; Bai, J.; Zhang, Y.; Li, F.; Guo, X. The bacterial community and metabolome dynamics and their interactions modulate fermentation process of whole crop corn silage prepared with or without inoculants. *Microb. Biotechnol.* **2021**, *14*, 561–576. [CrossRef]
26. Dong, Z.; Li, J.; Wang, S.; Dong, D.; Shao, T. Diurnal Variation of Epiphytic Microbiota: An Unignorable Factor Affecting the Anaerobic Fermentation Characteristics of Sorghum-Sudangrass Hybrid Silage. *Microbiol. Spectr.* **2023**, *11*, e0340422. [CrossRef] [PubMed]
27. Ni, K.; Zhao, J.; Zhu, B.; Su, R.; Pan, Y.; Ma, J.; Zhou, G.; Tao, Y.; Liu, X.; Zhong, J. Assessing the fermentation quality and microbial community of the mixed silage of forage soybean with crop corn or sorghum. *Bioresour. Technol.* **2018**, *265*, 563–567. [CrossRef] [PubMed]
28. Wang, M.; Franco, M.; Cai, Y.; Yu, Z. Dynamics of fermentation profile and bacterial community of silage prepared with alfalfa, whole-plant corn and their mixture. *Anim. Feed. Sci. Technol.* **2020**, *270*, 114702. [CrossRef]
29. Zeng, T.; Li, X.; Guan, H.; Yang, W.; Liu, W.; Liu, J.; Du, Z.; Li, X.; Xiao, Q.; Wang, X.; et al. Dynamic microbial diversity and fermentation quality of the mixed silage of corn and soybean grown in strip intercropping system. *Bioresour. Technol.* **2020**, *313*, 123655. [CrossRef] [PubMed]
30. Li, M.; Lv, R.; Zhou, H.; Zi, X. Dynamics and correlations of chlorophyll and phytol content with silage bacterial of different growth heights Pennisetum sinese. *Front. Plant Sci.* **2022**, *13*, 996970. [CrossRef] [PubMed]
31. Liu, Q.; Chen, M.; Zhang, J.; Shi, S.; Cai, Y. Characteristics of isolated lactic acid bacteria and their effectiveness to improve stylo (*Stylosanthes guianensis* Sw.) silage quality at various temperatures. *Anim. Sci. J.* **2012**, *83*, 128–135. [CrossRef]
32. Masaki, S. *Silage-no-Hinshitsu-Hyouka (V-SCORE); Nihon-Sochichikusan-Shushi-Kyokai*; CiNii: Tokyo, Japan, 2001; pp. 93–94.
33. Quast, C.; Pruesse, E.; Yilmaz, P.; Gerken, J.; Schweer, T.; Yarza, P.; Peplies, J.; Glöckner, F.O. The SILVA ribosomal RNA gene database project: Improved data processing and web-based tools. *Nucleic Acids Res.* **2013**, *41*, D590–D596. [CrossRef]
34. Segata, N.; Izard, J.; Waldron, L.; Gevers, D.; Miropolsky, L.; Garrett, W.S.; Huttenhower, C. Metagenomic biomarker discovery and explanation. *Genome Biol.* **2011**, *12*, R60. [CrossRef] [PubMed]
35. Langille, M.G.; Zaneveld, J.; Caporaso, J.G.; McDonald, D.; Knights, D.; Reyes, J.A.; Clemente, J.C.; Burkepile, D.E.; Vega Thurber, R.L.; Knight, R.; et al. Predictive functional profiling of microbial communities using 16S rRNA marker gene sequences. *Nat. Biotechnol.* **2013**, *31*, 814–821. [CrossRef] [PubMed]
36. Ni, K.; Wang, F.; Zhu, B.; Yang, J.; Zhou, G.; Pan, Y.; Tao, Y.; Zhong, J. Effects of lactic acid bacteria and molasses additives on the microbial community and fermentation quality of soybean silage. *Bioresour. Technol.* **2017**, *238*, 706–715. [CrossRef]
37. Li, M.; Zhang, L.; Zhang, Q.; Zi, X.; Lv, R.; Tang, J.; Zhou, H. Impacts of Citric Acid and Malic Acid on Fermentation Quality and Bacterial Community of Cassava Foliage Silage. *Front. Microbiol.* **2020**, *11*, 595622. [CrossRef]
38. Du, Z.; Sun, L.; Lin, Y.; Chen, C.; Yang, F.; Cai, Y. Use of Napier grass and rice straw hay as exogenous additive improves microbial community and fermentation quality of paper mulberry silage. *Anim. Feed. Sci. Technol.* **2022**, *285*, 115219. [CrossRef]
39. Li, X.; Chen, F.; Wang, X.; Xiong, Y.; Liu, Z.; Lin, Y.; Ni, K.; Yang, F. Innovative utilization of herbal residues: Exploring the diversity of mechanisms beneficial to regulate anaerobic fermentation of alfalfa. *Bioresour. Technol.* **2022**, *360*, 127429. [CrossRef] [PubMed]
40. Thébault, E.; Fontaine, C. Stability of ecological communities and the architecture of mutualistic and trophic networks. *Science* **2010**, *329*, 853–856. [CrossRef]
41. Toju, H.; Yamamichi, M.; Guimarães, P.R., Jr.; Olesen, J.M.; Mougi, A.; Yoshida, T.; Thompson, J.N. Species-rich networks and eco-evolutionary synthesis at the metacommunity level. *Nat. Ecol. Evol.* **2017**, *1*, 24. [CrossRef]

agronomy

Article

Exploring Microbial Rhizosphere Communities in Asymptomatic and Symptomatic Apple Trees Using Amplicon Sequencing and Shotgun Metagenomics

Zilia Y. Muñoz-Ramírez [1], Román González-Escobedo [2,*], Graciela D. Avila-Quezada [3], Obed Ramírez-Sánchez [4], Victor M. Higareda-Alvear [4], Emiliano Zapata-Chávez [1], Alejandra Borrego-Loya [1] and Laila N. Muñoz-Castellanos [1,*]

1 Facultad de Ciencias Químicas, Universidad Autónoma de Chihuahua, Campus II Circuito Universitario s/n, Chihuahua 31125, Chihuahua, Mexico; zramirez@uach.mx (Z.Y.M.-R.); ezapata@uach.mx (E.Z.-C.); aborrego@uach.mx (A.B.-L.)
2 Facultad de Zootecnia y Ecología, Universidad Autónoma de Chihuahua, Periférico Francisco R. Almada km 1, Chihuahua 31453, Chihuahua, Mexico
3 Facultad de Ciencias Agrotecnológicas, Universidad Autónoma de Chihuahua, Campus I s/n, Chihuahua 31350, Chihuahua, Mexico; gdavila@uach.mx
4 Soil Genomics & Discovery Department, Solena Inc. Av. Olímpica 3020-D, Villas de San Juan, León 37295, Guanajuato, Mexico; oramirez@solena.ag (O.R.-S.); vhigareda@solena.ag (V.M.H.-A.)
* Correspondence: rgescobedo@uach.mx (R.G.-E.); lmunoz@uach.mx (L.N.M.-C.)

Abstract: The rhizosphere is a dynamic and highly interactive habitat where diverse microbial communities are established, and it plays crucial roles in plant health and disease dynamics. In this study, microbial communities and functional profiles in the rhizosphere of both asymptomatic and symptomatic apple trees were investigated through amplicon sequencing and shotgun metagenomics. The research was conducted at a location in the municipality of Cuauhtemoc, Chihuahua State, Mexico, and a total of 22 samples were collected, comprising 12 for amplicon sequencing and 10 for shotgun metagenomic sequencing. Symptomatic trees were identified based on reddish branches and internal necrosis in the trunk and root, while asymptomatic trees exhibited a healthy physiology. The findings showed that the dominant bacterial phyla included Proteobacteria, Actinobacteria, and Bacteroidetes, with prevalent genera such as *Streptomyces*, *Pseudomonas*, and *Rhodanobacter*. The fungal communities featured Ascomycota, Mortierellomycota, and Basidiomycota, which were dominated by *Fusarium*, *Penicillium*, and *Mortierella*. In the fungal communities, Mortierellomycota, notably abundant in asymptomatic trees, holds potential as a biocontrol agent, as seen in other studies on the suppression of *Fusarium* wilt disease. The application of shotgun metagenomic sequencing revealed significant differences in alpha and beta diversities in bacterial communities, suggesting a health-dependent change in species composition and abundance. Functional profile analysis highlighted enzymatic activities associated with lipid synthesis/degradation, amino acid biosynthesis, carbohydrate metabolism, and nucleotide synthesis, which have been documented to participate in symbiotic relationships between plants. These insights not only contribute to understanding the dynamics of rhizosphere microbial activity but also provide valuable perspectives on the potential application of microbial communities for tree health and implications for the management of apple orchards.

Keywords: *Malus domestica*; rhizosphere; bacteria; fungi; microbiome; metagenomics

check for updates

Citation: Muñoz-Ramírez, Z.Y.; González-Escobedo, R.; Avila-Quezada, G.D.; Ramírez-Sánchez, O.; Higareda-Alvear, V.M.; Zapata-Chávez, E.; Borrego-Loya, A.; Muñoz-Castellanos, L.N. Exploring Microbial Rhizosphere Communities in Asymptomatic and Symptomatic Apple Trees Using Amplicon Sequencing and Shotgun Metagenomics. *Agronomy* **2024**, *14*, 357. https://doi.org/10.3390/agronomy14020357

Academic Editors: Yong-Xin Liu and Peng Yu

Received: 19 November 2023
Revised: 8 December 2023
Accepted: 9 December 2023
Published: 9 February 2024

1. Introduction

The rhizosphere, the soil region surrounding plant roots, is a dynamic and highly interactive habitat where diverse microbial communities are established. These communities play a fundamental role in numerous ecological and agricultural processes, including organic matter decomposition, nutrient cycling, and the promotion of plant growth [1–3]. Among the microorganisms present in the rhizosphere, bacteria and fungi stand out as key components of the microbial community. These microorganisms not only exist in close proximity to plant roots but also interact intimately with them, influencing plant health and productivity [1].

The apple tree (*Malus domestica*), a fruit species of great economic and agricultural importance, heavily relies on the interaction between tree roots and the microbial communities present in the rhizosphere. This microbial diversity and its functions have been the subject of increasing interest in the rhizosphere of several agricultural crops [4–8]. For instance, certain bacterial species can fix atmospheric nitrogen, making it available to plants, while others produce phytohormones that stimulate root development and enhance nutrient uptake [9–11]. Additionally, microbial communities can protect plants from pathogens through resource competition or by producing antimicrobial compounds [12,13]. Understanding the composition and function of these microbial communities is crucial to improving tree health, increasing the quality and quantity of apple production, and developing sustainable agricultural practices [14,15].

Although numerous studies have been conducted on the microbiological aspects associated with apple trees, the generated data due to next-generation sequencing (NGS) techniques are crucial for the generation of knowledge benefiting this agriculturally important crop. The use of metagenomic analysis in research has been applied to solve existing gaps by identifying new genetic variants and trying to explain emerging diseases or diseases caused by various pathogens, as is the case in this study. In the same way, progress in these matters has revolutionized our understanding of microbial communities in the rhizosphere, providing insights into the taxonomic composition, functional potential, and ecological roles of microbial communities [16,17].

Apple crown and root rot diseases have scarcely been studied in apple orchards in the apple-growing region of Chihuahua, Mexico [18]. They induce symptoms such as necrosis in the feeder root system and in the trunk, and finally, they lead to the death of the tree. Around the world, fungi (e.g., *Cylindrocarpon* spp., *Rhizoctonia* spp., and *Fusarium proliferatum*) and oomycetes (e.g., *Phytophthora* spp. and *Pythium* spp.) have been reported as the pathogens causing apple root and crown rot diseases [19–21]. However, it has been observed that in many instances, it is not just an individual pathogen but rather the involvement of many other groups of microorganisms that affect the outcome of the infection in plants [22]. In fact, the diseases known as apple replant disorders are associated with various causal agents but differ among countries and remain unclear [23]. In Chihuahua, Mexico, the incidence of root diseases in apple trees was assessed, revealing a rate of 17%. Various isolates of fungi and oomycetes were identified, and their pathogenicity was subsequently determined. In in vitro antagonistic activity tests, *Trichoderma* and *Bacillus* emerged as promising alternatives for biological control against the evaluated phytopathogens [18]. It is crucial to study the microbial community in the rhizosphere of both asymptomatic and symptomatic trees affected by root and crown rot to gain a better understanding of the possible microbial interactions that exist in the rhizosphere and, consequently, to enhance our overall comprehension of the disease dynamics.

In light of the existing gaps in knowledge, our hypothesis suggests that changes in the rhizosphere microbial community composition of symptomatic apple trees are structured by the presence and activities of pathogens and saprophytes, as well as beneficial microorganisms exhibiting antagonistic properties against the pathogens. The aim of this study was to characterize and compare the composition and diversity of the microbial communities found in rhizosphere samples of both conditions at a location in the municipality of Cuauhtemoc, Chihuahua State, Mexico. In pursuit of this objective, metagenomic shotgun sequencing and amplicon sequencing targeting the 16S rRNA gene and internal transcribed spacer (ITS1) region were employed to generate new knowledge. Additionally, the functional profile was analyzed to elucidate the ecological functions of microbial communities in the rhizosphere. Thus, this study significantly contributes to understanding rhizosphere microbial dynamics, providing insights that could have potential implications for the effective management of apple orchards.

2. Materials and Methods

2.1. Site Description and Sample Collection

Soil samples from an apple orchard area containing asymptomatic and symptomatic crown and root rot of *Malus domestica* trees (golden glory apples grafted onto Emla7 rootstock) were collected at a location in the municipality of Cuauhtemoc, Chihuahua State, Mexico (28°42′54.9″ N, 106°55′06.6″ W; 2,018 masl). The first sampling took place in June 2020, during the early summer, and involved collecting three rhizospheric soil samples from both asymptomatic and symptomatic apple trees. These samples were used for amplicon sequencing. A subsequent sampling was conducted in February 2022, during late winter, where six asymptomatic and four symptomatic trees were sampled for metagenomic sequencing.

Rhizosphere soil samples from 7-year-old trees under each condition were sampled: first, rhizosphere soil samples from asymptomatic and symptomatic apple trees were selected, with a minimum distance of 40 m between samples, based on visual symptoms from the roots and foliage, such as dead areas at the base of the tree. These dead areas started in the bark between the soil line and the crown roots, comprising darkened and collapsed tissue. The foliage symptoms were reduced vegetative growth, small leaves, and little budding. The uppermost 20 cm of soil was removed using a shovel, and approximately 10 g of rhizosphere soil was collected per individual plant by employing a dry sterile toothbrush to brush off the soil around the surface of the apple root, which was then placed into sterile bags and stored on ice for transport to the laboratory, where the samples were then stored in an ultralow temperature freezer at −80 °C until processing. In addition, the physical and chemical properties of the soil at a depth of approximately 20 cm from asymptomatic and symptomatic trees were as follows: soil texture = loam and sandy loam; pH = 6.24 and 2.91; electrical conductivity (EC) = 2.71 and 3.75 mS/cm; organic matter = 1.42 and 1.50%.

2.2. Total DNA Extraction and Sequencing

Total DNA was extracted from each sample using a ZymoBIOMICS^{TM} DNA Miniprep Kit (Zymo Research, Irvine, CA, USA) following the manufacturer's instructions. The DNA quality and quantity were determined by using a NanoDrop spectrophotometer (Thermo Scientific, Wilmington, DE, USA) based on its A260/280 ratio, and the DNA was observed with a 1.0% agarose gel electrophoresis. The preparation and sequencing of amplicon libraries (16S rRNA gene and ITS1 region) and shotgun metagenomics, following the manufacturer's protocol, were carried out at Novogene (Beijing, China) and Illumina (San Diego, CA, USA), respectively. For amplicons, the NovaSeq sequencing platform was used alongside a paired-end 2 × 250 bp strategy carried out on an Illumina sequencer (Illumina Inc., San Diego, CA, USA). In the case of bacteria, a fragment of the 16S rRNA gene was amplified using the primers 341F (5′-CCTAYGGGRBGCASCAG-3′) and 806R (5′-GGACTACNNGGGTATCTAAT-3′) flanking the V3 and V4 regions [24]. For fungi, the ITS1 region was amplified using the primer pairs ITS5-1737F (5′-GGAAGTAAAAGTCGTAACA AGG-3′) and ITS2-2043R (5′-GCTGCGTTCTTCATCGATGC-3′) [25]. For metagenomic library preparation and sequencing, the total DNA was shipped to Illumina (San Diego, CA, USA). Briefly, 100 ng of total DNA was processed following the instructions of the Illumina DNA prep (M) tagmentation kit (# Cat. 20018705). Because many samples were run in the same flow cell, specific indexes were added to each DNA library with IDT for Illumina DNA/RNA UD Indexes Set A Tagmentation (# Cat. 20027213). Library concentration was quantified with the Qubit dsDNA HS Assay kit (Thermo Fisher), and the integrity of the DNA libraries was assessed with the Bioanalyzer 2100 Agilent system (NGS 1-6000 kit). Libraries were sequenced in an S4 flow cell using a 2 × 150 bp strategy on an Illumina NovaSeq 6000 sequencer (Illumina Inc., San Diego, CA, USA).

2.3. Amplicon Data Analysis

Sequence data were obtained as fastq files in the CASAVA v1.8 paired-end demultiplexed format. Forward and reverse sequences were merged using FLASH v1.2.11 with default settings [26]. Merged sequences were imported into Quantitative Insights Into Microbial Ecology 2 v2020.2 (QIIME 2) [27]. Firstly, quality filtering, trimming, dereplication, and denoising, as well as the removal of chimeric sequences from the merged data using DADA2 v1.16, were performed [28] to obtain representative amplicon sequence variants (ASVs). Each ASV was assigned a taxonomy with a trained naïve Bayesian classifier using the Greengenes v13.8 database [29] for bacteria and the UNITE v8.99 database [30] for fungi. Bar plots based on relative abundance were generated to show the taxonomic distribution.

A multiple sequence alignment and phylogenetic reconstruction were produced from the ASVs using MAFFT v7 [31] and FastTree v2 [32], respectively, to generate a rooted phylogenetic tree and conduct subsequent analyses. In order to analyze the α- and β-diversity of bacterial and fungal communities and conduct related statistical tests, the samples from the library were rarefied with the lowest number of reads, and the metrics were calculated using RStudio [33] with the packages vegan [34], tidyverse [35], and qiime2R [36]. To explore the α-diversity within these communities under both asymptomatic and symptomatic conditions, the Chao1, Shannon, Simpson, and evenness indexes were estimated, and the diversity was compared among groups using the Kruskal–Wallis test followed by the post hoc Dunn's test, considering the difference to be statistically significant at p-values < 0.05.

To assess differences in the bacterial and fungal communities' composition between the asymptomatic and symptomatic conditions, a principal coordinate analysis (PCoA) was performed based on Jaccard distances. Then, a permutational multivariate analysis of variance (PERMANOVA) with 999 permutations was used for testing, followed by the post hoc Benjamini–Hochberg FDR test to determine significant differences in bacterial and fungal communities. Additionally, a functional inference analysis was performed with PICRUSt2 v2.5.1 [37] to explore the possible functional profiles of the microbial communities through their enzymatic activities, and the accuracy of the analysis was assessed through the weighted Nearest Sequences Taxon Index (NSTI).

2.4. Metagenomic Data Analysis

Raw fastq reads of each sample were quality filtered with FASTP v0.20.0 [38]. The filtered quality scores were set to 25, and a trimming of the first and last five bases was carried out. The taxonomic assignment of trimmed reads was performed with Kraken2 v2.1.2 [39] using the –download-library option from Kraken2-build to download the complete bacterial and fungal genomes from the RefSeq database. In addition, fungal genomes from *Fusarium* spp. and *Pythium* spp. were included. The Kraken2 files from all samples were combined to generate a biom file with the Kraken-biom utility. Two main datasets were generated from the biom file through filtering: Kingdom == "Bacteria" and Kingdom == "Eukaryota". Analysis of the microbial diversity, including measurements of α- and β-diversity, was performed with the Phyloseq package v1.40.0 [40] for both datasets independently. In addition, Venn diagrams were made to visualize which bacterial and fungal genera were exclusive or shared between the symptomatic and asymptomatic conditions. All metrics presented in this work are based on relative abundances and calculated using RStudio [33] with the packages vegan [34], tidyverse [35], and qiime2R [36]. The statistical analysis of α-diversity was evaluated with the Kruskal–Wallis test, and β-diversity was determined using PERMANOVA with 999 permutations to test for significant differences in bacterial and fungal communities, considering the difference to be statistically significant at p-values < 0.05.

Finally, functional predictions for the shotgun metagenomic data were determined using the MG-RAST pipeline v4.0.3 [41]. The sequences in fastq format were uploaded to MG-RAST, and the default settings were used. MG-RAST classifies sequences into subsystems, which are grouped into hierarchical categories and can be used to construct a heatmap with the top 50 most abundant functions.

The datasets generated and analyzed during this current study are available in the NCBI Bioproject database under the accession numbers PRJNA1003089 (amplicon sequences) and PRJNA1003562 (metagenomic sequences).

3. Results

3.1. Composition and Abundance of Bacterial and Fungal Communities

A total of 837,294 (16S rRNA) and 1,074,605 (ITS1 region) high-quality sequences were obtained from all rhizosphere soil samples under both asymptomatic and symptomatic conditions (Table S1). The samples with the lowest number of sequences were rarefied, homogenizing all bacterial samples to a total of 126,057 sequences and fungal samples to a total of 166,349 sequences, which were the totals for subsequent analyses. In the case of shotgun metagenomics, a total of 826,840,106 raw sequences were generated. After a quality check analysis, 708,403,234 high-quality sequences were obtained and used for further analysis.

The community composition derived from both amplicon and metagenomic sequences exhibited robust similarity in relation to the most prevalent phyla. Nonetheless, metagenomic sequencing enabled a more comprehensive retrieval of information across all taxonomic levels, but especially that related to the phyla. The amplicon sequences allowed for the identification of a total of 51 phyla, 196 families, and 507 bacterial genera, as well as a total of 11 phyla, 143 families, and 268 fungal genera. The most abundant prokaryotic phyla in the rhizosphere of apple trees consisted of Proteobacteria (48.91%), Bacteroidetes (11.19%), Actinobacteria (10.46%), Acidobacteria (9.48%), Gemmatimonadetes (6.43%), Firmicutes (3.93%), Verrucomicrobia (3.25), and Chloroflexi (3.11%), followed by 43 genera with <1.00% relative abundance (Figure 1a). In the fungal communities, the phyla comprised Ascomycota (77.69%), Mortierellomycota (9.32%), Basidiomycota (9.19), Rozellomycota (1.59%), and Blastocladiomycota (1.47%), followed by eight genera with <1.0% relative abundance (Figure 2a). At the genus level, the only bacterial genus that presented a relative abundance >10% was *Rhodanobacter* (10.6%), followed by 25 genera (e.g., *Kaistobacter*, *Streptomyces*, *Rhodoplanes*, *Dechloromonas*, and *Pseudomonas*) that had a relative abundance ranging between 1 and 10% and 481 genera that had a relative abundance of <1.00% (Figure 1b). In fungi, at the genus level, *Penicillium* (28.62%) and *Mortierella* (11.14%) presented an abundance >10%, 18 genera (e.g., *Acremonium*, *Fusarium*, *Ilyonectria*, *Setophaeosphaeria*, and *Apiotrichum*) had a relative abundance ranging between 1 and 10%, and 248 genera had a relative abundance of <1.0%. (Figure 2b).

On the other hand, in the metagenomic analysis, 39 phyla, 451 families, and 1741 bacterial genera were recovered. The most abundant bacterial phyla were Proteobacteria (57.25%) and Actinobacteria (33.17%), followed by phyla with <5.00% relative abundance such as Bacteroidetes (2.22%), Planctomycetes (1.88%), Acidobacteria (1.63%), and Firmicutes (1.24%), and 33 phyla had <1.0% relative abundance (Figure 1c). At the genus level, *Streptomyces* was the most abundant, with a relative abundance of 6.81%, followed by *Bradyrhizobium* (3.87%), *Pseudomonas* (3.13%), *Nocardioides* (2.83%), *Sphingomonas* (2.67%), *Rhodanobacter* (2.63%), and nine other genera with relative abundances of up to 1.0%; the 1726 remaining genera showed relative abundances of <1.0% (Figure 1d). For fungi, the annotated sequences accounted for 0.10–0.30% of the total sequences, suggesting potential annotation biases that resulted in the underestimation of eukaryotic communities and led to the recovery of a total of 2 phyla, 32 families, and 55 genera. The phyla Ascomycota (91.87%) was the most dominant, followed by Basidiomycota (8.13%) (Figure 2c); *Fusarium* (30.00%) and *Trichoderma* (11.21%) were the most abundant genera, followed by *Metarhizium* (5.73%), *Beauveria* (4.48%), *Pseudozyma* (3.58%), *Pyricularia* (3.52%), *Clonostachys* (3.22%), *Penicillium* (3.11%), and 11 other genera with relative abundances of up to 1.00%, while the remaining 36 genera exhibited a relative abundance of <1.0% (Figure 2d). In addition, the comparison using Venn diagrams showed a core microbiome at the genus level. In the case of bacteria, 198 genera were shared between both health conditions analyzed, while 12 genera were prevalent for both conditions in the case of fungi (Figure S1).

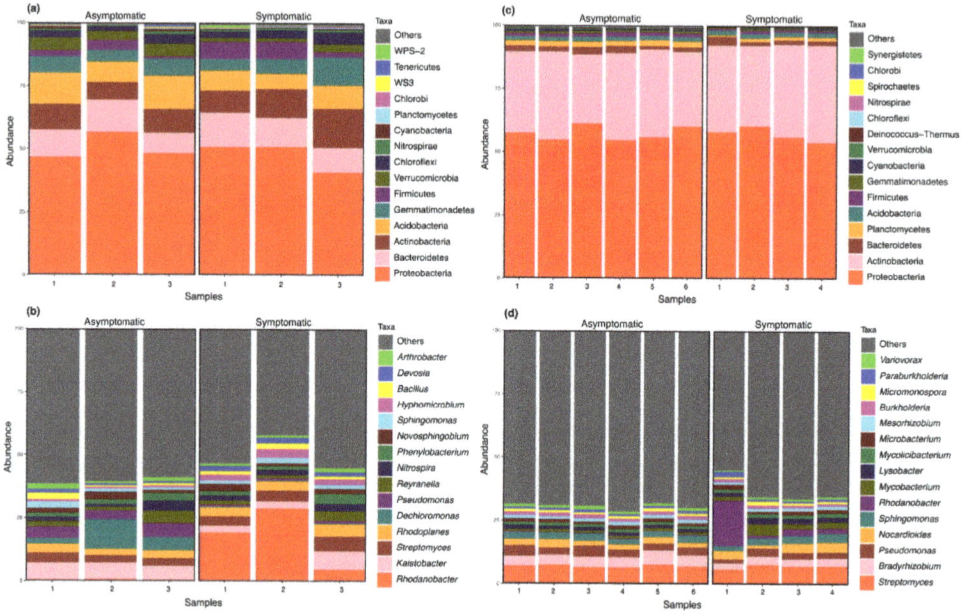

Figure 1. Bar plots depicting the relative abundance of bacterial communities at different taxonomic levels: (**a**) phyla using amplicon data, (**b**) phyla using metagenomic data, (**c**) genera using amplicon data, and (**d**) genera using metagenomic data.

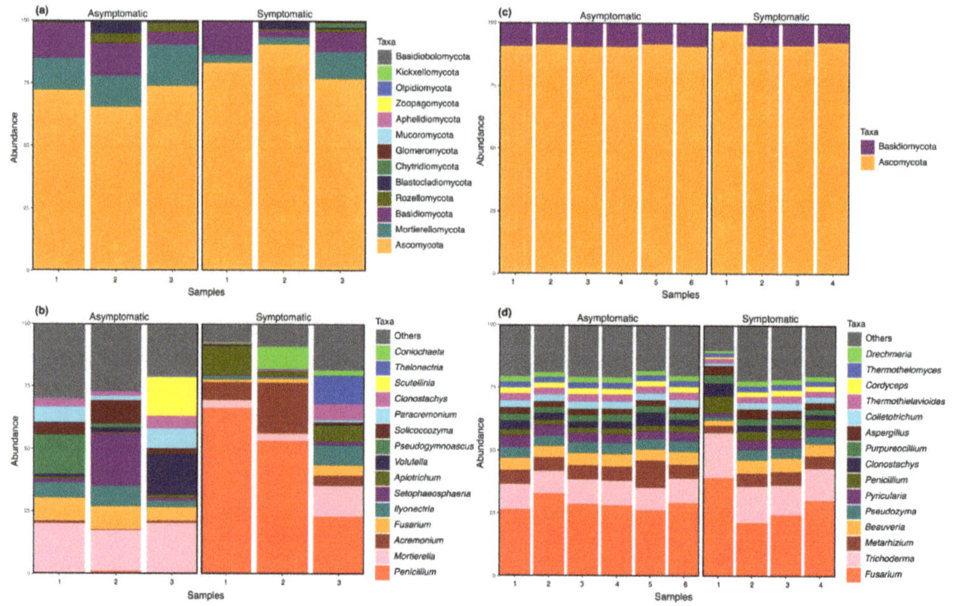

Figure 2. Bar plots depicting the relative abundance of fungal communities at different taxonomic levels: (**a**) phyla using amplicon data, (**b**) phyla using metagenomic data, (**c**) genera using amplicon data, and (**d**) genera using metagenomic data.

3.2. α- and β-Diversity of Asymptomatic and Symptomatic Apple Trees

The taxonomic distinctiveness of the rhizosphere in asymptomatic and symptomatic apple trees with regard to α-diversity was investigated. The analysis of amplicon sequences from bacteria and fungi revealed no significant differences ($p > 0.05$) in the Chao1, Shannon, Evenness, and Simpson indexes (Figures 3a and 4a). In the case of metagenomic data for bacteria, with the exception of the Simpson index ($p > 0.05$), the Chao1, Shannon, and Evenness indexes showed significant differences ($p < 0.05$) (Figure 3b). For fungi, no significant differences were observed in any index ($p > 0.05$) (Figure 4b).

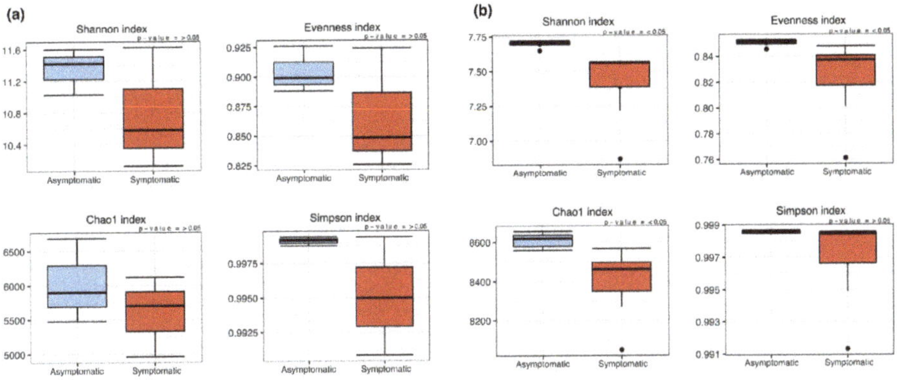

Figure 3. Box plots depicting the analysis of alpha diversity in the bacterial community using Shannon, Evenness, Chao1, and Simpson indexes, evaluated through (**a**) amplicon data and (**b**) metagenomic data.

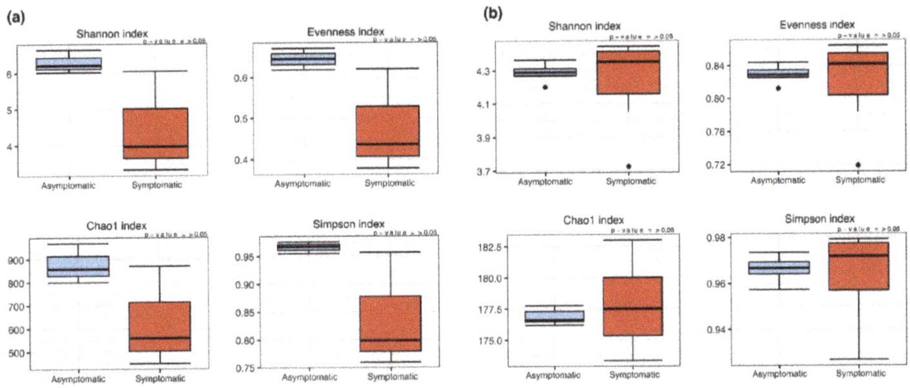

Figure 4. Box plots depicting the analysis of alpha diversity in the fungal community using Shannon, Evenness, Chao1, and Simpson indexes, evaluated through (**a**) amplicon data and (**b**) metagenomic data.

The β-diversity, as determined through the variation in microbial communities between asymptomatic and symptomatic conditions, was examined using PCoA analysis based on Jaccard distances. With bacterial and fungal amplicon data, the PCoAs showed 80.5% and 83.8% dissimilarity, respectively, and revealed that the microbial communities did not differ (PERMANOVA; $p > 0.05$) (Figure 5a,b). In contrast, the metagenomic data showed a dissimilarity of 81.11% and 90.18% in the PCoA analysis, unveiling significant differences in the bacterial communities (PERMANOVA; $p < 0.05$) (Figure 5c) but not in the fungal communities (PERMANOVA; $p > 0.05$) (Figure 5d).

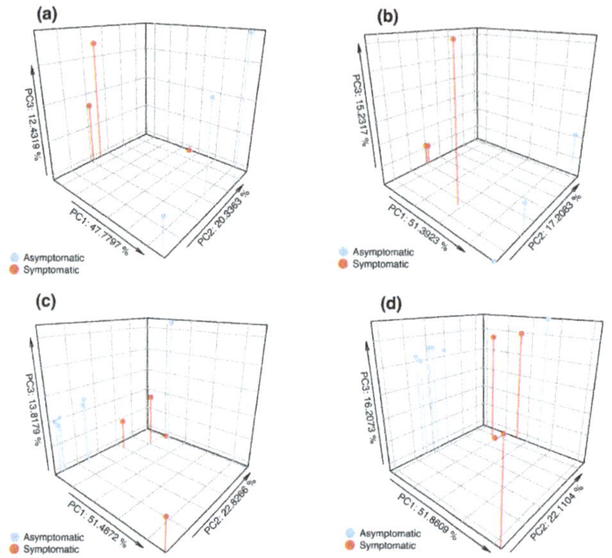

Figure 5. PCoA based on Jaccard distance shows the microbial community dissimilarity. (**a**) Bacterial 16S rRNA gene amplicon data; (**b**) fungal ITS region amplicon data; (**c**) bacterial metagenomic data; and (**d**) fungal metagenomic data.

3.3. Functional Profiles of Microbial Communities

PICRUSt2 analysis of the amplicon data was applied to identify the possible functional profiles of the microbial communities. A list of 2380 enzymatic functions was obtained, and the top 50 dominating enzymes were displayed in a heatmap (Figure 6a). The average NSTI value was 0.23, which is considered to be a low value; it indicates a satisfactory quality in the functional predictions made due to its proximity to the nearest reference genomes in the used database. In general, the top 50 enzymatic functions participate in essential processes for cellular functioning and the regulation of numerous metabolic pathways, such as DNA replication and transcription, protein synthesis and modification, lipid and carbohydrate metabolism, energy generation through cellular respiration, the synthesis and repair of nucleic acids, as well as the biosynthesis of secondary metabolites.

Based on the processed sequencing data from MG-RAST in 2019, genes/enzymes were identified using KEGG pathway annotation. Among these genes, a heatmap was generated for the top 50 most abundant enzymatic functions in the analyzed samples (Figure 6b). The analysis revealed a significant abundance of enzymes involved in various processes, including fatty acid, nucleotide, and protein synthesis, as well as amino acid biosynthesis, carbohydrate metabolism, and lipid metabolism. Additionally, enzymes related to energy generation and the degradation of toxic compounds were also highly prevalent. When comparing the top 50 enzymatic functions detected through PICRUSt2 and metagenomics, several enzymatic activities were found to be exactly the same. These activities are associated with various metabolic processes and biological functions related to lipid synthesis and degradation, amino acid biosynthesis, carbohydrate metabolism, and nucleotide synthesis. The enzymes involved in these processes were as follows: EC:1.1.1.100—3-oxoacyl-[acyl-carrier-protein] reductase; EC:1.8.1.9—thioredoxin-disulfide reductase; EC:2.2.1.6—acetolactate synthase; EC:2.3.1.9—acetyl-CoA C-acetyltransferase; EC:4.2.1.33—3-isopropylmalate dehydratase; EC:5.1.3.2—UDP-glucose 4-epimerase; EC:6.2.1.3—long-chain fatty acid CoA ligase; and EC:6.3.5.3—phosphoribosylformylglycinamidine synthase.

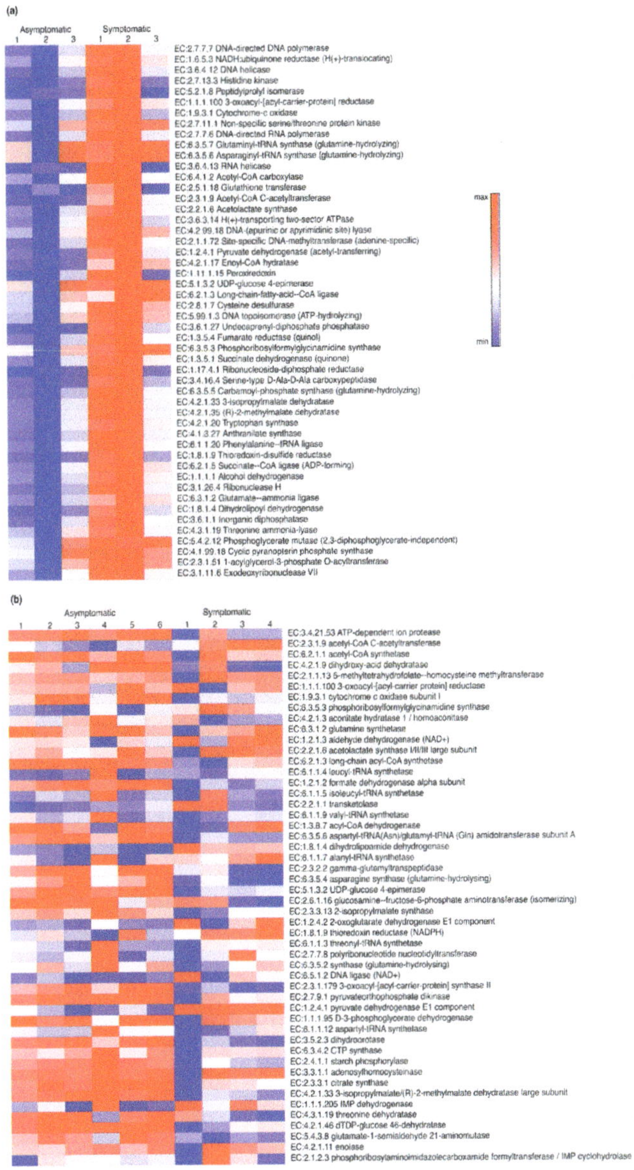

Figure 6. Heatmaps reveal the top 50 functional profiles predicted from (**a**) amplicon data using PICRUSt2 and (**b**) metagenomic shotgun data using MG-RAST.

4. Discussion

In this study, a thorough analysis of the taxonomic and functional traits of microbial communities in the rhizosphere of apple trees was conducted over two non-consecutive years. By using both shotgun metagenomic and amplicon sequencing, our aim was to enhance our understanding of the diversity and functional capabilities within these ecologically vital communities, which are linked to both asymptomatic and symptomatic trees. While previous studies on apple tree microbial communities have predominantly utilized amplicon sequencing approaches [42–45], only a limited number of shotgun metagenomic studies have been conducted on rhizosphere microbiomes [46].

In the current study's orchard, the incidence of apple tree crown and root rot is less than 10% (Omar Quintana, personal communication); however, it remains a persistent and growing problem that has not been completely eradicated despite continuous monitoring and management. Some studies have reported on the associated microorganisms inhabiting the rhizospheric soil of apple trees affected by crown disease, mainly through cultivable characterization [19] and non-cultivable techniques such as DGGE [47]. Therefore, exploring this through NGS techniques such as amplicon and metagenomics represents a baseline for further investigation.

Metagenomic shotgun sequencing provided greater depth in the analyses performed (39 phyla, 451 families, and 1741 bacterial genera), whereas amplicon sequencing identified a total of 23 phyla, 196 families, and 507 bacterial genera. It is important to note that different factors can influence changes in the communities, such as sampling over two non-consecutive years, different seasons, and possible variations in climatic conditions. Studies on the soil rhizosphere have shown that, despite being different samplings, the composition of microbial communities remains largely consistent when using either amplicon-based or shotgun sequencing technologies [48]. However, it is important to emphasize that understanding microbial communities and their functional capabilities provides an advantage and is a reason for choosing metagenomic technology over amplicon sequencing. Furthermore, it is important to highlight that, despite using different technologies, the most abundant phyla were successfully recovered with both high-throughput sequencing approaches. These phyla, including Proteobacteria, Actinobacteria, and Bacteroidetes, have been previously reported as the most abundant in the rhizosphere, particularly in some studies on soil in which *M. domestica* is grown [49–51].

In fungi, as mentioned earlier, an underestimation in composition occurred when using shotgun sequencing; however, two out of the three most abundant phyla (Ascomycota and Basidiomycota) remained consistent in both analyses. A possible technical issue could have been related to the Kraken database used, which contained 36,246 bacterial species and 455 fungal species, resulting in a sub-estimation of fungal diversity. The phyla Ascomycota, Basidiomycota, and Mortierellomycota have been reported in other agriculturally important crops. Notably, the phylum Mortierellomycota was found to be more abundant in asymptomatic trees; previous studies have reported the presence of this phylum in vanilla orchards, highlighting its ability to produce antibiotics and act as a potential antagonist against various plant pathogens [52]. Additionally, Basidiomycota is another abundant phylum in the rhizosphere of both asymptomatic and symptomatic trees, as it has been reported as the most abundant in apple orchard soils in China [53].

At the genus level, microbial diversity was high and heterogeneous, regardless of the health conditions of the trees. The greater sequencing depth of the shotgun technology allowed for the detection of more genera at a ratio of 1:3 compared to amplicon sequencing. Genera such as *Streptomyces*, *Pseudomonas*, and *Rhodanobacter* were prevalent, regardless of the sequencing technology used, and maintained a similar proportion despite health status. In particular, *Rhodanobacter* exhibited a higher relative abundance in the rhizosphere of symptomatic trees. While the available literature on *Rhodanobacter* is limited, it has been reported that it has the capacity to act as a denitrifying bacterium, which, if it occurs excessively, can result in the loss of nitrogen, an essential nutrient for plant growth, and reduce nitrogen availability for crops in agricultural soils [54,55]. Furthermore, in the case of *Streptomyces* and *Pseudomonas*, these genera have been widely reported in agricultural soils due to their biocontrol properties and their ability to act as antagonists against plant pathogens [56,57], their capacity to degrade and metabolize a wide range of organic compounds [58–60], and their ability to produce substances that promote plant growth, such as phytohormones, enzymes, and siderophores, which can enhance nutrient absorption and stimulate plant growth [61–63]. In the case of less common genera, it is important to highlight some low-abundance genera that, despite their scarcity, may play significant roles in the rhizosphere. One such is *Hydrogenobaculum*, a genus that participates in nitrate reduction and could be involved in modulating the distribution of microbial communities [64]. Another infrequent and low-abundance genus is *Metakosakonia*, which has been previously reported as a promoter of plant growth through in vitro assays, demonstrating increases in both the shoot and root growth of potatoes [65]. Therefore, it is of interest to investigate its interactions with other rhizosphere microorganisms and with the plant, which may lead to new discoveries for the development of biocontrol strategies and biostimulants for crop production.

In the case of fungi, *Fusarium* and *Penicillium* were the most abundant genera, followed by *Mortierella*, which was mainly abundant in samples from the rhizosphere of asymptomatic trees. Different *Fusarium* species have been documented as saprophytes [66], opportunists [67], and phytopathogens, meaning they are capable of causing diseases in plants such as vascular wilting, root rot, and stem decay, which results in a negative impact on the health and yield of crops [68–71]. Moreover, *Penicillium* is a common necrotrophic–saprophytic genus that might play an important role in diseased roots since it has been able to exhibit a variety of lifestyles, including mutualism, commensalism, and parasitism [72]. In fact, certain strains can also be phytopathogenic, leading to diseases and damage to crops [73,74]. On the other hand, in the rhizosphere of asymptomatic trees, the *Mortierella* genus presented the highest abundance, which was similar to the results reported by Xiong et al. [52], who found that suppressive soil was dominated by the fungal *Mortierella*, accounting for 37% of the total fungal sequences in a study of the suppression of vanilla *Fusarium* wilt disease; it seems that *Mortierella* produces fatty acid ethyl esters that contain arachidonic acid, which under greenhouse conditions reduced the development of tomato late blight and rhizoctonoise of potato tubers [75].

Significant differences in α-diversity were detected only in bacteria, exclusively through the shotgun metagenomic approach. Studies evaluating the α-diversity of rhizosphere communities in agricultural crops have reported variations in diversity due to different biotic factors (e.g., fungi, oomycetes, bacteria, and nematodes) and abiotic factors (e.g., temperature, tree age, sampling season, and physical and chemical soil properties) that influence the growth and distribution of microbes [76,77]. In fact, similar to the results from this study, greater microbial diversity has been reported in the rhizosphere of healthy plants, which decreases in diseased plants [78–80]. This was also demonstrated in sesame rhizosphere soil, where a positive correlation was found between the alpha diversity of the microbial community in the rhizosphere soil of crops and their health status. This correlation led to alterations in the dynamics of bacterial communities and their associated soil functions as part of a plant disease response mechanism [80].

The β-diversity results showed that the bacterial and fungal communities in the rhizosphere of asymptomatic trees were grouped separately from those of symptomatic trees. However, statistical analysis revealed no significant difference in the microbial community structure. In contrast, shotgun analysis showed a statistically significant difference, indicating a health-dependent shift in community structure. Regarding the amplicon-sequenced samples, several key observations were made when reviewing the β-diversity results, which may be interconnected. First, there was genuinely no significant difference in the analyzed samples, as observed in other studies evaluating microbial communities between asymptomatic and symptomatic avocado trees affected by root rot [81]. Secondly, the number of biological replicates used for amplicon sequencing is crucial. It is advised that at least three biological replicates be performed to obtain a more accurate representation of microbial diversity and reduce result variability. However, in some cases, additional replicates may be required for more robust and reliable results [82]. Thirdly, sampling for the first year was conducted in March 2020. Continuous monitoring was carried out in the orchard to detect trees with disease symptoms and apply appropriate management. It was found that out of the three rhizosphere samples collected from symptomatic trees, the apple tree, from which the third biological replication of bacteria and fungi was obtained, recovered, unlike the other two trees. This tree was treated with compost tea, which possibly promoted greater competition among soil microorganisms [83] due to the exudates produced by the roots and microorganisms and due to the self-regulation of microbial communities [84].

On the other hand, the observed variations in microbial communities between asymptomatic and symptomatic trees through the use of the two sequencing methodologies employed should be interpreted in the context of the seasonal differences during sample collection. Seasonal dynamics, such as temperature and moisture fluctuations, are known to influence microbial composition in soil ecosystems [85]. Recognizing the potential influence of seasonal variations on our results, it is crucial to emphasize that our data analysis utilized rigorous statistical methods to identify patterns related to health conditions. Indeed, similar to other studies such as those conducted by Bei et al. [86], despite the seasonal variations, the overall diversity of the rhizosphere microbiome remained relatively stable, as observed in the Venn diagrams. However, future studies with more frequent and extended sampling across seasons could offer a more comprehensive understanding of the interaction between seasonal dynamics and rhizosphere microbial communities.

Regarding the functional profiles of rhizosphere microbial communities, PICRUSt and MG-RAST have been used in several studies to identify enzymatic activities involved in the metabolic processes in the rhizosphere bacterial communities [87–89]. Considering the top 50 most abundant enzymatic functions from both approaches, several common metabolic functions were identified, including some directly associated with genes/enzymes involved in various metabolic processes and biological functions related to lipid synthesis and degradation, amino acid biosynthesis, carbohydrate metabolism, and nucleotide synthesis. For example, one of the important enzymes detected was 3-oxoacyl-[acyl-carrier-protein] reductase, which has been reported as an enzyme that participates in symbiotic relationships between plants and microorganisms, promoting the formation of nodules in plant roots [90]. In agricultural soil, the mentioned functions are of utmost importance as they perform activities that involve soil nutrient cycling, organic matter degradation, chemical compound transformation, secondary metabolite production, and interaction with plant roots [91,92]. Comprehending these metabolic profiles associated with the microbiome could serve as evidence of dependence on environmental factors and provide valuable insights into the interactions between the microbial community and the host plant [93].

Interestingly, shotgun sequencing allowed for the identification of the oomycetes *Phytophthora* and *Pythium*, which were found in the rhizosphere of both asymptomatic and symptomatic trees (Figure S2). It has been documented that this phytopathogen infects approximately 200 plant species, including economically important plants such as strawberries, pears, walnuts, and apples [94]. A higher relative abundance of *Phytophthora* and *Pythium* was observed in the rhizospheric soil samples from asymptomatic trees than those from symptomatic ones. *Phytophthora*, despite being known to cause diseases in apple trees, can also be present in healthy soil samples as part of its life cycle due to nutrient availability or the presence of related non-pathogenic species. It is also important to consider that the presence or abundance of a pathogen in soil does not always directly correlate with disease in plants. Other factors, such as plant defense mechanisms, the susceptibility of the host plants, or the interaction with other microorganisms present in the soil, can influence disease expression [95]. In fact, in studies conducted on the soils of various oak species showing nonspecific symptoms of branch death and canopy decline [96] and on those of asymptomatic *Eucalyptus coccifera* [97], several species of *Phytophthora* were recovered regardless of the health status of the trees, suggesting a possible ecological role as saprophytes. It is important to note that further analyses involving the isolation of these phytopathogens are required to understand the processes of pathogenicity or saprophytism. On the other hand, *Pythium* species are extensively spread as plant pathogens, including several apple pathogens [98]. However, other members of this genus are prevalent as soil saprophytes [99] and have been identified as exhibiting saprophytic behavior in soil samples, with distribution influenced by factors such as soil type, precipitation, and temperature. Hence, it is crucial to emphasize that not every Pythium species is pathogenic.

5. Conclusions

In summary, this study utilized amplicon sequencing and shotgun metagenomic sequencing to investigate the composition and diversity of bacterial and fungal communities in the rhizosphere of apple trees, revealing dominant taxa with similar relative abundances irrespective of the sequencing technology used. The findings showed significant variations in alpha and beta diversities within bacterial communities, indicating a shift in species composition and abundance influenced by both biotic and abiotic interactions in the dynamic ecological niche of rhizospheric soil, depending on tree health conditions. Additionally, microorganisms and their enzymes, which were previously identified in other studies as important agents of biological control due to their metabolic functions in the rhizosphere, particularly those involved in plant–microorganism interactions, collectively form the baseline for practical strategies in the development of sustainable agricultural practices.

Supplementary Materials: The following supporting information can be downloaded at: https://www.mdpi.com/article/10.3390/agronomy14020357/s1, Table S1: Summary of Illumina Data. Figure S1: Venn diagrams depicting the shared and exclusive (a) bacterial and (b) fungal genera under both asymptomatic and symptomatic conditions. Figure S2: Barplots of oomycetes community composition at the genus level.

Author Contributions: Conceptualization, R.G.-E., G.D.A.-Q. and L.N.M.-C.; methodology, G.D.A.-Q., A.B.-L., E.Z.-C. and L.N.M.-C.; validation, R.G.-E. and L.N.M.-C.; formal analysis, Z.Y.M.-R., O.R.-S., V.M.H.-A. and R.G.-E.; investigation, Z.Y.M.-R. and R.G.-E.; resources, G.D.A.-Q. and L.N.M.-C.; data curation, Z.Y.M.-R., O.R.-S., V.M.H.-A. and R.G.-E.; writing—original draft preparation, R.G.-E.; writing—review and editing, Z.Y.M.-R., G.D.A.-Q., O.R.-S., V.M.H.-A., A.B.-L., E.Z.-C., R.G.-E. and L.N.M.-C.; visualization, Z.Y.M.-R.; supervision, Z.Y.M.-R. and R.G.-E.; project administration, G.D.A.-Q. and L.N.M.-C. All authors have read and agreed to the published version of the manuscript.

Funding: This research received no external funding.

Data Availability Statement: The datasets generated and/or analyzed during the current study are available in the NCBI Bioproject database under the accession numbers PRJNA1003089 (amplicon sequences) and PRJNA1003562 (metagenomic sequences).

Acknowledgments: The authors would like to thank Omar Quintana for his assistance in the collection of samples and for logistical support. This paper was published with the support of the Instituto de Innovación y Competitividad of the Secretaría de Innovación y Desarrollo Económico del Estado de Chihuahua.

Conflicts of Interest: The authors declare no conflict of interest.

References

1. Neori, A.; Agami, M. The functioning of rhizosphere biota in wetlands—A review. *Wetlands* **2017**, *37*, 615–633. [CrossRef]
2. Kumar, S.; Diksha; Sindhu, S.S.; Kumar, R. Biofertilizers: An ecofriendly technology for nutrient recycling and environmental sustainability. *Curr. Res. Microb. Sci.* **2021**, *3*, 100094. [CrossRef]
3. Abdelfattah, A.; Freilich, S.; Bartuv, R.; Zhimo, V.Y.; Kumar, A.; Biasi, A.; Salim, S.; Feygenberg, O.; Burchard, E.; Dardick, C.; et al. Global analysis of the apple fruit microbiome: Are all apples the same? *Environ. Microbiol.* **2021**, *23*, 6038–6055. [CrossRef]
4. De Souza, R.S.; Okura, V.K.; Armanhi, J.S.; Jorrín, B.; Lozano, N.; da Silva, M.J.; González-Guerrero, M.; de Araújo, L.M.; Verza, N.C.; Bagheri, H.C.; et al. Unlocking the bacterial and fungal communities assemblages of sugarcane microbiome. *Sci. Rep.* **2016**, *30*, 28774. [CrossRef] [PubMed]
5. Yang, P.; Luo, Y.; Gao, Y.; Gao, X.; Gao, J.; Wang, P.; Feng, B. Soil properties, bacterial and fungal community compositions and the key factors after 5-year continuous monocropping of three minor crops. *PLoS ONE* **2020**, *15*, e0237164. [CrossRef] [PubMed]
6. Sun, A.; Jiao, X.Y.; Chen, Q.; Wu, A.L.; Zheng, Y.; Lin, Y.X.; He, J.Z.; Hu, H.W. Microbial communities in crop phyllosphere and root endosphere are more resistant than soil microbiota to fertilization. *Soil Biol. Biochem.* **2021**, *153*, 108113. [CrossRef]
7. Sun, R.; Zhang, W.; Liu, Y.; Yun, W.; Luo, B.; Chai, R.; Zhang, C.; Xian, X.; Su, X. Changes in phosphorus mobilization and community assembly of bacterial and fungal communities in rice rhizosphere under phosphate deficiency. *Front. Microbiol.* **2022**, *13*, 953340. [CrossRef]
8. Bokszczanin, K.Ł.; Przybyłko, S.; Molska-Kawulok, K.; Wrona, D. Tree root-associated microbial communities depend on various floor management systems in an intensive apple (*Malus* × *domestica* Borkh.) orchard. *Int. J. Mol. Sci.* **2023**, *24*, 9898. [CrossRef]
9. Jog, R.; Nareshkumar, G.; Rajkumar, S. Enhancing soil health and plant growth promotion by Actinomycetes. In *Plant Growth Promoting Actinobacteria*; Subramaniam, G., Arumugam, S., Rajendran, V., Eds.; Springer: Singapore, 2016. [CrossRef]

10. Prasad, S.; Malav, L.C.; Choudhary, J.; Kannojiya, S.; Kundu, M.; Kumar, S.; Yadav, A.N. Soil microbiomes for healthy nutrient recycling. In *Current Trends in Microbial Biotechnology for Sustainable Agriculture*; Springer: Singapore, 2021; pp. 1–21. [CrossRef]
11. Yadav, A.N.; Kour, D.; Kaur, T.; Devi, R.; Yadav, A.; Dikilitas, M.; Abdel-Azeem, A.M.; Ahluwalia, A.S.; Saxena, A.K. Biodiversity, and biotechnological contribution of beneficial soil microbiomes for nutrient cycling, plant growth improvement and nutrient uptake. *Biocatal. Agric. Biotechnol.* **2021**, *33*, 102009. [CrossRef]
12. Peralta, A.L.; Sun, Y.; McDaniel, M.D.; Lennon, J.T. Crop rotational diversity increases disease suppressive capacity of soil microbiomes. *Ecosphere* **2018**, *9*, e02235. [CrossRef]
13. Köhl, J.; Kolnaar, R.; Ravensberg, W.J. Mode of action of microbial biological control agents against plant diseases: Relevance beyond efficacy. *Front. Plant Sci.* **2019**, *10*, 845. [CrossRef]
14. Wang, L.; Li, J.; Yang, F.; Yao, E.; Raza, W.; Huang, Q.; Shen, Q. Application of bioorganic fertilizer significantly increased apple yields and shaped bacterial community structure in orchard soil. *Microb. Ecol.* **2017**, *73*, 404–416. [CrossRef]
15. Vida, C.; de Vicente, A.; Cazorla, F.M. The role of organic amendments to soil for crop protection: Induction of suppression of soilborne pathogens. *Ann. Appl. Biol.* **2020**, *176*, 1–15. [CrossRef]
16. Knief, C. Analysis of plant microbe interactions in the era of next generation sequencing technologies. *Front. Plant Sci.* **2014**, *5*, 216. [CrossRef] [PubMed]
17. Kalam, S.; Anirban, B.; Podile, A.R. Difficult-to-culture bacteria in the rhizosphere: The underexplored signature microbial groups. *Pedosphere* **2022**, *32*, 75–89. [CrossRef]
18. Ruiz-Cisneros, M.F.; Rios-Velasco, C.; Berlanga-Reyes, D.I.; Ornelas-Paz, J.D.J.; Acosta-Muñiz, C.H.; Romo-Chacón, A.; Zamudio-Flores, P.B.; Pérez-Corral, D.A.; Salas-Marina, M.Á.; Ibarra-Rendón, J.E.; et al. Incidence and causal agents of root diseases and its antagonists in apple orchards of Chihuahua, México. *Rev. Mex. Fitopatol.* **2017**, *35*, 437–462.
19. Tewoldemedhin, Y.T.; Mazzola, M.; Botha, W.J.; Spies, C.F.; McLeod, A. Characterization of fungi (*Fusarium* and *Rhizoctonia*) and oomycetes (*Phytophthora* and *Pythium*) associated with apple orchards in South Africa. *Eur. J. Plant Pathol.* **2011**, *130*, 215–229. [CrossRef]
20. Kelderer, M.; Manici, L.M.; Caputo, F.; Thalheimer, M. Planting in the 'inter-row' to overcome replant disease in apple orchards: A study on the effectiveness of the practice based on microbial indicators. *Plant Soil* **2012**, *357*, 381–393. [CrossRef]
21. Duan, Y.N.; Jiang, W.T.; Zhang, R.; Chen, R.; Chen, X.S.; Yin, C.M.; Mao, Z.Q. Discovery of *Fusarium proliferatum* f. sp. malus domestica causing apple replant disease in China. *Plant Dis.* **2022**, *106*, 2958–2966. [CrossRef] [PubMed]
22. Raaijmakers, J.M.; Paulitz, T.C.; Steinberg, C.; Alabouvette, C.; Moënne-Loccoz, Y. The rhizosphere: A playground and battlefield for soilborne pathogens and beneficial microorganisms. *Plant Soil* **2009**, *321*, 341–361. [CrossRef]
23. Manici, L.M.; Kelderer, M.; Franke-Whittle, I.H.; Rühmer, T.; Baab, G.; Nicoletti, F.; Caputo, F.; Topp, A.; Insam, H.; Naef, A. Relationship between root-endophytic microbial communities and replant disease in specialized apple growing areas in Europe. *Appl. Soil Ecol.* **2013**, *72*, 207–214. [CrossRef]
24. Yu, Y.; Lee, C.; Kim, J.; Hwang, S. Group-specific primer and probe sets to detect methanogenic communities using quantitative real-time polymerase chain reaction. *Biotechnol. Bioeng.* **2005**, *89*, 670–679. [CrossRef] [PubMed]
25. Bellemain, E.; Carlsen, T.; Brochmann, C.; Coissac, E.; Taberlet, P.; Kauserud, H. ITS as an environmental DNA barcode for fungi: An in silico approach reveals potential PCR biases. *BMC Microbiol.* **2010**, *10*, 189. [CrossRef] [PubMed]
26. Magoč, T.; Salzberg, S.L. FLASH: Fast length adjustment of short reads to improve genome assemblies. *Bioinformatics* **2011**, *27*, 2957–2963. [CrossRef]
27. Bolyen, E.; Rideout, J.R.; Dillon, M.R.; Bokulich, N.A.; Abnet, C.C.; Al-Ghalith, G.A.; Alexander, H.; Alm, E.J.; Arumugam, M.; Asnicar, F.; et al. Reproducible, interactive, scalable and extensible microbiome data science using QIIME 2. *Nat. Biotechnol.* **2019**, *37*, 852–857. [CrossRef] [PubMed]
28. Callahan, B.J.; McMurdie, P.J.; Rosen, M.J.; Han, A.W.; Johnson, A.J.; Holmes, S.P. DADA2: High-resolution sample inference from Illumina amplicon data. *Nat. Methods* **2016**, *13*, 581–583. [CrossRef] [PubMed]
29. McDonald, D.; Price, M.N.; Goodrich, J.; Nawrocki, E.P.; DeSantis, T.Z.; Probst, A.; Andersen, G.L.; Knight, R.; Hugenholtz, P. An improved Greengenes taxonomy with explicit ranks for ecological and evolutionary analyses of bacteria and archaea. *ISME J.* **2012**, *6*, 610–618. [CrossRef]
30. Nilsson, R.H.; Larsson, K.H.; Taylor, A.F.S.; Bengtsson-Palme, J.; Jeppesen, T.S.; Schigel, D.; Kennedy, P.; Picard, K.; Glöckner, F.O.; Tedersoo, L.; et al. The UNITE database for molecular identification of fungi: Handling dark taxa and parallel taxonomic classifications. *Nucleic Acids Res.* **2019**, *47*, D259–D264. [CrossRef] [PubMed]
31. Katoh, K.; Standley, D.M. MAFFT multiple sequence alignment software version 7: Improvements in performance and usability. *Mol. Biol. Evol.* **2013**, *30*, 772–780. [CrossRef] [PubMed]
32. Price, M.N.; Dehal, P.S.; Arkin, A.P. FastTree 2—Approximately maximum-likelihood trees for large alignments. *PLoS ONE* **2010**, *5*, e9490. [CrossRef]
33. RStudio Team. *RStudio: Integrated Development Environment for R*; RStudio—PBC: Boston, MA, USA, 2021; Available online: http://www.rstudio.com/ (accessed on 10 November 2023).
34. Oksanen, J.; Blanchet, F.G.; Friendly, M.; Kindt, R.; Legendre, P.; McGlinn, D.; Minchin, P.R.; O'Hara, R.B.; Simpson, G.L.; Solymos, P.; et al. Vegan: Community Ecology Package. R Package Version 2.5-7. 2021. Available online: https://CRAN.R-project.org/package=vegan (accessed on 10 November 2023).

35. Wickham, H.; Averick, M.; Bryan, J.; Chang, W.; McGowan, L.D.A.; François, R.; Grolemund, G.; Hayes, A.; Henry, L.; Hester, J.; et al. Welcome to the Tidyverse. *J. Open Source Softw.* **2019**, *4*, 1686. [CrossRef]
36. Bisanz, J.E. *qiime2R: Importing QIIME2 Artifacts and Associated Data into R Sessions (v0.99)*; GitHub: San Francisco, CA, USA, 2018.
37. Douglas, G.M.; Maffei, V.J.; Zaneveld, J.R.; Yurgel, S.N.; Brown, J.R.; Taylor, C.M.; Huttenhower, C.; Langille, M.G.I. PICRUSt2 for prediction of metagenome functions. *Nat. Biotechnol.* **2020**, *38*, 685–688. [CrossRef] [PubMed]
38. Chen, S.; Zhou, Y.; Chen, Y.; Gu, J. fastp: An ultra-fast all-in-one FASTQ preprocessor. *J. Bioinform.* **2018**, *34*, i884–i890. [CrossRef] [PubMed]
39. Wood, D.E.; Lu, J.; Langmead, B. Improved metagenomic analysis with Kraken 2. *Genome Biol.* **2019**, *20*, 257. [CrossRef] [PubMed]
40. McMurdie, P.J.; Holmes, S. Phyloseq: An R package for reproducible interactive analysis and graphics of microbiome census data. *PLoS ONE* **2013**, *8*, e61217. [CrossRef]
41. Meyer, F.; Paarmann, D.; D'Souza, M.; Olson, R.; Glass, E.M.; Kubal, M.; Paczian, T.; Rodriguez, A.; Stevens, R.; Wilke, A.; et al. The metagenomics RAST server—A public resource for the automatic phylogenetic and functional analysis of metagenomes. *BMC Bioinform.* **2008**, *9*, 386. [CrossRef] [PubMed]
42. Deakin, G.; Tilston, E.L.; Bennett, J.; Passey, T.; Harrison, N.; Fernández-Fernández, F.; Xu, X. Spatial structuring of soil microbial communities in commercial apple orchards. *Appl. Soil Ecol.* **2018**, *130*, 1–12. [CrossRef] [PubMed]
43. Deakin, G.; Fernández-Fernández, F.; Bennett, J.; Passey, T.; Harrison, N.; Tilston, E.L.; Xu, X. The effect of rotating apple rootstock genotypes on apple replant disease and rhizosphere microbiome. *Phytobiomes J.* **2019**, *3*, 273–285. [CrossRef]
44. Radl, V.; Winkler, J.B.; Kublik, S.; Yang, L.; Winkelmann, T.; Vestergaard, G.; Schröder, P.; Schloter, M. Reduced microbial potential for the degradation of phenolic compounds in the rhizosphere of apple plantlets grown in soils affected by replant disease. *Environ. Microbiome* **2019**, *14*, 8. [CrossRef]
45. Singh, J.; Silva, K.J.P.; Fuchs, M.; Khan, A. Potential role of weather, soil and plant microbial communities in rapid decline of apple trees. *PLoS ONE* **2019**, *14*, e0213293. [CrossRef]
46. Bartuv, R.; Berihu, M.; Medina, S.; Salim, S.; Feygenberg, O.; Faigenboim-Doron, A.; Zhimo, V.Y.; Abdelfattah, A.; Piombo, E.; Wisniewski, M.; et al. Functional analysis of the apple fruit microbiome based on shotgun metagenomic sequencing of conventional and organic orchard samples. *Environ. Microbiol.* **2023**, *25*, 1728–1746. [CrossRef] [PubMed]
47. Caputo, F.; Nicoletti, F.; Picione, F.D.L.; Manici, L.M. Rhizospheric changes of fungal and bacterial communities in relation to soil health of multi-generation apple orchards. *Biol. Control* **2015**, *88*, 8–17. [CrossRef]
48. Deng, S.; Wipf, H.M.; Pierroz, G.; Raab, T.K.; Khanna, R.; Coleman-Derr, D. A plant growth-promoting microbial soil amendment dynamically alters the strawberry root bacterial microbiome. *Sci. Rep.* **2019**, *9*, 17677. [CrossRef]
49. Bulgari, D.; Bozkurt, A.I.; Casati, P.; Cağlayan, K.; Quaglino, F.; Bianco, P.A. Endophytic bacterial community living in roots of healthy and 'Candidatus *Phytoplasma mali*'-infected apple (*Malus domestica*, Borkh.) trees. *Antonie Leeuwenhoek* **2012**, *102*, 677–687. [CrossRef] [PubMed]
50. Sun, J.; Zhang, Q.; Zhou, J.; Wei, Q. Illumina amplicon sequencing of 16S rRNA tag reveals bacterial community development in the rhizosphere of apple nurseries at a replant disease site and a new planting site. *PLoS ONE* **2014**, *9*, e111744. [CrossRef] [PubMed]
51. Mahnkopp-Dirks, F.; Radl, V.; Kublik, S.; Gschwendtner, S.; Schloter, M.; Winkelmann, T. Molecular barcoding reveals the genus *Streptomyces* as associated root endophytes of apple (*Malus domestica*) plants grown in soils affected by apple replant disease. *Phytobiomes J.* **2021**, *5*, 177–189. [CrossRef]
52. Xiong, W.; Li, R.; Ren, Y.; Liu, C.; Zhao, Q.; Wu, H.; Jousset, A.; Shen, Q. Distinct roles for soil fungal and bacterial communities associated with the suppression of vanilla *Fusarium* wilt disease. *Soil Biol. Biochem.* **2017**, *107*, 198–207. [CrossRef]
53. Ding, T.; Yan, Z.; Zhang, W.; Duan, T. Green manure crops affected soil chemical properties and fungal diversity and community of apple orchard in the Loess Plateau of China. *J. Soil Sci. Plant Nutr.* **2021**, *21*, 1089–1102. [CrossRef]
54. Lycus, P.; Lovise Bøthun, K.; Bergaust, L.; Peele Shapleigh, J.; Reier Bakken, L.; Frostegård, Å. Phenotypic and genotypic richness of denitrifiers revealed by a novel isolation strategy. *ISME J.* **2017**, *11*, 2219–2232. [CrossRef]
55. Qi, G.; Chen, S.; Ke, L.; Ma, G.; Zhao, X. Cover crops restore declining soil properties and suppress bacterial wilt by regulating rhizosphere bacterial communities and improving soil nutrient contents. *Microbiol. Res.* **2020**, *238*, 126505. [CrossRef]
56. Lahlali, R.; Ezrari, S.; Radouane, N.; Kenfaoui, J.; Esmaeel, Q.; El Hamss, H.; Belabess, Z.; Barka, E.A. Biological control of plant pathogens: A global perspective. *Microorganisms* **2022**, *10*, 596. [CrossRef] [PubMed]
57. Khatri, S.; Sazinas, P.; Strube, M.L.; Ding, L.; Dubey, S.; Shivay, Y.S.; Sharma, S.; Jelsbak, L. *Pseudomonas* is a key player in conferring disease suppressiveness in organic farming. *Plant Soil* **2023**, 1–20. [CrossRef]
58. Sehrawat, A.; Phour, M.; Kumar, R.; Sindhu, S.S. Bioremediation of pesticides: An eco-friendly approach for environment sustainability. In *Microbial Rejuvenation of Polluted Environment*; Springer: Singapore, 2021; Volume 25, pp. 23–84. [CrossRef]
59. Tao, J.; Chen, Q.; Chen, S.; Lu, P.; Chen, Y.; Jin, J.; Li, J.; Xu, Y.; He, W.; Long, T.; et al. Metagenomic insight into the microbial degradation of organic compounds in fermented plant leaves. *Environ. Res.* **2022**, *214*, 113902. [CrossRef]
60. Grgas, D.; Rukavina, M.; Bešlo, D.; Štefanac, T.; Crnek, V.; Šikić, T.; Habuda-Stanić, M.; Landeka Dragičević, T. The bacterial degradation of lignin—A review. *Water J.* **2023**, *15*, 1272. [CrossRef]
61. Etesami, H.; Glick, B.R. Halotolerant plant growth–promoting bacteria: Prospects for alleviating salinity stress in plants. *Environ. Exp. Bot.* **2020**, *178*, 104124. [CrossRef]

62. Omar, A.F.; Abdelmageed, A.H.A.; Al-Turki, A.; Abdelhameid, N.M.; Sayyed, R.Z.; Rehan, M. Exploring the plant growth-promotion of four *Streptomyces* strains from rhizosphere soil to enhance cucumber growth and yield. *Plants* **2022**, *11*, 3316. [CrossRef] [PubMed]
63. Rehan, M.; Al-Turki, A.; Abdelmageed, A.H.; Abdelhameid, N.M.; Omar, A.F. Performance of plant-growth-promoting rhizobacteria (PGPR) isolated from sandy soil on growth of tomato (*Solanum lycopersicum* L.). *Plants* **2023**, *12*, 1588. [CrossRef]
64. Mu, H.M.; Wan, Y.Y.; Wu, B.C.; Tian, Y.; Dong, H.L.; Xian, C.G.; Li, Y. A rapid change in microbial communities of the shale gas drilling fluid from 3548 m depth to the above-ground storage tank. *Sci. Total Environ.* **2021**, *784*, 147009. [CrossRef] [PubMed]
65. Higdon, S.M.; Pozzo, T.; Tibbett, E.J.; Chiu, C.; Jeannotte, R.; Weimer, B.C.; Bennett, A.B. Diazotrophic bacteria from maize exhibit multifaceted plant growth promotion traits in multiple hosts. *PLoS ONE* **2020**, *15*, e0239081. [CrossRef] [PubMed]
66. Davis, R.L.; Hayter, J.T.; Marlino, M.L.; Isakeit, T.; Chappell, T.M. Pathogenic and saprophytic growth rates of *Fusarium oxysporum* f. sp. *vasinfectum* interact to affect variation in inoculum density and interannual infection risk. *J. Phytopathol.* **2023**, *113*, 1447–1456. [CrossRef]
67. Navasca, A.M.; Singh, J.; Rivera-Varas, V.; Geddes, B.; Secor, G.; Gill, U.; Baldwin, T. First report and draft genome resource of a unique opportunistic *Fusarium solani* pathogen associated with unknown dark galls in sugarbeet. *PhytoFrontiers* **2023**, *3*, 704–707. [CrossRef]
68. El Hazzat, N.; Adnani, M.; Msairi, S.; El Alaoui, M.A.; Mouden, N.; Chliyeh, M.; Boughribil, S.; Selmaoui, K.; Ouazzani Touhami, A.; Douira, A. *Fusarium equiseti* as one of the main *Fusarium* species causing wilt and root rot of chickpeas in Morocco. *Acta Mycol.* **2022**, *57*, 1–10. [CrossRef]
69. Ezrari, S.; Radouane, N.; Tahiri, A.; El Housni, Z.; Mokrini, F.; Özer, G.; Lazraq, A.; Belabess, Z.; Amiri, S.; Lahlali, R. Dry root rot disease, an emerging threat to citrus industry worldwide under climate change: A review. *Physiol. Mol. Plant Pathol.* **2022**, *117*, 101753. [CrossRef]
70. Liu, Y.; Tian, Y.; Zhao, X.; Yue, L.; Uwaremwe, C.; Zhou, Q.; Wang, Y.; Zhang, Y.; Dun, Z.; Cui, Z.; et al. Identification of pathogenic *Fusarium* spp. responsible for root rot of *Angelica sinensis* and characterization of their biological enemies in Dingxi, China. *Plant Dis.* **2022**, *106*, 1898–1910. [CrossRef] [PubMed]
71. Zakaria, L. *Fusarium* species associated with diseases of major tropical fruit crops. *Horticulturae* **2023**, *9*, 322. [CrossRef]
72. Mazzola, M.; Manici, L.M. Apple replant disease: Role of microbial ecology in cause and control. *Annu. Rev. Phytopathol.* **2012**, *50*, 45–65. [CrossRef] [PubMed]
73. Kanashiro, A.M.; Akiyama, D.Y.; Kupper, K.C.; Fill, T.P. *Penicillium italicum*: An underexplored postharvest pathogen. *Front. Microbiol.* **2020**, *11*, 606852. [CrossRef]
74. Patel, D.; Andhare, P.; Menon, S.; Vadakan, S.; Goswami, D. Eccentricity in the behavior of *Penicillium* spp. as phytopathogen and phytoaugmentor. In *Beneficial Microbes for Sustainable Agriculture and Environmental Management*; CRC Press: Boca Raton, FL, USA, 2020; pp. 115–138.
75. Morgunov, I.G.; Kamzolova, S.V.; Dedyukhina, E.G.; Chistyakova, T.I.; Lunina, J.N.; Mironov, A.A.; Stepanova, N.N.; Shemshura, O.N.; Vainshtein, M.B. Application of organic acids for plant protection against phytopathogens. *Appl. Microbiol. Biotechnol.* **2017**, *101*, 921–932. [CrossRef]
76. Girvan, M.S.; Bullimore, J.; Pretty, J.N.; Osborn, A.M.; Ball, A.S. Soil type is the primary determinant of the composition of the total and active bacterial communities in arable soils. *Appl. Environ. Microbiol.* **2003**, *69*, 1800–1809. [CrossRef]
77. Yang, M.; Yang, D.; Yu, X. Soil microbial communities and enzyme activities in sea-buckthorn (*Hippophae rhamnoides*) plantation at different ages. *PLoS ONE* **2018**, *13*, e0190959. [CrossRef]
78. Zheng, X.; Liu, B.; Zhu, Y.; Wang, J.; Zhang, H.; Wang, Z. Bacterial community diversity associated with the severity of bacterial wilt disease in tomato fields in southeast China. *Can. J. Microbiol.* **2019**, *65*, 538–549. [CrossRef] [PubMed]
79. González-Escobedo, R.; Muñoz-Castellanos, L.N.; Muñoz-Ramirez, Z.Y.; Guigón-López, C.; Avila-Quezada, G.D. Rhizosphere bacterial and fungal communities of healthy and wilted pepper (*Capsicum annuum* L.) in an organic farming system. *Ciênc. Rural* **2023**, *53*, 0220072. [CrossRef]
80. Wang, R.; Lv, F.; Lv, R.; Lin, H.; Zhang, Z.; Wei, L. Sampling period and disease severity of bacterial wilt significantly affected the bacterial community structure and functional prediction in the sesame rhizosphere soil. *Rhizosphere* **2023**, *26*, 100704. [CrossRef]
81. Solís-García, I.A.; Ceballos-Luna, O.; Cortazar-Murillo, E.M.; Desgarennes, D.; Garay-Serrano, E.; Patiño-Conde, V.; Guevara-Avendaño, E.; Méndez-Bravo, A.; Reverchon, F. *Phytophthora* root rot modifies the composition of the avocado rhizosphere microbiome and increases the abundance of opportunistic fungal pathogens. *Front. Microbiol.* **2021**, *11*, 574110. [CrossRef] [PubMed]
82. Vestergaard, G.; Schulz, S.; Schöler, A.; Schloter, M. Making big data smart—How to use metagenomics to understand soil quality. *Biol. Fertil. Soils* **2017**, *53*, 479–484. [CrossRef]
83. Martin, C.C.S. Potential of compost tea for suppressing plant diseases. *CABI Rev.* **2014**, *9*, 1–38. [CrossRef]
84. Ladygina, N.; Hedlund, K. Plant species influence microbial diversity and carbon allocation in the rhizosphere. *Soil Biol. Biochem.* **2010**, *42*, 162–168. [CrossRef]
85. Shigyo, N.; Umeki, K.; Hirao, T. Seasonal dynamics of soil fungal and bacterial communities in cool-temperate montane forests. *Front. Microbiol.* **2019**, *10*, 1944. [CrossRef]
86. Bei, Q.; Moser, G.; Müller, C.; Liesack, W. Seasonality affects function and complexity but not diversity of the rhizosphere microbiome in European temperate grassland. *Sci. Total. Environ.* **2021**, *784*, 147036. [CrossRef]

87. Toole, D.R.; Zhao, J.; Martens-Habbena, W.; Strauss, S.L. Bacterial functional prediction tools detect but underestimate metabolic diversity compared to shotgun metagenomics in southwest Florida soils. *Appl. Soil Ecol.* **2021**, *168*, 104129. [CrossRef]
88. Ayiti, O.E.; Ayangbenro, A.S.; Babalola, O.O. Relationship between nitrifying microorganisms and other microorganisms residing in the maize rhizosphere. *Arch. Microbiol.* **2022**, *204*, 246. [CrossRef] [PubMed]
89. Sivaram, A.K.; Panneerselvan, L.; Mukunthan, K.; Megharaj, M. Effect of pyroligneous acid on the microbial community composition and plant growth-promoting bacteria (PGPB) in soils. *Soil Syst.* **2022**, *6*, 10. [CrossRef]
90. Baz, L.; Abulfaraj, A.A.; Tashkandi, M.A.; Baeissa, H.M.; Refai, M.Y.; Barqawi, A.A.; Shami, A.; Abuauf, H.W.; Ashy, R.A.; Jalal, R.S. Predicted functional shifts due to type of soil microbiome and watering of two wild plants in western region of Saudi Arabia. *Phyton* **2022**, *91*, 2249–2268. [CrossRef]
91. Kumar, A.; Dubey, A. Rhizosphere microbiome: Engineering bacterial competitiveness for enhancing crop production. *J. Adv. Res.* **2020**, *24*, 337–352. [CrossRef]
92. Zhao, X.; Dong, Q.; Han, Y.; Zhang, K.; Shi, X.; Yang, X.; Yuan, Y.; Zhou, D.; Wang, K.; Wang, X.; et al. Maize/peanut intercropping improves nutrient uptake of side-row maize and system microbial community diversity. *BMC Microbiol.* **2022**, *22*, 14. [CrossRef] [PubMed]
93. Mohan, S.M.; Sudhakar, P. Metagenomic Approaches for Studying Plant–Microbe Interactions. In *Understanding the Microbiome Interactions in Agriculture and the Environment*; Veera Bramhachari, P., Ed.; Springer: Singapore, 2022. [CrossRef]
94. Wijayawardene, N.N.; Hyde, K.D.; Dai, D.Q. Outline of Ascomycota. *Encycl. Mycol.* **2021**, *88*, 246–254. [CrossRef]
95. Gorshkov, V.; Tsers, I. Plant susceptible responses: The underestimated side of plant–pathogen interactions. *Biol. Rev.* **2022**, *97*, 45–66. [CrossRef]
96. Hansen, E.; Delatour, C. Phytophthora species in oak forests of north-east France. *Ann. For. Sci.* **1999**, *56*, 539–547. [CrossRef]
97. Khaliq, I.; Hardy, G.E.S.J.; McDougall, K.L.; Burgess, T.I. *Phytophthora* species isolated from alpine and sub-alpine regions of Australia, including the description of two new species; *Phytophthora cacuminis* sp. nov and *Phytophthora oreophila* sp. nov. *Fungal Biol.* **2019**, *123*, 29–41. [CrossRef] [PubMed]
98. Mannai, S.; Benfradj, N.; Boughalleb-M'Hamdi, N. Regional and seasonal variation of *Fusarium* and Oomycetes species associated with apple seedlings decline in Tunisian nurseries. *Nov. Res. Microbiol. J.* **2023**, *7*, 2015–2033. [CrossRef]
99. Wannicke, N.; Brust, H. Inactivation of the plant pathogen *Pythium ultimum* by plasma-processed air (PPA). *Appl. Sci.* **2023**, *13*, 4511. [CrossRef]

agronomy

Article

Paddy-*Lilium* Crop Rotation Improves Potential Beneficial Soil Fungi and Alleviates Soil Acidification in *Lilium* Cropping Soil

Li Wen [1], Fengqiu Huang [1], Zhongxiu Rao [1], Kaikai Cheng [1], Yong Guo [2] and Haiming Tang [1,*]

[1] Hunan Soil and Fertilizer Institute, Changsha 410125, China; wenli@hunaas.cn (L.W.); huangfengqiu1243@hunaas.cn (F.H.); raozhongxiu@hunaas.cn (Z.R.); ckk0306@126.com (K.C.)
[2] College of Agronomy, Hunan Agricultural University, Changsha 410128, China; guoyong2000@outlook.com
* Correspondence: tanghaiming@hunaas.cn

Abstract: *Lilium* growth is severely impeded by continuous cropping, and crop rotation is essential to reducing the detrimental effects of monocultures. Soil (0–20 cm) was collected in three *Lilium* cropping patterns in Longshan County, Hunan Province, including continuous *Lilium* cropping (Lilium), corn upland rotation with *Lilium* (Corn), and paddy rotation with *Lilium* (Rice). Using Illumina high-throughput sequencing technology, the fungal ribosomal DNA internal-transcribed spacer 1 (ITS1) was examined to evaluate the features of soil fungi communities among three cropping patterns. Crop rotation has an impact on soil properties and the microbial community. Rice soil has a significantly higher pH than *Lilium* and corn soil, while corn and rice soil have a greater total nitrogen and total phosphorus content than *Lilium* soil. Rotation cropping clearly shifted the fungi community diversity based on the results of principal coordinate analysis (PCoA) and nonmetric multidimensional scaling (NMDS). *Ascomycota* was the most prevalent phylum, with the highest levels in *Lilium* soil. Genetic analysis revealed that paddy rotation led to a clear reduction in or non-detection of eight potentially pathogenic fungal genera and a noticeable accumulation of eight beneficial fungal genera compared to *Lilium* continuous cropping. Fungi communities and their abundant taxa were correlated with soil pH and nutrients. Altogether, we propose that rice rotation, with its ability to mitigate soil acidification, reducing pathogenic and accumulating beneficial communities, may be an effective strategy for alleviating the continuous cropping barrier.

Keywords: rotation cropping; fungi diversity; beneficial community; soil pH; high-throughput sequencing

Citation: Wen, L.; Huang, F.; Rao, Z.; Cheng, K.; Guo, Y.; Tang, H. Paddy-*Lilium* Crop Rotation Improves Potential Beneficial Soil Fungi and Alleviates Soil Acidification in *Lilium* Cropping Soil. *Agronomy* **2024**, *14*, 161. https://doi.org/10.3390/agronomy14010161

Academic Editor: Fuyong Wu

Received: 13 November 2023
Revised: 22 December 2023
Accepted: 22 December 2023
Published: 11 January 2024

1. Introduction

Continuous or monoculture cropping, which is driven by production demands, involves cultivating the same crop over a long period of time [1,2]. This practice often causes soil nutrient imbalances, stunted plant growth, and increased crop disease and pest [3,4]. In addition, with the continuous cropping, crop roots secrete certain identical exudates, such as phenolic acids, which may alter the soil microbial community and enhance the prevalence of the pathogenic microbiome [5]. As a result, continuous or monoculture cultivation ultimately poses a serious threat to yield reductions and the healthy advancement of agriculture [5,6]. Previous studies have shown that continual cropping with a single crop can diminish crop yields by 20 to 50 percent [7,8]. Therefore, it is imperative to look into feasible approaches for alleviating the continuous cropping barrier and achieving green and sustainable development.

Soil microbes play a vital role in agroecosystems, involving nutrient cycling, organic matter decomposition, and the promotion or suppression of crop diseases [1,3]. The modification of soil microbe abundance and species can be used to predict soil health and fertility in agroecosystems [9]. It has been extensively demonstrated that continuous cropping can readily disrupt the balance of soil microbial communities, resulting in a decline in microbial diversity and homogenization of microbial community structure [1,4,5,10]. In soils, loss and simplification of soil community composition impair soil ecosystem function, further affecting plant growth [11]. Generally, soil microorganisms are frequently classified into pathogenic and beneficial groups based on their function in the agroecosystem [5,12]. Continuous cropping tends to cause a detectable increase in populations of pathogenic fungi and a decrease in beneficial fungi. Li et al. [5] and Gao et al. [10], for instance, demonstrated that continuous cropping of peanuts or potatoes significantly inhabited the populations of *Fusarium*, *Phoma*, *Verticillium*, *Colletotrichum*, and *Bionectria* and decreased some plant-beneficial fungal groups (like *Trichoderma* and *Mortierella*). The accumulation of a load of pathogenic fungi in the soil at the expense of plant-beneficial fungi appears to be a likely explanation for the decline in yield and quality as a consequence of continuous cropping [5]. Thus, knowledge of the variability in important functional microbial communities might shed light on the way certain cultivation strategies perform to eliminate barriers to continuous cropping.

Rotational cropping has been gradually implemented in agriculture to boost crop yields and eliminate the continuous cropping barrier [13,14]. A meta-analysis found that, in China, rotation cultivation increased crop yields by an average of 15–25% more than continuous monoculture [8,15]. In particular, the corn yield from the corn–soybean–wheat rotation was 11.8 Mg ha^{-1}, which was 15% greater than the yield obtained from continuous corn rotation [15]. However, the quantitative synthesis of magnitude and variability in yields has not been reported. This is mostly caused by the broad variation in climate factor, crop species, field management, and soil nutrient delivery associated with various rotation patterns [1,8,15–17]. In addition, research shows that crop rotation has a variety of consequences on the soil microbial population, including positive, negative, and neutral effects [6,18–22]. Therefore, it is necessary to evaluate the impact of various rotation cropping systems on the variation in soil microbial community structure.

Lilium tigrinum (*Lilium lancifolium* Ker Gawl.) is a valuable medicinal herb of the *Liliaceae* family and is distributed in Eastern and Central China. In addition, *L. tigrinum* is an ornamental horticultural crop and food with high economic value [4]. In China, the herb has been cultivated commercially for over 1400 years, and its bulbs are widely used in a variety of Chinese medicinal products due to their lung-moistening and cough-relieving properties [23]. As a result, there has been a growing push for farmers to plant *Lilium* continuously in growing areas. Nonetheless, continuous cropping with *Lilium* leads to severe consecutive replant problems, including extreme soil degradation, crop yield and quality decline, and frequent soil-borne diseases, e.g., wilt disease and black spot disease, which has become an important factor limiting the sustainable development of the *L. tigrinum* industry [4]. A data survey indicates that constant replanting around 6–9 years results in a decline in the dry weight of *Lilium* bulb of 13.25–15.54%, losing farmers an estimated USD 50 million [4]. Thus, it is imperative to investigate the most efficient ways to mitigate the barriers to continuous cropping. To avoid these issues, it is common practice to plant *Lilium* in previously uncultivated fields. Despite this, there is a restricted quantity of land available for producing *Lilium*, and it is anticipated that such detrimental planting strategies would pose a severe danger to the *Lilium* industry in the near future. Recently, certain rotational patterns, such as paddy–*Lilium* rotation, soybean–*Lilium* rotation, and potato–*Lilium* rotation, have been conducted to replace continuous cropping regions with a single crop [24–26]. However, there are distinct differences between the land management practices, crop species employed, and associated root quantity or quality, and other soil parameters among these types that may impact soil quality and fertility [6]. Additionally, soil microorganisms may be affected by these factors. Thus, the objectives of this study were to (1) assess how different patterns of *Lilium* cropping affect the abundance and composition of soil community; (2) compare the pattern of fungal diversity and structure among three different *Lilium* cropping patterns; and (3) examine how potentially pathogenic and beneficial fungal communities change when *Lilium* monoculture cropping is converted into corn–*Lilium* cropping and paddy–*Lilium* cropping. Understanding the influence of

Lilium planting on the soil fungus community would give theoretical guidelines for rational farming mode adoption and sustainable farmland usage.

2. Material and Methods

2.1. Field Description and Sampling

The study sites were located in Longshan County (109°13′–109°48′ E, 28°46′–29°38′ N) of Hunan Province. This region has a typical subtropical monsoon climate with mean annual temperature of 15.8 °C and annual precipitation of 1400 mm. The soil type is categorized as Gleysol in Chinese soil classification, which is derived from sand and shale [27]. At the beginning of this field experiment, soil chemical characteristics at 0–20 cm layer were as follows: pH 5.26, soil organic carbon (SOC) 10.38 g kg^{-1}, total nitrogen (TN) 1.36 g kg^{-1}, total phosphorous (TP) 0.88 g kg^{-1}, total potassium (TK) 23.90 g kg^{-1}, available P 126.7 mg kg^{-1}, available K 139 mg kg^{-1}, NH_4^+-N 0.90 mg kg^{-1}, and NO_3^--N 8.01 mg kg^{-1}. The experiment had three treatments, including continuous cropping of *Lilium* (Lilium), corn upland rotation with *Lilium* (corn), and paddy rotation with *Lilium* (rice). Each treatment set triple plots (20 m × 15 m). The agricultural land management was practiced with typical management of the region. Detailed information, including the crop species, tillage frequency, planting time, harvest frequency, fertilization rate, and replanting frequency, is presented in Table S1.

Lanceleaf *Lilium* bulb was grown in these research region for more than 40 years. The *Lilium* was generally managed as follows: the soil was tilled to a depth of 25 cm, and the plant was fertilized three times throughout the growth period. The experimental bulbs were uniformly sized, second-generation, and planted 7–10 cm deep. The soil was kept moist but not soggy during the planting procedure. The bulbs weighed roughly 100 g and were spaced 15–20 cm apart, with a planting density of roughly 100,000 plants/hm^2.

2.2. Soil Sampling and Determination of Soil Physiochemical Properties

Soil sampling (0–20 cm) was conducted in 2022, at the crop-harvesting stage for *Lilium*, corn, and rice, respectively. A random collection of 5 points were chosen in each plot and then mixed as one sample. After removing stones and roots, a portion of fresh soil sample was stored at −80 °C for DNA extraction; the remaining portion of the soil sample was air-dried and passed through a 0.15 mm mesh to measure soil physicochemical properties. Soil pH was measured using a pH meter (FE20K, Mettler Toledo, Greifensee, Switzerland) at a soil/water ratio of 1:2.5 [28]. The SOC and TN contents were analyzed by $K_2Cr_2O_7$-H_2SO_4 digestion and semi-micro Kjeldahl digestion, respectively [28]. Soil TP was measured using the molybdenum blue colorimetric method, and soil TK was determined using inductively coupled plasma-atomic emission spectrometry [28]. In addition, methods for detecting soil-based properties have also been performed previously [28].

2.3. DNA Extraction, PCR Amplification, and Amplicon High-Throughput Sequencing

ITS rRNA amplification sequencing was performed by Genesky Biotechnologies Inc. (Shanghai, China). Briefly, DNA was isolated from 0.25 g soil using a FastDNA® SPIN Kit for soil (MP Biomedicals, Santa Ana, CA, USA) according to the manufacturer's instructions. The integrity of genomic DNA was detected through agarose gel electrophoresis, and the concentration and purity of genomic DNA were detected through the NanoDrop 2000 and Qubit 3.0 Spectrophotometer (Thermo Fisher Scientific, Waltham, MA, USA). DNA from the soil samples was used as a template for PCR amplification. As positive control, standard genomic DNA of fungi was amplified in triplicate. The PCR products were purified with Agencourt AMPure XPPCR Purification Beads (Beckman Coulter, Brea, CA, USA) to assess their specificity. The primers used for the amplification of the ITS1 region were ITS1F(5′-CTTGGTCATTTAGAGGAAGTAA−3′)/ITS2R(5′-GCTGCGTTCTTCATCGATGC−3′) and then sequenced using Illumina NovaSeq 6000 sequencing (Illumina, San Diego, CA, USA) [29–32]. The PCR mixture consisted of 1 uL of 10×Toptaq buffer, 0.8uL of 2.5 mM dNTPs, 0.3 uL of F/R primer (10 uM), 0.2 uL of

Toptaq DNA polymerase (Transgen, Beijing, China), 3 uL template DNA, and up to 10 uL ddH_2O. The thermal cycling parameters for PCR amplification were as follows: initial denaturation at 94 °C for 3 min, 94 °C for 30 s, 55 °C for 30 s, 72 °C for 1 min; 25~27 cycles of denaturing at 72 °C for 10 min, and hold at 4 °C. The purified products were then indexed in the 16S V4–V5 library. The library quality was assessed on the Qubit®3.0 Fluorometer (Thermo Scientific, Wilmington, DE, USA) and Agilent 2100 Bioanalyzer (Agilent Technologies, Santa Clara, CA, USA) systems. Finally, the purified amplifications from all the samples were pooled in equimolar concentrations, and then the pooled library was sequenced on a NovaSeq 6000 platform (Illumina, San Diego, CA, USA), SP-Xp (PE250) using a 2 × 250-paired-end sequencing kit (250 bp).

Paired-end reads were assembled using FLASH to obtain raw tags. The raw read sequences were processed in the Quantitative Insights Into Microbial Ecology (QIIME) toolkit [29]. The adaptor and primer sequences were trimmed using the cutadapt plugin. High-throughput sequencing data were clustered into operational taxonomic units (OTUs) at a 97% similarity level using UPARSE pipeline, and the chimeric sequences were identified and removed. DADA2 plugin was used for quality control and to identify amplicon sequence variants (ASVs) [33]. Taxonomic assignments of ASV representative sequences were performed with confidence threshold 0.6 by a pre-trained Naive Bayes classifier, which was trained on the UNITE (version 8.2).

2.4. Data Analysis

Before statistical analyses, all the dependent variables were tested for normality and natural log-transformed when necessary. One-way ANOVA with LSD test was conducted using SPSS ver.16.0 for Windows (SPSS Inc., Chicago, IL, USA) to determine significant differences among three cropping types ($p < 0.05$). The relative abundance was determined by the number of sequences affiliated with the same phylogenetic groups divided by the total number of the target phyla or genera per sample. The Shannon, ACE, Chao 1, and Simpson index were calculated to compare the fungi community alpha diversity for each sample. The Venn analysis was performed to compare the fungal composition. Principal coordinate analysis (PCoA) and nonmetric multidimensional scaling (NMDS) were performed to detect the beta diversity between individual samples. Additionally, the similarity of community composition in the abovementioned communities was evaluated sequentially using the "Adonis" and "AOSIMN" functions (999 permutations) of the R package "phyloseq" (version 1.26.1) and "Vegan" (version 2.5.6). Linear discriminant analysis (LDA) effect size (LEfSe) was used to elucidate the biomarkers in each treatment. Those with an LDA $\geq$ 2.0 were considered to be important biomarkers in each treatment. Redundancy analysis (RDA) with a forward selection option and Monte Carlo permutation test (999 permutations) was conducted to select the significant explanatory variables (soil properties) contributing significantly ($p < 0.05$) to soil fungi microbial community variation. The RDA was performed using the R package "vegan" (version 2.4.5).

3. Results

3.1. Changes in Soil Properties Underlie Three Land Types

Except soil TK, soil pH, SOC, TN, TP, C:N, C:P, and N:P significantly differed among three cropping patterns (Table 1). The soil pH was significantly higher in rice soil than in *Lilium* and corn soils (Table 1). The content of TN and TP was higher in corn and rice soil than that in *Lilium* soil, and the highest content of SOC was observed in corn soil (Table 1). Inversely, the ratio of soil C:N and C:P was significantly higher in *Lilium* soil than that in the other two rotation patterns (Table 1, $p < 0.05$).

Table 1. Soil properties under three cropping patterns.

	Lilium	Corn	Rice
pH	5.12 (0.06) ^{b}	5.08 (0.02) ^{b}	5.51 (0.01) ^{a}
SOC, g/kg	11.57 (0.22) ^{b}	14.00 (0.06) ^{a}	10.71 (0.21) ^{c}
TN, g/kg	1.24 (0.02) ^{c}	1.73 (0.06) ^{a}	1.45 (0.02) ^{b}
TP, g/kg	0.72 (0.02) ^{c}	1.71 (0.01) ^{a}	0.91 (0.00) ^{b}
TK, g/kg	21.37 (0.03)	21.77 (0.15)	21.80 (0.17)
C:N	9.36 (0.20) ^{a}	8.13 (0.32) ^{b}	7.39 (0.20) ^{b}
C:P	16.19 (0.78) ^{a}	8.17 (0.02) ^{c}	11.73(0.25) ^{b}
N:P	1.59 (0.01) ^{a}	1.01 (0.04) ^{b}	1.73 (0.08) ^{a}

Values are presented as means (standard error). Lilium: continuous cropping with *Lilium*, Corn: corn upland rotation with *Lilium*, Rice: paddy rotation with *Lilium*. Different low letters present significant difference among three cropping patterns at $p < 0.05$.

3.2. Fungal Abundance and Diversity

Fungal composition and diversity were measured based on OTUs. The OTU distribution Venn analysis demonstrated that there were a total of 2412 OTUs, and 610, 453 and 608 unique OTUs in *Lilium*, corn and rice soil, respectively (Figure S1). These unique OTUs accounted for about 63.36–73.36% for all OTUs in three cropping patterns (Figure S1). In addition, there were 91 shared fungal OTUs, accounting for about 10.66–12.73% among three cropping patterns (Figure S1).

The alpha diversity indices were evaluated based on the OTUs. The ACE, Chao1, Shannon, and Simpson diversity reveal an depletion of OTU similarity in *Lilium* and rice soil compared to *Lilium* and corn soil (Table 2). The beta diversity of fungal community over all samples is illustrated using a PCoA and NMDS plot (Figure 1). At the OTU level, the distribution pattern revealed that the fungal community structure was distinctly different among three cropping patterns (Figure 1). The first two principal components explained approximately 76.58% of the variance in the fungal composition based on the result PCoA (Figure 1A). The aggregation degrees of the sample points from the rice soil were distant from those of the sample points from corn soil, and both soils were extremely far away from the sample points in *Lilium* soil (Adonis: $R^2 = 0.762$, $p < 0.01$; ANOSIM: statistic R = 1, $p < 0.01$; Figure 1B). Based on cluster tree analysis, all samples are grouped into two main groups: samples of rice soil are separated from samples of the other two soils, and samples of *Lilium* and corn soils are grouped into one group (Figure S2).

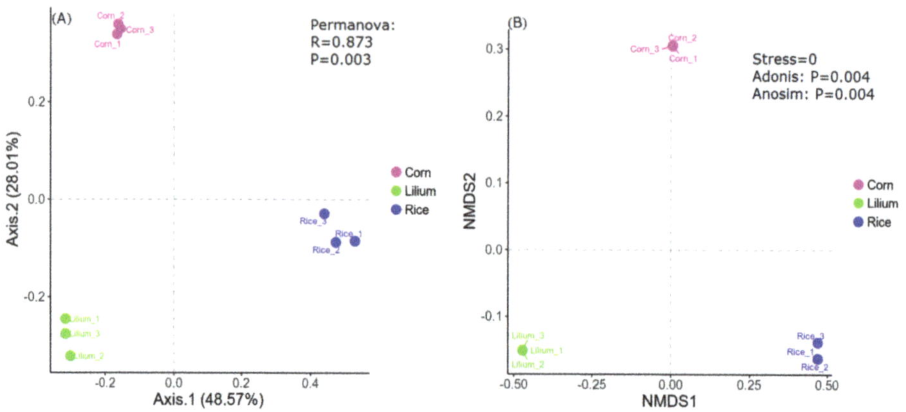

Figure 1. Beta diversity of soil fungal communities under three cropping patterns. PCoA (**A**) and NMDS (**B**) based on the Bray–Curtis distance dissimilarity. PCoA: principal coordinate analysis, NMDS: nonmetric multidimensional scaling. Lilium: continuous cropping with *Lilium*, Corn: corn upland rotation with *Lilium*, Rice: paddy rotation with *Lilium*.

Table 2. Soil fungi diversity under three rotation patterns.

	Alpha Diversity			
	ACE	**Chao 1**	**Shannon**	**Simpson**
Lilium	445.74 (26.70)	444.88 (26.38)	4.25 (0.09)	0.044 (0.00)
Corn	395.38 (31.76)	394.52 (31.44)	3.99 (0.05)	0.055 (0.01)
Rice	440.27 (26.01)	439.09 (26.28)	4.11 (0.18)	0.046 (0.01)

Value are presented as means (standard error). ACE: abundance-based coverage estimator metric. Lilium: continuous cropping with *Lilium*, Corn: corn upland rotation with *Lilium*, Rice: paddy rotation with *Lilium*.

3.3. Fungal Community Structure and Composition

The OTUs identified in all samples were divided into eight phyla, 18 classes, 40 orders, 64 families, and 72 genes. Among three cropping types, the overall fungal community was dominated by the phyla (relative abundance > 1% all samples) *Ascomycota* (average relative abundance > 65%), followed by *Basidiomycota* (14.82%), *Mortierellomycota* (7.26%), and *unassigned* fungi (6.72%) (Figure S3A). The relative abundance of these dominant fungal phyla responded differently to cropping patterns. For detail, the relative abundance of *Ascomycota* was highest in *Lilium* soil, followed by rice soil, in contrast to the corn soil (Figures 2A and S3B). Nevertheless, the relative abundance of *Basidiomycota* and *Mortierellomycota* was significantly higher in corn soil and lowest in rice or *Lilium* soil ($p < 0.05$, Figures 2A and S3B). Further taxonomic classification at the gene level revealed that 258 genera were detected, 79 of which differed between the three cropping patterns ($p < 0.05$). *Fusarium, Mortierella, Gibberella, Trichobolus,* and *Humicola* were the predominant fungal genera in *Lilium* and corn soils, while *Mortierella, Fusarium, Talaromyces, Lecanicillium,* and *Colletotrichum* were the dominant fungal genera in rice soil (Figure 2B). The relative abundances of *Fusarium, Gibberella, Humicola,* and *Trichobolus* were significantly reduced or not detected, while the relative abundances of *Mortierella, Pencillium, Talaromyces, Lecanicillium, Trichoderma,* and *Colletotrichum* increased in rice soil compared to the *Lilium* soil (Figure 2B).

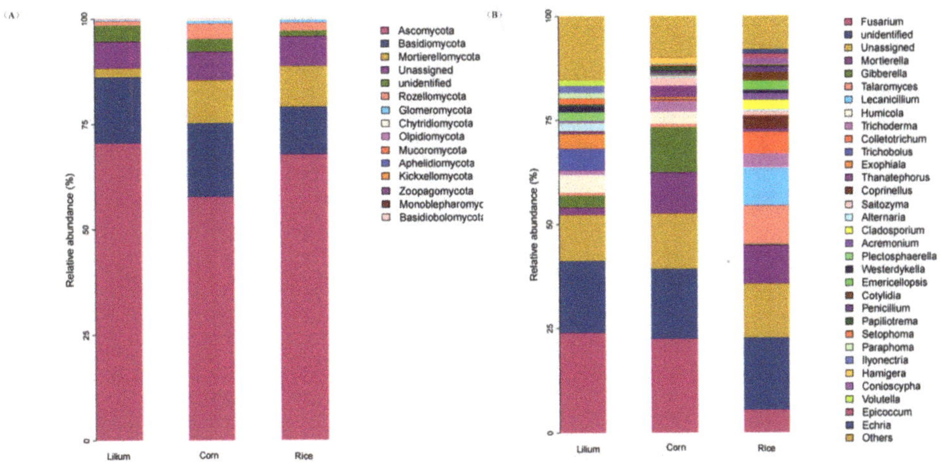

Figure 2. Relative abundance of dominant fungal phyla (**A**,**B**) among three cropping patterns. Lilium: continuous cropping with *Lilium*, Corn: corn upland rotation with *Lilium*, Rice: paddy rotation with *Lilium*.

The linear discriminant analysis effects size (LEfSe) was analyzed for discerning the abundant taxa among three cropping types (Figure 3). There are a total of one, seven, and five biomarkers in *Lilium*, corn, and rice soil, with all LDA values over 3.0. Furthermore, all biomarkers were completely different amongst soils, demonstrating that rotation cultivation induced alterations in the fungal community structure of *Lilium* soil (Figure 3).

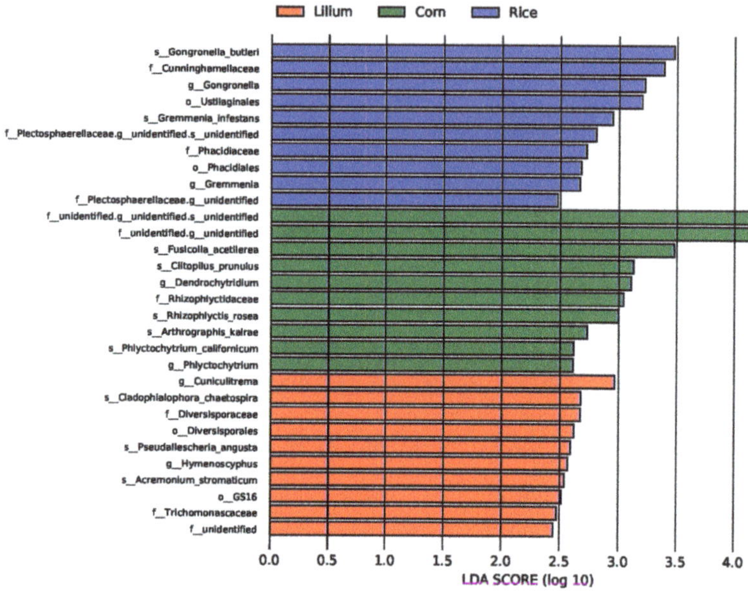

Figure 3. LDA (linear discriminant analysis) discriminant results. Taxa enriched in *Lilium* are shown in red with a positive LDA score, corn in green with positive LDA score and rice in blue with a positive LDA score ($p < 0.05$; LDA score ≥ 2.0).

3.4. Changes in Potential Pathogenic and Beneficial Fungi

Based on a subsequent fungal guild analysis, eight potentially pathogenic fungi and eight beneficial fungi were found to be considerably distinct in corn and rice soils in comparison to *Lilium* soils (Table 3). The relative abundance of potentially pathogenic fungus with names, like *Fusarium, Gibberella, Volutella, Alternaria, Humicola, Aspergillus, Trichobolus,* and *Exophiala,* was substantially reduced or not observed in rice soil compared to *Lilium* soil (Table 3). Meanwhile, certain plant-beneficial fungi, such as *Mortierella, Lecanicilium, Trichodema, Acremonium, Clonostachys, Metarhizium, Metacordyceps,* and *Penicillium,* expanded more in rice soil than that in *Lilium* soil (Table 3).

Table 3. The relative abundance of potentially pathogenic fungi and potentially beneficial fungi.

	Lilium	Corn	Rice
	Potentially pathogenic fungi		
Fusarium	23.97(2.09) ^{a}	22.46(2.37) ^{a}	5.47(2.01) ^{b}
Gibberella	2.83(0.49) ^{b}	10.88(1.43) ^{a}	0.25(0.11) ^{c}
Volutella	1.30(0.24)	0.02(0.01)	0
Humicola	4.28(1.18) ^{a}	2.90 (0.15) ^{a}	0.07 (0.02) ^{b}
Alternaria	1.91 (0.31) ^{a}	0.35 (0.07) ^{b}	0.44 (0.04) ^{b}
Aspergillus	0.48 (0.28)	0.03 (0.01)	0.01(0.01)
Trichobolus	5.28 (1.34) ^{a}	0.15(0.07) ^{b}	0
Exophiala	3.55(0.64) ^{a}	0.50(0.10) ^{b}	0.06(0.01) ^{b}

Table 3. *Cont.*

	Lilium	Corn	Rice
	Potentially beneficial fungi		
Mortierella	1.80(0.38) [b]	9.89(1.10) [a]	9.18(0.66) [a]
Lecanicillium	0	0	8.97(3.82)
Trichoderma	1.01(0.20)	2.19(0.89)	3.46(1.34)
Acremonium	0.43(0.14) [b]	0.34(0.07) [b]	1.58(0.20) [a]
Clonostachys	0.77(0.16) [a]	0.08(0.03) [ab]	0.35(0.19) [b]
Metarhizium	0.08(0.00) [b]	0.71(0.25) [a]	0.12(0.07) [b]
Penicillium	0.04(0.01) [b]	0.50(0.38) [b]	1.39(0.06) [a]
Metacordyceps	0	0	0.07

Values are presented as means (standard error). Different low letters present significant difference among three patterns at $p < 0.05$. Lilium: continuous cropping with *Lilium*, Corn: corn upland rotation with *Lilium*, Rice: paddy rotation with *Lilium*.

3.5. Relation between Fungal Community and Soil Properties

A mantel test analysis revealed a strong correlation between fungal community structures and soil pH, SOC, TP, TN, C:P, and N:P, and soil pH was the most substantial explanatory variable (r = 0.7863, $p < 0.01$, Table S2). Likewise, RDA was applied to analyze the relationships between the fungal community compositions and soil properties (Figure 4). RDA1 and RDA2 contributed to 60.93% of the total variances at the phylum level (Figure 4). The RDA plot of the fungal community composition rendered it abundantly evident that the samples from *Lilium*, corn, and rice varied considerably from one another (Figure 4). The first axis was positively correlated with soil pH, SOC, TN, TK, and negatively correlated with soil pH, C:P, and N:P. The second axis was positively correlated with soil TK, C:N, C:P, and N:P and negatively correlated with soil pH, SOC, TN, TK, and TP. Soil pH was the most crucial variable in shaping the relative abundance of *Basidiomycota* (r = −0.95, $p < 0.001$, Table S3) at the phylum level.

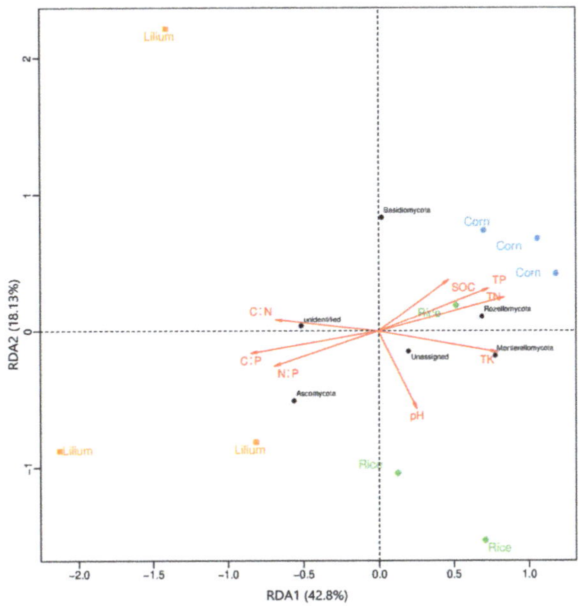

Figure 4. Redundancy analysis (RDA) diagram illustrating the relationship between the soil fungal community composition at the phyla-level from different sampling sites and environmental variables.

The explanatory variables are showed by different red arrows, soil fungal community composition by black star circle. pH: salinity, SOC: soil organic carbon, TN: soil total nitrogen, TP: soil total phosphorus, TK: soil total potassium, C:N: the ratio of SOC to TN, C:P: the ratio of SOC to TP, N:P: the ratio of TN to TP. The orange squares represent *Lilium* cropping (Lilium), the blue squares represent corn upland rotation with *Lilium* (Corn), the green squares present paddy rotation with *Lilium* (Rice).

We also used Spearman's rank correlation to evaluate the associations between the abundance of fungus phyla and soil physicochemistry properties (Figure 5). *Ascomycoto* had a negative correlation with SOC, TN, and TP but a substantial positive correlation with soil C:P and N:P. There was a significant and positive correlation observed within *Mortierellomycota* and *Mucoromycota* and SOC, TP, and/or TN. While *Olpidiomycota*, *Aphelidiomycota*, and *Kickxellomycota* were significantly and positively associated with pH and negatively correlated with TK, *Glomeromycota* and *Rozellomycota* were considerably and positively correlated with TK (Figure 5).

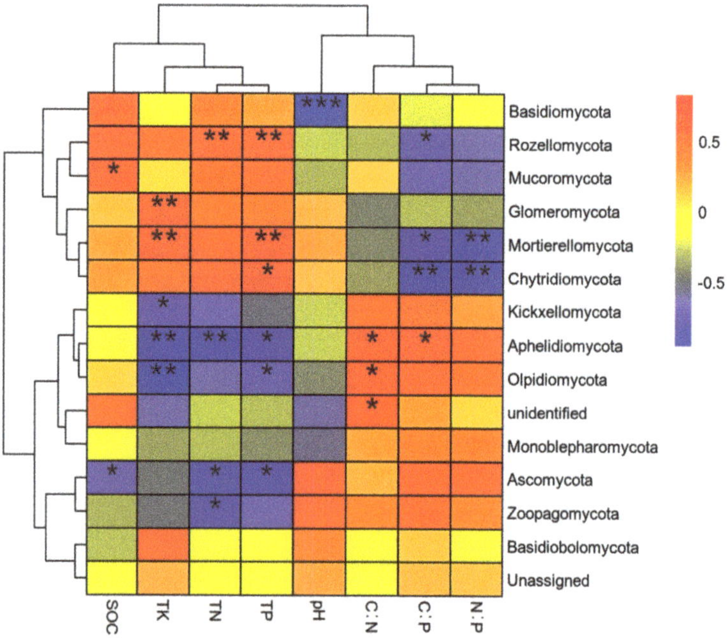

Figure 5. Spearman's rank correlations between soil nutrients and the relative abundances of the different fungal phyla. pH: salinity, SOC: soil organic carbon, TN: soil total nitrogen, TP: soil total phosphorus, TK: soil total potassium, C:N: the ratio of SOC to TN, C:P: the ratio of SOC to TP, N:P: the ratio of TN to TP. "*", "**" and "***"present significant difference at $p < 0.05$, $p < 0.01$ and $p < 0.001$, respectively.

4. Discussion

4.1. The Influence of Rotation Cropping Patterns on Slowing Soil Acidification and Improving Soil Nutrients

Long-term continuous cultivation with a single crop can deteriorate the soil environment, causing nutrient imbalances and acidity [34–36]. Rotation cropping, which effectively improves soil physical and chemical properties, regulates soil fertility and increases crop yields by altering the soil microenvironment, and it has been recognized as a suitable practice to protect crop production [6,17]. Among these practices, lowland paddy and upland crop rotations are accompanied by an anaerobic and aerobic microenvironment, which jointly modify soil physicochemical features and nutrient conversion, most likely facilitating the growth of microorganisms and plants [14,37]. In acidic soils, pH tends to move towards neutral after inundation [38]. Consequently, paddy-upland rotation has a noticeable effect on slowing soil acidification. The results of this study indicate that paddy-upland rotation may help to mitigate soil acidification since the average pH of rice soil increased significantly overall, as compared to Lilium soil by 0.39 units. The increased value (0.25 unit) of soil pH in rice soil, as compared to base soil (mean value = 5.26), further supports this result. This finding is in line with that of Zhang et al. [14], who demonstrated that the paddy-upland rotation can produce comparatively higher soil pH and diminish the tendency of soil acidification than continuous cultivation with a single upland crop. Different rotation types also regulated the soil nutrient content. Compared with *Lilium* monoculture cultivation, the content of SOC, TN, and TP increased by 21.00%, 16.94%, and 137.5%, respectively, in corn soil. Meanwhile, the TN and TP contents of rice soil were comparatively greater, suggesting that these two crop rotation strategies could improve the potential for sequestering soil organic matter. In contrast, the paddy-upland rotation model distinctly reduced SOC content, which is consistent with the results published by Wang et al. [16]. This is presumably due to water leaching reducing the excess SOC during the rice rotation phase. As taken together, rice–*Lilium* rotation cropping can minimize soil acidity, improve soil nutrients, and possibly modify the soil microbial community.

4.2. Impact of Different Cropping Patterns on Fungi Community Diversity

An indicator of soil microbial community diversity is critical for the productivity and stability of agricultural soil ecosystems [39]. The findings of the alpha diversity analysis showed that the fungal community's richness in rice soil was comparable to that in *Lilium* soil; however, it decreased substantially in corn soil, as compared to that in *Lilium* soil. Additional data indicated that fungi in *Lilium* and rice soils possessed more variety, as revealed by the Ace and Shannon index. Gao et al. [10] and Liu et al. [2] also proposed that the diversity of fungi was markedly increased by the continuous cropping of sweet potato and soybean. This observation, however, was contrary to the findings published by Shi et al. [4], who found that continuous cropping with *Lilium* reduced fungal richness. This might be due to inconsistencies in detection methodologies or differences in crop and soil conditions, as various crops have diverse root exudates that alter microbial communities [40]. For beta diversity, both PCoA and NMDS analysis indicated that crop rotation clearly had a significant impact on the matrix fungal community structure [22,41,42], which was confirmed by the hierarchical clustering (Figure S2).

It is widely acknowledged that the structure and composition of soil microbial community are determined by the soil physicochemical properties [1,41,43]. Similarly, this discovery showed that the structure and diversity of soil fungus are greatly affected by soil pH, TN, SOC, TP, and TK. The findings can be attributed to differences in soil properties brought about by various agricultural practices, which affect the availability of microbial nutritional elements. Consequently, these elements serve as cofactors in the synthesis of numerous organic compounds, including acids, proteins, and carbohydrates. Of these, soil pH appears to be one of the key factors regulating fungal community structure under different cropping patterns [35,41,43–45]. Shen et al. [46] and Shi et al. [47] showed that soil pH can alter the bioavailability of soil nutrients, substrates, and hazardous metals, which can alter the composition of the microbial community. Furthermore, the overly acidic soil may limit most soil fungal growth and propagation, and these fungal communities will thrive when the soil pH limitation is removed [45]. Thus, soil fungi growth improved in rice soil with relatively high pH.

4.3. Effects of Cropping Pattern on Fungal Community Taxonomic Classification and Structure

The alternation in soil fungal community structure and composition under continuous cropping has been well documented [1,2,44]. When OTUs are grouped at the phylum level, *Ascomycota*, *Basidiomycetes*, and *Mortiomycota* were the most dominant phyla that appear frequently in soil. The phylum *Mortierellomycota* was more conducive to enrichment in healthy soil, whereas *Ascomycota* was more likely to be formed in diseased soil, according to the findings presented by Yuan et al. [48]. *Ascomycota* has been found in association with a wide range of crop monoculture systems [3,10,44] and the phyla *Ascomycota* was detected in *Lilium* soil with higher relative abundance (average 70.45%) related to that in corn and rice soil. Moreover, we discovered that the phylum *Mortrellomycota* was considerably more prevalent in rice soil compared to *Liliun* soil, indicating that crop rotation is an improved strategy to defend crops against specific diseases like *Fusarium* wilt disease. Additionally, the *Ascomycota* class *Sordariomycetes* is the most dominant (top 1 class, average RA = 43.65%), which is consistent with numerous studies, showing *Sordariomycetes* to be the most prevalent fungal class in a variety of agricultural systems [2]. Members of this class, including *Fusarium, Gibberlla, Volutella,* and *Humicola* gene, have a broad distribution as plant endophytes and pathogens in almost all ecosystems [2]. At the genus level, *Fusarium, Mortierella, Gibberella, Humicola, Trichobolus,* and *Exophiala* were the dominating genes in *Lilium* soil at the genus level. One of them, *Mortierella,* has been proven to be a beneficial fungus with biocontrol characteristics [2,12,49], and its abundance increased when *Lilium* continuous cropping was converted into rice–*Lilium* rotation cropping. In addition, the existence of putative pathogenic fungi, such as *Fusarium, Gibberella, Humicola, Trichobolus,* and *Exophiala,* can cause soil-borne disease, affecting crop growth and production [16,50–52].

4.4. Changes in Potentially Pathogenic and Beneficial Fungi in Response to Different Crop Rotations

The emergence of crop diseases caused by the formation or expansion of pathogenic communities, such as *Fusarium, Alternaria, Gibberella,* and *Colletotrichum,* seems to be the main reason for the decline in crop output and the stunted growth of monoculture crops [5,44,50,51,53]. These pathogenic fungal communities, including *Fusarium, Alternaria,* and *Gibberella,* were the most prevalent in the current study. Their relative abundance was, likewise, significantly higher in Lilium monocultures and lower in rice–*Lilium* rotations, demonstrating that certain pathogenic fungi thrive in *Lilium* monocultures. Specifically, *Fusarium* is generally recognized as a pathogenic fungus that causes devastating crop diseases, such as *Lilium* wilt disease, *Fusarium* root rot, and banana *Fusarium* wilt disease [4,10,50,54,55]. As the largest genus, the relative abundance of *Fusarium* was distinctly significantly higher in *Lilium* soil and lower in rice soil (Table 3). *Alternaria* and *Gibberella* can infect various crops and cause foolish seeding disease of rice, soybean *Altenaris* leaf spot, citrus *Alternaria* brown spot, and Sanqing ginseng root rot [51,53]. In our study, other certain fungal genera, such as *Humicola, Trichobolus, Exophiala, Volutella,* and *Aspergillus,* were found to be more abundant in *Lilium* soil and reduced or not detected in rice soil (Table 3). These five fungal genera have been identified as potentially pathogenic fungi based on other research [52,56,57]. Thus, our findings imply that a rise in the abundance of the aforementioned pathogenic fungal community is a potential cause of the continuous cropping barrier, which restricts the establishment of *Lilium* monocultures.

The reduction in pathogenic and the accumulation of potential beneficial communities might help to alleviate the continuous cropping barrier [4,5,13]. Our findings also revealed that rotation cropping cultivation increased or some potential beneficial fungi species appeared, e.g., *Mortierella, Penicillium, Trichoderma, Lecanicillium,* and *Acremonium.* Certain species of *Mortierella* have the ability to shield crops from *Fusarium* wilt disease and *Fusarium* root rot [2,49]. As the major dominant genus, the relative abundance of *Mortierella* was four-times more abundant in rice soil (9.18% on average) than in *Lilium* soil (1.80% on average). Previous studies have illustrated substantial correlation between higher *Mortierella* abundance and good soil health during continuous replanting [1]. *Penicillium* can act as a fungal antagonist and plant growth promoter due to its function in producing a vast variety of physiologically active chemicals [58]. *Trichoderma, Lecanicillium,* and *Acremonium* also have biocontrol or plant growth-promoting ability based on their significant and positive correlation with crop physiological characteristics [14]. Here, the relative abundance of *Trichoderma*, *Lecanicillium*, and *Acremonium* was distinctly higher in rice soil compared with in *Lilium* soil. Meanwhile, other certain genera, i.e., *Metarhizium* and *Metacordyceps*, have been identified as parasitic fungi of insects and susceptible to rotation cropping [59,60]. In conclusion, compared to *Lilium* monoculture cropping, rice–*Lilium* rotation can improve the quantify of beneficial fungi and lessen the abundance of pathogenic fungi. As a consequence, it may be extremely crucial in alleviating the barrier associated with continuous cropping in agricultural production.

5. Conclusions

In this study, different cropping patterns had divergent effects on soil physicochemical properties and microbial community structure. Rice–*Lilium* rotation cropping considerably reduced the development of soil acidification and increased soil nutrient content, as compared to *Lilium* continuous cropping. Soil fungal communities in rice and corn soil were noticeably distinct from those in *Lilium* soil. *Ascomycota* was found to be the dominant group, with the highest abundance in *Lilium* soil. Additionally, we observed a higher relative abundance of eight beneficial fungi and a decline in or non-detection of eight potentially pathogenic fungi in rice soil compared to *Lilium* soil. Overall, this study provides evidence that paddy-upland rotation cropping can improve soil fungi structure by reducing soil acidity and improving some beneficial fungi communities.

Supplementary Materials: The following supporting information can be downloaded at: https://www.mdpi.com/article/10.3390/agronomy14010161/s1, Table S1: Management information for the three cropping patterns. Table S2: Mantel test result for the correlation between soil fungal community composition and environmental variables. Table S3: The correlation between fungal phyla and soil properties variables. Figure S1: Venn of soil fungi under different cropping patterns. Lilium: continuous cropping with Lilium, Corn: corn upland rotation with Lilium, Rice: paddy rotation with Lilium. Figure S2: Hierarchical clustering of fungal communities underlying three different cropping patterns. Lilium: continuous cropping with *Lilium*, Corn: corn upland rotation with *Lilium*, Rice: paddy rotation with *Lilium*. Figure S3: The barplot showed the relative abundance of the main fungal phyla(A) and the heatmap showed their variations among soil samples at phylum level (B) present in three cropping soils. Lilium: continuous cropping with *Lilium*, Corn: corn upland rotation with *Lilium*, Rice: paddy rotation with *Lilium*.

Author Contributions: Conceptualization, F.H. and H.T.; methodology, L.W., K.C. and H.T.; software, L.W. and Y.G.; formal analysis, L.W., K.C. and H.T.; investigation, Z.R. and Y.G.; writing—original draft preparation, L.W.; writing—review and editing, H.T.; funding acquisition, F.H. All authors have read and agreed to the published version of the manuscript.

Funding: This study was funded by the special funds for the construction of innovative provinces in Hunan Province (2021NK2028), Natural Science Foundation of Hunan Province (2022JJ40232) and Director Fund of Hunan Soil and Fertilizer Institute (2022tfs201).

Institutional Review Board Statement: Not applicable.

Informed Consent Statement: Not applicable.

Data Availability Statement: The datasets used and/or analyzed during the current study are available from the corresponding author on reasonable request.

Conflicts of Interest: The authors declare that the research was conducted in the absence of any commercial or financial relationships that could be construed as a potential conflict of interest.

References

1. Liu, Z.X.; Liu, J.J.; Yu, Z.H.; Yao, Q.; Li, Y.S.; Liang, A.Z.; Zhang, W.; Mi, G.; Jin, J.; Liu, X.B.; et al. Long-term continuous cropping of soybean is comparable to crop rotation in mediating microbial abundance, diversity and community composition. *Soil. Tillage Res.* **2020**, *197*, 104503. [CrossRef]
2. Liu, H.; Pan, F.; Han, X.; Song, F.; Zhang, Z.; Yan, J.; Xu, Y. Response of Soil Fungal Community Structure to Long-Term Continuous Soybean Cropping. *Front. Microbiol.* **2018**, *9*, 3316. [CrossRef] [PubMed]

3. Liu, J.J.; Yao, Q.; Li, Y.S.; Zhang, W.; Mi, G.; Chen, X.L.; Yu, Z.H.; Wang, G.H. Continuous cropping of soybean alters the bulk and rhizospheric soil fungal communities in a Mollisol of Northeast PR China. *Land. Degrad. Dev.* **2019**, *30*, 1725–1738. [CrossRef]
4. Shi, G.Y.; Sun, H.Q.; Calderon-Urrea, A.; Jia, X.X.; Yang, H.Y.; Su, G.L. Soil Fungal Diversity Loss and Appearance of Specific Fungal Pathogenic Communities Associated With the Consecutive Replant Problem (CRP) in Lily. *Front. Microbiol.* **2020**, *11*, 1649. [CrossRef] [PubMed]
5. Li, X.G.; Ding, C.F.; Zhang, T.L.; Wang, X.X. Fungal pathogen accumulation at the expense of plant-beneficial fungi as a consequence of consecutive peanut monoculturing. *Soil. Biol. Biochem.* **2014**, *72*, 11–18. [CrossRef]
6. Yang, T.; Siddique, K.H.M.; Liu, K. Cropping systems in agriculture and their impact on soil health—A review. *Glob. Ecol. Conserv.* **2020**, *23*, e01118. [CrossRef]
7. Chang, C.L.; Fu, X.P.; Zhou, X.G.; Guo, M.Y.; Wu, F.Z. Effects of seven different companion plants on cucumber productivity, soil chemical characteristics and community. *J. Integr. Agric.* **2017**, *16*, 2206–2214. [CrossRef]
8. Zhao, J.; Yang, Y.; Zhang, K.; Jeong, J.; Zeng, Z.; Zang, H. Does crop rotation yield more in China? A meta-analysis. *Field Crops Res.* **2020**, *245*, 107659. [CrossRef]
9. Hartman, W.H.; Richardson, C.J.; Vilgalys, R.; Bruland, G.L. Environmental and anthropogenic controls over bacterial communities in wetland soils. *Proc. Natl. Acad. Sci. USA* **2008**, *105*, 17842–17847. [CrossRef]
10. Gao, Z.Y.; Han, M.K.; Hu, Y.Y.; Li, Z.Q.; Liu, C.F.; Wang, X.; Tian, Q.; Jiao, W.J.; Hu, J.M.; Liu, L.F.; et al. Effects of Continuous Cropping of Sweet Potato on the Fungal Community Structure in Rhizospheric Soil. *Front. Microbiol.* **2019**, *10*, 2269. [CrossRef]
11. Wagg, C.; Bender, S.F.; Widmer, F.; van der Heijden, M.G.A. Soil biodiversity and soil community composition determine ecosystem multifunctionality. *Proc. Natl. Acad. Sci. USA* **2014**, *111*, 5266–5270. [CrossRef] [PubMed]
12. Meents, A.K.; Furch, A.C.U.; Almeida-Trapp, M.; Ozyurek, S.; Scholz, S.S.; Kirbis, A.; Lenser, T.; Theissen, G.; Grabe, V.; Hansson, B.; et al. Beneficial and Pathogenic Arabidopsis Root-Interacting Fungi Differently Affect Auxin Levels and Responsive Genes During Early Infection. *Front. Microbiol.* **2019**, *10*, 380. [CrossRef] [PubMed]
13. Chen, Y.; Du, J.; Li, Y.; Tang, H.; Yin, Z.; Yang, L.; Ding, X. Evolutions and Managements of Soil Microbial Community Structure Drove by Continuous Cropping. *Front. Microbiol.* **2022**, *13*, 839494. [CrossRef] [PubMed]
14. Zhang, M.M.; Liang, G.Y.; Ren, S.; Li, L.P.; Li, C.; Li, Y.J.; Yu, X.L.; Yin, Y.P.; Liu, T.; Liu, X.J. Responses of soil microbial community structure, potential ecological functions, and soil physicochemical properties to different cultivation patterns in cucumber. *Geoderma* **2023**, *429*, 116237. [CrossRef]
15. Kazula, M.J.; Lauer, J.G. The influence of crop rotation on corn total biomass production. *J. Soil. Water Conserv.* **2018**, *73*, 541–548. [CrossRef]
16. Wang, P.; Yan, S.; Zhang, W.S.; Xie, X.D.; Li, M.J.; Ren, T.B.; Gu, L.; Zhang, Z.Y. Effect of soil management systems on the rhizosphere bacterial community structure of tobacco: Continuous cropping vs. paddy-upland rotation. *Front. Plant Sci.* **2022**, *13*, 996858. [CrossRef] [PubMed]
17. Jahan, M.A.H.S.; Hossain, A.; Sarkar, M.A.R.; Teixeira da Silva, J.A.; Ferdousi, M.N.S. Productivity impacts and nutrient balances of an intensive potato-mungbean-rice crop rotation in multiple environments of Bangladesh. *Agric. Ecosyst. Environ.* **2016**, *231*, 79–97. [CrossRef]
18. Venter, Z.S.; Jacobs, K.; Hawkins, H.-J. The impact of crop rotation on soil microbial diversity: A meta-analysis. *Pedobiologia* **2016**, *59*, 215–223. [CrossRef]
19. Navarro-Noya, Y.E.; Gómez-Acata, S.; Montoya-Ciriaco, N.; Rojas-Valdez, A.; Suárez-Arriaga, M.C.; Valenzuela-Encinas, C.; Jiménez-Bueno, N.; Verhulst, N.; Govaerts, B.; Dendooven, L. Relative impacts of tillage, residue management and crop-rotation on soil bacterial communities in a semi-arid agroecosystem. *Soil. Biol. Biochem.* **2013**, *65*, 86–95. [CrossRef]
20. Yin, C.; Jones, K.L.; Peterson, D.E.; Garrett, K.A.; Hulbert, S.H.; Paulitz, T.C. Members of soil bacterial communities sensitive to tillage and crop rotation. *Soil. Biol. Biochem.* **2010**, *42*, 2111–2118. [CrossRef]
21. Peralta, A.L.; Sun, Y.; McDaniel, M.D.; Lennon, J.T. Crop rotational diversity increases disease suppressive capacity of soil microbiomes. *Ecosphere* **2018**, *9*, e02235. [CrossRef]
22. Azeem, M.; Sun, D.; Crowley, D.; Hayat, R.; Hussain, Q.; Ali, A.; Tahir, M.I.; Jeyasundar, P.G.S.A.; Rinklebe, J.; Zhang, Z. Crop types have stronger effects on soil microbial communities and functionalities than biochar or fertilizer during two cycles of legume-cereal rotations of dry land. *Sci. Total Environ.* **2020**, *715*, 136958. [CrossRef] [PubMed]
23. Zhou, J.; An, R.; Huang, X. Genus Lilium: A review on traditional uses, phytochemistry and pharmacology. *J. Ethnopharmacol.* **2021**, *270*, 113852. [CrossRef] [PubMed]
24. Zhou, L.; Wang, Y.; Xie, Z.; Zhang, Y.; Malhi, S.S.; Guo, Z.; Qiu, Y.; Wang, L. Effects of lily/maize intercropping on rhizosphere microbial community and yield of *Lilium davidii* var. unicolor. *J. Basic. Microbiol.* **2018**, *58*, 892–901. [CrossRef] [PubMed]
25. Mouqiang, L.; Wenzhu, W.; Shengfeng, F.; Dawuda, M.M.; Lipeng, Z.; Hongyu, Y.; Guiying, S. Evaluation for the rotation crops suitability of food lily by soil extract solution irrigation treatments in Western China. *Pak. J. Bot.* **2021**, *53*, 1645–1653. [CrossRef]
26. Borchers, A.; Pieler, T. Programming pluripotent precursor cells derived from Xenopus embryos to generate specific tissues and organs. *Genes* **2010**, *1*, 413–426. [CrossRef]
27. Shi, X.Z.; Yu, D.S.; Warner, E.D.; Pan, X.Z.; Petersen, G.W.; Gong, Z.G.; Weindorf, D.C. Soil Database of 1:1,000,000 Digital Soil Survey and Reference System of the Chinese Genetic Soil Classification System. *Soil Surv. Horiz.* **2004**, *45*, 129–136. [CrossRef]
28. Wen, L.; Li, D.; Xiao, X.; Tang, H. Alterations in soil microbial phospholipid fatty acid profile with soil depth following cropland conversion in karst region, southwest China. *Environ. Sci. Pollut. Res. Int.* **2023**, *30*, 1502–1519. [CrossRef]

29. Bolyen, E.; Rideout, J.R.; Dillon, M.R.; Bokulich, N.A.; Abnet, C.C.; Al-Ghalith, G.A.; Alexander, H.; Alm, E.J.; Arumugam, M.; Asnicar, F.; et al. Reproducible, interactive, scalable and extensible microbiome data science using QIIME 2. *Nat. Biotechnol.* **2019**, *37*, 852–857. [CrossRef]
30. Usyk, M.; Zolnik, C.P.; Patel, H.; Levi, M.H.; Burk, R.D. Novel ITS1 Fungal Primers for Characterization of the Mycobiome. *mSphere* **2017**, *2*, e00488-17. [CrossRef]
31. Yan, K.; Zhou, J.; Feng, C.; Wang, S.; Haegeman, B.; Zhang, W.; Chen, J.; Zhao, S.; Zhou, J.; Xu, J.; et al. Abundant fungi dominate the complexity of microbial networks in soil of contaminated site: High-precision community analysis by full-length sequencing. *Sci. Total Environ.* **2023**, *861*, 160563. [CrossRef] [PubMed]
32. Chang, C.; Zhang, J.; Liu, T.; Song, K.; Xie, J.; Luo, S.; Qu, T.; Zhang, J.; Tian, C.; Zhang, J. Rhizosphere fungal communities of wild and cultivated soybeans grown in three different soil suspensions. *Appl. Soil. Ecol.* **2020**, *153*, 103586. [CrossRef]
33. Callahan, B.J.; McMurdie, P.J.; Rosen, M.J.; Han, A.W.; Johnson, A.J.A.; Holmes, S.P. DADA2: High-resolution sample inference from Illumina amplicon data. *Nat. Methods* **2016**, *13*, 581–583. [CrossRef] [PubMed]
34. Behnke, G.D.; Zabaloy, M.C.; Riggins, C.W.; Rodriguez-Zas, S.; Huang, L.; Villamil, M.B. Acidification in corn monocultures favor fungi, ammonia oxidizing bacteria, and nirK-denitrifier groups. *Sci. Total Environ.* **2020**, *720*, 137514. [CrossRef] [PubMed]
35. Zhao, Y.; Qin, X.M.; Tian, X.P.; Yang, T.; Deng, R.; Huang, J. Effects of continuous cropping of *Pinellia ternata* (Thunb.) Breit. on soil physicochemical properties, enzyme activities, microbial communities and functional genes. *Chem. Biol. Technol. Agric.* **2021**, *8*, 43. [CrossRef]
36. Tan, W.J.; Wang, J.M.; Bai, W.Q.; Qi, J.J.; Chen, W.M. Soil bacterial diversity correlates with precipitation and soil pH in long-term maize cropping systems. *Sci. Rep.* **2020**, *10*, 6012. [CrossRef] [PubMed]
37. Wang, Q.; Xu, J.M.; Lin, H.; Zou, P.; Jiang, L. Effect of rice planting on the nutrient accumulation and transfer in soils under plastic greenhouse vegetable-rice rotation system in southeast China. *J. Soils Sediments* **2017**, *17*, 204–209. [CrossRef]
38. Lee, C.H.; Hong, C.O.; Kim, S.Y.; Schumacher, T.; Kim, P.J. Reduction of phosphorus release by liming from temporary flooded rice rotational system in greenhouse upland soil. *Ecol. Eng.* **2011**, *37*, 1239–1243. [CrossRef]
39. van der Heijden, M.G.; Bardgett, R.D.; van Straalen, N.M. The unseen majority: Soil microbes as drivers of plant diversity and productivity in terrestrial ecosystems. *Ecol. Lett.* **2008**, *11*, 296–310. [CrossRef]
40. Rengel, Z. Root exudation and microflora populations in rhizosphere of crop genotypes differing in tolerance to micronutrient deficiency. *Plant Soil.* **1997**, *196*, 255–260. [CrossRef]
41. Xia, Q.; Rufty, T.; Shi, W. Soil microbial diversity and composition: Links to soil texture and associated properties. *Soil. Biol. Biochem.* **2020**, *149*, 107953. [CrossRef]
42. Ai, C.; Zhang, S.; Zhang, X.; Guo, D.; Zhou, W.; Huang, S. Distinct responses of soil bacterial and fungal communities to changes in fertilization regime and crop rotation. *Geoderma* **2018**, *319*, 156–166. [CrossRef]
43. Fernandez-Calvino, D.; Baath, E. Interaction between pH and Cu toxicity on fungal and bacterial performance in soil. *Soil. Biol. Biochem.* **2016**, *96*, 20–29. [CrossRef]
44. Li, Y.; Shi, C.Q.; Wei, D.; Gu, X.J.; Wang, Y.F.; Sun, L.; Cai, S.S.; Hu, Y.; Jin, L.; Wang, W. Soybean continuous cropping affects yield by changing soil chemical properties and microbial community richness. *Front. Microbiol.* **2022**, *13*, 1083736. [CrossRef] [PubMed]
45. Rousk, J.; Bååth, E.; Brookes, P.C.; Lauber, C.L.; Lozupone, C.; Caporaso, J.G.; Knight, R.; Fierer, N. Soil bacterial and fungal communities across a pH gradient in an arable soil. *ISME J.* **2010**, *4*, 1340–1351. [CrossRef] [PubMed]
46. Shen, C.C.; Xiong, J.B.; Zhang, H.Y.; Feng, Y.Z.; Lin, X.G.; Li, X.Y.; Liang, W.J.; Chu, H.Y. Soil pH drives the spatial distribution of bacterial communities along elevation on Changbai Mountain. *Soil Biol. Biochem.* **2013**, *57*, 204–211. [CrossRef]
47. Shi, Y.; Li, Y.T.; Yang, T.; Chu, H.Y. Threshold effects of soil pH on microbial co-occurrence struc-ture in acidic and alkaline arable lands. *Sci. Total Environ.* **2021**, *800*, 149592. [CrossRef] [PubMed]
48. Yuan, J.; Wen, T.; Zhang, H.; Zhao, M.; Penton, C.R.; Thomashow, L.S.; Shen, Q. Predicting disease occurrence with high accuracy based on soil macroecological patterns of Fusarium wilt. *ISME J.* **2020**, *14*, 2936–2950. [CrossRef]
49. Shen, Z.Z.; Penton, C.R.; Lv, N.; Xue, C.; Yuan, X.F.; Ruan, Y.Z.; Li, R.; Shen, Q.R. Banana Fusarium Wilt Disease Incidence Is Influenced by Shifts of Soil Microbial Communities Under Different Monoculture Spans. *Microb. Ecol.* **2018**, *75*, 739–750. [CrossRef]
50. Lv, N.A.; Tao, C.Y.; Ou, Y.N.; Wang, J.B.; Deng, X.H.; Liu, H.J.; Shen, Z.Z.; Li, R.; Shen, Q.R. Root-Associated Antagonistic *Pseudomonas* spp. Contribute to Soil Suppressiveness against Banana Fusarium Wilt Disease of Banana. *Microbiol. Spectr.* **2023**, *11*, e03525-22. [CrossRef]
51. Zhao, J.; Li, Y.; Wang, B.; Huang, X.; Yang, L.; Lan, T.; Zhang, J.; Cai, Z. Comparative soil microbial communities and activities in adjacent Sanqi ginseng monoculture and maize-Sanqi ginseng systems. *Appl. Soil. Ecol.* **2017**, *120*, 89–96. [CrossRef]
52. Thitla, T.; Kumla, J.; Khuna, S.; Lumyong, S.; Suwannarach, N. Species Diversity, Distribution, and Phylogeny of Exophiala with the Addition of Four New Species from Thailand. *J. Fungi* **2022**, *8*, 766. [CrossRef] [PubMed]
53. Wang, B.Y.; Xia, Q.; Li, Y.L.; Zhao, J.; Yang, S.Z.; Wei, F.G.; Huang, X.Q.; Zhang, J.B.; Cai, Z.C. Root rot-infected Sanqi ginseng rhizosphere harbors dynamically pathogenic microbiotas driven by the shift of phenolic acids. *Plant Soil.* **2021**, *465*, 385–402. [CrossRef]
54. Bai, L.; Cui, J.; Jie, W.; Cai, B. Analysis of the community compositions of rhizosphere fungi in soybeans continuous cropping fields. *Microbiol. Res.* **2015**, *180*, 49–56. [CrossRef] [PubMed]

55. Hong, S.; Jv, H.L.; Lu, M.; Wang, B.B.; Zhao, Y.; Ruan, Y.Z. Significant decline in banana *Fusarium* wilt disease is associated with soil microbiome reconstruction under chilli pepper-banana rotation. *Eur. J. Soil. Biol.* **2020**, *97*, 103154. [CrossRef]
56. Cannon, P.; Buddie, A.; Bridge, P.; de Neergaard, E.; Lübeck, M.; Askar, M.J.M. Lectera, a new genus of the Plectosphaerellaceae for the legume pathogen *Volutella colletotrichoides*. *MycoKeys* **2012**, *3*, 23–36. [CrossRef]
57. Tawfik, S.M.; Zaky, M.F.; Mohammad, T.G.M.; Attia, H.A.E. Synthesis, characterization, and in vitro antifungal activity of anionic and nonionic surfactants against crop pathogenic fungi. *J. Ind. Eng. Chem.* **2015**, *29*, 163–171. [CrossRef]
58. Franke-Whittle, I.H.; Manici, L.M.; Insam, H.; Stres, B. Rhizosphere bacteria and fungi associated with plant growth in soils of three replanted apple orchards. *Plant Soil.* **2015**, *395*, 317–333. [CrossRef]
59. Baydar, R.; Guven, O.; Karaca, I. Occurrence of Entomopathogenic Fungi in Agricultural Soils from Isparta Province in Turkey and Their Pathogenicity to *Galleria mellonella* (L.) (Lepidoptera: Pyralidae) Larvae. *Egypt. J. Biol. Pest. Control* **2016**, *26*, 323–327.
60. Kepler, R.M.; Sung, G.H.; Ban, S.; Nakagiri, A.; Chen, M.J.; Huang, B.; Spatafora, Z.L.W. New teleomorph combinations in the entomopathogenic genus Metacordyceps. *Mycologia* **2012**, *104*, 182–197. [CrossRef]

agronomy

MDPI

Article

The Preliminary Research on Shifts in Maize Rhizosphere Soil Microbial Communities and Symbiotic Networks under Different Fertilizer Sources

Lidong Ji [1,2], Xing Xu [1,*], Fengju Zhang [1], Haili Si [2], Lei Li [1] and Guilian Mao [1]

1 School of Agriculture, Ningxia University, Yinchuan 750021, China; jilidong_0429@163.com (L.J.); tonyzfj@163.com (F.Z.); lilei867@163.com (L.L.); fransis-0327@163.com (G.M.)
2 Institute of Agricultural Resources and Environment, Ningxia Academy of Agriculture and Forestry Sciences, Yinchuan 750002, China; sihaili_0427@163.com
* Correspondence: xuxingscience@126.com

Abstract: The use of chemical fertilizer along with organic fertilizer is an important agricultural practice that improves crop yield but also affects soil biogeochemical cycles. In this study, a maize field experiment was conducted to investigate the effects of NPK fertilizer (NPK), organic fertilizer (OF), and their combination (NPK+OF) on soil chemical properties, bacterial and fungal community structures, and diversity compared the control (CK, without any fertilizer). The results showed that the application of OF and NPK-combined OF increased soil organic matter (OM), total N, total P, available N, available P, and available K levels. For alpha diversity analyses, the application of fertilizers led to decreases in soil bacterial and fungal Shannon indices (except for NPK in fungi). Compared with CK, NPK, OF, and NPK+OF fertilization treatments significantly increased the abundances of Acidobacteriota, Gemmatimonadota, and Basidiomycota. Network analysis showed that fertilization produced fewer connections among microbial taxa, especially in the combination of NPK and OF. A redundancy analysis combined with Mantel test further found that the soil OM, available N and P were the main soil-fertility factors driving microbial community variations. Therefore, using organic fertilizer or biological fertilizer combined with chemical fertilizer to improve the status of soil C, N, and P is a promising method to maintain the balance of soil microorganisms in maize field.

Keywords: organic fertilizer; soil properties; soil quality; microbial community; co-occurrence network

Citation: Ji, L.; Xu, X.; Zhang, F.; Si, H.; Li, L.; Mao, G. The Preliminary Research on Shifts in Maize Rhizosphere Soil Microbial Communities and Symbiotic Networks under Different Fertilizer Sources. *Agronomy* **2023**, *13*, 2111. https://doi.org/10.3390/agronomy13082111

Academic Editor: Nikolaos Monokrousos

Received: 27 June 2023
Revised: 1 August 2023
Accepted: 4 August 2023
Published: 11 August 2023

1. Introduction

In recent decades, the maize yield per unit area significantly increased, and was mainly driven by the large amount of chemical fertilizer input [1,2]. However, long-term and excessive chemical fertilizer input, especially inorganic fertilizer, had a negative impact on soil quality [3], leading to a series of environmental problems, such as water eutrophication, soil erosion and biodiversity loss [3–5]. The ministry of agriculture and rural affairs of China came up with a plan in 2015 characterized by the policy of "zero growth of chemical fertilizer in 2020" to solve the environmental problems caused by the excessive application of chemical fertilizer [6], which stressed the need to improve fertilization management in crop production. In short, sustainable management practices are essential to reduce the negative impact of agriculture on the environment.

Although there are still some disputes on organic agriculture, it is generally believed that organic fertilizer is superior to chemical fertilizer in improving soil biological fertility, such as animal manure [7]. The use of organic fertilizer combined with chemical fertilizer can not only reduce the use of synthetic fertilizers, but also improve soil fertility and nutrient circulation, further achieving the long-term stability of crop production by reshaping more active microbial soil–plant interactions [8,9]. However, as a typical organic fertilizer, manure is rich in nutrients and binders; thus, it is widely used to improve soil structure and function [10,11], such as improving soil nutrient availability loss and phosphorus fixation for crop utilization by reducing nitrogen [12]. Additionally, since the alkaline substances released from manure can also lead to an increase in soil pH, they are closely related to various soil physicochemical and biological processes determining crop productivity [13,14]. Moreover, animal manure also contains fresh organic matter, and its application to the soil will cause a significant increase in the soil organic carbon [15,16].

Soil microorganisms are critical to the function and sustainability of agroecosystems [17,18], not only playing an important role in driving soil–plant rhizosphere interactions, but also directly affecting soil properties and crop yields [19]. In addition, soil microorganisms not only play a key role in nutrient cycling and organic matter decomposition, but can also metabolize and transform organic matter by secreting specific extracellular enzymes [20]. However, soil ecological processes are mainly driven by soil microorganisms, and microbial metabolic activity and functional diversity are closely related to biogeochemical cycles and are considered more important indicators for assessing soil ecological functions and productivity [5,21]. Rhizosphere microorganisms grow together with plant roots and are critical for plant productivity because they play a key role in circulating soil nutrients [22], as well as in abiotic stress tolerance [23] and soil-borne pathogen inhibition [24]. Previous studies have provided comprehensive evidence that plant rhizosphere bacterial communities can comprise many characteristics, including soil physicochemical characteristics and nutrient availability [25], plant root exudates and other metabolites [26], and plant genotypes [27]. Ding et al. [28] also found that, compared with chemical fertilizer, organic fertilizer increased microbial functional diversity in paddy soil and changed bacterial community composition. Previous studies showed that the long-term application of organic fertilizer changed the activity of functional microorganisms related to carbon fixation and decomposition, increased the abundance of soil carbon fixation and refractory C compound degradation genes [29], and reduced the abundance of the degradation of unstable C compounds. It has been reported that manure application can significantly increase the number of key microorganisms related to nitrogen immobilization and mineralization, ammonification and nitrification [30]. The combined application of organic and inorganic fertilizers can accelerate the growth of microorganisms, change the structure of the soil's microbial community and improve enzyme activity [31,32]. The application of organic fertilizer will have a positive impact on bacterial and fungal diversity, regardless of whether chemical fertilizer is applied [33]. Although the impact of fertilization on soil microorganisms has been revealed in many studies, research on the impact mechanism of different fertilizer sources on the maize soil microbial community and the main factors affecting them is relatively weak. Further research would be helpful to evaluate the fertility and health status of maize soil and further promote the sustainable development of the maize industry and environment.

In addition, the ecological network of microbial communities is now a priority area in soil ecology research. In recent years, it has been widely used to study the interaction between soil organisms in different ecological environments, in which species (i.e., nodes) are connected by pairwise interactions (links) [34,35]. Based on this analysis, we can not only determine the symbiotic patterns of characteristic microbial species at different taxonomic levels [36,37], but also identify the key species affecting the stability of the microbial community [38,39]. However, to date, the establishment of the soil's microbial community network in maize fields under different fertilizer sources is still in the primary stage; thus, it is of great significance to further establish the network of the soil's microbial community to determine the interaction between microorganisms. Moreover, the effects of fertilizer sources on maize soil microbial communities and their symbiotic networks in arid irrigated areas of Ningxia have not been revealed. Here, a field trial was carried out to investigate the impacts of fertilizer sources on rhizosphere soil bacterial and fungal communities in maize field using amplicon sequencing and network analysis. The objectives were to: (i) investigate the effects of the use of chemical fertilizer/organic fertilizer/chemical fertilizer combined with organic fertilizer on the soil chemical properties; (ii) compare the differences in the effects of different fertilizer sources on the soil bacterial and fungal community abundances; and (iii) explore the soil fertility factors driving soil microbial community changes.

2. Materials and Methods

2.1. Study Site and Field Experiment Design

The field experiment site is located in Jinmaoyuan family farm, Huangquqiao Town, Yellow River Diversion Irrigation Area in the north of Yinchuan, Ningxia (38°30′ N, 106°18′ E) (Figure S1). The altitude and average annual temperature are about 1090 m and 8.8 °C, and there is a large temperature difference between day and night. In addition, the average annual precipitation is about 200 mm, and the evaporation is high. The experimental region is low-lying and the east–west trend is slightly undulating. The soil pH in the area is alkaline (alkalinity 18.67%), soil type is irrigation silted soil, and the soil texture is silty clay loam (according to international classification standards), which belongs to chloride sulfate saline soil and is irrigated by the Yellow River water.

The experiment was carried out with a completely randomized block design in 2019, which was being repeated for the third year. The maize variety was Xianyu No. 987. The wide and narrow rows (70 cm wide and 50 cm narrow) were used for planting, and the hole spacing was 18–20 cm. The roller seeder was used for sowing. The maize was irrigated twice every year at the jointing stage (mid-June) and tasseling stage (mid-July), and the single irrigation amount was 1500 m^3 ha^{-1}. The four fertilization treatments included (1) CK: with none fertilizer (control); (2) NPK: with chemical NPK fertilizer (360 kg N ha^{-1}, 120 kg P_2O_5 ha^{-1}, and 45 kg K_2O ha^{-1}); (3) OF: with organic fertilizer 9000 kg ha^{-1}; (4) NPK+OF: mineral NPK fertilizer with organic fertilizer (360 kg CH_4N_2O ha^{-1}, 120 kg P_2O_5 ha^{-1}, 45 kg K_2O ha^{-1}, and 9000 kg ha^{-1} organic fertilizer). Each treatment was replicated three times. The organic fertilizer adopted for this study was fermented by chicken manure, which contained 1.95% of total TN, 0.92% of total TP, 0.85% of total K, and 26.25% of organic C. The chemical fertilizers, including N (urea, 46% N), P (superphosphate, 12% P_2O_5), and K (potassium sulfate, 50% K_2O) fertilizers, were applied as a basic fertilizer before sowing, and organic fertilizer was applied once at the beginning of April every year. Additionally, regular weed and pest management was performed during maize growth stages.

2.2. Soil Sampling

At the harvest stage of maize in October, 2019, the rhizosphere soil samples were collected using a 5 cm diameter auger according to the five-point sampling method. Five fresh samples were taken from each plot and mixed thoroughly as one composite sample for further study. After the obtained roots were removed, all soil samples were passed through a 2 mm sieve, and each sample was separated into two subsamples. One subsample was used for the measurement of chemical properties, while the other was immediately stored at −80 °C for later DNA extraction.

2.3. Soil Chemical Analysis

The soil chemical properties included soil total nitrogen (TN), alkali-hydro nitrogen (AN), total phosphorus (TP), available P (AP), total potassium (TK), available K (AK), and organic matter (OM). After being air-dried, soil chemical properties were determined based on the methods described in Ji et al. [40]. Soil OM was determined by the potassium dichromate volumetric method. Soil TN, TP, and TK were determined by the Kjeldahl method, Vanado–Molybdate phosphoric yellow colorimetric procedure, and atomic absorption spectrophotometry, respectively. Soil AN, AP, and AK were determined by the alkali hydrolysis diffusion method, molybdenum antimony colorimetric method, and flame photometry, respectively.

2.4. DNA Extraction, PCR Amplification and Pyrosequencing

The Qiagen DNA Isolation Kit (Shanghai Tongyuan Biotechnology Co. Ltd, Shanghai, China) was used to extract soil genomic DNA according to the manufacturer's instructions. Then, DNA purity was detected by 1% agarose gel electrophoresis. For bacteria, the V4 region of the 16S rRNA gene was amplified using the primer set 515F/806R [41], and ITS region of fungi was amplified using the primer set ITS1/ITS2 [42]. Amplifications were performed using the KAPA 2G Robust Hot Start Ready Mix (KAPA, UK). The obtained products were purified using the Agencourt® AMPure® XP kit (Beckman Coulter, Bria, California, USA). High-throughput sequencing was performed on the Illumina Miseq PE300 platform at Allwegene Technology Co., Ltd. (Beijing, China). The raw sequence data were deposited in the NCBI small read archive dataset under the study number PRJNA880499 and PRJNA880503.

2.5. Data Analysis

After removing the barcode and primer sequences, the raw sequences were assembled for each sample according to a unique barcode. The paired-end sequences for each sample were merged using the FLASH v1.2.0 tool. Quality filtering of the raw tags was performed under specific filtering conditions to obtain high-quality clean tags according to the quality control process of QIIME v1.8.0 (http://qiime.org). The effective tags were obtained after removing the chimeric sequences obtained. The sequence analysis was conducted using the UPARSE v7.0.1001 software, and the sequences with a similarity of ≥97% were assigned to the same operational taxonomic unit (OTU) to generate rarefaction curves (Figure S2) and calculate the richness and diversity indices. For each representative sequence, the RDP (bacteria) and UNITE (fungi) databases were used to annotate taxonomic information. Principal component analysis (PCA) was used to examine the similarity and differences among all soil samples based on the OTUs level.

Network analysis was performed to explore the ecological connections within and between bacterial and fungal taxa at the genus level in the soil microbial communities. Bacterial and fungal genera with the relative abundances greater than 0.1% were used to construct the networks in all treatments. The networks were visualized by the Gephi software (v0.9.2) combined with the 'psych' package in R (V4.0.3).

2.6. Statistical Analysis

All analysis results were reported as the means standard deviation (SD) for the three replications. A two-way analysis of variance (ANOVA) was performed to compare the data, including soil chemical properties, the α-diversity indices, and the relative abundances of soil microbial taxa among four treatments using the IBM SPSS v22.0 (IMB Corp., Armonk, USA). Three diversity indices (observed OTUs species, Shannon, and Chao1) were used to assess the diversity of microbial communities. The differences in microbial community compositions among the four fertilization treatments were analyzed with the Kruskal–Wallis in R, and p-values < 0.05 were considered significant. The differential abundant taxa were identified by the linear discriminant analysis (LDA) effect size (LEfSe) method. The LEfSe algorithm used the nonparametric factorial Kruskal–Wallis test ($\alpha = 0.05$) to analyze the differential taxa among the four fertilization treatments (LDA score > 3). The redundancy analysis (RDA) was performed using the vegan package in R with 999 permutations to identify the major factors driving microbial distribution in maize field with the four fertilization treatments. Spearman correlation analysis was performed using the IBM SPSS Statistics (v 25.0) (SPSS, Chicago, IL, USA).

3. Results

3.1. Soil Chemical Properties

The influences of different fertilizer sources on soil chemical properties are shown in Table 1. Except for total potassium (TK), significant differences ($p < 0.05$) in the soil organic matter (OM), total nitrogen (TN), total phosphorus (TP), available nitrogen (AN), available phosphorus (AP), and available potassium (AK) were observed among the four fertilization treatments. Compared with the CK, NPK, OF, and NPK+OF significantly increased soil OM, TN, AN, and AP. The NPK+OF was a better fertilization method than the CK, increasing the soil OM, TP, AN, and AP by 58.03%, 8.22%, 55.22%, and 162.70%, respectively, while the highest values of TN and AK occurred in OF treatments. TK was decreased by all fertilization treatments, but there were no significant differences among the four fertilization treatments ($p > 0.05$).

Table 1. Soil chemical properties analysis.

Treatments	OM (g kg^{-1})	TN (g kg^{-1})	TP (g kg^{-1})	TK (g kg^{-1})	AN (mg kg^{-1})	AP (mg kg^{-1})	AK (mg kg^{-1})
CK	11.8 ± 0.21 d	0.75 ± 0.04 c	0.73 ± 0.04 ab	19.3 ± 0.92 a	32.5 ± 0.12 c	10.7 ± 0.06 c	243 ± 7 b
NPK	15.8 ± 0.35 c	0.86 ± 0.01 bc	0.71 ± 0.02 b	18.7 ± 0.61 a	39.5 ± 0.40 b	14.1 ± 2.70 b	222 ± 7 c
OF	17.9 ± 0.40 b	1.04 ± 0.13 a	0.74 ± 0.03 ab	18.7 ± 0.06 a	49.9 ± 0.70 a	26.2 ± 0.32 a	304 ± 9 a
NPK+OF	18.6 ± 0.36 a	0.98 ± 0.09 ab	0.79 ± 0.05 a	18.9 ± 0.40 a	50.4 ± 1.25 a	28.0 ± 0.12 a	253 ± 3 b
Source of variance							
NPK	142 ***	0.27	0.79	0.29	76.8 ***	11.2 *	87.2 ***
OF	529 ***	20.1 ***	4.06	0.54	1074 ***	351 ***	139 ***
NPK+OF	72.4 ***	3.37	2.34	1.28	57.8 ***	0.95	15.5 ***

Values are mean ± standard deviation (n = 3). Different letters in the same column represent significant differences at the $p = 0.05$ level. * $p < 0.05$ and *** $p < 0.001$. OM, organic matter; TN, total N; TP, total P; TK, total K; AN, alkali-hydrolyzed N; AP, available P; AK, available K.

3.2. Impacts of Fertilization on Bacterial and Fungal Alpha and Beta Diversity

Similar to the soil chemical properties, bacterial and fungal alpha diversity indices were also affected by the the four fertilizer sources. The observed species, Shannon, and Chao 1 indices showed that the changes in soil bacterial and fungal alpha diversity observed in the various fertilization treatments were significant (Figure S3 and Table 2). The observed species, Shannon, and Chao 1 indices showed that fertilizer sources markedly decreased soil bacterial diversity, and the lowest value of bacteria occurred in NPK+OF treatments. Regarding soil fungal alpha diversity indices, NPK and NPK+OF fertilization treatments significantly decreased the observed species and Chao1 indices, while NPK significantly increased fungal Shannon index, which was significantly decreased by OF and NPK+OF treatments. Spearman correlation analysis suggested that some bacterial diversity indices were significantly negatively correlated with soil OM, N, and P, while fungal diversity indices were only closely negatively correlated with soil OM and P (Table 3). In addition, the numbers for the shared bacterial and fungal OTUs in all four fertilization treatments were 359 and 18, respectively (Figure S4).

Table 2. Bacterial and fungal alpha-diversity.

Treatments	Bacteria			Fungi		
	Observed_Species	Shannon	Chao1	Observed_Species	Shannon	Chao1
CK	4639 ± 155 a	10.4 ± 0.09 a	6388 ± 150 a	687 ± 46 a	5.19 ± 0.38 ab	977 ± 98.3 a
NPK	4114 ± 111 b	10.1 ± 0.19 ab	6132 ± 83 b	589 ± 50 b	5.64 ± 0.28 a	792 ± 16.49 b
OF	4074 ± 86 b	10.2 ± 0.09 bc	5968 ± 91 bc	685 ± 7 a	4.99 ± 0.36 bc	922 ± 1.14 a
NPK+OF	3870 ± 42 c	9.98 ± 0.06 c	5867 ± 87 c	451 ± 10 c	4.46 ± 0.22 c	622 ± 7.86 c
Source of variance						
NPK		14.15 ***	8.43 *	69.8 ***	0.05	70.8 ***
OF	19.18 ***	6.33 *	31.2 ***	12.43 ***	14.49 ***	15.29 ***
NPK+OF	0.17	0.02	1.60	11.72 ***	7.32 *	3.90

Values are mean ± standard deviation (n = 3). Different letters in the same column represent significant differences at the $p = 0.05$ level. * $p < 0.05$ and *** $p < 0.001$.

Table 3. Correlation analysis between bacterial and fungal diversity indexes and soil chemical properties.

	Indices	OM	TN	TP	TK	AN	AP	AK
Bacteria	Observed species	−0.81 **	−0.60 *	−0.54	0.19	−0.71 *	−0.88 **	−0.27
	Shannon	−0.63 *	−0.39	−0.40	0.26	−0.52	−0.76 **	−0.10
	Chao 1	−0.90 *	−0.75 *	−0.61 *	0.02	−0.81 **	−0.93 **	−0.47
Fungi	Observed species	−0.43	−0.11	−0.35	0.21	−0.27	−0.48	0.22
	Shannon	−0.69 *	−0.55	−0.71 *	−0.22	−0.49	−0.62 *	−0.53
	Chao 1	−0.59 *	−0.37	−0.28	0.22	−0.51	−0.65 *	0.15

* $p < 0.05$, ** $p < 0.01$. OM, organic matter; TN, total N; TP, total P; TK, total K; AN, alkali-hydrolyzed N; AP, available P; AK, available K.

To evaluate the differences in bacterial and fungal beta diversity, principal component analysis (PCA) was performed based on the OTU level. The first two main coordinates of PCA (PC1 = 25.89% and PC2 = 14.54%) explained 40.43% of the variation in bacterial beta diversity (Figure 1a), similarly, which also explained 34.57% of the variation in fungal beta diversity (Figure 1b). This suggested that the influences of fertilization on bacterial beta diversity were larger than that of fungi. Additionally, the bacterial and fungal communities in the three fertilization treatments (NPK, OF, and NPK+OF) were separated from the CK (Anosim analysis, p = 0.001) (Table S1), indicating that the fertilization treatments had significant effects on the bacterial and fungal communities.

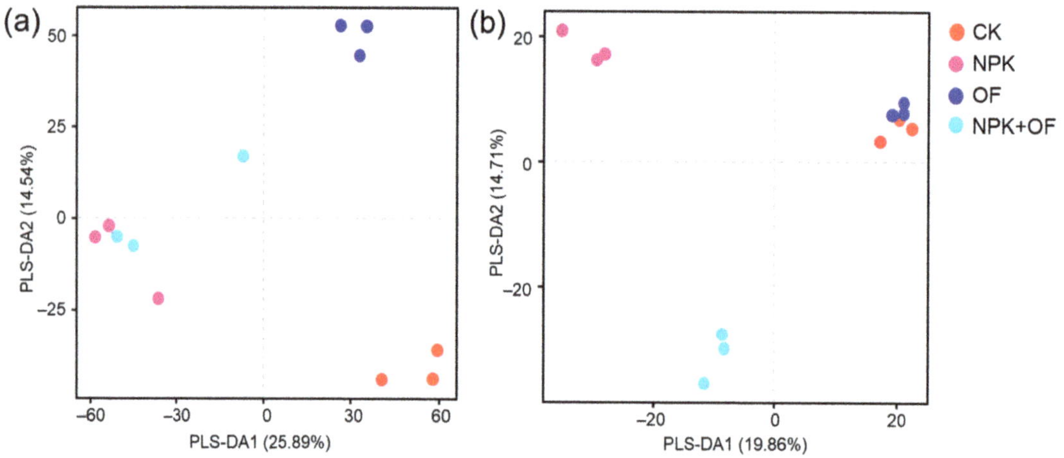

Figure 1. Principal component analysis (PCA) of bacterial (**a**) and fungal (**b**) communities in soils of four treatments.

3.3. Relative Abundance of Major Bacterial and Fungal Taxa

A total of 59 bacterial phyla and 848 genera were detected in all soil samples, and the average relative abundance of 11 phyla and 10 genera exceeded 1%. Among the top bacterial community at the phylum level, Proteobacteria (23.15%) were clearly dominant, followed by Acidobacteriota (13.15%), Gemmatimonadota (10.14%), Actinobacteriota (9.72%), Chloroflexi (9.25%), Planctomycetota (7.74%), Bacteroidota (6.74%), Myxococcota (4.58%), Crenarchaeota (3.09%), and Desulfobacterota (3.07%) (Figure 2a and Table S2). There were significant differences in the relative abundances of Proteobacteria, Acidobacteriota, Gemmatimonadota, Bacteroidota and Desulfobacterota. The relative abundances of Acidobacteriota and Gemmatimonadota were significantly increased by all fertilization treatments, while those of Proteobacteria and Desulfobacterota were markedly decreased. The top 10 dominant genera included *MND1* (3.09%), *RB41* (3.09%), *Pseudomonas* (3.09%), *Haliangium* (3.09%), *Citrifermentans* (3.09%), *metagenome* (3.09%), *Subgroup_10* (3.09%), *Lysobacter* (3.09%), *Sphingomonas* (3.09%) and *Defluviicoccus* (3.09%) (Figure 2b and Table S2). These genera belong to five phyla (Proteobacteria, Acidobacteriota, Myxococcota, Desulfobacterota, and Actinobacteriota). All fertilization treatments significantly decreased the relative abundances of *MND1*, *Citrifermentans*, and *Lysobacter*, and increased that of *RB41*. The relative abundance of genus *metagenome* was significantly reduced only in NPK+OF treatment, while that of *Sphingomonas* was significantly reduced by NPK and NPK+OF treatments.

Regarding soil fungal community, a total of 12 phyla and 347 genera were detected in all soil samples, and the average relative abundance of 5 phyla and 5 genera exceeded 1%. All these samples were dominated at the phylum level by Ascomycota, which comprised 57.45% of the total sequences on average. Only NPK+OF caused an increase in the relative abundance of Ascomycota. The other dominant fungal phyla were Basidiomycota (6.84%), Mortierellomycota (5.52%), Glomeromycota (1.77%) and Chytridiomycota (1.55%). In addition, among the top 10 dominant fungal phyla, only the relative abundance of Basidiomycota was significantly different in all treatments, fertilization treatment significantly increased its relative abundance, and the maximum abundance appeared in NPK treatment (Figure 2c and Table S3). At the genus level, *Botryotrichum* (19.66%), *Mortierella* (5.51%), *Fusarium* (3.10%), *Acremonium* (2.66%), *Chaetomium* (2.06%), *Clitocybe* (0.93%), *Neopaxillus* (0.90%), *Stachybotrys* (0.86%), *Lophotrichus* (0.71%), and *Plectosphaerella* (0.65%), detected in all samples, were the top 10 dominant fungal genera (Figure 2d and Table S3). These genera belong to three phyla (Ascomycota, Mortierellomycota, and Basidiomycota). Among the fungal genera, compared to the CK, the NPK treatment caused a remarkable increase in the relative abundances of *Fusarium*, *Neopaxillus*, and *Stachybotrys*, while a similar result occurred in the OF treatment regarding the relative abundance of *Clitocybe*. Furthermore, the relative abundances of *Acremonium* and *Plectosphaerella* markedly decreased following all fertilization treatments relative to the CK. These results indicated that fertilizer sources can cause significant changes in the composition of the soil microbial community.

Additionally, linear discriminant biomarker analysis (LEfSe) was performed to determine the classified bacterial and fungal taxa with significant abundance differences among the fertilization and control groups (Figures S5 and S6). The biomarkers differ depending on the fertilizer sources. Specifically, the number of bacterial biomarkers in the NPK, OF, and NPK+OF treatment groups was 10, 2, and 7, respectively. Similarly, the number of fungal biomarkers in the NPK, OF, and NPK+OF treatment groups was 39, 8, and 8, respectively. Bacterial biomarkers were mostly distributed in Acidobacteria, Thaumarchaeota, Nitrospirae, Gemmatimonadetes, Bacteroidetes, Proteobacteria, Actinobacteria, and Chloroflexi, and fungal biomarkers were mostly distributed in Ascomycota, Basidiomycota, Mortierellomycota, and Glomeromycota. In addition, although there were no common bacterial biomarkers between the three fertilization treatments at the genus level, the common fungal biomarkers mostly included genera *Lecanicillium*, *Mortierella*, *Filobasidium*, *Vishniacozyma*, *Cladosporium*, *Hannaella*, *Mycosphaerella*, *Coprinellus*, and *Gliomastix*.

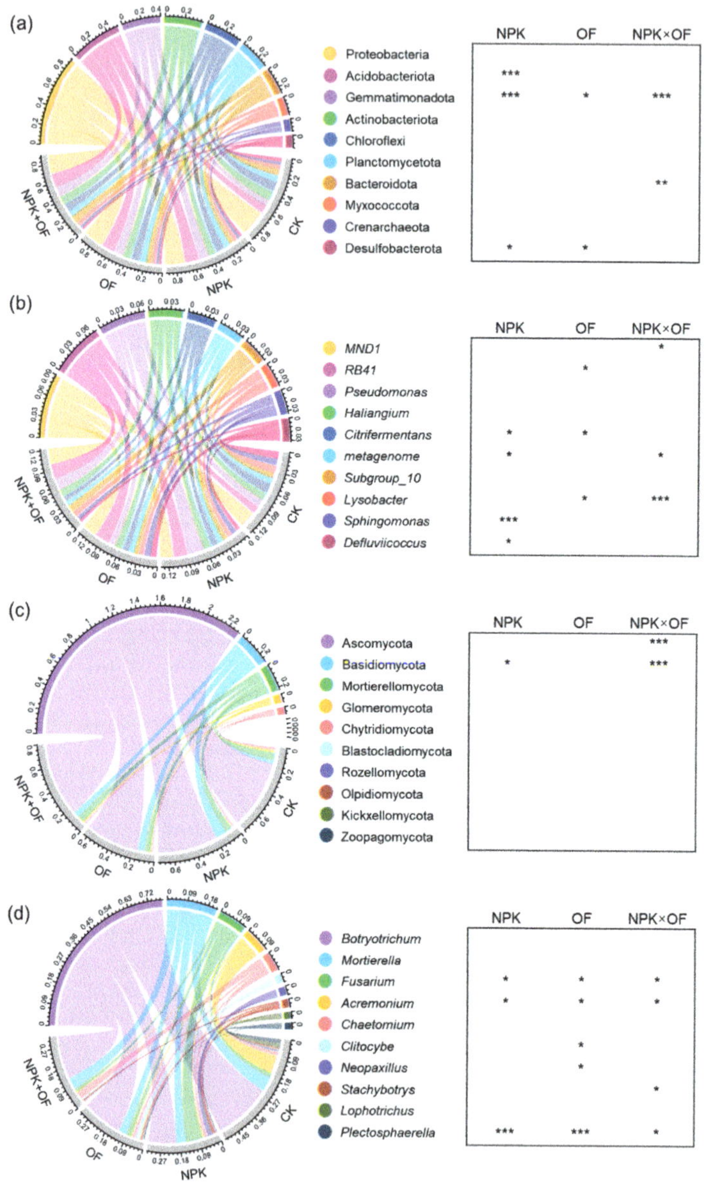

Figure 2. Relative abundances of the top 10 bacterial (**a**) and fungal (**c**) phyla, and bacterial (**b**) and fungal (**d**) genera in corn soils under four fertilization treatments. CK = no fertilization, NPK = chemical fertilizer only, OF = organic fertilizer only, and NPK+OF = chemical plus organic fertilizer. * $p < 0.05$, ** $p < 0.01$, and *** $p < 0.001$.

3.4. The Influences of Fertilization Treatment on the Complexity of the Microbial Co-Occurrence Network

Microorganisms usually form complex networks to deal with the interference of foreign substances. The co-occurrence networks of soil bacterial and fungal communities were constructed to explore the co-occurrence patterns of microbes under fertilization treatments (Figure 3). The soil microbial network of each treatment showed a different co-occurrence pattern. Network topology parameters (node number and edge number) are used to evaluate the complexity of the soil microbial network. The results showed that there were significant differences in the topological structure of co-occurrence networks between the control and the fertilization treatment groups. In general, fertilizer sources interfered with the network complexity of microbial communities. The numbers of nodes and edges showed that the fertilization treatments significantly reduced the network complexity of microbial communities, and network density showed that fertilization treatments significantly increased the density of microbes compared with the CK. In addition, the clustering coefficients suggested that the three fertilization treatments did not change the tightness of microbial community connections, and the average degree in NPK+OF treatment was also significantly lower than the control, indicating that NPK+OF weakened the relationships between bacterial fungal taxa. The complexity of the soil microbial community networks proved that fertilizer sources can interfere with the complexity of the soil microbial community co-occurrence networks, further reducing microbial relationships.

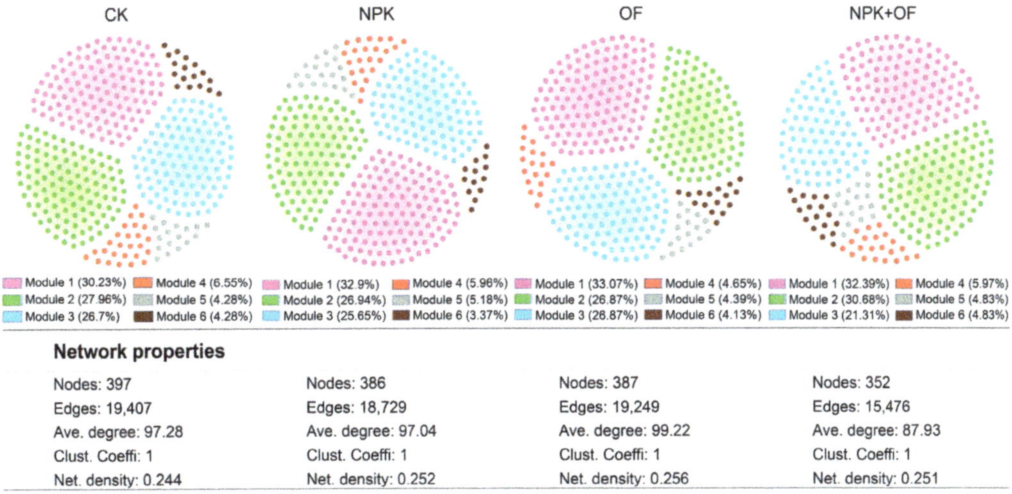

Figure 3. Effect of different fertilization regimes on the co-occurrence patterns of soil bacterial and fungal community. Networks were constructed at the genera level. The size of nodes is scaled to the degree of nodes, and the nodes are colored according to modules. Edges indicate correlations among nodes ($r > 0.9$, $p < 0.01$). Ave. degree: average degree; Clust. coeff.: clustering coefficient; Net. density: network density.

3.5. Driving Soil Fertility Factors of Microbial Community Variations

The relationships among soil chemical variables and microbial communities were found using a redundancy analysis (RDA) (Figure 4). Similar to beta diversity, RDA results also revealed clear separations among all fertilization treatments. The environmental characteristics associated with the bacterial and fungal communities showed that the selected soil fertility factors explained 53.15% of the bacterial community structure variation and 69.09% of the fungal community structure variation (Figure 4a,b), and permutation tests showed a significant correlation between soil OM, TN, AN, AP, and AK and bacterial community structure, and soil OM, AN, AP, AK and fungal community structure (Table S4). Additionally, the Mantel test revealed that the soil OM, AN, and AP were significantly correlated with bacterial community composition, while soil OM, TP, AN, and AP were remarkably correlated with fungal community composition (Figure 4c). These results indicated that soil OM, and available N and P, were the main soil fertility factors driving microbial community variations.

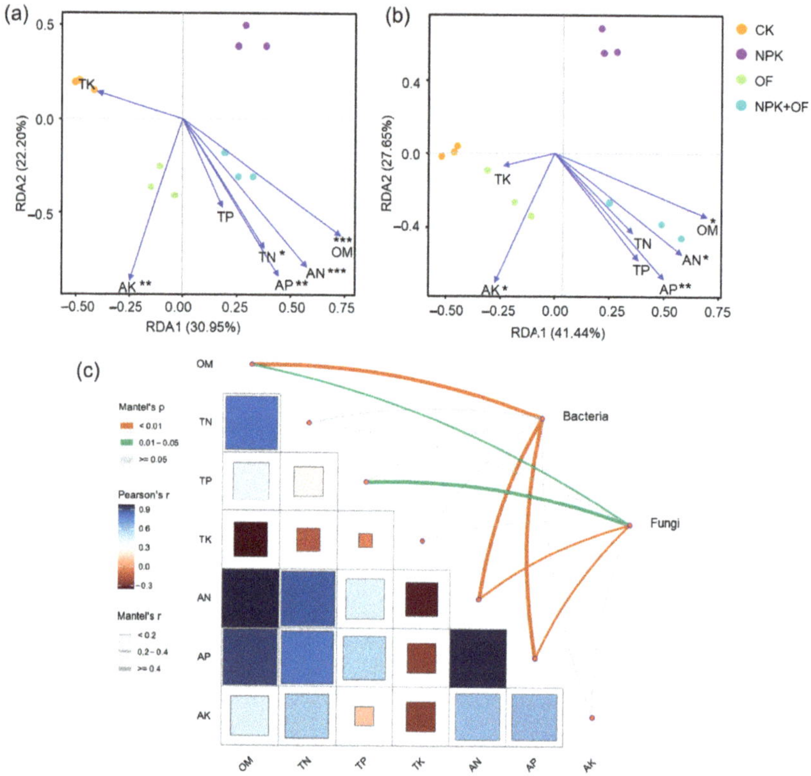

Figure 4. Environmental drivers of soil microbial community compositions. Redundancy analysis (RDA) of the bacterial (**a**) and fungal (**b**) communities and soil properties for individual samples. The direction of the arrows indicates correlations with the first two canonical axes, and the length of the arrows represents the strength of the correlations. Pairwise comparisons of soil fertility factors with soil bacterial and fungal communities with color gradient denoting Spearman's correlation coefficients (**c**). Edge width corresponds to Mantel's r statistic for the corresponding distance correlations, and edge color denotes the statistical significance based on 9999 permutations. Soil properties included soil organic matter (OM), total N (TN), total P (TP), total K (TK), dissolved organic C (AN), available P (AP), available K (AK). * $p < 0.05$, ** $p < 0.01$, and *** $p < 0.001$.

4. Discussion

Many studies have explored the responses of the soil microbial community to different fertilization systems and fertilizer sources. However, we studied the changes in bacterial and fungal communities in maize field under different fertilizer sources after irrigation. PCA analysis showed significant differences in microbial community compositions among different fertilization treatments. A large number of studies suggested that microbes showed different biogeographic patterns, which were continuously and permanently affected by the local climate, crop diversification, tillage, fertilization type, and other environmental factors [43], while our results showed that the application of organic and inorganic fertilizers could cause significant changes in microbial communities in maize fields after irrigation.

Soil physicochemical properties are not only considered to be the key to soil quality, but are also generally defined as an important indicator to maintain soil environmental quality and improve crop productivity [44,45]. Previous studies have confirmed that fertilization may interfere with the soil's physicochemical properties, mainly due to the soil type, and fertilization dosage [7]. Other studies have shown that the application of chemical fertilizers can lead to soil acidification; however, adding the organic improvers can not only alleviate soil acidification, but can increase organic matter input, which is helpful to promote soil nutrients' transformation and cycling [46]. In the present study, we found that many soil nutrients were significantly altered by the chemical and organic fertilizer sources. In addition, the application of an organic fertilizer can also lead to improvements in nutrient utilization efficiency and further reduce losses, such as N and P [47]. Similarly, our study showed that the application of organic fertilizer and its combination with NPK led to increases in the soil OM, TN, TP, AN, AP, and AK, further improving soil fertility, which also confirmed the results of many studies [31,48]. Other research also found that organic and N fertilizer significantly increased the topsoil total N, PON, MBN, DON, and NO_3^- [49], mainly due to the nature of the organic fertilizer containing certain nutrients, and can improve soil fertility and crop yield [50]. Moreover, the combination of chemical fertilizer and organic fertilizer not only improves the soil nutrients [17], but also facilitates the growth of crop roots and improves crop yield [51,52].

Microorganisms are not only an important indicator of soil health [53], but their activities regulate the accumulation of soil organic carbon and the nutrient cycles. Some studies have shown that long-term chemical fertilization will reduce the diversity of the soil bacterial community, but the use of an organic fertilizer caused it to increase [54]. Additionally, it has also been found that long-term NPK fertilization lead to a decrease in soil bacterial community diversity, while the input of organic fertilizer significantly stabilized bacterial diversity, further restoring it to the level of unfertilized soil [17]. Our results showed that the soil bacterial and fungal diversity in maize farmland were significantly reduced with different fertilization treatments, which was inconsistent with many research results. For example, most studies found that the diversity of soil bacteria was not significantly increased by fertilization [11,46,55], but other studies found that the application of a bio-fertilizer lead to a decrease in soil fungal diversity [46,56]. In addition, the useo f NPK combined with organic fertilizer can not only increase the diversity and richness of soil bacterial population, but also can promote the root exudates [57]. More interestingly, soil organic matter and other nutrients were significantly correlated with soil diversity indices, and the same results were found in other studies [46,58], which are basically consistent with previous studies showing that the fertilization system leads to changes in soil microbial communities by changing soil chemistry [55,59].

Additionally, soil health depends on not only the diversity and richness of soil microbes, but also the community compositions [60]. Firstly, bacteria are the most important decomposers in soil, and long-term organic fertilization leads to the enrichment of specific bacteria that can effectively utilize nutrients [31]. The dominant bacterial phyla detected in the kiwifruit orchard soil were Proteobacteria, Acidobacteria, Bacteroidetes, and Actinobacteria [46,61]. Some researchers also found that the dominant bacteria after long-term fertilization were Actinobacteria, Proteobacteria, Chloroflexi, and Acidobacteria [62]. Actinobacteria are a group of co-trophic organisms suitable to enhance plant growth in high-C environments. Biological manure promoted the proliferation of Actinomycetes, because they can create an environment that is rich in nutrients and carbon, and their abundances significantly increase under the treatment of biological manure [63]. Although Proteobacteria, Acidobacteriota, Gemmatimonadota, and Actinobacteriota were detected to be the dominant bacterial phyla in our study, the relative abundance of Actinobacteriota decreased under organic fertilizer treatment, which may be caused by the difference between manure and biological manure. The relative abundance of Acidobacteriota significantly increased while that of Proteobacteria significantly decreased, but no significant differences were observed among the three fertilization treatments. However, in other studies, the results were the opposite [46], which may be the reason that Acidobacteriota belongs to a class of oligotrophs (k-strategists), which can degrade relatively stable carbon, grows slowly, dominates the microbial community in a dystrophic environment, and is negatively correlated with C level [17]. At the same time, studies found that high C/N and N limitations caused by long-term organic or inorganic fertilization affected the structure of soil microbial communities and their dominant SOC decomposition [64].

Followed by bacteria, the dominant fungal phyla were also detected in the soil samples in our study, such as Ascomycota, Basidiomycota, and Mortierellomycota, which is consistent with the findings in a kiwifruit orchard soil under organic and inorganic fertilization, but the results were different from the findings in soybean rotation system under organic and inorganic fertilization [65]. Moreover, Ascomycota is a kind of saprophytic fungi, which can thrive in an arid environment, has strong environmental adaptability, degrades organic matter, and is the main decomposer of soil organic matter containing cellulose, lignin, and pectin [66]. Mortierellomycota are saprophytic and widely exist, which can dissolve P, increase crop yield, and form symbiotic relationships with plants [67]. In the present study, the abundances of Ascomycota and Mortierellomycota decreased in NPK and OF fertilization treatments, while in other studies, the opposite results were observed [62]. Importantly, the relative abundance of *Fusarium*, a soil-borne plant pathogenic fungus that can induce crop *Fusarium wilt*, was significantly reduced after the use of OF combined with NPK fertilizer [46,68], which showed that the application of organic fertilizer was beneficial to inhibit the growth of plant pathogenic fungi. Therefore, fungal communities are closely related to soil fertilizer sources.

The co-occurrence networks are not only an important manifestation of microbial community stability in response to external disturbances [69], but they also have the ability to maintain the relationships between soil microbial diversity and ecosystem multifunctionality [70]. At present, network analysis has been widely used in microbial ecology. For example, some studies indicated that the complexity of soil microbial networks was affected by environmental changes, such as agricultural management and the addition of other foreign substances [71]. The use of a bio-organic fertilizer enhanced stable network of soil microbial communities [46]. A high altitude not only reduced the diversity of microbial communities, the complexity of co-occurrence networks, and the versatility of ecosystems, but also revealed that the impacts of microbial community diversity on versatility was indirectly driven by the complexity of microbial networks [70]. Another study suggested that the molecular ecological network became more stable under warming conditions [72]. Organic fertilization drives changes in the complexity of microbial communities, and key groups increase the resistance of microbial-mediated functions to biodiversity loss [73]. Similarly, our study on the complexity of soil microbial community networks under different fertilization treatments showed that fertilizer sources interfered with the complexity of the co-occurrence networks in soil microbial communities, and reduced the interactions among them. Therefore, both organic and inorganic fertilizers will reveal its response to external interference by affecting the ability of the soil microbial co-occurrence pattern, and during this process, some synergistic microbial co-occurrence patterns may be generated to resist soil microbes.

Additionally, soil properties after fertilization can explain the changes in microbial communities in this experiment [74]. Among them, OM, AN, and AP were the main factors driving the changes in bacterial and fungal communities. Some studies have shown that most microbial parameters are mainly related to the soil OM [74]. Previous studies also showed that pH and SOC were the main indirect factors regulating the crop yield by regulating the structure and diversity of bacterial communities and even the abundances of potential functional genes of microorganisms [11]. In addition, soil organic carbon was not only closely related to nutrient availability, but was another key factor affecting soil microbial community composition and diversity [75]. In summary, our research showed that fertilizer sources (chemical fertilizer, organic fertilizer and their combination) changed the abundances, diversity and compositions of soil microbes by changing the soil chemical properties.

5. Conclusions

In this work, a field trial was carried out to investigate the impacts of different fertilizer sources on the rhizosphere soil bacterial and fungal communities in maize field by using amplicon sequencing and network analysis. Our results showed that different fertilizer sources significantly improved the soil chemical properties, changed the soil bacterial and fungal communities, and improved the soil microenvironment in maize field after irrigation. Network analysis also suggested that the fertilization treatments reduced interactions among bacterial and fungal taxa in the microbial community, especially the combination of NPK and OF, indicating that treatment can decrease microbial competition among microbes by improving soil fertility, etc. Moreover, the redundancy analysis combined with Mantel test further revealed that soil OM, available N and P were the main soil fertility factors driving microbial community variations. Above all, the results indicated that the application of NPK combined with OF is the most appropriate method for planting, which is beneficial for regulating soil biogeochemical cycles through significantly affecting the soil C, N, and P, as well as microbial communities in maize rhizosphere soil. This study can provide important methods and data support for further research on improvements in soil quality and yield in the future, as well as the green development of agriculture.

Supplementary Materials: The following Supporting information can be downloaded at: https://www.mdpi.com/article/10.3390/agronomy13082111/s1. Table S1. The differences in the bacterial and fungal community Structures among Samples; Table S2. The relative abundance of phyla and genera (top 10) for bacteria; Table S3. The relative abundance of phyla and genera (top 10) for fungi; Table S4. The results of Monte Carlo permutation test in the RDA in the vegan package of R; Figure S1. The geographical location of the experimental field; Figure S2. Rarefaction curves of bacteria (a) and fungi (b); Figure S3. Comparison of microbial diversity indices among all four treatments; Figure S4. Shared and unique OTUs in all treatments for bacteria (a) and fungi (b); Figure S5. Results of LEfSe analysis Showing bacterial taxa that Significantly differed in the four treatments (a). Cladogram plotted from LEfSe analysis Showing the Significant differences ($p < 0.05$) in the relative abundance of bacterial taxon among four treatments (b); Figure S6. Results of LEfSe analysis Showing fungal taxa that Significantly differed in the four treatments (a). Cladogram plotted from LEfSe analysis Showing the Significant differences ($p < 0.05$) in the relative abundance of fungal taxon among four treatments (b).

Author Contributions: X.X. contributed to the overall design of the experiment, and L.J. and F.Z. were responsible for the implementation of the field experiment and sample collection. The samples were collected by L.J., F.Z. and H.S., the determination of soil properties was analyzed by H.S. and L.L., and the writing and modification of the manuscript was implemented by X.X. and G.M. All authors have read and agreed to the published version of the manuscript.

Funding: This research was funded by the Foundation project: Project of national key research and development program of China (2021YFD1900605).

Data Availability Statement: The raw reads were deposited into the NCBI Sequence Read Archive (SRA) database.

Conflicts of Interest: The authors declare no conflict of interest.

References

1. Xiao, G.; Zhao, Z.; Liang, L.; Meng, F.; Wu, W.; Guo, Y. Improving nitrogen and water use efficiency in a wheat-maize rotation system in the North China Plain using optimized farming practices. *Agr. Water Manag.* **2019**, *212*, 172–180. [CrossRef]
2. Liu, X.; Wang, H.; Zhou, J.; Chen, Z.; Lu, D.; Zhu, D.; Deng, P. Effect of nitrogen root zone fertilization on rice yield, uptake and utilization of macronutrient in lower reaches of Yangtze River, China. *Paddy Water Environ.* **2017**, *15*, 625–638. [CrossRef]
3. Krauss, M.; Berner, A.; Perrochet, F.; Frei, R.; Niggli, U.; Mäder, P. Enhanced soil quality with reduced tillage and solid manures in organic farming—A synthesis of 15 years. *Sci. Rep.* **2020**, *10*, 4403. [CrossRef] [PubMed]
4. Mei, N.; Zhang, X.; Wang, X.; Peng, C.; Gao, H.; Zhu, P.; Gu, Y. Effects of 40 years applications of inorganic and organic fertilization on soil bacterial community in a maize agroecosystem in northeast China. *Eur. J. Agron.* **2021**, *130*, 126332. [CrossRef]

5. Song, D.; Dai, X.; Guo, T.; Cui, J.; Zhou, W.; Huang, S.; Shen, J.; Liang, G.; He, P.; Wang, X.; et al. Organic amendment regulates soil microbial biomass and activity in wheat-maize and wheat-soybean rotation systems. *Agr. Ecosyst. Environ.* **2022**, *333*, 107974. [CrossRef]
6. Zhai, L.; Wang, Z.; Zhai, Y.; Zhang, L.; Zheng, M.; Yao, H.; Lv, L.; Shen, H.; Zhang, J.; Yao, Y.; et al. Partial substitution of chemical fertilizer by organic fertilizer benefits grain yield, water use efficiency, and economic return of summer maize. *Soil Till. Res.* **2022**, *217*, 105287. [CrossRef]
7. Li, R.; Tao, R.; Ling, N.; Chu, G. Chemical, organic and bio-fertilizer management practices effect on soil physicochemical property and antagonistic bacteria abundance of a cotton field: Implications for soil biological quality. *Soil Till. Res.* **2017**, *167*, 30–38. [CrossRef]
8. Li, Y.; Wang, Z.; Li, T.; Zhao, D.; Han, J.; Liao, Y. Wheat rhizosphere fungal community is affected by tillage and plant growth. *Agr. Ecosyst. Environ.* **2021**, *317*, 107475. [CrossRef]
9. Zhao, Z.-B.; He, J.-Z.; Quan, Z.; Wu, C.-F.; Sheng, R.; Zhang, L.-M.; Geisen, S. Fertilization changes soil microbiome functioning, especially phagotrophic protists. *Soil Biol. Biochem.* **2020**, *148*, 107863. [CrossRef]
10. Huang, X.; Jia, Z.; Guo, J.; Li, T.; Sun, D.; Meng, H.; Yu, G.; He, X.; Ran, W.; Zhang, S.; et al. Ten-year long-term organic fertilization enhances carbon sequestration and calcium-mediated stabilization of aggregate-associated organic carbon in a reclaimed Cambisol. *Geoderma* **2019**, *355*, 113880. [CrossRef]
11. Li, J.; Yang, Y.; Wen, J.; Mo, F.; Liu, Y. Continuous manure application strengthens the associations between soil microbial function and crop production: Evidence from a 7-year multisite field experiment on the Guanzhong Plain. *Agr. Ecosyst. Environ.* **2022**, *338*, 108082. [CrossRef]
12. Cai, A.; Xu, M.; Wang, B.; Zhang, W.; Liang, G.; Hou, E.; Luo, Y. Manure acts as a better fertilizer for increasing crop yields than synthetic fertilizer does by improving soil fertility. *Soil Till. Res.* **2019**, *189*, 168–175. [CrossRef]
13. Liu, E.; Yan, C.; Mei, X.; He, W.; Bing, S.H.; Ding, L.; Liu, Q.; Liu, S.; Fan, T. Long-term effect of chemical fertilizer, straw, and manure on soil chemical and biological properties in northwest China. *Geoderma* **2010**, *158*, 173–180. [CrossRef]
14. Mi, W.; Sun, Y.; Xia, S.; Zhao, H.; Mi, W.; Brookes, P.C.; Liu, Y.; Wu, L. Effect of inorganic fertilizers with organic amendments on soil chemical properties and rice yield in a low-productivity paddy soil. *Geoderma* **2018**, *320*, 23–29. [CrossRef]
15. Hua, W.; Luo, P.; An, N.; Cai, F.; Zhang, S.; Chen, K.; Yang, J.; Han, X. Manure application increased crop yields by promoting nitrogen use efficiency in the soils of 40-year soybean-maize rotation. *Sci. Rep.* **2020**, *10*, 14882. [CrossRef] [PubMed]
16. Li, T.; Zhang, Y.; Bei, S.; Li, X.; Reinsch, S.; Zhang, H.; Zhang, J. Contrasting impacts of manure and inorganic fertilizer applications for nine years on soil organic carbon and its labile fractions in bulk soil and soil aggregates. *Catena* **2020**, *194*, 104739. [CrossRef]
17. Li, X.; Li, B.; Chen, L.; Liang, J.; Huang, R.; Tang, X.; Zhang, X.; Wang, C. Partial substitution of chemical fertilizer with organic fertilizer over seven years increases yields and restores soil bacterial community diversity in wheat–rice rotation. *Eur. J. Agron.* **2022**, *133*, 126445. [CrossRef]
18. Shi, Y.; Delgado-Baquerizo, M.; Li, Y.; Yang, Y.; Zhu, Y.-G.; Peñuelas, J.; Chu, H. Abundance of kinless hubs within soil microbial networks are associated with high functional potential in agricultural ecosystems. *Environ. Int.* **2020**, *142*, 105869. [CrossRef]
19. Semenov, M.V.; Krasnov, G.S.; Semenov, V.M.; van Bruggen, A.H.C. Long-term fertilization rather than plant species shapes rhizosphere and bulk soil prokaryotic communities in agroecosystems. *Appl. Soil Ecol.* **2020**, *154*, 103641. [CrossRef]
20. Domeignoz-Horta, L.A.; Shinfuku, M.; Junier, P.; Poirier, S.; Verrecchia, E.; Sebag, D.; DeAngelis, K.M. Direct evidence for the role of microbial community composition in the formation of soil organic matter composition and persistence. *ISME Commun.* **2021**, *1*, 64. [CrossRef]
21. Marinari, S.; Bonifacio, E.; Moscatelli, M.C.; Falsone, G.; Antisari, L.V.; Vianello, G. Soil development and microbial functional diversity: Proposal for a methodological approach. *Geoderma.* **2013**, *192*, 437–445. [CrossRef]
22. Zhang, J.; Liu, Y.-X.; Zhang, N.; Hu, B.; Jin, T.; Xu, H.; Qin, Y.; Yan, P.; Zhang, X.; Guo, X.; et al. *NRT1.1B* is associated with root microbiota composition and nitrogen use in field-grown rice. *Nat. Biotechnol.* **2019**, *37*, 676–684. [CrossRef] [PubMed]
23. Yang, J.; Kloepper, J.W.; Ryu, C.M. Rhizosphere bacteria help plants tolerate abiotic stress. *Trends Plant Sci.* **2009**, *14*, 1–4. [CrossRef] [PubMed]
24. Kwak, M.J.; Kong, H.G.; Choi, K.; Kwon, S.K.; Song, J.; Lee, J.; Lee, P.A.; Choi, S.Y.; Seo, M.; Lee, H.J.; et al. Rhizosphere microbiome structure alters to enable wilt resistance in tomato. *Nat. Biotechnol.* **2018**, *36*, 1117. [CrossRef] [PubMed]
25. Castrillo, G.; Teixeira, P.J.P.L.; Paredes, S.H.; Law, T.F.; de Lorenzo, L.; Feltcher, M.E.; Finkel, O.M.; Breakfield, N.W.; Mieczkowski, P.; Jones, C.D.; et al. Root microbiota drive direct integration of phosphate stress and immunity. *Nature.* **2017**, *543*, 513–518. [CrossRef]
26. Zhalnina, K.; Louie, K.B.; Hao, Z.; Mansoori, N.; da Rocha, U.N.; Shi, S.; Cho, H.; Karaoz, U.; Loqué, D.; Bowen, B.P.; et al. Dynamic root exudate chemistry and microbial substrate preferences drive patterns in rhizosphere microbial community assembly. *Nat. Microbiol.* **2018**, *3*, 470–480. [CrossRef]
27. Bulgarelli, D.; Garrido-Oter, R.; Münch, P.C.; Weiman, A.; Dröge, J.; Pan, Y.; McHardy, A.C.; Schulze-Lefert, P. Structure and function of the bacterial root microbiota in wild and domesticated barley. *Cell Host Microbe* **2015**, *17*, 392–403. [CrossRef]
28. Ding, W.; He, P.; Zhang, J.; Liu, Y.; Xu, X.; Ullah, S.; Cui, Z.; Zhou, W. Optimizing rates and sources of nutrient input to mitigate nitrogen, phosphorus, and carbon losses from rice paddies. *J. Clean. Prod.* **2020**, *256*, 120603. [CrossRef]
29. Liao, H.; Qin, F.; Wang, K.; Zhang, Y.; Hao, X.; Chen, W.; Huang, Q. Long-term chemical fertilization-driving changes in soil autotrophic microbial community depresses soil CO_2 fixation in a Mollisol. *Sci. Total Environ.* **2020**, *748*, 141317. [CrossRef]

30. Ouyang, Y.; Reeve, J.R.; Norton, J.M. Soil enzyme activities and abundance of microbial functional genes involved in nitrogen transformations in an organic farming system. *Biol. Fert. Soils* **2018**, *54*, 437–450. [CrossRef]
31. Han, J.; Dong, Y.; Zhang, M. Chemical fertilizer reduction with organic fertilizer effectively improve soil fertility and microbial community from newly cultivated land in the Loess Plateau of China. *Appl. Soil Ecol.* **2021**, *165*, 103966. [CrossRef]
32. Lazcano, C.; Gómez-Brandón, M.; Revilla, P.; Domínguez, J. Short-term effects of organic and inorganic fertilizers on soil microbial community structure and function. *Biol. Fert. Soils* **2013**, *49*, 723–733. [CrossRef]
33. Kamaa, M.; Mburu, H.; Blanchart, E.; Chibole, L.; Chotte, J.-L.; Kibunja, C.; Lesueur, D. Effects of organic and inorganic fertilization on soil bacterial and fungal microbial diversity in the Kabete long-term trial, Kenya. *Biol. Fert. Soils* **2011**, *47*, 315–321. [CrossRef]
34. Feng, K.; Zhang, Z.; Cai, W.; Liu, W.; Xu, M.; Yin, H.; Wang, A.; He, Z.; Deng, Y. Biodiversity and species competition regulate the resilience of microbial biofilm community. *Mol. Ecol.* **2017**, *26*, 6170–6182. [CrossRef]
35. Rui, Z.; Lu, X.; Li, Z.; Lin, Z.; Lu, H.; Zhang, D.; Shen, S.; Liu, X.; Zheng, J.; Drosos, M.; et al. Macroaggregates serve as micro-hotspots enriched with functional and networked microbial communities and enhanced under organic/inorganic fertilization in a paddy topsoil from Southeastern China. *Front. Microbiol.* **2022**, *13*, 909. [CrossRef]
36. Lupatini, M.; Suleiman, A.K.A.; Jacques, R.J.S.; Antoniolli, Z.I.; Ferreira, A.d.S.; Kuramae, E.E.; Roesch, L.F.W. Network topology reveals high connectance levels and few key microbial genera within soils. *Front. Env. Sci.* **2014**, *2*, 10. [CrossRef]
37. Fan, K.; Weisenhorn, P.; Gilbert, J.A.; Chu, H. Wheat rhizosphere harbors a less complex and more stable microbial co-occurrence pattern than bulk soil. *Soil Biol. Biochem.* **2018**, *125*, 251–260. [CrossRef]
38. Banerjee, S.; Schlaeppi, K.; van der Heijden, M.G.A. Keystone taxa as drivers of microbiome structure and functioning. *Nat. Rev. Microbiol.* **2018**, *16*, 567–576. [CrossRef]
39. Liu, S.; Yu, H.; Yu, Y.; Huang, J.; Zhou, Z.; Zeng, J.; Chen, P.; Xiao, F.; He, Z.; Yan, Q. Ecological stability of microbial communities in Lake Donghu regulated by keystone taxa. *Ecol. Indic.* **2022**, *136*, 108695. [CrossRef]
40. Ji, L.; Si, H.; He, J.; Fan, L.; Li, L. The shifts of maize soil microbial community and networks are related to soil properties under different organic fertilizers. *Rhizosphere* **2021**, *19*, 100388. [CrossRef]
41. Munyaka, P.M.; Eissa, N.; Bernstein, C.N.; Khafipour, E.; Ghia, J.E. Antepartum antibiotic treatment increases offspring susceptibility to experimental colitis: A role of the gut microbiota. *PLoS ONE* **2015**, *10*, e0142536.
42. Li, Z.; Zu, C.; Wang, C.; Yang, J.; Yu, H.; Wu, H. Different responses of rhizosphere and non-rhizosphere soil microbial communities to consecutive Piper nigrum L. monoculture. *Sci. Rep.* **2016**, *6*, 35825. [CrossRef] [PubMed]
43. Morugán, A.; Pérez-Rodríguez, P.; Insolia, E.; Gómez, D.; Fernández-Calviño, D.; Zornoza, R. The impact of crop diversification, tillage and fertilization type on soil total microbial, fungal and bacterial abundance: A worldwide meta-analysis of agricultural sites. *Agr. Ecosyst. Environ.* **2022**, *329*, 107867. [CrossRef]
44. Jeyamangalam, F. Improvement of soil physical properties with organic manure. *J. Res. Sci.* **2014**, *2*, 2278–9073.
45. Yang, G.; Li, F.; Tian, L.; He, X.; Gao, Y.; Wang, Z.; Ren, F. Soil physicochemical properties and cotton (*Gossypium hirsutum* L.) yield under brackish water mulched drip irrigation. *Soil Till. Res.* **2020**, *199*, 104592. [CrossRef]
46. Liu, Z.; Ma, X.; He, N.; Zhang, J.; Wu, J.; Liu, C. Shifts in microbial communities and networks are correlated with the soil ionome in a kiwifruit orchard under different fertilization regimes. *Appl. Soil Ecol.* **2020**, *149*, 103517. [CrossRef]
47. Zhang, Y.; Yan, J.; Rong, X.; Han, Y.; Yang, Z.; Hou, K.; Zhao, H.; Hu, W. Responses of maize yield, nitrogen and phosphorus runoff losses and soil properties to biochar and organic fertilizer application in a light-loamy fluvo-aquic soil. *Agr. Ecosyst. Environ.* **2021**, *314*, 107433. [CrossRef]
48. Tian, S.; Zhu, B.; Yin, R.; Wang, M.; Jiang, Y.; Zhang, C.; Li, D.; Chen, X.; Kardol, P.; Liu, M. Organic fertilization promotes crop productivity through changes in soil aggregation. *Soil Biol. Biochem.* **2022**, *165*, 108533. [CrossRef]
49. Hossain, E.; Mei, X.; Zhang, W.; Dong, W.; Yan, Z.; Liu, X.; Rachit, S.; Gopalakrishnan, S.; Liu, E. Substitution of chemical fertilizer with organic fertilizer affects soil total nitrogen and its fractions in Northern China. *Int. J. Env. Res. Pub. Health* **2021**, *18*, 12848.
50. Xu, F.; Liu, Y.; Du, W.; Li, C.; Xu, M.; Xie, T.; Yin, Y.; Guo, H. Response of soil bacterial communities, antibiotic residuals, and crop yields to organic fertilizer substitution in North China under wheat–maize rotation. *Sci. Total Environ.* **2021**, *785*, 147248.
51. Chen, M.; Zhang, S.; Liu, L.; Wu, L.; Ding, X. Combined organic amendments and mineral fertilizer application increase rice yield by improving soil structure, P availability and root growth in saline-alkaline soil. *Soil Till. Res.* **2021**, *212*, 105060. [CrossRef]
52. Holland, J.E.; White, P.J.; Glendining, M.J.; Goulding, K.W.T.; McGrath, S.P. Yield responses of arable crops to liming—An evaluation of relationships between yields and soil pH from a long-term liming experiment. *Eur. J. Agron.* **2019**, *105*, 176–188. [CrossRef]
53. van Bruggen, A.H.C.; Sharma, K.; Kaku, E.; Karfopoulos, S.; Zelenev, V.V.; Blok, W.J. Soil health indicators and Fusarium wilt suppression in organically and conventionally managed greenhouse soils. *Appl. Soil Ecol.* **2015**, *86*, 192–201. [CrossRef]
54. van der Bom, F.; Nunes, I.; Raymond, N.S.; Hansen, V.; Bonnichsen, L.; Magid, J.; Nybroe, O.; Jensen, L.S. Long-term fertilisation form, level and duration affect the diversity, structure and functioning of soil microbial communities in the field. *Soil Biol. Biochem.* **2018**, *122*, 91–103.
55. Liu, J.; Shu, A.; Song, W.; Shi, W.; Li, M.; Zhang, W.; Li, Z.; Liu, G.; Yuan, F.; Zhang, S.; et al. Long-term organic fertilizer substitution increases rice yield by improving soil properties and regulating soil bacteria. *Geoderma* **2021**, *404*, 115287.
56. Zhang, F.; Zhu, Z.; Yang, X.; Ran, W.; Shen, Q. Trichoderma harzianum T-E5 significantly affects cucumber root exudates and fungal community in the cucumber rhizosphere. *Appl. Soil Ecol.* **2013**, *72*, 41–48. [CrossRef]

57. Kumar, U.; Nayak, A.K.; Shahid, M.; Gupta, V.V.; Panneerselvam, P.; Mohanty, S.; Kaviraj, M.; Kumar, A.; Chatterjee, D.; Lal, B.; et al. Continuous application of inorganic and organic fertilizers over 47 years in paddy soil alters the bacterial community structure and its influence on rice production. *Agr. Ecosyst. Environ.* **2018**, *262*, 65–75. [CrossRef]
58. Beugnon, R.; Du, J.; Cesarz, S.; Jurburg, S.D.; Pang, Z.; Singavarapu, B.; Wubet, T.; Xue, K.; Wang, Y.; Eisenhauer, N. Tree diversity and soil chemical properties drive the linkages between soil microbial community and ecosystem functioning. *ISME Commun.* **2021**, *1*, 41. [CrossRef]
59. Mishra, A.; Singh, L.; Singh, D. Unboxing the black box—One step forward to understand the soil microbiome: A systematic review. *Microb. Ecol.* **2023**, *85*, 669–683. [CrossRef]
60. Zhang, S.; Song, X.; Li, N.; Zhang, K.; Liu, G.; Li, X.; Wang, Z.-Z.; He, X.-B.; Wang, G.-F.; Shao, H.-F. Influence of high-carbon basal fertiliser on the structure and composition of a soil microbial community under tobacco cultivation. *Res. Microbiol.* **2018**, *169*, 115–126. [CrossRef]
61. Wu, K.; Yuan, S.; Wang, L.; Shi, J.; Zhao, J.; Shen, B.; Shen, Q. Effects of bio-organic fertilizer plus soil amendment on the control of tobacco bacterial wilt and composition of soil bacterial communities. *Biol. Fert. Soils* **2014**, *50*, 961–971. [CrossRef]
62. Jin, L.; Jin, N.; Wang, S.; Li, J. Changes in the microbial structure of the root soil and the yield of Chinese baby cabbage by chemical fertilizer reduction with bio-organic fertilizer application. *Microbiol. Spectr.* **2022**, *10*, e0121522. [CrossRef] [PubMed]
63. Fierer, N.; Lauber, C.; Ramirez, K.; Zaneveld, J.; Bradford, M.; Knight, R. Comparative metagenomic, phylogenetic and physiological analyses of soil microbial communities across nitrogen gradients. *ISME J.* **2012**, *6*, 1007–1017. [CrossRef] [PubMed]
64. Cui, J.; Zhu, R.; Wang, X.; Xu, X.; Ai, C.; He, P.; Liang, G.; Zhou, W.; Zhu, P. Effect of high soil C/N ratio and nitrogen limitation caused by the long-term combined organic-inorganic fertilization on the soil microbial community structure and its dominated SOC decomposition. *J. Environ. Manag.* **2022**, *303*, 114155.
65. Ma, M.; Jiang, X.; Wang, Q.; Ongena, M.; Wei, D.; Ding, J.; Guan, D.; Cao, F.; Zhao, B.; Li, J. Responses of fungal community composition to long-term chemical and organic fertilization strategies in Chinese Mollisols. *Microbiologyopen* **2018**, *7*, e00597. [CrossRef]
66. Yelle, D.; Ralph, J.; Lu, F.; Hammel, K. Evidence for cleavage of lignin by a brown rot basidiomycete. *Environ. Microbiol.* **2008**, *10*, 1844–1849. [CrossRef]
67. Grządziel, J.; Gałązka, A. Fungal biodiversity of the most common types of polish soil in a long-term microplot experiment. *Front. Microbiol.* **2019**, *10*, 6. [CrossRef]
68. Wang, Q.; Song, R.; Fan, S.; Coleman, J.J.; Xu, X.; Hu, X. Diversity of Fusarium community assembly shapes mycotoxin accumulation of diseased wheat heads. *Mol. Ecol.* **2022**, *32*, 2504–2518. [CrossRef] [PubMed]
69. Jiao, S.; Wang, J.; Wei, G.; Chen, W.; Lu, Y. Dominant role of abundant rather than rare bacterial taxa in maintaining agro-soil microbiomes under environmental disturbances. *Chemosphere* **2019**, *235*, 248–259. [CrossRef]
70. Chen, W.; Wang, J.; Chen, X.; Meng, Z.; Xu, R.; Duoji, D.; Zhang, J.; He, J.; Wang, Z.; Chen, J.; et al. Soil microbial network complexity predicts ecosystem function along elevation gradients on the Tibetan Plateau. *Soil Biol. Biochem.* **2022**, *172*, 108766. [CrossRef]
71. Wu, C.; Ma, Y.; Wang, D.; Shan, Y.; Song, X.; Hu, H.; Ren, X.; Ma, X.; Cui, J.; Ma, Y. Integrated microbiology and metabolomics analysis reveal plastic mulch film residue affects soil microorganisms and their metabolic functions. *J. Hazard. Mater.* **2022**, *423*, 127258. [CrossRef] [PubMed]
72. Yuan, M.M.; Guo, X.; Wu, L.; Zhang, Y.; Xiao, N.; Ning, D.; Shi, Z.; Zhou, X.; Wu, L.; Yang, Y.; et al. Climate warming enhances microbial network complexity and stability. *Nat. Clim. Chang.* **2021**, *11*, 343–348. [CrossRef]
73. Luo, J.; Banerjee, S.; Ma, Q.; Liao, G.; Hu, B.; Zhao, H.; Li, T. Organic fertilization drives shifts in microbiome complexity and keystone taxa increase the resistance of microbial mediated functions to biodiversity loss. *Biol. Fert. Soils* **2023**, *59*, 441–458. [CrossRef]
74. Nakasaki, K.; Hirai, H.; Mimoto, H.; Quyen, T.N.M.; Koyama, M.; Takeda, K. Succession of microbial community during vigorous organic matter degradation in the primary fermentation stage of food waste composting. *Sci. Total Environ.* **2019**, *671*, 1237–1244. [CrossRef]
75. Zheng, Q.; Hu, Y.; Zhang, S.; Noll, L.; Böckle, T.; Dietrich, M.; Herbold, C.W.; Eichorst, S.A.; Woebken, D.; Richter, A.; et al. Soil multifunctionality is affected by the soil environment and by microbial community composition and diversity. *Soil Biol. Biochem.* **2019**, *136*, 107521. [CrossRef] [PubMed]

agronomy

Article

Microbial Communities and Soil Respiration during Rice Growth in Paddy Fields from Karst and Non-Karst Areas

Junbo Zhou [1], Zhenjiang Jin [1,2,3,*], Wu Yuan [1], Weijian Chen [1], Xuesong Li [1], Liyuan Xiong [1] and Guanwen Cheng [1,2,3]

1 Environmental Science and Engineering College, Guilin University of Technology, Guilin 541004, China; junbozhou0901@163.com (J.Z.)
2 Guangxi Key Laboratory of Environmental Pollution Control Theory and Technology, Guilin University of Technology, Guilin 541004, China
3 Collaborative Innovation Center for Water Pollution Control and Water Safety in Karst Area, Guilin University of Technology, Guilin 541004, China
* Correspondence: 2004041@glut.edu.cn; Tel.: +86-773-13647737636; Fax: +86-773-5891059

Abstract: Soil microorganism and their relationships with soil respiration in paddy systems in karst areas (KA) of southern China is important for understanding the mechanisms of greenhouse gas emission reduction. Soils were collected from the tillage layer (0–20 cm) during the rice growing season from KA and non-karst areas (NKA) (red soils) from the Guilin Karst Experimental Site in China. Community structures and inferred functionalities of bacteria and fungi were analyzed using the high-throughput sequencing techniques, FAPROTAX and FUNGuild. A bacterial–fungal co-occurrence network was constructed and soil respiration was measured using dark box-gas chromatography and built their relationships. The results indicated that soil respiration was significantly lower in KA than in NKA. Principal component analysis indicated that bacterial and fungal community structures significantly differed between KA and NKA. The OTU ratio of fungi to bacteria (F/B) was positively correlated with soil respiration ($p = 0.044$). Further, the key network microorganisms were OTU69 and OTU1133 and OTU1599 in the KA. Soil respiration negatively correlated with Acidobacteria Gp6, dung saprotroph-endophyte-litter saprotroph-undefined saprotroph, aerobic nitrite oxidizers and nitrifier in KA ($p < 0.05$). Overall, this study demonstrated that soil respiration was reduced when soil microorganisms shifted from bacterial to fungal dominance during the rice growing season in KA.

Keywords: high-throughput sequencing; FAPROTAX and FUNGuild; carbon dioxide emission flux; the ratio of fungi to bacteria; functional group; co-occurrence network

Citation: Zhou, J.; Jin, Z.; Yuan, W.; Chen, W.; Li, X.; Xiong, L.; Cheng, G. Microbial Communities and Soil Respiration during Rice Growth in Paddy Fields from Karst and Non-Karst Areas. *Agronomy* **2023**, *13*, 2001. https://doi.org/10.3390/agronomy13082001

Academic Editors: Yong-Xin Liu and Peng Yu

Received: 30 June 2023
Revised: 21 July 2023
Accepted: 24 July 2023
Published: 28 July 2023

1. Introduction

Karst ecosystem is an ecosystem restricted by geological background which accounts for one-third of the national land area in China [1]. In karst area (KA) with high temperature and humidity, the dissolution of carbonate brings a large amount of calcium ion (Ca^{2+}) and HCO_3^- into the soil. HCO_3^- combines with H^+ in the water to generate Carbon dioxide (CO_2), which increases soil CO_2 concentration [2]. The carbonate weathering consumes CO_2 and represent large sinks for atmospheric CO_2 that can influence global carbon balance [3,4]. Thereby, carbon cycles during the weathering of carbonate rocks are very important [5], and biological activities catalyze and regulate this process of weathering, thus resulting in alkaline soils rich in calcium [6]. This process also alters soil microbial communities involved in carbon cycling within KA.

CO_2 is the most important greenhouse gases (GHG) [7] and the global CO_2 imbalance is one of the most critical problems for the global carbon cycle [8]. Soil respiration was an important process of CO_2 emission and initially used to describe soil metabolic processes [9], and soil aeration, temperature, moisture, and microbial communities influence respiration [10]. Microbial respiration accounts for more than 80% of all soil respiration [11], which indicated that microbes were the main contributor to soil CO_2 emissions [12]. In general, long-term inundation slows down the degradation process of soil organic matter in paddy field, which is conducive to the accumulation of soil organic carbon (SOC). The degradable organic carbon in paddy soil increase aboveground and underground biomass of rice [13], thus enhancing the outflow of root exudates and increasing the microbial biomass [14]. Both decreasing the ratio of carbon to nitrogen [15] and increasing the exogenous organic matter [16] retard soil respiration and increase SOC accumulation in paddy soils. The soil characterized by neutral alkalinity, the metal ions (including Ca^{2+}) can not only combine with inorganic carbon [17,18], but also with carboxyl group of soil organic matter, which changed the microbial community and its utilization of organic matter, thus affecting soil respiration [14,19]. Consequently, soil respiration and its relation with microbe in rice fields of KA are critical for the accumulation and stability of SOC.

Both bacteria and fungi are important microbial taxa that synergistically interact with soil respiration during rice cultivation [20,21] and determine the carbon sequestration potential of paddy fields [22]. The inter-roots of rice colonization can rapidly limit oxygen and establish anaerobic zones that lead to some microbial populations performing the anaerobic respiration of nitrates and carbonates, thereby accelerating the release of photosynthetically fixed carbon through roots into the surrounding soils [23]. The input of carbon substrates drives the structural and functional changes in soil microbial communities, making the assimilated carbon of microorganisms more stable than aboveground and root-derived carbon [24].

SOC played an important pole in shaping the pattern of soil bacteria and fungi community structures [25], which indicated that increasing biomasses of root and litter stimulated microbial growth in KA [26]. In addition, bacterial and fungal populations are able to increase the stability of calcareous SOC [27] that might reduce the decomposition of soil organic matter. For example, *Flavobacterium* and *Lysobacter* dominated the karst area communities and exhibited relative abundances of 1.24–14.73%,which *Flavobacterium* can decompose organic matter and *Lysobacter* can synthesize organic matter using CO_2 as a carbon source [25]. Further, numerous bacterial and fungal taxa (such as *Bradyrhizobium, Herbaspirillum, Cellulomonas, Blastococcus,* some endophytes and ectomycorrhizae) that are symbiotic with moss plants in KA are able to increase the peroxidase activity of mosses, thereby enhancing the photosynthetic efficiency of the plants [28]. Thus, bacteria and fungi in karst ecosystem play critical roles in dynamic changes of carbon fixation and CO_2 emissions. In previous studies, we reported that soil bacterial communities of rice field in KA are considerably different with those in NKA at the same latitude, which is mainly attributed to the main ecological factors such as SOC, total nitrogen, pH and so on [25]. Our study also found that there are significant differences in the bacterial community structure, key groups, and functional groups across the three particle size aggregates between KA and NKA [29]. At the same time, we monitored the soil respiration of paddy fields during fallow and found that the soil respiration in KA was far lower than that in NKA (Supplementary Figure S1). During rice growth, soil microbial community structure may be more variable due to the influence of aboveground litter, roots and their secretions, which will affect soil respiration [30–32]. However, if there was difference in the soil respiration of paddy field during rice growth between KA and NKA? Which specific microorganisms are closely related to soil respiration? and which are the top microorganisms in the two different areas? The above problems are rarely reported.

In this study, the karst experimental site in Maocun village of Guangxi Zhuang Autonomous Region, China was as research site. The paddy fields across the entire rice growing period in KA and NKA were selected to investigate: (1) dynamic changes in bacterial and fungal abundances during rice growth using real-time PCR and high-throughput sequencing; (2) in situ soil respiration in rice fields across the growth season using static box-meteorological chromatography; (3) predicted functional groups of bacteria and fungi; and (4) co-occurrence networks of fungal and bacterial communities. These studies were used to analyze the relationship between major microbial groups and soil respiration and to help us understand the process and mechanism of microbial communitiy affecting soil respiration in KA during rice growth.

2. Materials and Methods

2.1. Soil Collection and Gas Sampling

2.1.1. Soil Sample Collection and Soil Physicochemical Properties Analysis

From June to September 2019, rice fields from a typical karst area (KA) (25°08′30″ N, 110°31′28″ E) and a non-karst area (NKA) (25°10′51″ N, 110°31′35″ E) were selected as studying fields within the karst experimental site in Maocun village of Guangxi Zhuang Autonomous Region, China. The soil type in the KA is a limestone brown soil, whereas the soil type in NKA is a silicate red soil (Soil forming parent material are mainly sandstone and granite). Each field was divided into three replicate plots with the same area (KA: 42.20 m^2 and NKA: 46.67 m^2) using by ridge. Each plot was subjected to uniform irrigation, fertilization, and management practices. Three random sampling points were used in each plot. Paddy fields were planted with single-season rice, and the rice growing period was 94 d. Compound fertilizers (N-P_2O_5-K_2O) containing 18% each of nitrogen, phosphorus and potassium were applied as base fertilizer to the fields at 7.56 kg (KA) and 8.53 kg (NKA) before transplanting seedlings. In addition, 5.70 kg (KA) and 6.40 kg (NKA) of compound fertilizer were applied in each plot as follow-up fertilizer on 14 July. A total of 1 kg of soil was collected from each sample point at a sampling depth of 0–20 cm; the soil samples from the three sampling points were equally and uniformly mixed into a single 3 kg sample. A total of 48 soil samples were collected on eight dates with intervals of about 14 d (with the exception of the first sampling interval of 7 d). The sample from 4 June represented the rice planting date, 12 June was from the seedling stage, 26 June and 10 July were from the tillering stage, 24 July was from the nodulation stage, 7 August was from the gestation stage, 21 August was from the tasseling stage, and 4 September was from the maturity stage.

Soil physicochemical properties were analyzed by routine methods [33]. Briefly, soil moisture content was measured using the drying method. pH was measured using CO_2-free distilled water as the leaching agent after mixing soil with water at a 1:2.5 ratio, followed by measurement with a precision pH meter (model: IS128C). SOC was measured using the sulfuric acid potassium dichromate external heating method. Soil total nitrogen (TN) and hydrolyzed nitrogen (AN) measured using the concentrated sulfuric acid digestion-Kjeldahl method and the alkali hydrolysis diffusion method. Total phosphorus (TP) and available phosphorus (AP) determined by the sodium carbonate melting method and the hydrochloric acid-ammonium fluoride method. Soluble organic carbon (DOC) was extracted by shaking a soil-water mix (at a ratio of 10:1), filtering through a 0.45 μm fiber membrane, and analysis on a multi N/C3100 total organic carbon analyzer (Analytik Jena, Jena, Germany). Cation exchange capacity (CEC) was measured using the EDTA-ammonium fast method. An optima 7000 DV inductively coupled plasma emission spectrometer (Perkin Elmer, Waltham, MA, USA) (ICP-OES) was used to measure the exchangeable calcium (Ca^{2+}) and magnesium (Mg^{2+}) ion concentrations.

2.1.2. Measurement of Soil Respiration

Static dark boxes with 50 cm × 50 cm × 50 cm were used as a field gas collection device, with the boxes raised as plant heights changed. The box was constructed polyethylene (PVC) covered with foam and reflective material and had a fan for mixing gases inside the box installed on the top along with a gas collection port. The gas was collected in a vacuum bag using a syringe from 9:00 to 9:30 on the above-mentioned soil sample collection date. Each sample collection point needs to be separated by 10 min for continuous gas collection. Temperature was measured with a thermometer (JM624, Guangzhou, China) at 5 cm depth. Three rice plants were randomly collected from each sample site during rice harvesting and then dried in oven at 105 °C until achieving. Gas samples were immediately brought to the laboratory for quantification of CO_2 using an Agilent 7890 B gas chromatograph-mass spectrometer (Agilent, Santa Clara, CA, USA) with a hydrogen flame ion detector (FID). An external standard working curve was plotted after running every 48 samples. The gas emission rate (*dc/dt*) was derived from the slope of four consecutive sample concentration values using linear regression analysis. Greenhouse gas emission flux was calculated using the following equation:

$$F = H\frac{MP}{R(273+T)}\frac{dc}{dt} \quad (1)$$

where F is the gas emission flux (units of mg·m^{-2}·h^{-1}); H is the height of the sampling box (m); M is the molar mass of gas (g·mol^{-1}); P is the air pressure at the sampling point (Pa); R is the universal gas constant (8.314, Pa·m^3·mol^{-1}·k^{-1}); T is the average temperature inside the box at the time of sampling (°C) and dc/dt is the gas emission rate (uL·L^{-1}·min^{-1}).

Heterotrophic respiration (R_h) was then calculated using the following equations:

$$NPP = NPP_{\text{seeds}} + NPP_{\text{straw}} + NPP_{\text{roots}} + NPP_{\text{apoplast}} + NPP_{\text{root sediment}} \quad (2)$$

$$GPP = NPP + R_{\text{a}} \quad (3)$$

$$R_{\text{h}} = R_{\text{e}} - R_{\text{a}} \quad (4)$$

where NPP is the total carbon added to the above- and below-ground portions of the plant during the entire growth cycle (kg·hm^{-2}); NPP_{seeds} and NPP_{straw} are the estimated biomass from the harvested plants after drying, while NPP_{root}, NPP_{apoplast} and $NPP_{\text{root sediment}}$ were estimated by referencing previous studies [30–32]. The equation assumed an above-ground/root system ratio for rice of 1.0/0.1, and apoplankton coverage of 5% and 8% of the upper and root dry biomass, respectively, and root sediment accounted for 15% of the total plant biomass. In Equation (3), GPP is total primary productivity, R_{a} is autotrophic respiration (unit: kg·hm^{-2}), and the NPP/GPP ratio is 0.58 [34]. In Equation (4), R_e is ecosystem respiration and R_{h} is heterotrophic respiration (kg·hm^{-2}).

The cumulative greenhouse gas (single-season rice) emissions were calculated as follows:

$$Ec = \left[\frac{F_1 + F_n}{2} + \sum_{i=1}^{n}\left(\frac{F_i + F_{i+1}}{2}\right) \times (t_{i+1} - t_i)\right] \times 24 \times 0.01 \times a \quad (5)$$

where E_{c} is cumulative greenhouse gas (single-season rice) emissions (kg·hm^{-2}); n is the number of observations during a single rice season; i is the sampling interval, F_i and F_{i+1} are the GHG emission fluxes (mg·m^{-2}·h^{-1}) at the ith and i + 1th sampling, respectively; F_1 and F_n are the GHG emission fluxes at the first and last sampling times, respectively; $t_{\text{i+1}}$ and t_{i} are the time intervals (d) between the i + 1th and ith sampling; and a is a conversion factor of 98/94.

The cumulative net CO_2 emission, F_{CO2}, was calculated as follows:

$$F_{CO2} = Ec - Ra \quad (6)$$

where F_{CO2} is the cumulative CO_2 emission minus plant autotrophic respiration (kg·hm^{-2}).

2.2. High-Throughput DNA Sequencing

The genomic DNA of total 48 soil samples was extracted using an EZNA Soil DNA extraction kit (Omega, Norcross, GA, USA) according to the manufacturer's instructions (0.5 g per sample). The V3-V4 and ITS3-ITS4 amplicons of 16S rRNA genes and an internal transcribed spacer between ribosomal genes were amplified using a KAPA HiFi Hot Start Ready Mix (2x, TaKaRa Bio Inc., Kusatsu City, Japan). Universal PCR primers (PAGE purified) were used including the bacterial PCR forward primer (CCTACGGGNGGCWGCAG) and reverse primer (GACTACHVGGGTATCTAATCC) in addition to the fungal forward primer (GCATCGATGAAGAACGCAGC) and fungal reverse primer (TCCTCCGCTTATTGATATGC) [35,36] (primer sequences provided by Sangon Biotech company(Shanghai, China)). Reactions included 2 μL of DNA (10 ng μL^{-1}); 1 μL each of PCR forward and reserve primers (10 μmol each); 15 μL of 2× KAPA HiFi Hot Start Ready Mix; and total reaction 30 μL. PCR plates were sealed and PCRs were performed in a thermal cycler (Applied Biosystems 9700, Waltham, MA, USA) using the following steps: pre-denaturation at 94 °C for 3 min, followed by denaturation at 95 °C for 20 s, 20 s of annealing at 55 °C, and a final extension at 72 °C for 30 s. The above denaturation, annealing, and extension steps were repeated for 20 cycles. The PCR products were detected gel electrophoresis using a 1% agarose gel in TBE (Tris-H^3BO^3-EDTA) buffer and stain with ethidium bromide (EB), followed by visualization with UV light. Samples were used to construct libraries using the universal Illumina adapters and indices. Sequencing was then performed on an Illumina MiSeq system (Illumina MiSeq, San Diego, CA, USA). The paired-end (PE) reads were obtained via paired-end sequencing and first combined using the fast length adjustment of short reads (FLASH) software package, which was also used to quality filter the datasets and obtain high-quality sequences for subsequent analysis. Using PRINSEQ to cut off the bases with quality value below 20 in the tail of reads, and set a 10 bp window. If the average quality value in the window is lower than 20, cut off the back-end bases from the window, filter the N-containing sequences and short sequences after quality control and finally filtered out the sequences with low complexity. After sequencing, chimeras were removed using the UCHIME programs. Quality-filtered sequences from each sample were taxonomically classified using the RDP classifier accessed on 28 June 2020. (RDP 16S database: http://rdp.cme.msu.edu/misc/resources.jsp and RDP ITS database: http://rdp.cme.msu.edu/misc/resources.jsp). A total of 211,930 bacterial sequences with 211,696 valid sequences and 67,106 fungal sequences with 66,988 valid sequences were detected. The datasets generated during the current study have been uploaded to the Sequence Read Archive (SRA), and had the accession number PRJNA763299.

2.3. Data Processing

Data processing was performed using the Excel 2016 software program. Correlation line plots were produced from fungal and bacterial OTUs in addition to fungal/bacterial OTU ratios in association with in situ CO_2 fluxes for the 8 sampling dates. Histograms of fungal and bacterial OTUs numbers, in addition to soil respiration rates, were also established for the 8 and 14 sampling dates, respectively. The above were all using Origin 2017 software. Classes with greater than 1% relative abundance were designated as dominant classes, and the Origin 2017 software established relative abundance histogram. OTUs with greater than 1% relative abundance were designated as dominant OTUs, and the CoNet plug-in for the Cytoscape 3.7.1 software was used to construct a bacterial-fungal co-occurrence network [37]. Four statistical algorithms were used: Pearson's correlation, Spearman correlation, Bray-Curtis dissimilarity and Kullback-Leibler dissimilarity. The Brown method was used to integrate the p values. The data with significant correlation ($p < 0.05$) were selected for subsequent analysis. The Benjamin–Hochberg method was used as the multiple test correction. The MCODE plug-in for Cytoscape 3.7.1 was used to analyze network modularity using default criteria, whereas the Network Analyzer for Cytoscape 3.7.1 was used to identify the number of nodes, number of connected edges, and the connectivity within the network. Functional predictions based on the dominant fungal and bacterial OTUs were inferred using the FUNGuild and FAPROTAX databases, respectively, and the resulting functional profiles were visualized as heat maps using the R Studio 5.3.1 program. All statistical analyses were carried out with SPSS 24.0. Correlation analysis was conducted using Pearson correlation (trigonometric function converted to normal distribution) and Spearman correlation [38]. The abundance difference was conducted with independent sample t-test [39].

3. Results

3.1. Physicochemical Properties of Paddy Soils in Karst and Non-Karst Areas

The 11 physicochemical indicators of paddy soils in KA were significantly higher than those in NKA [Table 1], which indicated that there was significantly difference between KA and NKA.

Table 1. Physicochemical properties of paddy soils in KAs and NKAs.

Site	pH (H_2O)	SOC /g·kg^{-1}	DOC /ug·L^{-1}	TN /g·kg^{-1}	AN /mg·kg^{-1}	TP /g·kg^{-1}	AP /mg·kg^{-1}	C/N /g·kg^{-1}	CEC /cmol·kg^{-1}	Ca^{2+} /cmol·kg^{-1}	Mg^{2+} /cmol·kg^{-1}
Karst area	7.40 ± 0.18 a	25.15 ± 1.03 a	261.62 ± 9.22 a	1.74 ± 0.09 a	90.21 ± 1.24 a	1.24 ± 0.02 a	22.04 ± 1.63 a	12..46 ± 0.11 a	13.79 ± 0.42 a	3.89 ± 0.04 a	1.20 ± 0.01 a
Non-karst area	5.76 ± 0.15 b	13.86 ± 1.61 b	202.78 ± 20.46 b	1.50 ± 0.04 b	85.24 ± 0.09 b	0.49 ± 0.01 b	18.63 ± 0.57 b	9.23 ± 0.47 b	6.09 ± 0.17 b	2.36 ± 0.01 b	0.48 ± 0.03 b

Note: Different lowercase letters in the same column indicate statistically significant differences between KA and NKA ($p < 0.05$). Data represent means ± standard deviation. KA, karst areas; NKA, non-karst areas; SOC, soil organic carbon; DOC, dissolved organic carbon; TN, total nitrogen; TP, total phosphorus; AP, Available phosphorus; C/N, ratio of SOC to TN; CEC, Cation exchange capacity; Ca^{2+}, exchangeable calcium; magnesium Mg^{2+}, exchangeable magnesium.

3.2. In Situ Soil Respiration and Cumulative CO_2 Emissions of Paddy Fields during Rice Growth in KA and NKA

Variation of in situ soil respiration in KA and NKA ranged from 26.86 to 1293.59 mg·m^{-2}·h^{-1} and 62.99 to 1501.21 mg·m^{-2}·h^{-1}, respectively, during rice growing where the temperature ranged from 27 to 36 °C [Figure 1]. Among the 14 sampling times evaluated across the growing period, eight of the in situ CO_2 flux values in KA were significantly lower than those in NKA, while one sample measurement was significantly higher than those in NKA, and five were not significantly different. The total cumulative CO_2 emissions across the entire growth period were 7632.82 kg·hm^{-2} and 9637.66 kg·hm^{-2} in KA and NKA, respectively. Thus, soil respiration of paddy field in KA was generally lower than that in NKA.

3.3. The Abundance of Bacteria and Fungi and the Ratio of Fungi to Bacteria (F/B) in Paddy Field Soils of KA and NKA

Among the 14 sampling times evaluated across the growing period, the bacterial abundance in the KA which was (87.52 ± 15.54) $\times 10^{10}$ ~ (242.79 ± 96.20) $\times$ 10^{10} copies·g^{-1} was significantly higher than that in the NKA which had an abundance of (2.51 ± 0.54) $\times$ 10^{10} copies·g^{-1} ~ (5.52 ± 0.87) $\times$ 10^{10} copies ·g^{-1} [Figure 2a]. The fungal abundance in the KA which was (17.26 ± 1.32) $\times$ 10^9 copies ·g^{-1} ~ (27.73 ± 0.61) $\times$ 10^9 copies·g^{-1} was significantly higher than that in the NKA which had an abundance of (1.90 ± 0.55) $\times$ 10^9 ~ (13.26 ± 1.47) $\times$ 10^9 copies g^{-1} on Jun-26 (tillering stage) [Figure 2b].

Across the entire rice growing period, the F/B was 5.76 $\times$ 10^{-2} ~ 31.1 $\times$ 10^{-2} and 1.26 $\times$ 10^{-2} ~ 2.51 $\times$ 10^{-2} in KA and NKA, respectively [Figure 3a]. The F/B in KA exhibited significantly higher than those in NKA. The F/B was also significantly negatively correlated with soil respiration when combining the data of both areas ($p = 0.044$) [Figure 3b]. Thus, soil respiration decreases with reducing F/B during the rice growing period.

3.4. Variation in Fungal and Bacterial Communities in Paddy Fields of KA and NKA

PCoA analysis based on Bray-Curtis distances revealed that 50.56% and 13.56% of the fungal community variation was explained by axes 1 and 2, respectively [Figure 4a], while 66.78% and 18.79% of bacterial community variation was explained by axes 1 and 2 [Figure 4b]. Bacteral and fungal communities generally segregated across the ordinations, indicating clear microbial differences between the KA and the NKA.

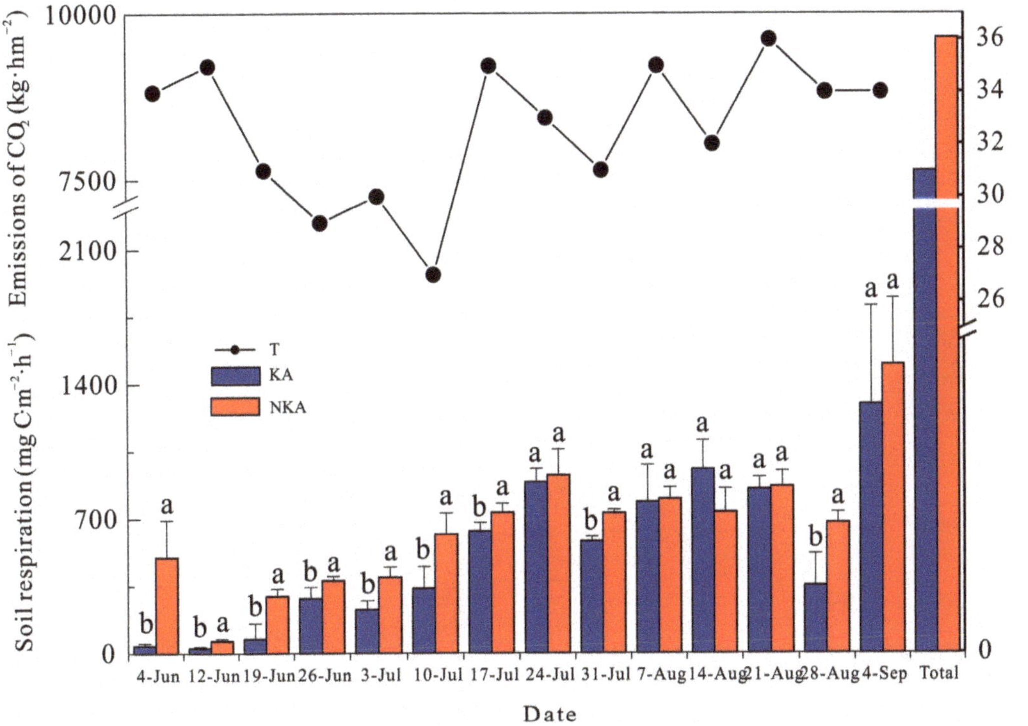

Figure 1. Situ soil respiration and cumulative CO_2 emissions of paddy fields. Note: Different lowercase letters in the same date indicate statistically significant differences between the KA and NKA ($p < 0.05$); KA, karst areas; NKA, non-karst areas.

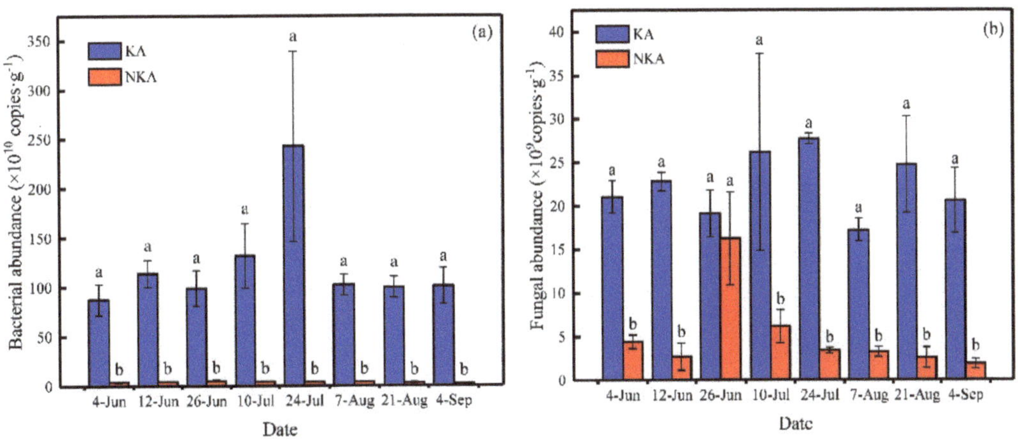

Figure 2. The abundance of soil bacteria (**a**) and fungi (**b**) based on real-time PCR during rice growth in the KA and NKA. Note: Different lowercase letters in the same date indicate statistically significant differences between KA and NKA ($p < 0.05$); KA, karst areas; NKA, non-karst areas.

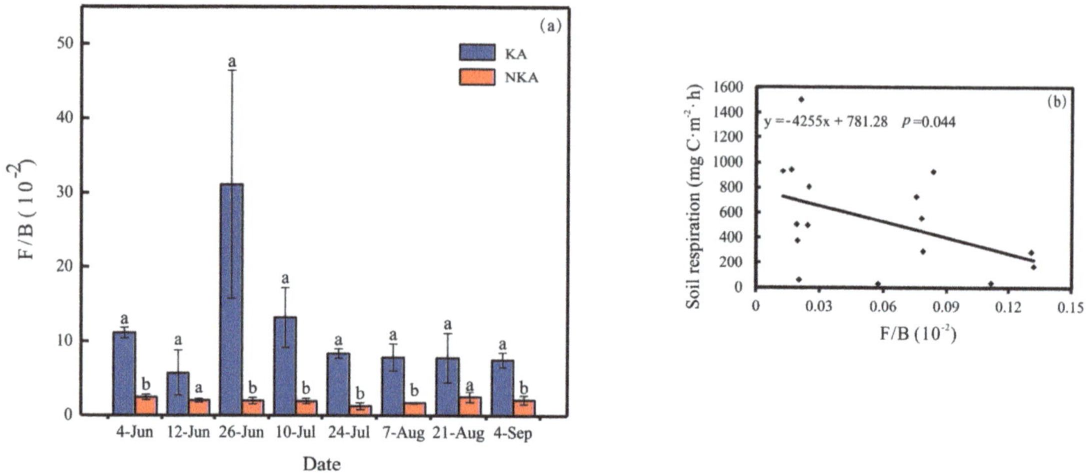

Figure 3. (**a**) The F/B during rice growth; (**b**) Correlation between soil respiration and the F/B. Note: Different lowercase letters in the same date indicate statistically significant differences between KA and NKA ($p < 0.05$); KA, karst areas; NKA, non-karst areas; F/B, ratio of fungi to bacteria.

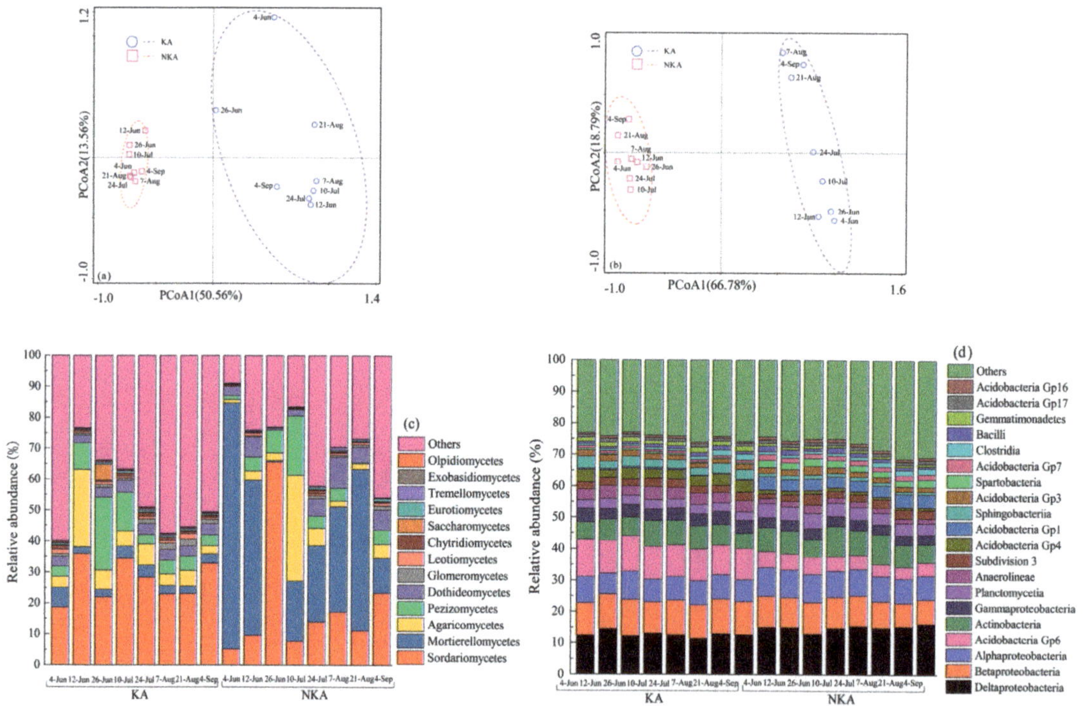

Figure 4. Principal coordinates analysis (PCoA) ordination of fungal (**a**) and bacterial (**b**) community variation based on Bray-Curtis distances; The relative abundances of the soil fungal (**c**) and bacterial (**d**) community at Class level. Note: KA, karst areas; NKA, non-karst areas.

At the class level, fungi and bacteria had a total of 13 and 20 dominant classes, respectively [Figure 4c,d]. The average relative abundances of fungal Glomeromycetes, Chytridiomycetes and Exobasidiomycetes (0.91%, 0.98% and 0.23%) in KA were significantly higher than those in NKA (0.47%, 0.28% and 0.04%). The average relative abundances of bacterial β-proteobacteria, Acidobacteria Gp6, γ-proteobacteria, Anaerolineae, Acidobacteria Gp4, Sphingobacteriia and Gemmatimonadetes (10.32%, 9.88%, 3.74%, 3.54%, 3.13%, 2.76% and 1.38%) in KA were significantly higher than those (9.07%, 4.54%, 3.18%, 1.78%, 1.01%, 1.10% and 0.73%). The above results showed that there was a significantly difference of microorganisms between KA and NKA.

3.5. Co-Occurrence Network Analysis of Fungal and Bacterial Taxa in Rice Field Soils

The distributions of dominant fungal (19 total) and bacterial (34 total) OTUs were used to construct fungal and bacterial co-occurrence networks [Figure 5a], With bacterial and fungal networks analyzed separately in [Figure 5b,c]. The characteristic path length (CPL) of the fungal-bacterial co-occurrence network was 1.59 cm (Figure 5), with a diameter of 5.00 cm, and a clustering coefficient (CC) of 0.35. The bacterial-only co-occurrence network also contained 53 nodes that were significantly correlated, comprising a total of 502 edges representing 293 and 209 positive and negative correlations, respectively. Thus, putative mutualistic or cooperative relationships between bacterial and fungal taxa (58.36%) were more apparent than putative competitive associations (41.64%). The OTUs with the highest connectivity in the fungal and bacterial networks were identified as top network taxa. The top taxa in KA soil networks included OTU69 (*Emericellopsis*) (connectivity of 31) in addition to OTUs 1599 (Acidobacteria Gp6) and 1133 (Acidobacteria Gp6) (both with a connectivity of 24), which were mutually exclusive. The top fungal and bacterial taxa in NKA were OTU139 (Sordariales) and OTU9 (*Nitrospira*) (connectivity of 26), respectively, which exhibited a positive correlation with one another.

The CPL of the bacterial correlation network was 1.38 cm [Figure 5b], while the diameter was 3.00 cm, and the clustering coefficient (CC) was 0.41. The network consisted of 34 nodes and 329 edges representing 229 positive and 100 negative correlations. The CPL of the network in module 1 was 1.254 cm, and the diameter was 3.00 cm, while the CC was 0.43, and the module comprised 29 nodes and 296 edges (196 positive and 100 negative correlations). The top bacterial OTUs, OTU1599 (Acidobacteria Gp6) and OTU9 (Nitrospira), were present in module 1. The top bacteria of KA was OTU1599 (Acidobacteria Gp6), which exhibited positive correlations with OTU1603 (Burkholderiales), OTU3528 (*Phycicoccus*), and OTU835 (ß-proteobacteria). The top bacteria in NKA included OTU9 (*Nitrospira*), which exhibited mutually exclusive relationships with OTU835 (ß-proteobacteria), OTU3528 (*Phycicoccus*), and OTU1603 (Burkholderiales).

The internal fungal correlation network CPL was 1.39 cm, its diameter was 3.00 cm, and its CC was 0.28 [Figure 5c]. The network consisted of 19 nodes and 38 edges representing 30 positive and 8 negative correlations. The network CPL for module 1 was 1.23 cm, its diameter was 2.00 cm, and its CC was 0.41. The network consisted of 9 nodes and 27 edges, with 22 positive and 5 negative correlations. The network CPL for module 2 was 1.17 cm, while its diameter was 2.00 cm, and its CC was 0.42. The network consisted of four nodes and four positively correlated edges. The top fungal taxa, OTU69 (*Emericellopsis*) and OTU139 (Sordariales), in KA were present in module 1. Both the OTUs exhibited positive correlations with OTU3379 (Gaeumannomyces), OTU81 (Sordariaceae), and OTU63 (*Fusarium*).

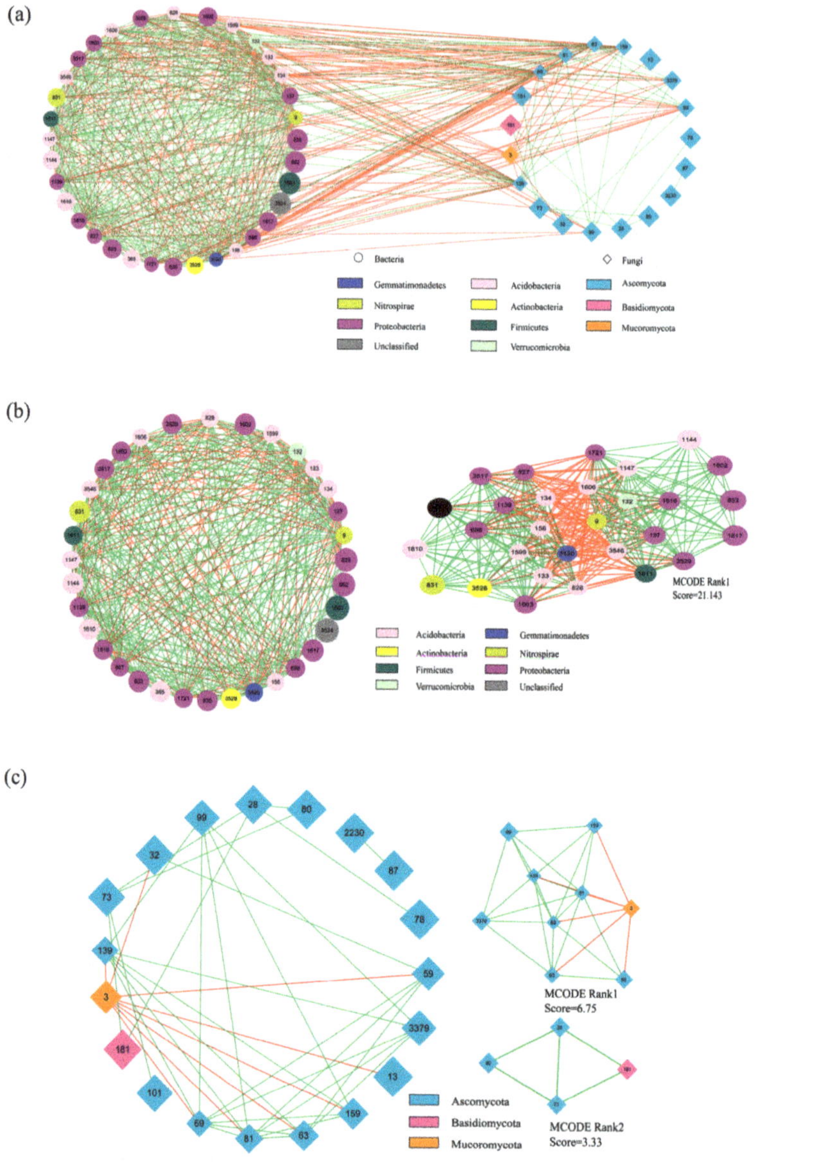

Figure 5. (**a**): Co-occurrence network analysis of dominant fungal and bacterial taxa; (**b**): Co-occurrence network and modules comprising dominant bacterial OTUs among KA and NKA soils; (**c**): Co-occurrence network and modules of dominant fungal OTUs among KA and NKA soils. Note: KA, karst areas; NKA, non-karst areas. Different colored nodes in [Figure 5a] represent different clades of bacteria and fungi, with the value on the node indicating the OTU number for the particular taxa, and node size representing the abundance of the taxa. Different colored nodes in [Figure 5b] represent the different phylum classifications for bacterial OTUs; and the value on the node represents the number of each bacterial OTU, while the size of the node indicates abundance. Different colored nodes [Figure 5c] represent different phylum classifications for fungal OTUs; and the value on the node represents the number of each OTU, while the size of the node indicates abundance. Green edges indicate positive correlations, while red edges indicate negative correlations.

3.6. Correlational Analysis of Soil Respiration with Fungal and Bacterial Functional Groups

The abundances of the dominant fungal and bacterial OTUs were subjected to functional group prediction using FAPROTAX and FUNGuild, respectively [Figure 6a,b], and their correlations with in situ CO_2 fluxes were analyzed [Table 2]. The mean abundance of *Stellatospora* in NKA accounted for 0.07% of that of all fungi and its abundances were positively correlated with in situ CO_2 fluxes. The mean abundance of Acidobacteria Gp6 in KA accounted for 7.05% of that of all bacteria and was negatively correlated with in situ CO_2 fluxes. Among the predicted functional groups, the average abundance of fungal plant pathogens in KA and endomycorrhizal-plant pathogen-undefined saprotrophs in NKA accounted for 2.61% and 0.26% of the average abundance of functional taxa, respectively, and the abundances of both were positively correlated with in situ CO_2 fluxes. In addition, the average abundance of dung saprotroph-endophyte-litter saprotroph–undefined saprotrophs in KA accounted for 0.88% of the total fungal functional groups and was negatively correlated with in situ CO_2 fluxes. The average abundance of aerobic nitrite oxidizers (nitrifiers) were 2.13% of all bacterial functional groups and negatively correlated with in situ CO_2 fluxes.

(a)

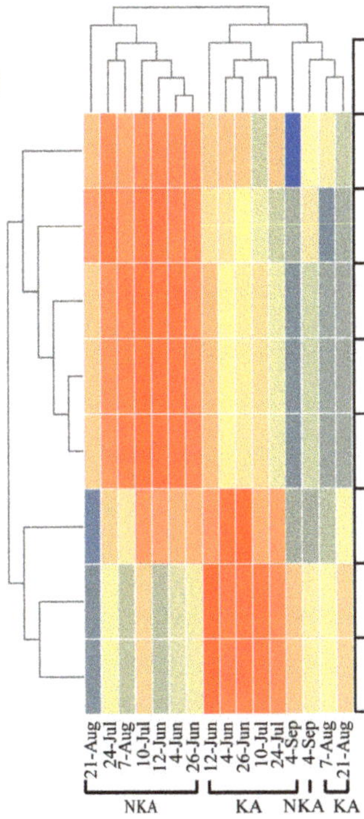

Function	Group	OTU ID	Rank
Nitrogen fixation	*Bradyrhizobium*	OTU1173	g
Iron respiration	*Geobacter*	OTU1721	g
Photoheterotrophy	*Rhodoplanes*	OTU1616	g
Denitrification	*Rhodoplanes*	OTU1616	g
Nitrite denitrification	*Rhodoplanes*	OTU1616	g
Chemoheterotrophy	*Phycicoccus*	OTU3528	g
Nitrification	*Nitrospira*	OTU831	g
Aerobic nitrite oxidation	*Nitrospira*	OTU831	g

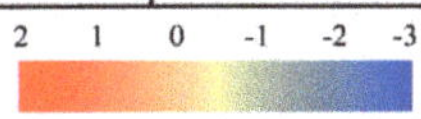

K-Kingdom, P-Phylum, C-Class, O-Order, F-Family, G-genus, S-Species

Figure 6. *Cont.*

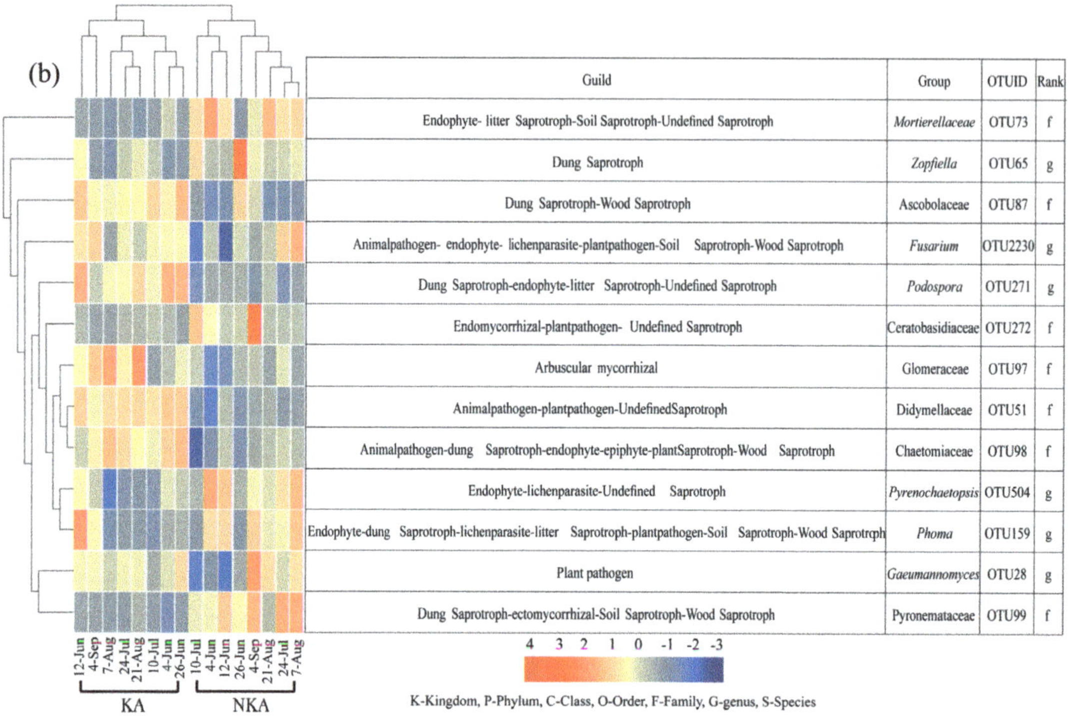

Figure 6. The heatmap clustering and distribution of bacterial (**a**) and fungal (**b**) functional groups. Note: KA, karst areas; NKA, non-karst areas.

Table 2. Correlation between soil respiration and bacterial and fungal functional.

Microbial Taxa	Groups	Region	Percentage	Correlation Coefficient
Bacteria	Acidobacteria Gp6	KA	7.05%	−0.565 *
Fungi	*Stellatospora*	NKA	0.07%	0.571 *
Functional groups of fungi	Plant pathogen	KA	2.61%	0.582 *
	Endomycorrhizal-plant pathogen-undefined saprotroph	NKA	0.26%	0.530 *
	Dung saprotroph-endophyte-litter saprotroph-undefined saprotroph	KA	0.88%	−0.568 *
Functional groups of bacteria	Aerobic nitrite oxidizers (Nitrifiers)	KA	2.13%	−0.545 *

Note: * Significant at the 0.05 level (two-tailed); KA, karst areas; NKA, non-karst areas.

4. Discussion

4.1. Effects of the F/B on In Situ CO_2 Fluxes

In present study, the abundance of bacteria and fungi in KA was higher than that in NKA, which indicated that higher soil nutrients shaped higher abundance of microorganisms. In KA, high levels of cations (such as Ca^{2+} and Mg^{2+}) tend to combine with easily decomposed organics to form more stable large aggregates and organic carbon pools [40,41]. Therefore, compared with fungi, bacteria are more susceptible to partly inhibition in KA environment.

The higher contents of chitin and chitosan in fungi cell wall could not be easily degraded, so that the total biomass of fungi was significantly higher than that of bacteria, which indicated that refractory organic matter form fungi is an important sources of SOC accumulation [42]. Long-term combination of organic and inorganic fertilization increased SOC accumulation of paddy fields in South China. The numbers of fungi and bacteria both also increased significantly, but fungi increased more than bacteria, thereby increasing the F/B [14]. Thiet argued that fungi didn't have greater growth efficiency than bacteria in greater C storage and slower C turnover in fungal-dominated soils [43]. Luo et al. [40] reported that there was a significant correlation between the SOC and F/B with the reason of increased fungal necromass resulting in a change in the SOC composition. The above result is in accordance with the findings of higher SOC content and F/B in KA in present study.

The KA with high temperature and humidity, the dissolution of carbonate brings a large amount of Ca^{2+} and HCO_3^- into the soil. HCO_3^- combines with H^+ in the water to generate CO_2, which increases soil CO_2 concentration [44]. Though some CO_2 in KA soil was consumed through reversibly combining with H_2O to form H_2CO_3 to a certain [3], CO_2 concentration of surface soil in KA is still higher than that in NKA. The higher concentration of soil CO_2 also change the quantity and quality of soluble sugars, organic acids, amino acids and other compounds secreted by rice roots. The bacterial abundance was significantly decreased and the fungal abundance and biomass were significantly increased, resulting in a significant increase in the F/B as previously reported [45,46].

More studies in the past 20 years have shown that the primary producers of CO_2 in soils are microorganism [13,14,47]. In present study, soil respiration in KA was lower than that in NKA, and the communities of both bacteria and fungi in KA were also significantly different with those in the NKA. Moreover, the F/B was higher in the KA than in the NKA. We also found that there was a significant negative correlation between soil respiration and the F/B across KA and NKA. In KA, soil carbon turnover rate would increase due to the changes on soil physical properties and the effect of fungi on physiology [43]. Keiblinger reported that more CO_2 will be released in bacterium-dominated soil because of lower carbon utilization efficient of bacteria [48]. Other researches have showed that fungal-derived soil processes and fungal biomass increased in the presence of elevated atmospheric CO_2 concentrations, which leading to increased organic C inputs into soils [46]. These results showed that fungi played an important role in the allocation of higher carbon and nitrogen nutrients in paddy soil, thus affecting soil respiration.

In our study, arbuscular mycorrhizal fungi were dominant in KA (showed in Figure 6b). The contribution of mycorrhizal fungi to the high carbon content of soil is reflected the conversion of floor C into mineral soil C [49] and the acceleration of microbial residues into soil [50]. Studies had shown that inoculation with arbuscular mycorrhizal fungi could promote the formation of large aggregates and ease their decomposition into aggregates with smaller particle sizes [51], which enhanced the formation of macroaggregates caused by sequestration of SOC [52]. Fungal hyphae and their cell wall residues act as binders by adsorption, cross-linking and adhesion of primary mineral particles, organic matter, and microaggregates, thereby enhancing the formation and stability of large aggregates [52]. Therefore, the distribution difference of mycorrhizal fungi in aggregates with different particle sizes in paddy soil between KA and NKA deserves further study.

4.2. Effect of Fungal and Background Community Composition On Paddy Field In Situ CO_2

Fungal-bacterial co-occurrence network analysis indicated a greater degree of putative mutualistic symbiotic relationships between fungi and bacteria (58.36%) compared with putative competitive relationships (41.64%) of KA and NKA. The presence of a typical "core" plant-microbe microbiota has been documented in several environmental studies [20,53]. These microbiomes may interact through direct or indirect mechanisms of carbon sequestration and can promote plant development [53,54]. Significantly higher relative abundance of

Acidobacteria Gp6 in KA of this study (average relative abundance of 7.05%) was negatively correlated with in situ CO_2 fluxes. Acidobacterium Gp6 may be the top microbial species in paddy soil in karst areas, as it is present in aggregate networks of different particle size grades [29]. Previous studies have demonstrated that the average relative abundance of Acidobacteria Gp6 was positively correlated with SOC and pH [55] in addition to amylase activity [56]. These results suggest the presence of coordinated activities of bacteria in KA with respect to SOC and pH. In addition, elevated CO_2 treatments have been shown to increase the relative abundance of Acidobacteria Gp6 [57], presumably because some Acidobacteria exhibit photosynthetic capacity. Some acidobacterial genomes contain *pscA* genes that encode the Fenna-Matthews-Olson (FMO) protein that binds to the bacterial chlorophyll molecule (BCh1) in photoresponse I, enabling chlorophyll-based photosynthetic capacity [58]. In this study, some strains of Acidobacteria Gp6 who absorb CO_2 for photosynthesis might have led to reduced CO_2 emissions in KA soils. Nevertheless, the carbon sequestration capacity of Acidobacteria Gp6 taxa in KA soils deserves our attention.

Nitrospira (OTU9) was also a top taxon in the internal bacterial correlation network and was dominant in NKA. *Nitrospira* abundance were negatively correlated with pH and TOC [59], and are highly resistant to high ammonia, high pH, salt, and SOC environments. In particular, the *Nitrospira* was most likely represented by the single-step nitrifying bacteria comammox *Nitrospira* that drives full nitrification [60]. Comammox *Nitrospira* encode a single gene cluster containing both *amo* and *hao* genes, but do not use Ca^{2+} (e.g., calcium nitrate and calcium nitrate) in nitrification reactions [61]. Instead, comammox *Nitrospira* require oxygen molecules to activate ammonia during nitrification [62], which are key adaptations for red soil environments with low carbon, nitrogen, and Ca^{2+} concentrations. Thus, OTU9 (comammox *Nitrospira*) is probably an important indicator in NKA rather than KA.

The internal fungal correlation network indicated that OTU69 (*Emericellopsis*) and OTU139 (Sordariales) as the core groups within the soil fungal networks in KA and NKA, respectively. *Emericellopsis* is an alkalophilic endophytic fungus with the ability to resist pathogenic peptide synthesis [63] and it enhances the plant host's ability to cope with environmental stresses, while also contributing to healthy host-plant growth and reproduction [64]. The neutral-alkaline pH in KA resulted in increased relative abundances of *Emericellopsis* that may then intermittently raise the quantity and quality of algae during the flooding periods within rice fields. Further, most water column algae in KA can use CO_2 and free HCO_3^- to conduct photosynthesis and produce organic carbon. The synergistic interaction between algae and their endophytic fungi in photosynthesis of KA therefore requires further investigation [65]. *Stellatospora* belongs to the Sordariales group, whose abundances were positively correlated with CO_2 emission. Some studies have shown that Sordariales are highly adapted to tropical soil environments that experience warming due to elevated atmospheric CO_2 [66]. Thus, *Stellatospora* might be regarded as a biomarker indicator to distinguish NKA from KA.

4.3. *Effects of Fungal and Bacterial Functional Groups on Paddy Field In Situ CO_2 Fluxes*

Among the predicted fungal functional groups, the endomycorrhizal-plant pathogen-undefined saprotroph functional group was largely represented by Ceratobasidiaceae, with dominant relative abundances in NKA. Ceratobasidiaceae are able to rice blight, where in stems and leaves become yellow and wilt, leading to weakened photosynthetic ability [67]. The plant pathogenic fungal functional group was largely represented by the genus *Gaeumannomyces*, which also exhibited higher abundances in NKA. Some fungi of this genus are capable of infesting wheat and resulted in wheat allozyme diseases. The plant pathogenic and endomycorrhizal-plant pathogen-undefined saprotroph abundances were positively correlated with CO_2 in situ emission. The infected plants transmit infection information to soil through root exudates, causing the bacterial quorum sensing [68] leading to compensatory increase in soil respiration. In addition, Ca^{2+} affect activity of the antifungal proteins released by bacteria [69], which in turn influences the resistance of crops to allozyme disease. Therefore, the interaction between pathogenic microbes and antagonistic microbes in KA and NKA should be further studied.

The abundance of dung saprotroph-endophyte-litter saprotroph–undefined saprotroph group were negatively correlated with in situ CO_2 emission. This functional group was mostly represented by *Podospora* that are dominant in KA. van Erven reported that *Podospora anserine*, a late colonizer of herbivorous dung, has high NADPH oxidase activity [70]. The oxidase promotes the formation of H_2O_2 and acts specifically on the more recalcitrant fraction of lignocellulose, which may promote the degradation of organic matter. Malagnac reported that NADPH oxidase can also promote the sexual reproduction and ascospore germination of the filamentous fungus *Podospora anserina* [71]. The products of lignocellulose degradation by fungi are likely to be used by bacteria, which improved utilization efficiency of organic matter and increased thermally stable SOC [72]. These results suggest that the relationship between soil microorganisms including *Podospora* and the thermal stability of SOC in KA is worth further study.

CO_2 in soils not only changed general microbial taxa, but also can be assimilated by autotrophic bacteria and turned into microbial biomass [73]. In the present study, autotrophic bacteria (such as aerobic nitrite oxidizers and nitrifiers) using CO_2 as the only carbon source were dominant in KA (as shown in Figure 6a). This was similar to the results of our previous investigation, which showed that high abundance of carbon fixing bacteria was in KA [25]. Among the predicted bacterial functional groups, the abundances of aerobic nitrite oxidizers and nitrifiers were negatively correlated with in situ CO_2 fluxes. These functional groups derive from the aerobic nitrite oxidation and nitrification activities associated with OTU831 (*Nitrospira*), which were dominant in KA. Ammonia-oxidizing bacteria (AOB) are autotrophic microorganisms using CO_2 as a carbon source in autotrophy [74]. Nitrification substrates are NH_3 molecules rather than NH_4^+ ions, and a neutral alkaline karst environment might make the chemical equilibrium of NH_3 and NH_4^+ tend towards NH_3 molecule production. Consequently, the abundances of AOB and their ability to fix CO_2 might be improved by increasing the abundances of NH_4^+-N [75,76]. Therefore, the lower in situ CO_2 emission in KA might also contribute to autotrophic nitrification by *Nitrospira* [77].

5. Conclusions

In this study, we observed higher abundance of bacteria and fungi and higher F/B but lower soil respiration in karst area compared with those in non-karst area. We found that there was a significant difference of bacterial and fungi community between karst area and non-karst area. We also found that there was a significantly negatively correlation between soil respiration and ratio of fungi to bacteria. On the one hand, higher abundances of fungi and their associated functional activities might enable better use of recalcitrance. On the other hand, bacterial abundances were decreased to adapt to environmental stresses from reduced easily available carbon supplies. Further, in response to the unique karst environments, microbial community structures and their associated functional groups were altered. In addition, the abundances of some autotrophic carbon-fixing bacteria and arbuscular mycorrhizal fungi increased and the abundance of pathogenic fungi decreased, which thereby also probably improved the utilization of SOC and reduces soil CO_2 emissions. In the future research, combining these microbial characteristics associated with soil respiration and through C and N isotopic labeling tracking technology and indoor incubation experiments can be provided more accurate soil carbon and nitrogen dynamics in KA.

Supplementary Materials: The following supporting information can be downloaded at: https://www.mdpi.com/article/10.3390/agronomy13082001/s1, Figure S1: The soil respiration of paddy fields during fallow.

Author Contributions: Conceptualization, Z.J. and J.Z.; software, J.Z.; formal analysis, G.C. and J.Z.; investigation, L.X., X.L., W.C., J.Z. and Z.J.; data curation, W.Y. and J.Z.; writing—original draft preparation, J.Z.; writing—review and editing, Z.J. and J.Z.; visualization, J.Z. All authors have read and agreed to the published version of the manuscript.

Funding: This work was supported by grants from the National Natural Science Foundation of China (41867008), the Key Research and Development Program of Guangxi (GuiKe-AB21196050, GuiKe-AB20297039), Guangxi Natural Science Foundation of China (2018GXNSFAA281247), Foundation of Guilin University of Technology (GUTQDJJ2004041), and Technology Planning Project (gxzz201903).

Data Availability Statement: The datasets generated during the current study were uploaded to the Sequence Read Archive (SRA) and have the accession number PRJNA763299.

Acknowledgments: Thank Xiaowen Zhang of Guilin University of Technology for providing qPCR experimental guidance.

Conflicts of Interest: The authors declare that they have no known competing financial interest or personal relationships that could have appeared to influence the work reported in this paper.

References

1. Wang, K.; Zhang, C.; Chen, H.; Yue, Y.; Zhang, W.; Zhang, M.; Qi, X.; Fu, Z. Karst Landscapes of China: Patterns, Ecosystem Processes and Services. *Landsc. Ecol.* **2019**, *34*, 2743–2763. [CrossRef]
2. Liu, C.; Huang, Y.; Wu, F.; Liu, W.; Ning, Y.; Huang, Z.; Tang, S.; Liang, Y. Plant Adaptability in Karst Regions. *J. Plant Res.* **2021**, *134*, 889–906. [CrossRef] [PubMed]
3. Liu, Z.; Dreybrodt, W. Significance of the Carbon Sink Produced by H_2O-Carbonate-CO_2-Aquatic Phototroph Interaction on Land. *Sci. Bull.* **2015**, *60*, 182–191. [CrossRef]
4. Abbaszadeh, M.; Nasiri, M.; Riazi, M. Experimental Investigation of the Impact of Rock Dissolution on Carbonate Rock Properties in the Presence of Carbonated Water. *Environ. Earth Sci.* **2016**, *75*, 791. [CrossRef]
5. Zeng, S.; Liu, Z.; Kaufmann, G. Sensitivity of the Global Carbonate Weathering Carbon-Sink Flux to Climate and Land-Use Changes. *Nat. Commun.* **2019**, *10*, 5749. [CrossRef]
6. Dass, P.; Houlton, B.Z.; Wang, Y.; Warlind, D.; Morford, S. Bedrock Weathering Controls on Terrestrial Carbon-Nitrogen-Climate Interactions. *Glob. Biogeochem. Cycle* **2021**, *35*, e2020GB006933. [CrossRef]
7. Zhao, B.; Su, Y. Process Effect of Microalgal-Carbon Dioxide Fixation and Biomass Production: A Review. *Renew. Sustain. Energ. Rev.* **2014**, *31*, 121–132. [CrossRef]
8. Sha, Z.; Bai, Y.; Li, R.; Lan, H.; Zhang, X.; Li, J.; Liu, X.; Chang, S.; Xie, Y. The Global Carbon Sink Potential of Terrestrial Vegetation Can Be Increased Substantially by Optimal Land Management. *Commun. Earth Environ.* **2022**, *3*, 8. [CrossRef]
9. Huang, W.; Han, T.; Liu, J.; Wang, G.; Zhou, G. Changes in Soil Respiration Components and Their Specific Respiration along Three Successional Forests in the Subtropics. *Funct. Ecol.* **2016**, *30*, 1466–1474. [CrossRef]
10. Bond-Lamberty, B.; Thomson, A. Temperature-Associated Increases in the Global Soil Respiration Record. *Nature* **2010**, *464*, 579–582. [CrossRef]
11. Su, Y.G.; Huang, G.; Lin, Y.J.; Zhang, Y.M. No Synergistic Effects of Water and Nitrogen Addition on Soil Microbial Communities and Soil Respiration in a Temperate Desert. *Catena* **2016**, *142*, 126–133. [CrossRef]
12. Jian, J.; Steele, M.K.; Thomas, R.Q.; Day, S.D.; Hodges, S.C. Constraining Estimates of Global Soil Respiration by Quantifying Sources of Variability. *Glob. Change Biol.* **2018**, *24*, 4143–4159. [CrossRef]
13. Moinet, G.Y.K.; Cieraad, E.; Hunt, J.E.; Fraser, A.; Turnbull, M.H.; Whitehead, D. Soil Heterotrophic Respiration Is Insensitive to Changes in Soil Water Content but Related to Microbial Access to Organic Matter. *Geoderma* **2016**, *274*, 68–78. [CrossRef]
14. Ming, L.; Ekschmitt, K.; Bin, Z.; Holzhauer, S.I.J.; Zhong-pei, L.; Tao-lin, Z.; Rauch, S. Effect of Intensive Inorganic Fertilizer Application on Microbial Properties in a Paddy Soil of Subtropical China. *Agric. Sci. China* **2011**, *10*, 1758–1764. [CrossRef]
15. Chen, X.; Xia, Y.; Rui, Y.; Ning, Z.; Hu, Y.; Tang, H.; He, H.; Li, H.; Kuzyakov, Y.; Ge, T.; et al. Microbial Carbon Use Efficiency, Biomass Turnover, and Necromass Accumulation in Paddy Soil Depending on Fertilization. *Agric. Ecosyst. Environ.* **2020**, *292*, 106816. [CrossRef]
16. Jia, X.; Zha, T.; Wu, B.; Zhang, Y.; Chen, W.; Wang, X.; Yu, H.; He, G. Temperature Response of Soil Respiration in a Chinese Pine Plantation: Hysteresis and Seasonal vs. Diel Q_{10}. *PLoS ONE* **2013**, *8*, e57858. [CrossRef] [PubMed]
17. Ji, W.; Yang, Z.; Yu, T.; Yang, Q.; Wen, Y.; Wu, T. Potential Ecological Risk Assessment of Heavy Metals in the Fe-Mn Nodules in the Karst Area of Guangxi, Southwest China. *Bull. Environ. Contam. Toxicol.* **2021**, *106*, 51–56. [CrossRef]
18. Miao, X.; Hao, Y.; Liu, H.; Xie, Z.; Miao, D.; He, X. Effects of Heavy Metals Speciations in Sediments on Their Bioaccumulation in Wild Fish in Rivers in Liuzhou—A Typical Karst Catchment in Southwest China. *Ecotox. Environ. Saf.* **2021**, *214*, 112099. [CrossRef]
19. Liu, S.; Zhang, Y.; Zong, Y.; Hu, Z.; Wu, S.; Zhou, J.; Jin, Y.; Zou, J. Response of Soil Carbon Dioxide Fluxes, Soil Organic Carbon and Microbial Biomass Carbon to Biochar Amendment: A Meta-Analysis. *GCB Bioenergy* **2016**, *8*, 392–406. [CrossRef]

20. Bulgarelli, D.; Rott, M.; Schlaeppi, K.; van Themaat, E.V.L.; Ahmadinejad, N.; Assenza, F.; Rauf, P.; Huettel, B.; Reinhardt, R.; Schmelzer, E.; et al. Revealing Structure and Assembly Cues for Arabidopsis Root-Inhabiting Bacterial Microbiota. *Nature* **2012**, *488*, 91–95. [CrossRef]
21. Pierce, E.C.; Morin, M.; Little, J.C.; Liu, R.B.; Tannous, J.; Keller, N.P.; Pogliano, K.; Wolfe, B.E.; Sanchez, L.M.; Dutton, R.J. Bacterial-Fungal Interactions Revealed by Genome-Wide Analysis of Bacterial Mutant Fitness. *Nat. Microbiol.* **2021**, *6*, 87–102. [CrossRef] [PubMed]
22. Wu, M.; Feng, Q.; Sun, X.; Wang, H.; Gielen, G.; Wu, W. Rice (*Oryza Sativa* L.) Plantation Affects the Stability of Biochar in Paddy Soil. *Sci. Rep.* **2015**, *5*, 10001. [CrossRef] [PubMed]
23. Lecomte, S.M.; Achouak, W.; Abrouk, D.; Heulin, T.; Nesme, X.; Haichar, F.e.Z. Diversifying Anaerobic Respiration Strategies to Compete in the Rhizosphere. *Front. Environ. Sci.* **2018**, *6*, 139. [CrossRef]
24. Zhu, Z.; Ge, T.; Hu, Y.; Zhou, P.; Wang, T.; Shibistova, O.; Guggenberger, G.; Su, Y.; Wu, J. Fate of Rice Shoot and Root Residues, Rhizodeposits, and Microbial Assimilated Carbon in Paddy Soil—Part 2: Turnover and Microbial Utilization. *Plant Soil* **2017**, *416*, 243–257. [CrossRef]
25. Zhou, J.; Jin, Z.; Leng, M.; Xiao, X.; Wang, X.; Pan, F.; Li, Q. Investigation of Soil Bacterial Communities and Functionalities Within Typical Karst Paddy Field Soils in Southern China. *Fresenius Environ. Bull.* **2021**, *30*, 3537–3548.
26. Xu, H.; Du, H.; Zeng, F.; Song, T.; Peng, W. Diminished Rhizosphere and Bulk Soil Microbial Abundance and Diversity across Succession Stages in Karst Area, Southwest China. *Appl. Soil Ecol.* **2021**, *158*, 103799. [CrossRef]
27. Yang, H.; Xie, Y.; Zhu, T.; Zhou, M. Reduced Organic Carbon Content during the Evolvement of Calcareous Soils in Karst Region. *Forests* **2021**, *12*, 221. [CrossRef]
28. Cao, W.; Xiong, Y.; Zhao, D.; Tan, H.; Qu, J. Bryophytes and the Symbiotic Microorganisms, the Pioneers of Vegetation Restoration in Karst Rocky Desertification Areas in Southwestern China. *Appl. Microbiol. Biotechnol.* **2020**, *104*, 873–891. [CrossRef]
29. Xiao, X.; Jin, Z.; Leng, M.; Li, X.; Xiong, L. Comparison of Bacterial Community Structure and Functional Groups of Paddy Soil Aggregates Between Karst and Non-Karst Areas. *Huan Jing Ke Xue = Huanjing Kexue* **2022**, *43*, 3865–3875. [CrossRef] [PubMed]
30. Huang, Y.; Zhang, W.; Sun, W.; Zheng, X. Net Primary Production of Chinese Croplands from 1950 to 1999. *Ecol. Appl.* **2007**, *17*, 692–701. [CrossRef] [PubMed]
31. Kimura, M.; Murase, J.; Lu, Y.H. Carbon Cycling in Rice Field Ecosystems in the Context of Input, Decomposition and Translocation of Organic Materials and the Fates of Their End Products (CO_2 and CH_4). *Soil Biol. Biochem.* **2004**, *36*, 1399–1416. [CrossRef]
32. Mandal, B.; Majumder, B.; Adhya, T.K.; Bandyopadhyay, P.K.; Gangopadhyay, A.; Sarkar, D.; Kundu, M.C.; Choudhury, S.G.; Hazra, G.C.; Kundu, S.; et al. Potential of Double-Cropped Rice Ecology to Conserve Organic Carbon under Subtropical Climate. *Glob. Change Biol.* **2008**, *14*, 2139–2151. [CrossRef]
33. Li, X.; Jin, Z.; Xiong, L.; Tong, L.; Zhu, H.; Zhang, X.; Qin, G. Effects of Land Reclamation on Soil Bacterial Community and Potential Functions in Bauxite Mining Area. *Int. J. Environ. Res. Public Health* **2022**, *19*, 16921. [CrossRef] [PubMed]
34. Zhang, L.; Yu, D.; Shi, X.; Weindorf, D.C.; Zhao, L.; Ding, W.; Wang, H.; Pan, J.; Li, C. Simulation of Global Warming Potential (GWP) from Rice Fields in the Tai-Lake Region, China by Coupling 1:50,000 Soil Database with DNDC Model. *Atmos. Environ.* **2009**, *43*, 2737–2746. [CrossRef]
35. Claesson, M.J.; O'Sullivan, O.; Wang, Q.; Nikkilae, J.; Marchesi, J.R.; Smidt, H.; de Vos, W.M.; Ross, R.P.; O'Toole, P.W. Comparative Analysis of Pyrosequencing and a Phylogenetic Microarray for Exploring Microbial Community Structures in the Human Distal Intestine. *PLoS ONE* **2009**, *4*, e6669. [CrossRef]
36. Schmieder, R.; Edwards, R. Quality Control and Preprocessing of Metagenomic Datasets. *Bioinformatics* **2011**, *27*, 863–864. [CrossRef]
37. Cannon, A.J. Multivariate Quantile Mapping Bias Correction: An N-Dimensional Probability Density Function Transform for Climate Model Simulations of Multiple Variables. *Clim. Dyn.* **2018**, *50*, 31–49. [CrossRef]
38. Doncheva, N.T.; Assenov, Y.; Domingues, F.S.; Albrecht, M. Topological Analysis and Interactive Visualization of Biological Networks and Protein Structures. *Nat. Protoc.* **2012**, *7*, 670–685. [CrossRef]
39. Zallot, R.; Oberg, N.; Gerlt, J.A. The EFI Web Resource for Genomic Enzymology Tools: Leveraging Protein, Genome, and Metagenome Databases to Discover Novel Enzymes and Metabolic Pathways. *Biochemistry* **2019**, *58*, 4169–4182. [CrossRef]
40. Luo, Y.; Xiao, M.; Yuan, H.; Liang, C.; Zhu, Z.; Xu, J.; Kuzyakov, Y.; Wu, J.; Ge, T.; Tang, C. Rice Rhizodeposition Promotes the Build-up of Organic Carbon in Soil via Fungal Necromass. *Soil Biol. Biochem.* **2021**, *160*, 108345. [CrossRef]
41. Song, Y.; Liu, C.; Wang, X.; Ma, X.; Jiang, L.; Zhu, J.; Gao, J.; Song, C. Microbial Abundance as an Indicator of Soil Carbon and Nitrogen Nutrient in Permafrost Peatlands. *Ecol. Indic.* **2020**, *115*, 106362. [CrossRef]
42. Ananyeva, N.D.; Polyanskaya, L.M.; Stolnikova, E.V.; Zvyagintzev, D.G. Fungal to Bacterial Biomass Ratio in the Forests Soil Profile. *Biol. Bull* **2010**, *37*, 254–262. [CrossRef]
43. Thiet, R.K.; Frey, S.D.; Six, J. Do Growth Yield Efficiencies Differ between Soil Microbial Communities Differing in Fungal: Bacterial Ratios? Reality Check and Methodological Issues. *Soil Biol. Biochem.* **2006**, *38*, 837–844. [CrossRef]
44. Jianhua, C.; Daoxian, Y.; Groves, C.; Fen, H.; Hui, Y.; Qian, L. Carbon Fluxes and Sinks: The Consumption of Atmospheric and Soil CO_2 by Carbonate Rock Dissolution. *Acta Geol. Sin.-Engl. Ed.* **2012**, *86*, 963–972. [CrossRef]

45. Zhong, L.; Bowatte, S.; Newton, P.C.D.; Hoogendoorn, C.J.; Luo, D. An Increased Ratio of Fungi to Bacteria Indicates Greater Potential for N_2O Production in a Grazed Grassland Exposed to Elevated CO_2. *Agric. Ecosyst. Environ.* **2018**, *254*, 111–116. [CrossRef]
46. Laughlin, R.J.; Rutting, T.; Mueller, C.; Watson, C.J.; Stevens, R.J. Effect of Acetate on Soil Respiration, N_2O Emissions and Gross N Transformations Related to Fungi and Bacteria in a Grassland Soil. *Appl. Soil Ecol.* **2009**, *42*, 25–30. [CrossRef]
47. Li, W.; Yu, L.J.; Yuan, D.X.; Xu, H.B.; Yang, Y. Bacteria Biomass and Carbonic Anhydrase Activity in Some Karst Areas of Southwest China. *J. Asian Earth Sci.* **2004**, *24*, 145–152. [CrossRef]
48. Keiblinger, K.M.; Hall, E.K.; Wanek, W.; Szukics, U.; Haemmerle, I.; Ellersdorfer, G.; Boeck, S.; Strauss, J.; Sterflinger, K.; Richter, A.; et al. The Effect of Resource Quantity and Resource Stoichiometry on Microbial Carbon-Use-Efficiency. *FEMS Microbiol. Ecol.* **2010**, *73*, 430–440. [CrossRef] [PubMed]
49. Lin, G.; McCormack, M.L.; Ma, C.; Guo, D. Similar Below-Ground Carbon Cycling Dynamics but Contrasting Modes of Nitrogen Cycling between Arbuscular Mycorrhizal and Ectomycorrhizal Forests. *New Phytol.* **2017**, *213*, 1440–1451. [CrossRef]
50. Craig, M.E.; Turner, B.L.; Liang, C.; Clay, K.; Johnson, D.J.; Phillips, R.P. Tree Mycorrhizal Type Predicts Within-Site Variability in the Storage and Distribution of Soil Organic Matter. *Glob. Change Biol.* **2018**, *24*, 3317–3330. [CrossRef] [PubMed]
51. Tian, X.; Wang, C.; Bao, X.; Wang, P.; Li, X.; Yang, S.; Ding, G.; Christie, P.; Li, L. Crop Diversity Facilitates Soil Aggregation in Relation to Soil Microbial Community Composition Driven by Intercropping. *Plant Soil* **2019**, *436*, 173–192. [CrossRef]
52. Xiao, L.; Zhang, W.; Hu, P.; Xiao, D.; Yang, R.; Ye, Y.; Wang, K. The Formation of Large Macroaggregates Induces Soil Organic Carbon Sequestration in Short-Term Cropland Restoration in a Typical Karst Area. *Sci. Total Environ.* **2021**, *801*, 149588. [CrossRef] [PubMed]
53. Lundberg, D.S.; Lebeis, S.L.; Paredes, S.H.; Yourstone, S.; Gehring, J.; Malfatti, S.; Tremblay, J.; Engelbrektson, A.; Kunin, V.; del Rio, T.G.; et al. Defining the Core Arabidopsis Thaliana Root Microbiome. *Nature* **2012**, *488*, 86–90. [CrossRef] [PubMed]
54. Bulgarelli, D.; Schlaeppi, K.; Spaepen, S.; van Themaat, E.V.L.; Schulze-Lefert, P. Structure and Functions of the Bacterial Microbiota of Plants. *Annu. Rev. Plant Biol.* **2013**, *64*, 807–838. [CrossRef] [PubMed]
55. Han, Z.; Ma, J.; Yang, C.-H.; Ibekwe, A.M. Soil Salinity, PH, and Indigenous Bacterial Community Interactively Influence the Survival of *E. Coli* O157:H7 Revealed by Multivariate Statistics. *Environ. Sci. Pollut. Res.* **2021**, *28*, 5575–5586. [CrossRef] [PubMed]
56. Zhang, Y.; Cong, J.; Lu, H.; Li, G.; Qu, Y.; Su, X.; Zhou, J.; Li, D. Community Structure and Elevational Diversity Patterns of Soil Acidobacteria. *J. Environ. Sci.* **2014**, *26*, 1717–1724. [CrossRef]
57. Dunbar, J.; Gallegos-Graves, L.V.; Steven, B.; Mueller, R.; Hesse, C.; Zak, D.R.; Kuske, C.R. Surface Soil Fungal and Bacterial Communities in Aspen Stands Are Resilient to Eleven Years of Elevated CO_2 and O_3. *Soil Biol. Biochem.* **2014**, *76*, 227–234. [CrossRef]
58. Bryant, D.A.; Costas, A.M.G.; Maresca, J.A.; Chew, A.G.M.; Klatt, C.G.; Bateson, M.M.; Tallon, L.J.; Hostetler, J.; Nelson, W.C.; Heidelberg, J.F.; et al. Candidatus Chloracidobacterium Thermophilum: An Aerobic Phototrophic Acidobacterium. *Science* **2007**, *317*, 523–526. [CrossRef]
59. Mehrani, M.-J.; Sobotka, D.; Kowal, P.; Ciesielski, S.; Makinia, J. The Occurrence and Role of Nitrospira in Nitrogen Removal Systems. *Bioresour. Technol.* **2020**, *303*, 122936. [CrossRef]
60. Hu, H.-W.; He, J.-Z. Comammox-a Newly Discovered Nitrification Process in the Terrestrial Nitrogen Cycle. *J. Soils Sediments* **2017**, *17*, 2709–2717. [CrossRef]
61. Palomo, A.; Pedersen, A.G.; Fowler, S.J.; Dechesne, A.; Sicheritz-Pontén, T.; Smets, B.F. Comparative Genomics Sheds Light on Niche Differentiation and the Evolutionary History of Comammox Nitrospira. *ISME J.* **2018**, *12*, 1779–1793. [CrossRef]
62. Koch, H.; van Kessel, M.A.H.J.; Lucker, S. Complete Nitrification: Insights into the Ecophysiology of Comammox Nitrospira. *Appl. Microbiol. Biotechnol.* **2019**, *103*, 177–189. [CrossRef]
63. Baranova, A.A.; Rogozhin, E.A.; Georgieva, M.L.; Bilanenko, E.N.; Kul'ko, A.B.; Yakushev, A.V.; Alferova, V.A.; Sadykova, V.S. Antimicrobial Peptides Produced by Alkaliphilic Fungi Emericellopsis Alkalina: Biosynthesis and Biological Activity Against Pathogenic Multidrug-Resistant Fungi. *Appl. Biochem. Microbiol.* **2019**, *55*, 145–151. [CrossRef]
64. Goncalves, M.F.M.; Vicente, T.F.L.; Esteves, A.C.; Alves, A. Novel Halotolerant Species of Emericellopsis and Parasarocladium Associated with Macroalgae in an Estuarine Environment. *Mycologia* **2020**, *112*, 154–171. [CrossRef] [PubMed]
65. Zaitseva, L.V.; Orleanskii, V.K.; Gerasimenko, L.M.; Ushatinskaya, G.T. The Role of Cyanobacteria in Crystallization of Magnesium Calcites. *Paleontol. J.* **2006**, *40*, 125–133. [CrossRef]
66. de Oliveira, T.B.; de Lucas, R.C.; de Almeida Scarcella, A.S.; Contato, A.G.; Pasin, T.M.; Martinez, C.A.; Teixeira de Moraes Polizeli, M.d.L. Fungal Communities Differentially Respond to Warming and Drought in Tropical Grassland Soil. *Mol. Ecol.* **2020**, *29*, 1550–1559. [CrossRef] [PubMed]
67. Selim, H.M.M.; Gomaa, N.M.; Essa, A.M.M. Application of Endophytic Bacteria for the Biocontrol of *Rhizoctonia solani* (Cantharellales: Ceratobasidiaceae) Damping-off Disease in Cotton Seedlings. *Biocontrol Sci. Technol.* **2017**, *27*, 81–95. [CrossRef]
68. Qu, T.; Du, X.; Peng, Y.; Guo, W.; Zhao, C.; Losapio, G. Invasive Species Allelopathy Decreases Plant Growth and Soil Microbial Activity. *PLoS ONE* **2021**, *16*, e0246685. [CrossRef]
69. Wang, Z.; Wang, Y.; Zheng, L.; Yang, X.; Liu, H.; Guo, J. Isolation and Characterization of an Antifungal Protein from Bacillus Licheniformis HS10. *Biochem. Biophys. Res. Commun.* **2014**, *454*, 48–52. [CrossRef]
70. van Erven, G.; Kleijn, A.F.; Patyshakuliyeva, A.; Di Falco, M.; Tsang, A.; de Vries, R.P.; van Berkel, W.J.H.; Kabel, M.A. Evidence for Ligninolytic Activity of the Ascomycete Fungus Podospora Anserina. *Biotechnol. Biofuels* **2020**, *13*, 75. [CrossRef]

71. Malagnac, F.; Lalucque, H.; Lepere, G.; Silar, P. Two NADPH Oxidase Isoforms Are Required for Sexual Reproduction and Ascospore Germination in the Filamentous Fungus Podospora Anserina. *Fungal Genet. Biol.* **2004**, *41*, 982–997. [CrossRef]
72. Domeignoz-Horta, L.A.; Pold, G.; Liu, X.-J.A.; Frey, S.D.; Melillo, J.M.; DeAngelis, K.M. Microbial Diversity Drives Carbon Use Efficiency in a Model Soil. *Nat. Commun.* **2020**, *11*, 3684. [CrossRef] [PubMed]
73. Zhao, X.; Zhao, C.; Stahr, K.; Kuzyakov, Y.; Wei, X. The Effect of Microorganisms on Soil Carbonate Recrystallization and Abiotic CO_2 Uptake of Soil. *Catena* **2020**, *192*, 104592. [CrossRef]
74. Luecker, S.; Wagner, M.; Maixner, F.; Pelletier, E.; Koch, H.; Vacherie, B.; Rattei, T.; Damste, J.S.S.; Spieck, E.; Le Paslier, D.; et al. A Nitrospira Metagenome Illuminates the Physiology and Evolution of Globally Important Nitrite-Oxidizing Bacteria. *Proc. Natl. Acad. Sci. USA* **2010**, *107*, 13479–13484. [CrossRef] [PubMed]
75. Di, H.J.; Cameron, K.C.; Shen, J.P.; Winefield, C.S.; O'Callaghan, M.; Bowatte, S.; He, J.Z. Nitrification Driven by Bacteria and Not Archaea in Nitrogen-Rich Grassland Soils. *Nat. Geosci.* **2009**, *2*, 621–624. [CrossRef]
76. Kemmitt, S.J.; Wright, D.; Goulding, K.W.T.; Jones, D.L. PH Regulation of Carbon and Nitrogen Dynamics in Two Agricultural Soils. *Soil Biol. Biochem.* **2006**, *38*, 898–911. [CrossRef]
77. Zhang, Q.; Li, Y.; He, Y.; Brookes, P.C.; Xu, J. Elevated Temperature Increased Nitrification Activity by Stimulating AOB Growth and Activity in an Acidic Paddy Soil. *Plant Soil* **2019**, *445*, 71–83. [CrossRef]

agronomy

Article

Effects of Transplantation and Microhabitat on Rhizosphere Microbial Communities during the Growth of American Ginseng

Fan Chang [1,2], Feng-An Jia [1], Min Guan [3], Qing-An Jia [4], Yan Sun [2,*] and Zhi Li [2,*]

1 Agricultural and Environmental Microbiology Research Center, Shaanxi Institute of Microbiology, Xi'an 710043, China; fox387@163.com (F.C.); jiafenganwsw@163.com (F.-A.J.)
2 College of Life Science, Shaanxi Normal University, Xi'an 710062, China
3 Shaanxi Agricultural Machinery Research Institute, Xianyang 712000, China; guanmin387@163.com
4 Institute of Medical Research, Northwestern Polytechnical University, Xi'an 710072, China; jiaqingan666@163.com
* Correspondence: sunyan23@163.com (Y.S.); jiaoli3lou@163.com (Z.L.); Tel.: +86-153-5355-4537 (Y.S.); +86-135-7290-0787 (Z.L.)

Abstract: Transplanting has been widely used in American ginseng (*Panax quinquefolium* L.) cultivation in Northwest China to mitigate the negative effects of continuous cropping obstacles. Because of the accumulation of pathogenic microorganisms and the change in soil properties, transplanting American ginseng to newly cultivated fields after two years of growth has become a major planting pattern. Despite transplanting improving the quality of American ginseng, the effects of soil properties and microbiota on growth during the transplanting process are poorly understood. In the present study, microbial communities, soil physico-chemical properties and morpho-physiological parameters were analyzed to investigate the effects of microbiota and soil characteristics on American ginseng growth in both soil and ginseng root microhabitats. Results indicated that the structure and species of bacterial and fungal communities changed significantly in different microhabitats before and after transplantation. Moreover, the assemblage process of the bacterial community was dominated by deterministic processes. The stochastic process ratio increased and niche breadth decreased significantly after transplanting. While the assembly of the fungal community was dominated by stochastic process, and there was no significant difference in NST, βNTI or niche breadth before and after transplanting. Bacterial co-occurrence networks demonstrated a higher connectivity but a lower aggregation in soil microhabitat, while the fungal community networks remained stable before and after transplantation. Gammaproteobacteria was the biomarker in the soil microhabitat, while Alphaproteobacteria, Betaproteobacteria and Gemmatimonadetes were biomarkers in the ginseng root microhabitat. Sordariomycetes was a biomarker with high relative abundance in the fungal community before and after transplanting. The bacterial functional and important ASVs were significantly correlated with pH, organic matter, total nitrogen, available phosphorus, total potassium root fresh weight, taproot diameter and stem height of American ginseng. Partial least squares path modeling showed that soil properties significantly affected the formation of different microbial specific ASVs. The important functional ASVs in ginseng root microhabitat had a positive effect on American ginseng growth, while the rare taxa had a negative effect. Our results provide a good starting point for future studies of microbial community succession in different microhabitats influenced by the transplantation pattern of American ginseng.

Keywords: American ginseng; transplanting; microhabitats; microbiota; soil properties

check for updates

Citation: Chang, F.; Jia, F.-A.; Guan, M.; Jia, Q.-A.; Sun, Y.; Li, Z. Effects of Transplantation and Microhabitat on Rhizosphere Microbial Communities during the Growth of American Ginseng. *Agronomy* **2023**, *13*, 1876. https://doi.org/10.3390/agronomy13071876

Academic Editors: Yong-Xin Liu and Peng Yu

Received: 5 June 2023
Revised: 5 July 2023
Accepted: 13 July 2023
Published: 16 July 2023

1. Introduction

American ginseng (*Panax quinquefolius* L.) has been used as a herbal medicine with a "cool" property in China for nearly 300 years. It is known for its pharmacological properties such as anti-oxidation, anti-cancer activity, stimulation of blood flow, and enhancement of the central nervous system [1,2]. Originally native to southeastern Canada and northern United States, American ginseng was introduced to China in the 1980s [3,4]. Because of the imbalance in soil nutrients and continuous cropping obstacles caused by long-term growth in the same field, American ginseng is generally cultivated in Northwest China using a "two-year land-changing planting" pattern. This pattern involves transplanting American ginseng after two years of growth into newly cultivated land for another two years [5,6].

The reduction in crop yield and quality caused by long-term monoculture, known as continuous cropping obstacle, is an pressing challenge to the worldwide cultivation of Chinese herbal medicines, including American ginseng [4]. Transplanting can mitigate, but cannot avoid, the degradation in American ginseng quality and soil deterioration caused by continuous cropping [5,6]. Therefore, it is important to pay attention to the dynamic changes of soil microenvironment during American ginseng planting. Continuous cropping obstacles are influenced by various factors. Previous studies have shown that soil physicochemical properties, such as physical characteristics, nutrient, pH and allelopathic autotoxicity of plants [7–9], can impact the growth of ginseng plants [7,9] and also increase the incidence of diseases [10]. The biotic factors that cause continuous cropping obstacles in the American ginseng soil commonly include the microbiome, protozoa and insect pests [11,12]. While soil-borne pathogenic microorganisms are mainly responsible for biological diseases, other beneficial microbes exhibit resistance to such ailments as well [13,14]. The plant–soil feedback mechanism proposes that alterations in soil properties resulting from plant growth can impact both plant populations and microbial communities [15]. Typically, these mechanisms are usually studied independently; however, given their potential for interaction, it is unlikely that any single mechanism can fully account for plant–soil feedback.

Previous studies have suggested that the rhizosphere bridges the plant–soil relationship and facilitates the exchange of substances between roots and soil [16]. Many rhizosphere activities, including nutrient cycling and bioremediation [6], are mediated by microbiota in the soil–rhizosphere microhabitat, and roots provide a favorable environment for the enrichment of rhizosphere microbiota [17,18]. Moreover, mounting evidence suggests that plant species and sampling time exert significant impacts on the soil and rhizosphere microbial community [19,20]. The advancement of high-throughput sequencing technology and the expansion of bioinformatics methodology [21] has facilitated our comprehension of the interrelationships and interactions among the microbiome, soil properties and plant growth in agricultural practices. However, there is currently no available information on how transplantation pattern affects the rhizosphere microhabitat and microbiota of American ginseng roots. This knowledge is helpful for investigating the causes of obstacles in continuous cropping and their impact on soil environment, thereby promoting the scientific cultivation of American ginseng and improving land use efficiency.

Here, we conducted field trials to investigate the potential impacts of transplanting patterns on soil properties and American ginseng growth by assessing the succession of bacteria and fungi in the soil and rhizosphere microhabitat. We hypothesized that the matter exchange and information transfer between the ginseng root and the soil would greatly disturb the microbiota in the ginseng habitat. We established a randomized experimental transplanting of American ginseng. The changes of bacterial and fungal communities were analyzed, and their interactions with soil properties and the growth of American ginseng was discussed.

2. Materials and Methods

2.1. Experimental Setup and Soil Collection

The whole experiment was conducted in Liuba (latitude 33°40′ N, longitude 106°52′ E) village, Shaanxi province, the only cultivated American ginseng area in Northwest China [5]. The altitude is 1540 m, the mean annual temperature is 11.5 °C, the average annual amount of sunshine is about 1800 h, the annual precipitation is 840–880 mm, the frost-free period is about 210 days and the soil type is yellow brown soil. In this planting field, the American ginseng was planted in 1.5 m wide and 50 m long seedbeds before and after transplantation. The distance between the seedbeds was approximately 0.5 m, the plant spacing was about 10 cm, and the row spacing was about 20 cm. The field blocks were arranged in a completely randomized block design, with five replicate plots (1.5 × 10 m). American ginseng was cultivated and transplanted on newly cultivated farmland with no previous agricultural tillage. The organic fertilizer used in the experiment was aerobically composted cow manure and mushroom residue, where the organic matter (OM) was 25.10%, total nitrogen (TN) was 11.32 $g \cdot kg^{-1}$, available phosphorus (AP) was 275.15 $mg \cdot kg^{-1}$, available potassium (AK) was 2657.13 $mg \cdot kg^{-1}$ and water content (WC) was 1.02%. Organic fertilizer was collected from the experimental base of Shaanxi Institute of Microbiology and was applied as base fertilizer (4.0 $kg \cdot m^{-2}$) in March before the cultivation of American ginseng, followed by an equal amount of topdressing in March every year. Soil water content was adjusted to a range of 40–50% (w/w). Good Agricultural Practice (GAP) was followed during the cultivation process [22]. The whole American ginseng plants would be transplanted into new soil blocks after 2 years of cultivation.

Ginseng rhizosphere and non-rhizosphere soil was collected at the beginning of September 2021. The ginseng rhizosphere soil was defined as tightly attached to plant roots. Ten to fifteen healthy plants with green leaves and stems without spots, and ginseng roots without markings and rot were randomly selected. The taproot was vigorously shaken to collect the adhering soil from non-rhizosphere soil samples, followed by vortex oscillation for collection of tightly attached soil for ginseng rhizosphere samples. In this study, ginseng rhizosphere and non-rhizosphere soil was referred as "microhabitats". In total, 20 samples were collected and each sample was subjected to three rounds of liquid nitrogen treatment before being kept on ice and immediately stored at −80 °C until further analysis.

2.2. DNA Extraction, Amplification and Sequence Processing

DNA was extracted from 0.5 g soil using the Fast DNA SPIN Kit for Soil (MP Biomedicals, Solon, OH, USA), and the isolated microbial DNA was used as a template for subsequent sequencing. The yield and quality were assessed using an Agilent 2100 Bioanalyzer (Agilent Technologies, Santa Clara, CA, USA). Bacterial 16S rRNA (V3-V4) gene amplifications were amplified using primers 341F (5′-CCTACGGGNGGCWGCAG-3′) and 805R (5′-GACTACHVGGGTATCTAATCC-3′) [23], while fungal internal transcribed spacer region ITS2 amplification employed primers ITS1FI2 (5′-GTGARTCATCGAATCTTTG-3′) and ITS2 (5′-TCCTCCGCTTATTGATATGC-3′) [24]. Next-generation sequencing (NGS) preparation, Illumina HiSeq NGS library preparations and Illumina HiSeq sequencing were performed by LC-Bio Technology Co., Ltd, Hangzhou, China. DNA samples were quantified using a Qubit 2.0 fluorometer (Invitrogen, Carlsbad, CA, USA), followed by amplicon generation from 30–50 ng of DNA using the NEBNext® UltraTM DNA Library Prep Kit for Illumina® (New England Biolabs, Beverly, MA, USA), according to the manufacturer's protocol. The total DNA was eluted in 50 μL of Elution buffer and stored at −80 °C until PCR measurement.

The Usearch10 [25] and Vsearch 2.8.1 [26] pipelines were employed for sequence analysis. Forward and reverse sequences were joined, assigned to respective samples based on barcodes and truncated through the removal of the barcode and primer sequences. Quality filtering on joined sequences was performed, and sequences with ambiguous bases and expected errors per base rate ≤0.01 were discarded. Subsequently, the sequences were dereplicated, and singletons (with a minuniquesize < 8) were removed. The sequences were clustered into Amplicon sequence variants (ASVs) using the exact sequence variants algorithm [27,28] (Unoise3), and chimeric sequences were simultaneously removed. The effective sequences were used in the final analysis. The taxonomic identities of the bacterial and fungal ASVs were determined via the clustering program VSEARCH 2.8.1 against the Ribosomal Database Program (RDP, http://rdp.cme.msu.edu/) (accessed on 13 January 2022) and UNITE (https://unite.ut.ee) (accessed on 28 January 2022) at 97% sequence identity, and the confidence threshold for the RDP classifier for ASVs is 0.8 [29]. The raw data of bacterial and fungal sequences were available at the National Center for Biotechnology Information under BioProject ID PRJNA975712.

2.3. Soil Physico-Chemical Properties and Morpho-Physiological Parameters

Physico-chemical properties of soil were assessed using five replicates per group. For each sample, soil was collected from the top layer (0–30 cm) at five random locations. Then, the soils were thoroughly mixed and transported from field to laboratory in sterile ice containers. In the laboratory, samples were sieved (2 mm mesh) to remove plant debris and air-dried in a designated soil drying room and analyzed for organic matter (OM), total nitrogen (TN), amino nitrogen (AN), total phosphorus (TP), Olsen-P (OP), total potassium (TK) and available potassium (AK) [30]. Dried samples were mixed with deionized water (volume ratio 1:2.5), shaken at 200 rpm for 30 min, and then centrifuged at 12,000 rpm for 5 min to determine pH and electrical conductivity (EC) [31]. OM was quantified through sulfuric acid-potassium dichromate wet oxidation, followed by titration with ferrous sulfate according to the Walkley–Black procedure [30]. TN was determined using the Kjeldahl method [32]. Soil AN was measured via diffusion methods [33]. Soil TP was assessed using the Mo-Sb anti spectrophotometric method [34]. The OP in the soil was measured using the Olsen method [35]. The TK and AK in the soil were determined via ammonium acetate extraction followed by flame photometry [36]. Concurrently, the morpho-physiological parameters of American ginseng in each group were measured. The fresh weight of ginseng root (GW) and stem (SW) was weighed. The length and diameter of taproot (TL, TD), as well as the height and diameter of stem (SH, SD), were measured. The number of branches (BN), stems (SN) and leaves (LN) per plant was recorded.

2.4. Statistical Analysis

Statistical analysis and graphic display were performed using R software version 4.1.2, Auckland, NZ and ImageGP (version 2.0, https://www.bic.ac.cn/ImageGP/ (accessed on 14 July 2022), Chen T., Beijing, China) [37]. Alpha diversity was assessed using the Shannon index with the "vegan" package (version 2.5-6, https://CRAN.R-project.org/package=vegan (accessed on 28 June 2022), Helsinki, Finland) [38], and Faith's phylogenetic diversity index (Faith's PD index) with the "picante" package (version 1.8.2, https://CRAN.R-project.org/package=picante (accessed on 28 June 2022), Eugene, OR, USA) (accessed on 28 June 2022) [39]. Phylogenetic tree analysis of soil was carried out using *cluster_agg* of Usearch10. The alpha diversity indices were calculated through one-way analysis of variance (ANOVA) and Tukey's multiple comparison tests.

The beta diversity of bacteria and fungi was evaluated through principal coordinate analysis (PCoA) based on Bray–Curtis distance. The significance of the effect of transplanting and microhabitats on community dissimilarity was tested using PERMANOVA with the *adonis* function in the "vegan" package. To investigate the assembly processes and multifunctionality of soil and ginseng rhizosphere microbiota, we quantified the Normalized stochasticity ratio (NST) index, beta nearest taxon (βNTI) index and niche breadth under American ginseng transplanting. Nearest-taxon index (NST) [40] and β-nearest taxon index (βNTI) [41] were calculated with the "picante" package (version 1.8.2) to explore the community assembly processes and quantify phylogenetic structure. Niche breadth index was calculated according to Levin's niche breadth [42] equation using the "spaa" package (version 0.2.1, https://CRAN.R-project.org/package=spaa (accessed on 6 July 2022), Hong Kong, China).

To further characterize the impact of transplanting on microbiota in different microhabitats, we evaluated the composition and co-occurrence patterns of bacterial and fungal communities. For taxonomic analysis, a class-level stacking bar chart was generated to check relative abundance of species. The co-occurrence network was estimated using the "igraph" package (version 1.2.5, https://CRAN.R-project.org/package=igraph (accessed on 7 July 2022), Oxford, GB). A valid co-occurrence was considered a statistically robust correlation between ASVs when the correlation threshold exceeded 0.8 and the *p* value was below 0.01. The microbial co-occurrence network graphs were visualized using Gephi (version 0.9.2, https://gephi.org/ (accessed on 7 July 2022), Paris, France) [43].

To evaluate the impact of transplanting on microbial community structure and function, we employed two classifications: microbial biomarkers filtered with the random forest algorithm and functional microorganisms selected based on niche breadth. To identify the microbial biomarkers, a random forest classifier (RFC) model was constructed and a 10-fold cross-validation on the RFC model was performed to detect unique ASV-based microbial biomarkers. The filtering criteria for functional microorganisms were as follows: ASVs set with the niche breadth index larger than the mean value were selected; ASVs set with a contribution rate higher than the average and significant *p*-value according to SIMPER analysis were also selected; The intersection of these two sets of ASVs was considered as the functional microorganism. The random forest algorithm [44] of the "randomForest" package (version 4.6-14, https://CRAN.R-project.org/package=randomForest (accessed on 11 Augest 2022), Berkeley, CA, USA) was used to quantify classification of biomarker ASVs, and the percentage of increase in mean square error (increase in MSE (%)) was used to show the importance of the biomarker ASVs. Additionally, the similarity percentage (SIMPER) algorithm was calculated with the 'vegan' package. To investigate the interrelationships among soil properties and American ginseng growth, correlation analysis was employed to demonstrate the associations between these three variables sets. The correlation between the soil properties and growth variables was evaluated using Spearman's rank correlation test. Prior to correlation analysis, the "Hmisc" package (https://cran.r-project.org/package=Hmisc (accessed on 12 August 2022), Nashville, TN, USA) was used to eliminate collinear factors. To evaluate their associations with ASVs, factors were transformed and normalized before application of the Mantel test [45]. The ASVs with relative abundances below 0.01% of total sequences were defined as "rare" ASVs [46,47]. Partial least squares path modeling (PLS-PM) [48] was employed to quantify the relationships between soil properties, ASVs and American ginseng growth variables.

3. Results

3.1. Soil and Ginseng Rhizosphere Microbiota Harbor Distinct Communities after Transplanting

We first compared the differences in bacterial and fungal communities between soil and ginseng rhizosphere samples pre- and post- transplanting. After transplantation, there was a significant decrease in the phylogenetic diversity, richness and evenness of the bacterial community (Figure 1A, Faith's PD index: $F_{1,18} = 9.527$, $p < 0.01$, Shannon index: $F_{1,18} = 19.2$, $p < 0.001$), while no significant difference was observed in the alpha diversity indices of the fungal community (Figure 1B, Faith's PD index: $p = 0.613$, Shannon index: $p = 0.838$). Moreover, a significant reduction in the alpha diversity of bacterial communities within soil microhabitat was observed after transplantation (Figure 1A, Faith's PD index: $F_{1,8} = 6.19$, $p = 0.0376$, Shannon index: $F_{1,8} = 8.946$, $p = 0.0173$), while no such reduction was detected in ginseng rhizosphere microhabitat (Supplementary data: Table S1). These findings suggested that the transplanting process primarily impacted soil bacterial community diversity, and the bacterial community in ginseng rhizosphere microhabitat remained relatively stable post-transplantation. No discernible differences were detected in soil fungal communities in the transplanting process.

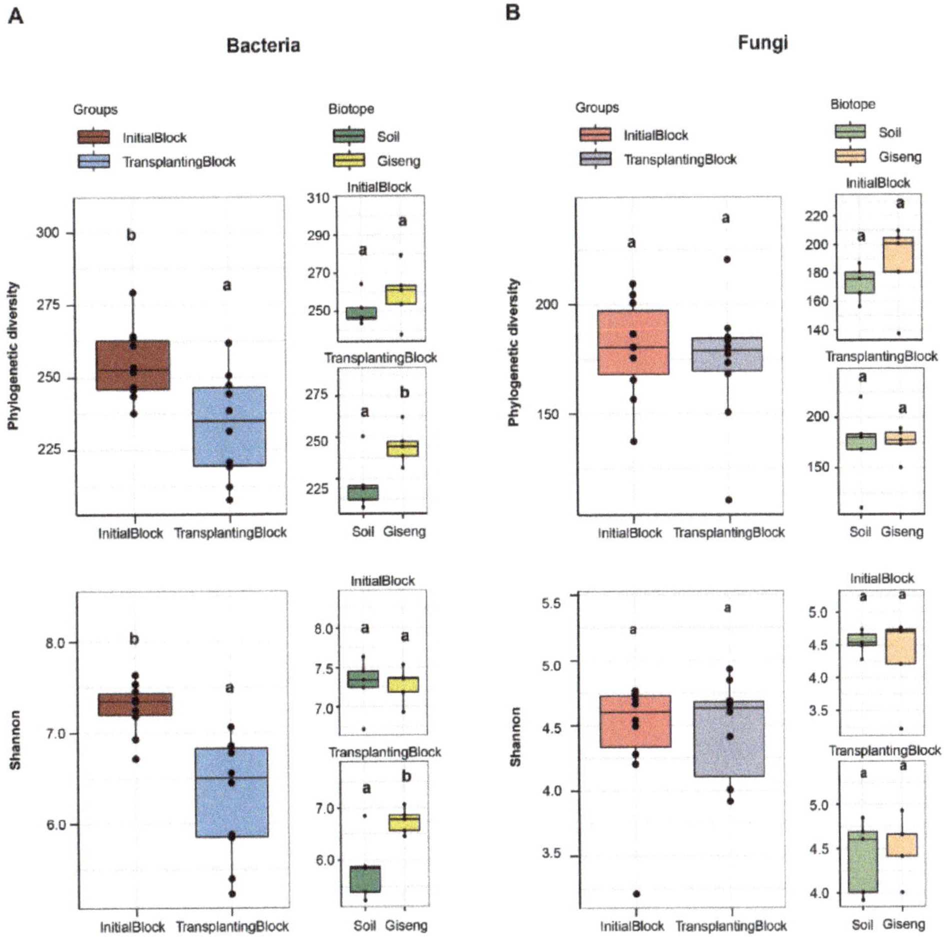

Figure 1. Comparison of microbial diversity and evenness in different microhabitats before and after transplanting. (**A**) Faith's PD and Shannon indices of bacterial community. (**B**) Faith's PD and Shannon indices of fungal community. ANOVA and Tukey's multiple comparison tests were carried out for each group, and the statistical significances ($p < 0.05$) were indicated by different letters.

3.2. Different Structures and Variation of Soil and Rhizosphere Microbiota in Transplanting Process

In principle coordinate analysis (PCoA) of bacterial Bray–Curtis distance across all samples, ginseng rhizosphere samples exhibited clustering while soil samples shifted away from the rhizosphere after transplanting in the second coordinate axis. Additionally, a significant decrease was observed in the Bray–Curtis distance between pre- and post-transplantation samples (Figure 2A). Interestingly, the fungal community structure exhibited the opposite trend, in that a significant separation was observed among ginseng rhizosphere fungi samples along the first coordinate axis after transplantation, while there was no significant alteration in the Bray–Curtis distance of soil samples before and after transplanting. Our study also found that the Bray–Curtis dissimilarity index of the bacterial communities in soil microhabitat decreased significantly after transplanting, as compared to fungal communities (Figure 2B).

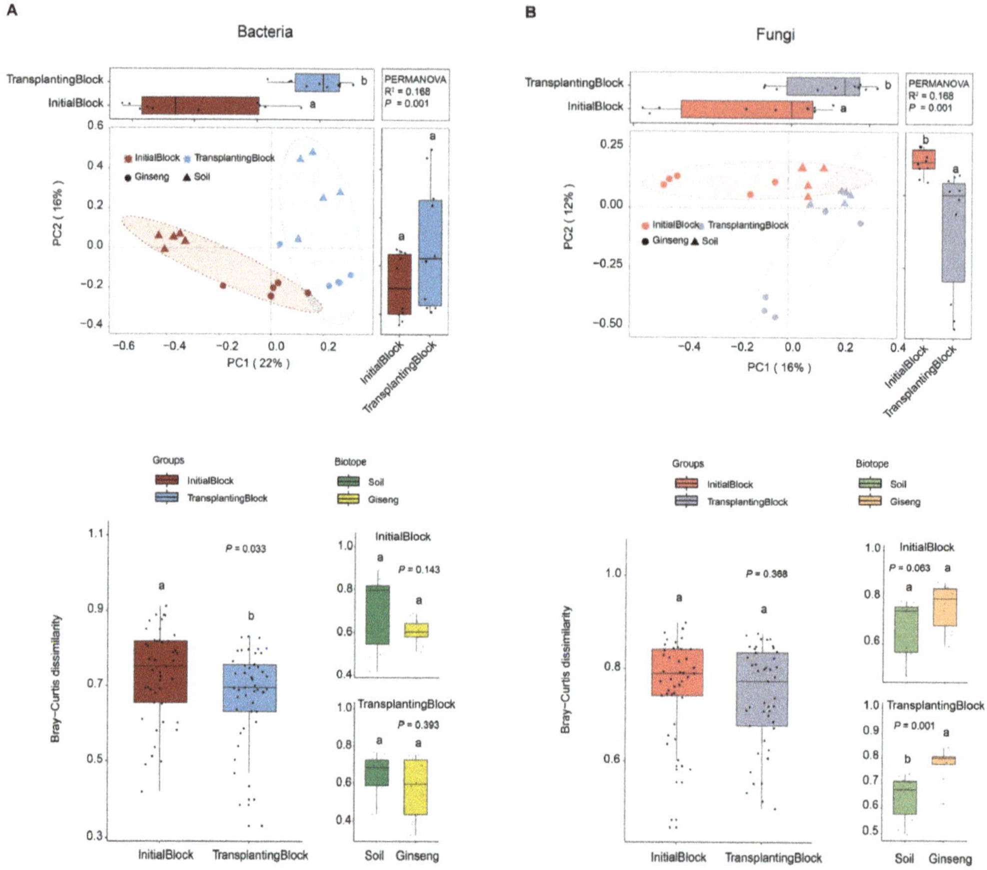

Figure 2. Comparison of community composition and variation degree between soil and ginseng rhizosphere microbiota before and after transplanting based on ASVs. (**A**) Beta diversity measurement with Principal coordinate analysis (PCoA; pairwise comparisons based on PERMANOVA). (**B**) Bray–Curtis distances before and after transplanting between soil and ginseng rhizosphere microbiota.

Generally, significant differences were found in NST, βNTI and niche breadth of bacterial communities before and after transplanting, but not in fungal communities. The NST of the bacterial community increased significantly after transplanting, indicating that the proportion of randomness increased, and the increase was mainly attributed to the contribution of soil microhabitat. The βNTI of the bacterial community was >2 before and after transplantation, suggesting that the community assembly was a deterministic process. However, the βNTI decreased after transplantation, revealing a trend towards stochasticity in the rhizosphere microhabitat bacterial community (Figure 3A). These results suggest that the assembly of the bacterial community after transplanting was stochastic, especially in the ginseng rhizosphere microhabitat. Additionally, the niche breadth of the bacterial community decreased significantly after transplanting, indicating that decreased environmental adaptability caused by this reduction could ultimately lead to a decline in deterministic processes of bacterial community assembly. Although there was no significant difference in the fungal community before and after transplanting, NST significantly decreased and βNTI significantly increased after transplanting in ginseng rhizosphere microhabitat (Figure 3B). This suggests an increase in the deterministic trend of fungal community assembly in ginseng rhizosphere microhabitat after transplanting. In addition, we also observed that the niche breadth of the fungi community in the ginseng rhizosphere microhabitat was significantly lower compared to that in the soil microhabitat, regardless of transplanting (Figure 3B). This result suggested that the fungi in the ginseng roots were composed of specific taxa and tended to be conservative in function.

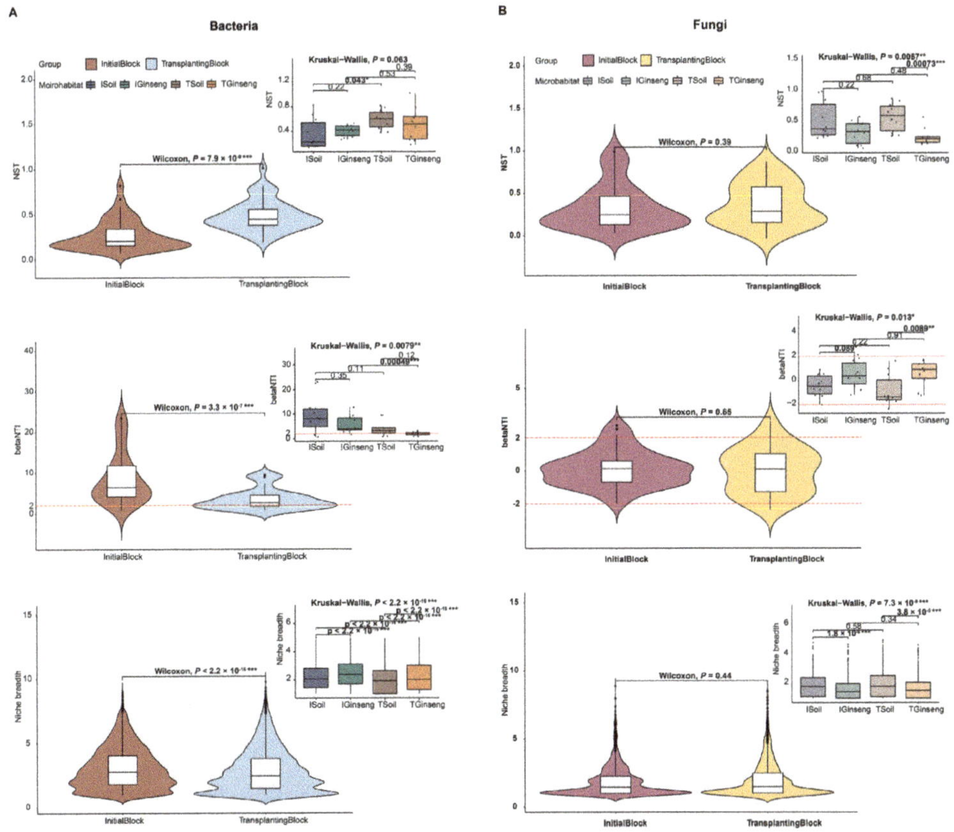

Figure 3. Comparison of assembly processes and multifunctionality of microbiota between soil and ginseng rhizosphere microbiota in transplanting process. (**A**) Normalized stochasticity ratio (NST) index, beta nearest taxon (βNTI) index and niche breadth of bacterial community. (**B**) NST index, βNTI index and niche breadth of fungal community.Black asterisks represented significant differences between groups with Wilcoxon or Kruskal–Wallis test. Asterisk (*) represented $p < 0.05$; double asterisk (**) represented $p < 0.01$; triple asterisk (***) represented $p < 0.001$.

3.3. Composition and Co-Occurrence Network of Microbial Communities in Different Microhabitats before and after Transplanting

Overall, the bacterial community was dominated by Alphaproteobacteria, Gammaproteobacteria, Actinobacteria, Gemmatimonadetes and Betaproteobacteria with the relative abundance exceeding 60% (Figure 4A). Furthermore, there were notable variations in the abundance of certain classes in different microhabitats after transplanting. Gammaproteobacteria and Sphingobacteriia exhibited a significant decrease, whereas Gemmatimonadetes and Betaproteobacteria demonstrated a marked increase in the ginseng rhizosphere microhabitat (Supplementary Materials Table S2). In co-occurrence networks, bacterial classes in soil microhabitat demonstrated a higher network connectivity (as characterized by the degree distribution) but a lower aggregation (as characterized by the clustering coefficient) than those observed in the ginseng rhizosphere, regardless of transplanting. Additionally, ASVs belonging to Gammaproteobacteria in soil microhabitat were the main hub nodes of the network, whereas classes with high abundance such as Alphaproteobacteria, Gemmatimonadetes and Betaproteobacteria in the ginseng rhizosphere microhabitat were not in the dense area of the network. Meanwhile, we observed a significant increase in the number of nodes (99 to 129) and edges (528 to 932) above the average degree of distribution in the ginseng rhizosphere microhabitat after transplantation, while there was an obvious decrease in the number of edges (1052 to 897) in the soil habitat (Figure 4B).

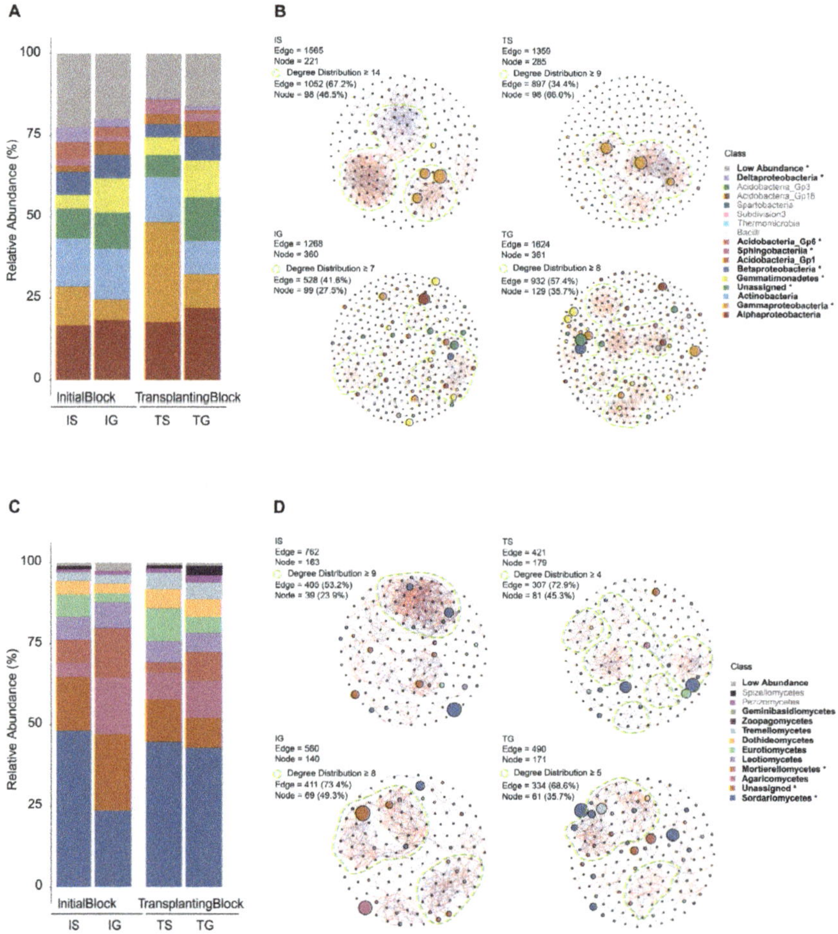

Figure 4. Relative abundance and correlation networks of microbiota before and after transplanting in different microhabitats at class level. (**A**) The relative abundance of the top 10 members of the bacterial community. (**B**) Different effects of transplanting on the networks of bacterial communities in soil and ginseng rhizosphere microhabitat. (**C**) The relative abundance of top 10 members of the fungal community. (**D**) Different effects of transplanting on the networks of fungal communities in soil and ginseng rhizosphere microhabitat. Black asterisks (*) represent classes with significant differences ($p < 0.05$) in relative abundance based on the ANOVA test. The lines between the points represent correlation coefficients (red, positively correlated; blue, negatively correlated). The points enclosed by the green dotted lines represent classes with higher than average degree distribution.

Overall, the fungal community and correlations remained stable before and after transplantation. The combined relative abundance of Sordariomycetes, Agaricomycetes, Mortierellomycetes and Leotiomycetes accounted for over 75% (Figure 4C). Sordariomycetes increased significantly in the ginseng rhizosphere microhabitat after transplanting (Supplementary Materials Table S2). The fungal correlation network had fewer edges and lower distribution after transplanting. Additionally, ASVs belonging to Sordariomycetes in soil microhabitat were the main hub nodes in the network, but were not distributed in the dense network of the ginseng rhizosphere microhabitat. We also found that the proportion of nodes (23.9% to 45.3%) and edges (53.2% to 72.9%) above the average distribution of soil microhabitats increased significantly after transplanting, while those of ginseng rhizosphere microhabitats decreased markedly (Figure 4D, nodes: 49.3% to 35.7%; edges: 73.4% to 68.6%). In summary, these results showed that the transplanting process reduced the connectivity of fungal communities in different microhabitats and weakened the relationship between taxa. Interestingly, our analysis also revealed certain taxa in the microbial correlation network that were not among the top 10 classes, including Bacilli, Thermomicrobia, Spartobacteria, Acidobacteria_Gp16 and Acidobacteria_Gp3 in bacteria, as well as Pezizomycetes and Spizellomycetes in fungi.

3.4. Association of Microbial Biomarkers and Functional Microorganisms with Soil Properties and Growth of American Ginseng

We observed that, although the proportion of ASVs exceeded 50% (55.3% in bacteria and 57.8% in fungi), functional microorganisms exhibited differential distribution before and after transplanting. In the fungal biomarkers screened, we opted for fewer ASVs (156 ASVs) rather than enhancing model accuracy. Finally, through the intersection of microbial markers and functional microorganisms, we identified three subgroups of ASVs (Figure 5): functional ASVs (Bacteria: 232 ASVs, 39.1%; Fungi: 28 ASVs, 15.2%), important ASVs (Bacteria: 255 ASVs, 43.0%; Fungi: 132 ASVs, 71.7%), and important functional ASVs (Bacteria: 106 ASVs, 17.9%; Fungi: 24 ASVs, 13.0%).

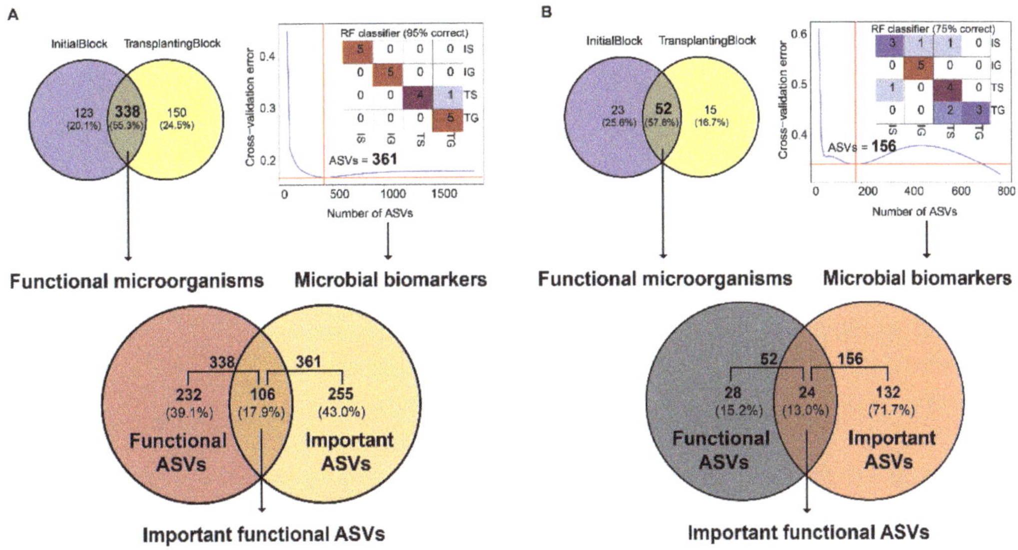

Figure 5. Filtering strategy of microbial biomarkers and functional microorganisms in different microhabitats and definitions of different types of ASVs. (**A**) Bacterial ASV filtering and definition. (**B**) fungal ASV filtering and definition. The diagonal numbers and colors represent the accuracy of the RF classifier with top features. The redder color represented higher correct rate.

Overall, a negative correlation was observed between soil properties and growth variables. In soil properties, pH exhibited a significant positive correlation with organic matter (OM), total nitrogen (TN), and total potassium (TK). Conversely, the growth weight (GW) of American ginseng displayed a significant negative correlation with soil OM, TN and TP. Height of stem (SH) was negatively correlated with both OM and TN. Diameter of taproot (TD) and TN were significantly negatively correlated. Only soil available phosphorus (AP) demonstrated a positive correlation with growth variables (Figure 6).

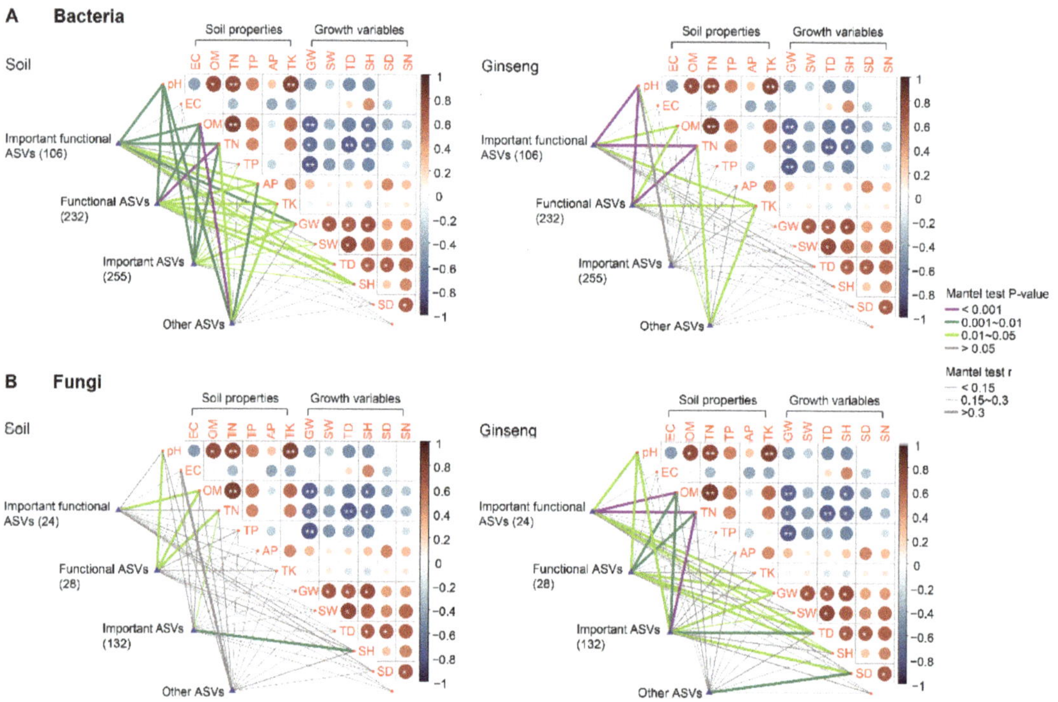

Figure 6. Relationships between soil properties, growth variables and subgroups of important functional ASVs, functional ASVs, important ASVs and other ASVs. (**A**) Bacterial subgroups in soil and ginseng rhizosphere microhabitats. (**B**) Fungal subgroups in soil and ginseng rhizosphere microhabitats. The upper right matrix represented the correlation coefficient matrix diagram between soil properties and growth variables, and the right bar was the correlation coefficient contrast color. Significant Spearman correlation coefficients were marked with an asterisks (*) ($p < 0.05$) and double asterisks (**) ($p < 0.01$).

To determine the relationship between microbiota in different microhabitats and American ginseng growth, we profiled the correlation between ASV subgroups and variables with the Mantel test. We observed that the bacterial ASVs in the soil microhabitat and the fungal ASVs in the ginseng rhizosphere microhabitat were more strongly correlated with soil properties and American ginseng growth. Within the bacterial ASVs subgroups, important functional ASVs and functional ASVs were significantly correlated with soil pH, OM, TN, AP and TK, as well as American ginseng GW and SH in the soil microhabitat. Additionally, functional ASVs were also significantly correlated with TD, and important ASVs were significantly correlated with pH, OM, TN and GW (Figure 6A). Interestingly, although other ASVs subgroup were significantly correlated with some soil properties (pH, OM, TN, AP and TK), this subgroup was not significantly correlated with growth variables of American ginseng. In the ginseng rhizosphere microhabitat, all bacterial subgroups were not significantly correlated with growth variables of American ginseng. Conversely, important functional ASVs and functional ASVs were strongly significantly correlated with soil pH and OM (Mantel's r > 0.3, *p*-value < 0.001). The results indicated that the bacterial communities in soil microhabitats mainly interacted with soil properties and had little effect on the growth of American ginseng.

In the fungal subgroups, important functional ASVs, functional ASVs and important ASVs were significantly correlated with soil pH, OM and TN as well as American ginseng GW and TD in the ginseng rhizosphere microhabitat (Figure 6B). Notably, OM and TN showed strong correlations with all subgroups of ASVs (Mantel's r > 0.3, *p*-value < 0.01). We observed that important ASVs were significantly strongly correlated with American ginseng TD, while other ASVs were strongly significantly correlated with SD (Mantel's r > 0.3, *p*-value < 0.01). In the soil microhabitat, functional ASVs showed significant correlations with pH, OM and TN, while important functional ASVs were significantly correlated only with OM. Only important ASVs were strongly significantly correlated with American ginseng SH (Mantel's r > 0.3, *p*-value < 0.01). In general, fungal communities mainly interacted with soil properties in the root rhizosphere microhabitat. It was also noteworthy that different subgroups of fungal ASVs, including important ASVs and other ASVs, had a greater impact on the growth of American ginseng.

3.5. Driving Forces for American Ginseng American Growth

Recent studies have increasingly emphasized the importance of rare taxa and studied the responses of rare sub-communities to the cropping process [47,49]. In this study, PLS-PM analysis was conducted to assess the direct and indirect effects of soil properties, bacterial ASVs and fungal ASVs on growth variables in soil and ginseng root microhabitat of American ginseng. According to the previous results, we combined three types of ASV into NW-RF ASVs, and divided the other ASVs into intermediate and rare ASVs (Figure 7). PLS-PM analysis showed that soil properties had significant positive effects on bacterial and fungal ASVs in the soil microhabitat (Figure 7A). In the ginseng root microhabitat, all soil properties had positive effects on bacterial ASVs and negative effects on fungal ASVs (Figure 7B). It was found that bacterial and fungal sub-group ASVs showed different effects in different microhabitat. Bacterial NW-RF ASVs showed negative total standardized effects in soil microhabitat, but a positive one in ginseng root microhabitat, while fungal NW-RF ASVs displayed positive total standardized effects in both microhabitats. Bacterial intermediate ASVs showed positive and rare ASVs showed negative total standardized effects on the American ginseng growth in both microhabitats (Figure 7A). In fungal ASVs, intermediate ASVs exhibited negative effects in soil microhabitats but positive effects in ginseng root microhabitats. Conversely, rare ASVs displayed the opposite trend and had the largest total standardized effects (Figure 7B).

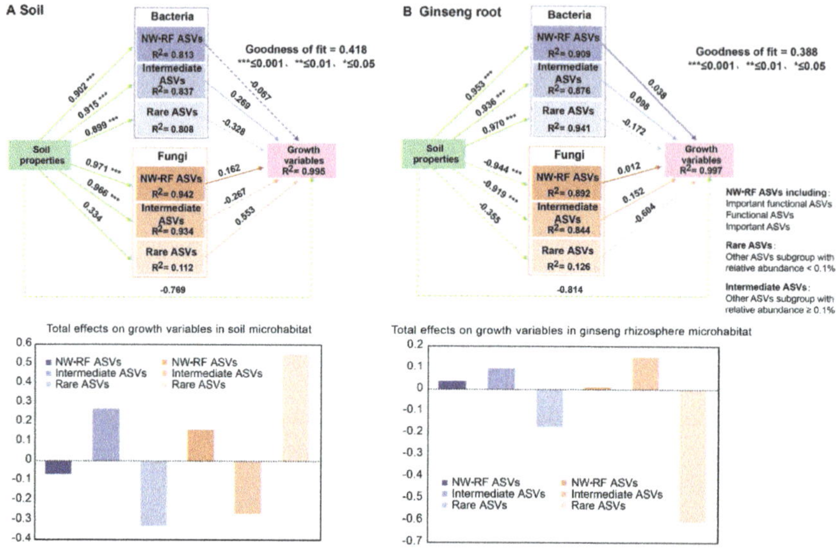

Figure 7. The partial least squares path models showing the effects of soil properties, bacterial ASVs, fungal ASVs on growth variables in soil (**A**) and ginseng root (**B**) microhabitat of American ginseng. Solid and dashed lines indicate positive and negative effects, respectively. Numbers adjacent to each arrow denote partial correlation coefficients (significance codes: *** $\leq$ 0.001, ** $\leq$ 0.01, * $\leq$ 0.05). R^2 values displayed the proportion of variance explained for each factor. The bar chart shows the standardized total effect of each factor on the American ginseng growth variables soil and ginseng root microhabitat.

4. Discussion

Our initial hypothesis was that transplanting pattern was an important factor affecting American ginseng rhizosphere soil properties and microbiota. To adapt to soil environmental disturbances, plants were able to recruit functional microorganisms and alter microbial interactions in the rhizosphere soil by changing chemical conditions and releasing root-derived compounds [18,50]. To verify this hypothesis, the rhizosphere–soil microbial community was investigated before and after transplanting, Further, their effects on soil properties and American ginseng growth was examined. We observed that the transplanting process mainly affected the diversity and structure of the bacterial community, and the fungal community was relatively stable. The bacterial community in soil microhabitat was more active and had a stronger correlation with soil properties, whereas the fungal community of the ginseng rhizosphere microhabitat was more closely related to soil properties and American ginseng growth.

4.1. Response of Soil and Rhizosphere Microhabitat Microbial Communities to Transplanting of American Ginseng

Changes in soil properties of American ginseng have been reported as being the main factor causing disturbance in microbial community structure and composition [51,52]. In this study, significant changes were observed in the structure, variation, assembly process, niche breadth and composition of bacteria compared to fungi before and after transplanting (Figures 1–4). These results suggest that transplanting had little effect on the diversity and structure of fungal communities, consistent with previous studies [10,53]. Moreover, microhabitats had more significant effects on the bacterial community. After transplanting, the Faith's PD and Shannon indices of the soil microhabitat bacterial community decreased significantly, and the Bray–Curtis distances of ginseng root microhabitat were larger (Figures 1 and 2). In agro-ecosystems soils, plant roots established interactions with rhizosphere microorganisms during growth [16,54]. This effect might contribute to the establishment of a stable bacterial community in the ginseng root microhabitat, while the bacterial community in the soil microhabitat was more dynamic and significantly associated with soil properties and American ginseng growth (Figure 6). Microbial community abundance distributions in the environment could provide a broad reservoir of ecological function and resiliency [55]. Significant changes in soil microhabitat bacteria might be due to the species migration of the adjacent bulk soil microbiota. However, this effect was not observed in fungi. The Bray–Curtis dissimilarity and deterministic process of fungal community in ginseng root microhabitat increased significantly after transplanting (Figures 2 and 3). These results indicated that the ginseng rhizosphere bacterial community structure tended to be conserved after transplantation, while the fungal community structure changed significantly because of the selectivity of American ginseng roots. The composition of plant-colonizing microbial communities was mainly influenced by the selective pressure of plants [56]. In general, these results indicated that the bacterial community was mainly affected by soil properties and mediated rhizosphere activities though soil microhabitats [6]. The fungal community was mainly affected by ginseng root growth, and the root microhabitat provided a favorable environment for the enrichment of rhizosphere fungi [17,18].

4.2. Microbial Composition Exhibited Distinct Correlation Network Patterns in Different Microhabitats before and after Transplanting

The soil physicochemical properties and soil microbial communities interactions shifted with the growth of American ginseng plants [53]. In this study, different microhabitat greatly altered the distribution, the clustering and interactions among hub nodes of different microbial classes of the co-occurrence network (Figure 4; Supplementary data: Table S3). One interesting finding was that Alphaproteobacteria, which exhibited the highest relative abundance, displayed a weak association with the bacterial network (Figure 4). Keystone species belonging to Gammaproteobacteria were more connected with other taxa in soil microhabitat than in ginseng root microhabitat, while Gemmatimonadetes in ginseng root microhabitat showed higher connectivity in the correlation network. The fungal keystone species belonging to Sordariomycetes were more conservative, maintaining high abundance and connectivity in the network (Figure 4), which implied the fungal co-occurrence network was more stable. This study further found that the connectivity of the fungal community network in the ginseng root microhabitat was greatly reduced after transplanting compared with bacteria (Supplementary data: Table S3). After transplanting, soil habitat microbes were more loosely connected, but fungal classes of hub nodes were more closely connected than bacteria (Figure 4). After transplanting, the bacterial network in ginseng root microhabitat was more complex and more connected, while fungal networks had shorter average paths length, more network modules and a higher degree of modularity (Supplementary data: Table S3). The findings suggested that the ginseng rhizosphere microhabitat had a recruitment effect on the bacterial community, promoting the aggregation of bacterial taxa, and maintained this feature even after transplanting. While the fungi in the ginseng root microhabitat showed small-world characteristics after transplanting, which allowed the effects of a disturbance to be transmitted quickly throughout the network, making the fungal system efficient [57].

4.3. Effects of Microbial Communities and Soil Properties on Growth Variables in Soil and Ginseng Root Microhabitats

The similarity percentage (SIMPER) could evaluate the contribution of different species to inter-group variations, while niche breadth was indicative of a species' utilization of environmental resources [58,59]. In this study, we defined functional ASVs based on the niche breadth and contribution rate, and identified important ASVs using a random forest algorithm. Our study revealed that distinct types of ASVs exhibited significant associations with diverse soil properties and growth variables (Figure 6; Supplementary data: Table S4). We observed that important functional bacterial ASVs with high relative abundance were bASV_11 (*Rhodanobacter*), bASV_31 (*Bradyrhizobium*) and bASV_10 (*Acidobacterium*). *Rhodanobacter* was reported to exhibit antagonistic effects against ginseng root rot pathogen *Fusarium solani* [60]. Additionally, it was previously positively correlated with the soil pH and N, which could impact soil denitrification [61]. *Bradyrhizobium* could protect plants growing in a natural soil against abiotic stress, as it is a known nitrogen-fixing bacteria commonly found in plant soils [62]. As a endophytic and rhizosphere bacterium, *Acidobacterium* could be used as a candidate microorganism to reflect soil fertility and plant health [63]. *Mizugakiibacter* and *Gemmatimonas* was observed in high relative abundance in important (bASV_16, bASV_75) and functional ASVs (bASV_2, bASV_40). *Mizugakiibacter* have commonly been found in abundance at low pH in crop fields [64], and *Gemmatimonas* has been reported to be abundant in healthy Panax ginseng soil, as it has a greater ability to use available C sources [50,65]. In fungal ASVs, *Fusarium*, the main soil-borne pathogen causing root rot disease in American ginseng [10,53], was highly abundant in important functional (fASV_31) and important ASVs (fASV_1), particularly after transplanting. Meanwhile, potential disease-suppressive fungi *Mortierella* was found in high relative abundance in all three types of ASVs (fASV_113 in important functional ASVs, fASV_70 and fASV_121 in important ASVs, fASV_14 and fASV_34 in functional ASVs) [52]. In general, different types of abundant bacterial ASVs primarily regulated soil physiological and biochemical properties like pH, OM and TN, providing a stable growth environment for American ginseng. Fungal ASVs primarily impacted ginseng root growth and played an important role in both the generation and antagonism of soil-borne diseases.

The composition of the microbial community is considered an important component of the plant–soil feedback process [47,66]. Below-ground biomass of plants could affect microbial biomass in soil and further affect microbial diversity and multifunctionality [67]. In our study, soil properties had negative effects on the fungal community in the ginseng root microhabitat. This might be due to the antagonistic effect of American ginseng on pathogenic fungi such as *Fusarium* (fASV_1 and fASV_31; Supplementary data: Table S4), which were highly ranked among NW-RF ASVs, during its growth. Previous studies also showed that root exudates like phenolic acids had antagonistic effects on fungi during the growth of American ginseng [4]. The effect of bacterial NW-RF ASVs on growth was microhabitat-dependent, suggesting that the American ginseng roots were enriched with functional bacteria, including nutrient uptake and pathogen antagonism [6]. Additionally, rare ASVs showed the largest total standardized effects on the growth of American ginseng and the effects of intermediate and rare ASVs varied across different habitats, indicating that fungal communities with low abundance could serve as a reservoir of diversity and function and be enriched in diverse environments [55]. It is worth noting that the influence of microbiota and soil properties on American ginseng growth was not significant, indicating that soil enzyme activity, climate conditions, above-ground plants, and other factors might have an impact on its growth. Overall, this study enhanced our comprehension of the dynamics of rhizosphere microorganisms in American ginseng roots during the transplanting process, thereby providing new insights into the regulation of continuous cropping obstacles in agro-ecosystems under a transplanting pattern.

5. Conclusions

This study revealed the succession of bacterial and fungal communities throughout the transplanting process of American ginseng, which showed different interactions on soil properties and growth in both soil and ginseng root microhabitats. In particular, significant differences were observed in bacterial diversity, NST, βNTI and niche breadth before and after transplantation. However, highly correlated bacterial networks were microhabitat-conserved. Specific bacterial taxa defined by different characteristics appeared to be driving changes in soil properties within the soil microhabitat. The response of fungal Bray–Curtis dissimilarity, NST and βNTI to transplanting was microhabitat-specific, particularly after transplanting. Moreover, specific fungal ASVs mainly affected the growth of American ginseng. Soil properties primarily influenced all microbial ASVs; however, the total effects on the fungal ASVs were opposite in soil and ginseng root microhabitats, showing the habitat specificity of soil properties on fungal community effects. In addition, it was found that rare ASVs had the largest total effect on American ginseng growth despite their low relative abundance. This study greatly expands our understanding of the interactions between microbial communities and soil properties during the American ginseng transplanting process, which also provides new insights into microbial regulation of rhizosphere microhabitat.

Supplementary Materials: The following supporting information can be downloaded at: https://www.mdpi.com/article/10.3390/agronomy13071876/s1, Table S1: Shannon and Faith's PD indices before and after transplanting in different microhabitats. All values were an average from all replicates ± standard deviations. Black asterisks represent significant differences based on the ANOVA test. Values with the same letters within a column did not significantly differ based on the Tukey's multiple comparisons test. Table S2: The relative abundance of top 10 class level before and after transplanting in different microhabitats. All values were an average from all replicates ± standard deviations. Black asterisks represent significant differences based on the ANOVA test. Values with the same letters within a column did not significantly differ based on the Tukey's multiple comparisons test. Table S3: Properties of the microbiota association networks before and after transplanting in different microhabitats. Table S4 Different types of ASVs, species classification, and average relative abundance in different groups. The bold numbers represented the relative abundance of ASV within the top 30 of the group.

Author Contributions: Conceptualization, F.C.; Validation, F.-A.J. and Q.-A.J.; Investigation, M.G.; Writing – original draft, F.C.; Writing – review & editing, Y.S. and Z.L.; Funding acquisition, F.C., Y.S. and Z.L. All authors have read and agreed to the published version of the manuscript.

Funding: This study was supported by Shaanxi Key Research and Development Program (2021ZDLNY05-08), Science and Technology Program of Shaanxi Academy of Science (Program No. 2018nk-08) and Shaanxi Normal University Central University Project (GK201604009), Science and Technology Program of Shaanxi Academy of Science (Program No. 2023k-08).

Data Availability Statement: The raw data presented in this study are available at the National Center for Biotechnology Information under BioProject ID PRJNA975712.

Conflicts of Interest: The authors declare no conflict of interest.

References

1. Xue, P.; Yao, Y.; Yang, X.-S.; Feng, J.; Ren, G.-X. Improved antimicrobial effect of ginseng extract by heat transformation. *J. Ginseng Res.* **2017**, *41*, 180–187. [CrossRef]
2. Kim, K.H.; Lee, D.; Lee, H.L.; Kim, C.-E.; Jung, K.; Kang, K.S. Beneficial effects of Panax ginseng for the treatment and prevention of neurodegenerative diseases: Past findings and future directions. *J. Ginseng Res.* **2018**, *42*, 239–247. [CrossRef]
3. Huang, X.; Liu, Y.; Zhang, Y.; Li, S.-P.; Yue, H.; Chen, C.-B.; Liu, S.-Y. Multicomponent assessment and ginsenoside conversions of *Panax quinquefolium* L. roots before and after steaming by HPLC-MSn. *J. Ginseng Res.* **2019**, *43*, 27–37. [CrossRef]
4. Jiao, X.-L.; Zhang, X.-S.; Lu, X.-H.; Qin, R.; Bi, Y.-M.; Gao, W.-W. Effects of maize rotation on the physicochemical properties and microbial communities of American ginseng cultivated soil. *Sci. Rep.* **2019**, *9*, 8615. [CrossRef]
5. Chang, F.; Jia, F.; Lv, R.; Guan, M.; Jia, Q.; Sun, Y.; Li, Z. Effects of American Ginseng Cultivation on Bacterial Community Structure and Responses of Soil Nutrients in Different Ecological Niches. *J. Microbiol. Biotechnol.* **2022**, *32*, 419–429. [CrossRef]
6. Xiao, C.; Yang, L.; Zhang, L.; Liu, C.; Han, M. Effects of cultivation ages and modes on microbial diversity in the rhizosphere soil of Panax ginseng. *J. Ginseng Res.* **2016**, *40*, 28–37. [CrossRef]
7. Chung, I.-M.; Lee, T.-J.; Oh, Y.-T.; Ghimire, B.K.; Jang, I.-B.; Kim, S.-H. Ginseng authenticity testing by measuring carbon, nitrogen, and sulfur stable isotope compositions that differ based on cultivation land and organic fertilizer type. *J. Ginseng Res.* **2017**, *41*, 195–200. [CrossRef]
8. Wang, F.; Suo, Y.; Wei, H.; Li, M.; Xie, C.; Wang, L.; Chen, X.; Zhang, Z. Identification and Characterization of 40 Isolated Rehmannia glutinosa MYB Family Genes and Their Expression Profiles in Response to Shading and Continuous Cropping. *IJMS* **2015**, *16*, 15009–15030. [CrossRef]
9. Wei, W.; Yang, M.; Liu, Y.; Huang, H.; Ye, C.; Zheng, J.; Guo, C.; Hao, M.; He, X.; Zhu, S. Fertilizer N application rate impacts plant-soil feedback in a sanqi production system. *Sci. Total Environ.* **2018**, *633*, 796–807. [CrossRef]
10. Zhang, J.; Fan, S.; Qin, J.; Dai, J.; Zhao, F.; Gao, L.; Lian, X.; Shang, W.; Xu, X.; Hu, X. Changes in the Microbiome in the Soil of an American Ginseng Continuous Plantation. *Front. Plant Sci.* **2020**, *11*, 572199. [CrossRef]

11. Dong, L.; Xu, J.; Zhang, L.; Yang, J.; Liao, B.; Li, X.; Chen, S. High-throughput sequencing technology reveals that continuous cropping of American ginseng results in changes in the microbial community in arable soil. *Chin. Med.* **2017**, *12*, 18. [CrossRef]
12. Zhu, B.; Wu, J.; Ji, Q.; Wu, W.; Dong, S.; Yu, J.; Zhang, Q.; Qin, L. Diversity of rhizosphere and endophytic fungi in *Atractylodes macrocephala* during continuous cropping. *PeerJ* **2020**, *8*, e8905. [CrossRef]
13. DesRochers, N.; Walsh, J.P.; Renaud, J.B.; Seifert, K.A.; Yeung, K.K.-C.; Sumarah, M.W. Metabolomic Profiling of Fungal Pathogens Responsible for Root Rot in American Ginseng. *Metabolites* **2020**, *10*, 35. [CrossRef]
14. Tao, C.; Li, R.; Xiong, W.; Shen, Z.; Liu, S.; Wang, B.; Ruan, Y.; Geisen, S.; Shen, Q.; Kowalchuk, G.A. Bio-organic fertilizers stimulate indigenous soil Pseudomonas populations to enhance plant disease suppression. *Microbiome* **2020**, *8*, 137. [CrossRef]
15. Bennett, J.A.; Klironomos, J. Mechanisms of plant–soil feedback: Interactions among biotic and abiotic drivers. *New Phytol.* **2019**, *222*, 91–96. [CrossRef]
16. Toal, M.E.; Yeomans, C.; Killham, K.; Meharg, A. A review of rhizosphere carbon flow modelling. *Plant Soil* **2000**, *222*, 263–281. [CrossRef]
17. Tong, A.-Z.; Liu, W.; Liu, Q.; Xia, G.-Q.; Zhu, J.-Y. Diversity and composition of the Panax ginseng rhizosphere microbiome in various cultivation modesand ages. *BMC Microbiol.* **2021**, *21*, 18. [CrossRef]
18. Wu, J.; Shi, Z.; Zhu, J.; Cao, A.; Fang, W.; Yan, D.; Wang, Q.; Li, Y. Taxonomic response of bacterial and fungal populations to biofertilizers applied to soil or substrate in greenhouse-grown cucumber. *Sci. Rep.* **2022**, *12*, 18522. [CrossRef]
19. Cao, X.; Liu, S.; Chen, L.; Wang, J.; Xiang, D.; Zhang, J.; Qiao, Z. Effects of different crops on rhizosphere bacterial diversity under immature soil conditions. *Arch. Agron. Soil Sci.* **2020**, *68*, 18–30. [CrossRef]
20. Lazcano, C.; Boyd, E.; Holmes, G.; Hewavitharana, S.; Pasulka, A.; Ivors, K. The rhizosphere microbiome plays a role in the resistance to soil-borne pathogens and nutrient uptake of strawberry cultivars under field conditions. *Sci. Rep.* **2021**, *11*, 3188. [CrossRef]
21. Liu, Y.; Chen, T.; Li, D.; Fu, J.; Liu, S. iMeta: Integrated Meta-Omics for Biology and Environments. iMeta [Internet] 2022; 1. Available online: https://onlinelibrary.wiley.com/doi/10.1002/imt2.15 (accessed on 5 July 2023).
22. Zhang, B.; Peng, Y.; Zhang, Z.; Liu, H.; Qi, Y.; Liu, S.; Xiao, P. GAP Production of TCM Herbs in China. *Planta Med.* **2010**, *76*, 1948–1955. [CrossRef]
23. Logue, J.B.; Stedmon, C.A.; Kellerman, A.M.; Nielsen, N.J.; Andersson, A.F.; Laudon, H.; Lindström, E.S.; Kritzberg, E.S. Experimental insights into the importance of aquatic bacterial community composition to the degradation of dissolved organic matter. *ISME J.* **2016**, *10*, 533–545. [CrossRef] [PubMed]
24. Karlsson, I.; Friberg, H.; Steinberg, C.; Persson, P. Fungicide Effects on Fungal Community Composition in the Wheat Phyllosphere. *PLoS ONE* **2014**, *9*, e111786. [CrossRef]
25. Edgar, R.C. Search and clustering orders of magnitude faster than BLAST. *Bioinformatics* **2010**, *26*, 2460–2461. [CrossRef] [PubMed]
26. Rognes, T.; Flouri, T.; Nichols, B.; Quince, C.; Mahé, F. VSEARCH: A versatile open source tool for metagenomics. *PeerJ* **2016**, *4*, e2584. [CrossRef]
27. Edgar, R.C. Updating the 97% identity threshold for 16S ribosomal RNA OTUs. Valencia, A., editor. *Bioinformatics* **2018**, *34*, 2371–2375. [CrossRef] [PubMed]
28. Knight, R.; Vrbanac, A.; Taylor, B.C.; Aksenov, A.; Callewaert, C.; Debelius, J.; Gonzalez, A.; Kosciolek, T.; McCall, L.-I.; McDonald, D.; et al. Best practices for analysing microbiomes. *Nat. Rev. Microbiol.* **2018**, *16*, 410–422. [CrossRef] [PubMed]
29. Bacci, G.; Bani, A.; Bazzicalupo, M.; Ceccherini, M.T.; Galardini, M.; Nannipieri, P.; Pietramellara, G.; Mengoni, A. Evaluation of the Performances of Ribosomal Database Project (RDP) Classifier for Taxonomic Assignment of 16S rRNA Metabarcoding Sequences Generated from Illumina-Solexa NGS. *J. Genom.* **2015**, *3*, 36–39. [CrossRef] [PubMed]
30. Qu, B.; Liu, Y.; Sun, X.; Li, S.; Wang, X.; Xiong, K.; Yun, B.; Zhang, H. Effect of various mulches on soil physico—Chemical properties and tree growth (Sophora japonica) in urban tree pits. *PLoS ONE* **2019**, *14*, e0210777. [CrossRef]
31. Fan, K.; Weisenhorn, P.; Gilbert, J.A.; Shi, Y.; Bai, Y.; Chu, H. Soil pH correlates with the co-occurrence and assemblage process of diazotrophic communities in rhizosphere and bulk soils of wheat fields. *Soil Biol. Biochem.* **2018**, *121*, 185–192. [CrossRef]
32. Bradstreet, R.B. Kjeldahl Method for Organic Nitrogen. *Anal. Chem.* **1954**, *26*, 185–187. [CrossRef]
33. Khan, S.A.; Mulvaney, R.L.; Mulvaney, C.S. Accelerated Diffusion Methods for Inorganic-Nitrogen Analysis of Soil Extracts and Water. *Soil Sci. Soc. Am. J.* **1997**, *61*, 936–942. [CrossRef]
34. Murphy, J.A.; Riley, J.P. A modified single solution method for the determination of phosphate in natural waters. *Anal. Chim. Acta* **1962**, *27*, 31–36. [CrossRef]
35. Wolf, A.M.; Baker, D.E. Comparisons of soil test phosphorus by Olsen, Bray P1, Mehlich I and Mehlich III methods. *Commun. Soil Sci. Plant Anal.* **1985**, *16*, 467–484. [CrossRef]
36. Attoe, O.J.; Truog, E. Rapid Photometric Determination of Exchangeable Potassium and Sodium. *Soil Sci. Soc. Am. J.* **1947**, *11*, 221–226. [CrossRef]
37. Chen, T.; Liu, Y.; Huang, L. ImageGP: An Easy-to-Use Data Visualization Web Server for Scientific Researchers. iMeta [Internet] 2022; 1. Available online: https://onlinelibrary.wiley.com/doi/10.1002/imt2.5 (accessed on 5 July 2023).
38. Oksanen, J.; Blanchet, F.G.; Kindt, R.; Legendre, P.; O'hara, R.B.; Simpson, G.L.; Solymos, P.; Stevens, M.H.H.; Wagner, H. Vegan: Community Ecology Package [Internet] 2019. Available online: https://CRAN.R-project.org/package=vegan (accessed on 5 July 2023).

39. Kembel, S.W.; Cowan, P.D.; Helmus, M.R.; Cornwell, W.K.; Morlon, H.; Ackerly, D.D.; Blomberg, S.P.; Webb, C.O. Picante: R tools for integrating phylogenies and ecology. *Bioinformatics* **2010**, *26*, 1463–1464. [CrossRef]
40. Webb, C.O.; Ackerly, D.D.; McPeek, M.A.; Donoghue, M.J. Phylogenies and Community Ecology. *Annu. Rev. Ecol. Syst.* **2002**, *33*, 475–505. [CrossRef]
41. Stegen, J.C.; Lin, X.; Konopka, A.E.; Fredrickson, J.K. Stochastic and deterministic assembly processes in subsurface microbial communities. *ISME J.* **2012**, *6*, 1653–1664. [CrossRef]
42. Levins, R. *Evolution in Changing Environments: Some Theoretical Explorations*; Monographs in Population Biology; Princeton University Press: Princeton, NJ, USA, 1968.
43. Bastian, M.; Heymann, S.; Jacomy, M. Gephi: An Open Source Software for Exploring and Manipulating Networks. *ICWSM* **2009**, *3*, 361–362. [CrossRef]
44. Breiman, L. Random forests. *Mach. Learn.* **2001**, *45*, 5–32. [CrossRef]
45. Diniz-Filho, J.A.; Soares, T.N.; Lima, J.S.; Dobrovolski, R.; Lemes Landeiro, V.; Pires de Campos Telles, M.; Rangel, T.F.; Bini, L.M. Mantel test in population genetics. *Genet. Mol. Biol.* **2013**, *36*, 475–485. [CrossRef]
46. Campbell, B.J.; Yu, L.; Heidelberg, J.F.; Kirchman, D.L. Activity of abundant and rare bacteria in a coastal ocean. *Proc. Natl. Acad. Sci. USA* **2011**, *108*, 12776–12781. [CrossRef]
47. Jiao, S.; Lu, Y. Soil pH and temperature regulate assembly processes of abundant and rare bacterial communities in agricultural ecosystems. *Environ. Microbiol.* **2020**, *22*, 1052–1065. [CrossRef]
48. Tenenhaus, M.; Vinzi, V.E.; Chatelin, Y.-M.; Lauro, C. PLS path modeling. *Comput. Stat. Data Anal.* **2005**, *48*, 159–205. [CrossRef]
49. Li, G.-L.; Wu, M.; Li, P.-F.; Wei, S.-P.; Liu, J.; Jiang, C.-Y.; Liu, M.; Li, Z.-P. Assembly and co-occurrence patterns of rare and abundant bacterial sub-communities in rice rhizosphere soil under short-term nitrogen deep placement. *J. Integr. Agric.* **2021**, *20*, 3299–3311. [CrossRef]
50. Wei, X.; Wang, X.; Cao, P.; Gao, Z.; Chen, A.J.; Han, J. Microbial Community Changes in the Rhizosphere Soil of Healthy and Rusty *Panax ginseng* and Discovery of Pivotal Fungal Genera Associated with Rusty Roots. *BioMed Res. Int.* **2020**, *2020*, 8018525. [CrossRef] [PubMed]
51. Liu, S.; Wang, Z.; Niu, J.; Dang, K.; Zhang, S.; Wang, S.; Wang, Z. Changes in physicochemical properties, enzymatic activities, and the microbial community of soil significantly influence the continuous cropping of Panax quinquefolius L. (American ginseng). *Plant Soil* **2021**, *463*, 427–446. [CrossRef]
52. Ji, L.; Tian, L.; Nasir, F.; Chang, J.; Chang, C.; Zhang, J.; Li, X.; Tian, C. Impacts of replanting American ginseng on fungal assembly and abundance in response to disease outbreaks. *Arch. Microbiol.* **2021**, *203*, 2157–2170. [CrossRef]
53. Ji, L.; Nasir, F.; Tian, L.; Chang, J.; Sun, Y.; Zhang, J.; Li, X.; Tian, C. Outbreaks of Root Rot Disease in Different Aged American Ginseng Plants Are Associated With Field Microbial Dynamics. *Front. Microbiol.* **2021**, *12*, 676880. [CrossRef]
54. Favela, A.; Bohn, M.O.; Kent, A.D. Maize germplasm chronosequence shows crop breeding history impacts recruitment of the rhizosphere microbiome. *ISME J.* **2021**, *15*, 2454–2464. [CrossRef] [PubMed]
55. Lynch, M.D.J.; Neufeld, J.D. Ecology and exploration of the rare biosphere. *Nat. Rev. Microbiol.* **2015**, *13*, 217. [CrossRef]
56. Dahlstrom, K.M.; McRose, D.L.; Newman, D.K. Keystone metabolites of crop rhizosphere microbiomes. *Curr. Biol.* **2020**, *30*, R1131–R1137. [CrossRef]
57. Yuan, M.M.; Guo, X.; Wu, L.; Zhang, Y.; Xiao, N.; Ning, D.; Shi, Z.; Zhou, X.; Wu, L.; Yang, Y.; et al. Climate warming enhances microbial network complexity and stability. *Nat. Clim. Chang.* **2021**, *11*, 343–348. [CrossRef]
58. Custodio, M.; Espinoza, C.; Peñaloza, R.; Peralta-Ortiz, T.; Sánchez-Suárez, H.; Ordinola-Zapata, A.; Vieyra-Peña, E. Microbial diversity in intensively farmed lake sediment contaminated by heavy metals and identification of microbial taxa bioindicators of environmental quality. *Sci. Rep.* **2022**, *12*, 80. [CrossRef]
59. Jiao, S.; Yang, Y.; Xu, Y.; Zhang, J.; Lu, Y. Balance between community assembly processes mediates species coexistence in agricultural soil microbiomes across eastern China. *ISME J.* **2020**, *14*, 202–216. [CrossRef]
60. Huo, Y.; Kang, J.-P.; Park, J.-K.; Li, J.; Chen, L.; Yang, D.-C. Rhodanobacter ginsengiterrae sp. nov., an antagonistic bacterium against root rot fungal pathogen Fusarium solani, isolated from ginseng rhizospheric soil. *Arch Microbiol.* **2018**, *200*, 1457–1463. [CrossRef]
61. De Clercq, D.; Van Trappen, S.; Cleenwerck, I.; Ceustermans, A.; Swings, J.; Coosemans, J.; Ryckeboer, J. Rhodanobacter spathiphylli sp. nov., a gammaproteobacterium isolated from the roots of Spathiphyllum plants grown in a compost-amended potting mix. *Int. J. Syst. Evol. Microbiol.* **2006**, *56*, 1755–1759. [CrossRef] [PubMed]
62. Schmitz, L.; Yan, Z.; Schneijderberg, M.; de Roij, M.; Pijnenburg, R.; Zheng, Q.; Franken, C.; Dechesne, A.; Trindade, L.M.; van Velzen, R.; et al. Synthetic bacterial community derived from a desert rhizosphere confers salt stress resilience to tomato in the presence of a soil microbiome. *ISME J.* **2022**, *16*, 1907–1920. [CrossRef] [PubMed]
63. Liu, J.; He, X.; Sun, J.; Ma, Y. A Degeneration Gradient of Poplar Trees Contributes to the Taxonomic, Functional, and Resistome Diversity of Bacterial Communities in Rhizosphere Soils. *IJMS* **2021**, *22*, 3438. [CrossRef]
64. Jiao, S.; Xu, Y.; Zhang, J.; Hao, X.; Lu, Y. Core Microbiota in Agricultural Soils and Their Potential Associations with Nutrient Cycling. *mSystems* **2019**, *4*, e00313–e00318. [CrossRef] [PubMed]
65. Guo, L.; Zheng, S.; Cao, C.; Li, C. Tillage practices and straw-returning methods affect topsoil bacterial community and organic C under a rice-wheat cropping system in central China. *Sci. Rep.* **2016**, *6*, 33155. [CrossRef] [PubMed]

66. Bever, J.D.; Platt, T.G.; Morton, E.R. Microbial Population and Community Dynamics on Plant Roots and Their Feedbacks on Plant Communities. *Annu. Rev. Microbiol.* **2012**, *66*, 265–283. [CrossRef] [PubMed]
67. Li, Z.; Liu, X.; Zhang, M.; Xing, F. Plant Diversity and Fungal Richness Regulate the Changes in Soil Multifunctionality in a Semi-Arid Grassland. *Biology* **2022**, *11*, 870. [CrossRef] [PubMed]

agronomy

MDPI

Article

A Meta-Analysis in Nine Different Continuous Cropping Fields to Find the Relationship between Plant Species and Rhizosphere Fungal Community

Qiuling Pang [1,†,‡], Mohammad Murtaza Alami [1,†], Weilong Yu [1], Zhen Ouyang [1], Shaohua Shu [1], Daiqun Tu [2], Mohammad Jawad Alami [3] and Xuekui Wang [1,*]

1 College of Plant Science and Technology, Huazhong Agricultural University, Wuhan 430070, China; pql1226@webmail.hzau.edu.cn (Q.P.); murtazaalami@webmail.hzau.edu.cn (M.M.A.); yuweilong@webmail.hzau.edu.cn (W.Y.); oyzzds@webmail.hzau.edu.cn (Z.O.); shushaohua@mail.hzau.edu.cn (S.S.)
2 Bureau of Agriculture and Rural Affairs of Lichuan City, Lichuan 445400, China; tudaiqun@163.com
3 Institute of Urban Environment (IUE), Chinese Academy of Sciences (CAS), Xiamen 361021, China; jawadalami459@gmail.com
* Correspondence: wang-xuekui@mail.hzau.edu.cn
† These authors contributed equally to this paper.
‡ Current Address: Guangxi Laboratory of Germplasm Innovation and Utilization of Specialty Commercial Crops in North Guangxi, Guangxi Academy of Specialty Crops, Guilin 541004, China.

Abstract: Plant species and cropping systems influence rhizospheric fungal communities' composition, diversity, and structure. The fungus community is one of the main factors behind soil health and quality. Yet, there is insufficient evidence and research on the effect of plant species with continuous cropping histories on the rhizospheric fungal community. In order to investigate how the fungal community responds to the various plant species and cropping systems, we have chosen one field that is left fallow along with eight continuously farmed areas to research. Among the eight phyla, the relative abundance of *Ascomycota* was significantly higher in *Polygonum multiflorum,* which was continuously cropped in fields for two years (P2). *Basidiomycota* was considerably higher in the fallow field (CK). Among the 1063 genera, the relative abundance of *Fusarium* was significantly higher in maize continuous-cropped fields for six years (M6), followed by *Fritillaria thunbergii* continuous-cropped fields for two years (F2), and found lower *Fusarium* abundance in CK. The alpha diversity observed in taxa, Chao1, and phylogenetic diversity indices were significantly higher in M2. β-diversity found that the fungal communities in the samples clustered from the fields in the same year were quite similar. In all the soil samples, the saprotrophic trophic type was the most common among the OTUs that had been given a function. Our studies have proved that continuous cropping and plant species changed the fungal community's composition, diversity, and structure. This research may serve as a guide for overcoming significant agricultural challenges and advancing the industry's sustainable growth.

Keywords: fungal community; composition; diversity; structure; cropping system; next generation sequencing

Citation: Pang, Q.; Alami, M.M.; Yu, W.; Ouyang, Z.; Shu, S.; Tu, D.; Alami, M.J.; Wang, X. A Meta-Analysis in Nine Different Continuous Cropping Fields to Find the Relationship between Plant Species and Rhizosphere Fungal Community. *Agronomy* **2023**, *13*, 1827. https://doi.org/10.3390/agronomy13071827

Academic Editor: Nikolaos Monokrousos

Received: 1 June 2023
Revised: 1 July 2023
Accepted: 4 July 2023
Published: 10 July 2023

1. Introduction

Due to limited land and rising market demand for food and cash crops, producers typically adopt continuous cropping systems [1]. Continuous cropping produced considerable plant production and quality changes, as well as the occurrence of different root rot diseases, for a long period of time [2]. In addition to plants and soil, soil microbial community composition and diversity are also altered by the continuous cropping system, as previously reported by the researcher [1–3]. However, in several studies, rotational diversity has been shown to provide below-ground advantages, such as increases in soil organic matter stocks. The advantages of plant diversity below ground have been connected to microbial community shifts in natural systems. This is also true for agroecosystems, which may function at many geographical and temporal dimensions. Microbial community composition, diversity, and structure vary significantly in fallow or continuous cropping soils [4–6]. Fallow or crop rotation has recently been used to fix continuous farming difficulties, lowering the presence of dangerous bacteria. By enhancing the soil microenvironment, technologies have been enhanced to reduce ongoing agricultural challenges [1,7–11].

Soil fungal communities are an integral factor in the soil ecosystem's processes of the material cycle and the transformation of energy, and they are different in bulk and rhizosphere soils due to a variety of biotic and abiotic variables. Changes in the composition, diversity, and organization of the soil fungus community directly impact soil health and quality. Fungal populations in the rhizosphere are linked to plant development, nutrient absorption, and soil-borne illnesses [8,11,12]. Soil fungi often vary in response to changing plant species, altering the composition and productivity of plant communities. Plant roots also create complex chemicals, generating resource-rich hotspots with unique properties from the bulk soil and recruiting microbial communities in the rhizosphere preferentially [1,7,9–11,13]. Other soil physical and chemical indices are less susceptible to changes in external variables such as plant rotation, land management, and cultivation methods than soil microorganisms. As a result, the biomass, composition, and diversity of microbial communities are often utilized as markers of changes in soil quality [14–16].

Previous studies have reported the effect of cropping systems on soil fungal communities in some specific crops and cash crops. However, less attention has been paid to comparing plant species with cropping histories, and it is still unclear whether plant species affect the composition and diversity of the fungal community or not. In this study, we have selected eight fields that have been continuously cropped with different plant species, such as *Coptis chinensis* Franch for six years, maize for 2 and 6 years, *P. multiflorum* for 2 and 6 years, sweet potato for 2 years, *F. thunbergia* for 2 years, cabbage for 2 years, and fallow fields as controls. This study aims to find the relationship between plant species and fungal community composition and diversity, find a suitable crop rotation for *C. chinensis* (one of the most important medicinal plants in the area) and other cash crops, and mitigate the continuous cropping obstacles.

2. Material and Methods

2.1. Soil Sampling and Soil Physicochemical Property Analysis

Soil samples were collected from eight distinct fields situated in Lichuan City, Hubei Province, China. These fields exhibited variations in terms of plant types, cropping years, and fallow areas. The predominant soil types encountered were sandy and clay, and the region experienced an average annual rainfall ranging from 1198 to 1650 mm with a yearly temperature of 12.7 °C. This particular research project focused on different types of fields: maize continuous-cropped fields referred to as M2 and M6, continuous-cropped *C. chinensis* fields known as C6, sweet potato continuous-cropped fields labeled as S2, *P. multiflorum* continuous-cropped fields termed P2 and P6, cabbage continuous-cropped fields denoted as CB2, *F. thunbergii* continuous-cropped fields identified as F2 and fallow fields simply called CK. Within each field, four soil samples were collected from the rhizospheres of ten randomly selected plants: *C. chinensis*, maize, *P. multiflorum*, sweet potato, *F. thunbergii*, and cabbage. Carefully, the roots of these plants were shaken to eliminate loosely attached soil particles. Additionally, four soil samples were obtained from the fallow fields, each consisting of five cores. All soil samples were then transported to the laboratory in ice boxes. To ensure the removal of debris and rocky components, a 2 mm sieve was employed to sieve each soil sample. Following the homogenization of the samples, 10 g of soil was placed in sterile tubes and preserved at −80 °C for subsequent DNA extraction [17].

In the analysis of soil physicochemical properties, the pH of the soil was measured using a Mettler-Toledo TE 20 instrument (Mettler-Toledo, Columbus, OH, USA). A soil suspension was prepared by mixing the soil with deionized distilled water in a ratio of 1:20 (weight/volume). The determination of soil organic matter (OM) was carried out using the potassium dichromate internal heating technique. For the assessment of total nitrogen (TN) and phosphorus (TP) levels, a Smartchem 200 discrete analyzer (Unity Scientific, Milford, MA, USA) was employed. The total potassium (TK) content was determined using an FP series multielement flame photometer (Xiang Yi, Hunan, China). To analyze the soil-available nutrients, specific extraction and measurement methods were utilized. The alkali hydrolyzed diffusion method was employed for determining soil-available nitrogen (AN). The sodium bicarbonate extraction molybdenum antimony anti-colorimetry method was used to assess soil-available phosphorus (OP). The ammonium acetate extraction flame photometer was utilized to measure soil-available potassium (AK). Lastly, the boiling water extraction curcumin colorimetry technique was employed to determine the soil-available boron (AB). These analytical techniques and instruments allowed for the comprehensive assessment of soil physicochemical properties, enabling the understanding of important nutrient levels and pH conditions in the soil [17].

2.2. Soil Enzyme Activities

Sucrase activity (S_SC) was assessed using the 3,5-Dinitrosalicylic acid colorimetry method, and the results were reported as the amount of glucose produced per gram of soil after 1 h. Urease activity (S_UE) was determined through indigo blue colorimetry, and the values were expressed as the quantity of NH3-N generated per gram of soil after 1 h. To measure phosphatase activity (S_ACP), disodium phenyl phosphate was utilized, and the activity was quantified accordingly [5].

2.3. 18S rRNA Gene Amplification and Sequencing

The PowerSoil Kit (MO BIO Laboratories, Carlsbad, CA, USA) was able to extract DNA from 0.5 g of dry soil samples (weight). The concentration of the DNA was measured using a Thermo Scientific Nanodrop 2000C Spectrophotometer (which is located in Wilmington, DE, USA). Before it was put to use, the soil DNA was kept at a temperature of −80 °C.

The primers SSU0817F (5′-TTAGCATGGAATAATRRAATAGGA-3′) and 1196R (5′-TCTGGACCTGGTGAGTTTCC-3′) of the V5–V7 regions of the 18S rRNA gene of fungus were used in this study. The following procedures were used to conduct the PCR reaction: 3 min of denaturation at 95 °C, followed by 30 cycles of 30 s at 95 °C, 30 s of annealing at 55 °C, 45 s of elongation at 72 °C, and then extension at 72 °C for 10 min and triplicated in a 20 μL mixture that included FastPfu Buffer 4 μL, dNTPs2.5 mM 2 μL, each primer (5M) 0.8 μL, FastPfu Polymerase 0.4 μL, BSA 0.2 μL, and template DNA 10 μg. Purified PCR products were measured using QuantiFluorTM-ST (Promega, Madison, WI, USA) and purified using the DNA gel extraction kit AxyPrep (Axygen Biosciences, Union City, CA, USA) [6].

We sequenced the pooled pure amplicons using an Illumina MiSeq (Illumina, San Diego, CA, USA) according to Illumina's standard procedures. The proportions of the purified amplicons in the pool were equal. The quality of the raw FASTQ was improved by using Trimomatic [18] and FLASH [19] to filter them according to the following criteria: (1) Any site that had an average quality score of 20 or below over a sliding window of 50 base pairs was used to truncate the 300 base pair reads, and any truncated reads that were shorter than 50 base pairs were removed from the analysis. Additionally, reads that included N-bases were thrown out. (2) Sequences were merged according to their overlaps, regardless of whether there was a mismatch of more than 2 base pairs, or if the overlaps were longer than 10 base pairs. (3) The sequences of each sample were identified by using barcodes (exactly matching), primers (allowing for two nucleotide mismatching), and readings with ambiguous bases. Barcodes and primers were both used to ensure that each sample's sequences were identical. Quantitative insights into microbial ecology (QIIME) [20] were used to develop operational taxonomic units (OTUs) by similarity using Mothur version 1.31.1 [21] with 97 percent cutoff points, and chimeric sequences were removed from the dataset. A degree of confidence of 70% was assigned to the results of comparing the taxonomy of each gene sequence to the Silva (SSU123) database.

For fungal functional prediction, the fungal OTUs were formatted into text and sent to the FUNGuild (https://github.com/UMNFuN/FUNGuild, (accessed on 10 December 2021)) [22].

2.4. Statistical Analyses

For the purpose of conducting multiple comparisons, a one-way analysis of variance (ANOVA) was done, followed by Tukey's (honestly significant difference) multiple-range tests. The Pearson correlation coefficients between soil parameters and fungal phylum abundances were computed using SPSS v20.0 (provided by SPSS Inc., Chicago, IL, USA). The OTU was used as a basis for establishing all of the other alpha diversity evaluations. The maximum amount of dissimilarity that may exist across clusters is 3%. Calculations of Chao, Simpson, Shannon, and phylogenetic diversity were performed on each sample in order to quantify its inherent richness and diversity. The Mothur Version 1.35.1 program was used to construct rarefaction curves by comparing the relative quantities of fungal OTUs based on the average number of OTUs that were found. The nine distinct field soils contributed to the overall variety. In order to create PCoA plots, both the weighted and unweighted UniFrac distance metrics were used, both of which were determined by the phylogenetic structure. Finally, Venn diagrams were used to illustrate species that were shared by more than one sample.

3. Results

3.1. Composition of Fungal Community

The rarefaction curves were generated by picking sequences at random from among the altered sequences and then graphing the sequences against the OTUs that each sequence represented (Figure S1). The curves became more uniform in shape as the number of sequences and OTUs increased. In addition to this, the number of reads found in M2 was the greatest. The nine soil samples included OTUs that belonged to 16 phyla, 70 classes, 179 orders, 432, 1063, 1963, and 9443 OTUs. Eight phyla and one unclassified phylum were classified among the phyla, which accounted for over 99.32% of the fungal sequences. The nine most dominant phyla were *Ascomycota* (46.50%), *Basidiomycota* (22.67%), *Mortierellomycota* (15.84%), unclassified fungi (9.41%), *Rozellomycota* (2.74%), *Chytridiomycota* (1.44%), *Glomeromycota* (0.72%), and *Olpidiomycota* (0.7%) (Figure 1).

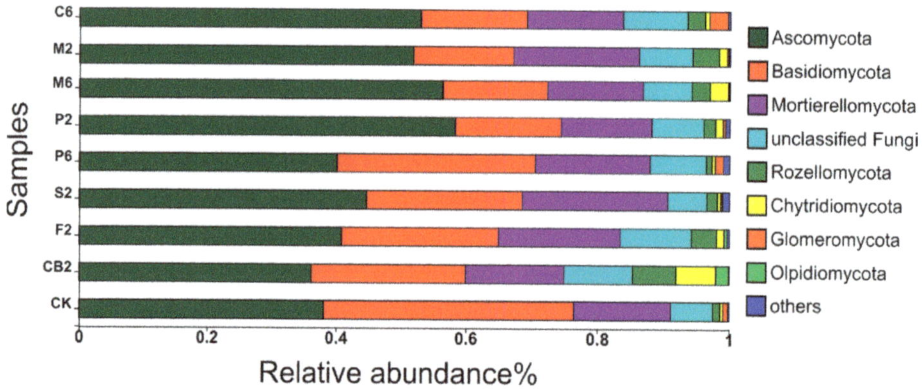

Figure 1. Fungal community composition at the phylum taxonomic level under different cropping systems C6; continuous cropping of *Coptis chinensis* Franch for 6 years, M2 and M6; maize for 2 and 6 years, P2 and P6; *Polygonum multiflorum* for 2 and 6 years, S2; sweet potato for 2 years, F2; *Fritillaria thunbergia* for 2 years, CB2; cabbage for 2 years, and CK; fallow field.

Mortierellaceae, *Piskurozymaceae*, *Nectriaceae*, and *Trichosporonaceae* were the most abundant fungal families among the 432 fungal families. The relative abundance of *Russulaceae* was higher in CK (23.50%), followed by CB2 (0.62%). At the genus taxonomic level, *Mortierella*, *Solicoccozyma*, *Saitozyma*, *Trichoderma*, *Exophiala*, *Apiotrichum*, *Penicillium*, *Trichocladium*, and *Fusarium* were identified in the soil samples. Among the 1063 genera, the relative abundance of *Fusarium* was significantly higher in M6 (4.67%), followed by F2 (2.96%), and lower in CK (0.01%) (Table S1).

The findings of several comparisons revealed that the mean percentage of *Ascomycota* was substantially more significant in (P2) (58.31%). The relative abundance of *Basidiomycota* was markedly higher in (CK) (38.37%). The *Mortierellomycota* phylum was significantly higher in S2 (22.04%) and lower in P2 (13.74%). *Rozellomycota and Chytridiomycota* were considerably more abundant in CB2 (6.03% and 6.03%, respectively). C6 had the highest mean proportion of *Glomeromycota* (2.80%), followed by P6 (1.22%) and CK (0.74%). The relative abundance of *Olpidiomycota* was considerably higher in CB2 (1.87%) than in the other fields (Figure 2 and Table S2).

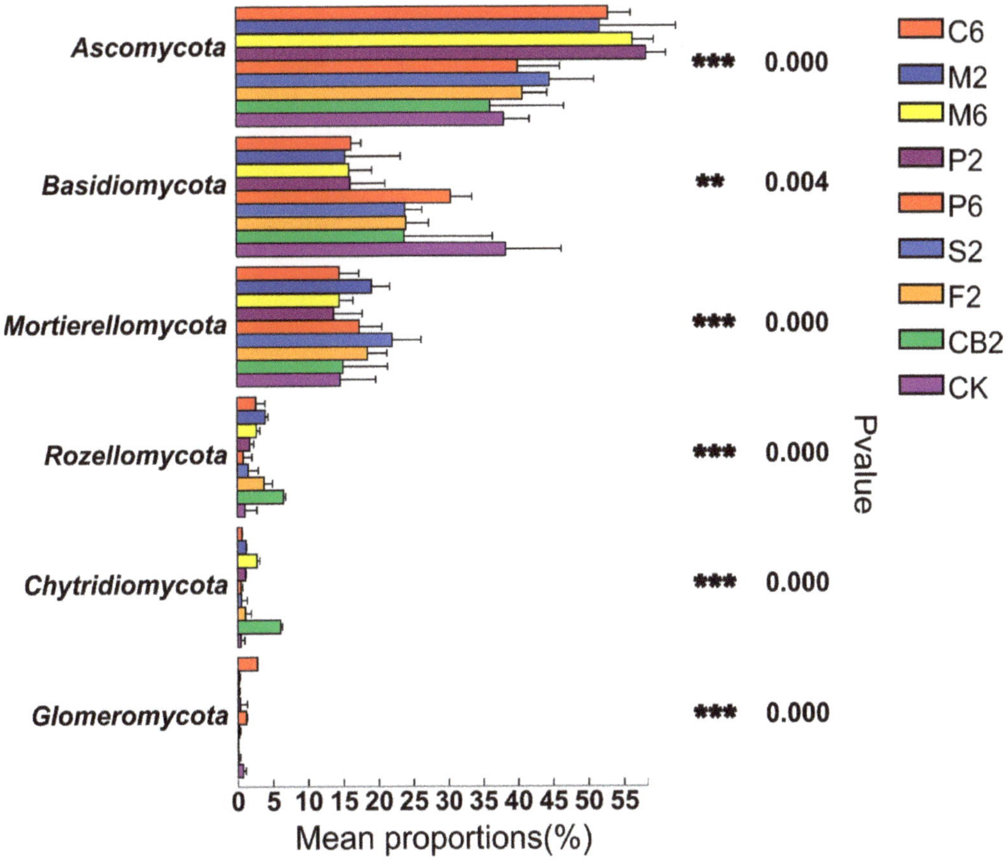

Figure 2. Multiple comparisons of soil fungal phyla in different cropping fields C6; continuous cropping of *Coptis chinensis* Franch for 6 years, M2 and M6; maize for 2 and 6 years, P2 and P6; *Polygonum multiflorum* for 2 and 6 years, S2; sweet potato for 2 years, F2; *Fritillaria thunbergia* for 2 years, CB2; cabbage for 2 years, and CK; fallow field. Asterisks (** and ***) show significant differences as p-values < 0.01, and 0.001, respectively.

Venn diagrams illustrate the degree of similarity and overlap between several samples of a species. There were 56, 46, 40, 40, 40, 37, 39, 26, and 111 unique species that were present in the C6, M2, M6, P2, P6, S2, F2, CB2, and CK fields, respectively. 148 (7.66%) species were common among the nine treatments, with CK (111) having the highest species number (Figure 3).

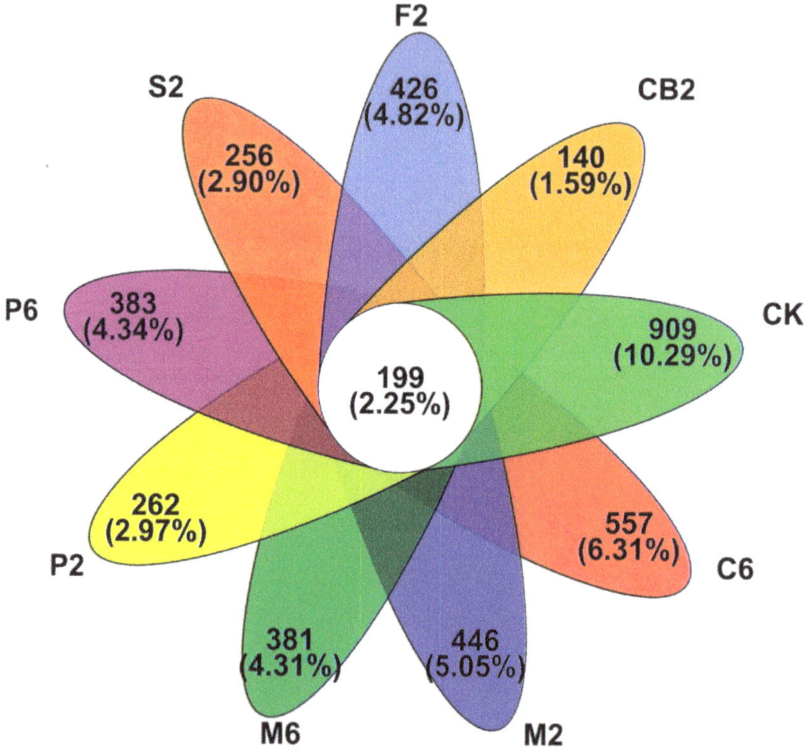

Figure 3. The Venn diagram shows the unique and shared fungal species numbers in different fields with different plant species. C6; continuous cropping of *Coptis chinensis* Franch for 6 years, M2 and M6; maize for 2 and 6 years, P2 and P6; *Polygonum multiflorum* for 2 and 6 years, S2; sweet potato for 2 years, F2; *Fritillaria thunbergia* for 2 years, CB2; cabbage for 2 years, and CK; fallow field.

3.2. Fungal Community Diversity

The alpha diversity provides a measurement of the variety that exists within the fungal community and may be used in the process of comparing the diversity that exists among fungal communities found in various fields that practice continuous cropping. Alpha diversity indicators such as Chao and Shannon, as well as observed species (Sobs) and Phylogenetic diversity (Pd), were used in order to conduct an analysis of the richness and diversity of the community contained within the samples. The alpha diversity of Sobs, Chao, and PD was significantly higher in M2. Compared with C6, the Chao and observed species' (Sobs) diversity was more elevated in M2, M6, and F2 and lower in S2, CB2, CK, P2, and P6. The Shannon index was higher in M2, P2, P6, and F2 and lower in M6, S2, CB2, and CK. It indicates that the plant's type and continuous cropping were the major factors affecting the fungal community's diversity (Figure 4).

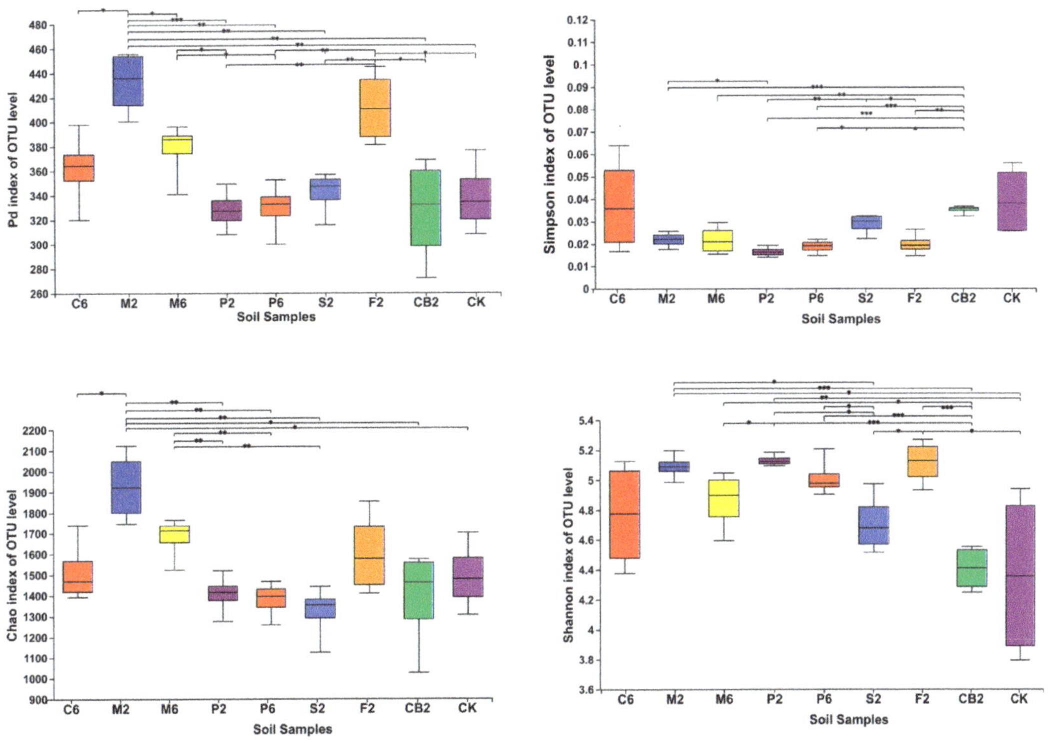

Figure 4. Alpha diversity of the rhizosphere fungal community under different cropping systems with different plant species C6; continuous cropping of *Coptis chinensis* Franch for 6 years, M2 and M6; maize for 2 and 6 years, P2 and P6; *Polygonum multiflorum* for 2 and 6 years, S2; sweet potato for 2 years, F2; *Fritillaria thunbergia* for 2 years, CB2; cabbage for 2 years, and CK; fallow field. Asterisks (*, **, and ***) show significant differences as p-values < 0.05, 0.01, and 0.001, respectively.

The fungal ß-diversity was evaluated based on an unweighted UniFrace matrix in order to provide a more accurate assessment of the distance connection between a variety of soil samples and plant species. The principal coordinate analysis (PCoA) provides a graphical representation of the fact that the fungal community compositions differed significantly across the various soil samples. The biological replicates clustered together; samples from the C6 soil clustered with M2, M6, and CK soil samples, which showed that the fungal species in these soil samples were more similar than other samples, and samples from the other sites (P2, P6, CB2, F2, and S2) clustered together. The first two axes explain 40.9% and 18.11% of the total variation. Importantly, significant divergences in β-diversity were identified between maize-cropped fields and other fields. Moreover, the plant types affected the community composition and were marginally influenced by the sampling sites (Figure 5).

β-diversity indicated that fungal communities in soil samples gathered from fields in the same year were more comparable to plant kinds in soil samples collected from different years. For instance, the fungal communities in soil from *C. chinensis* continuous cropping fields (C6), maize cropped fields (M2, M6), and fallow fields (CK) were more comparable within fields than between fields.

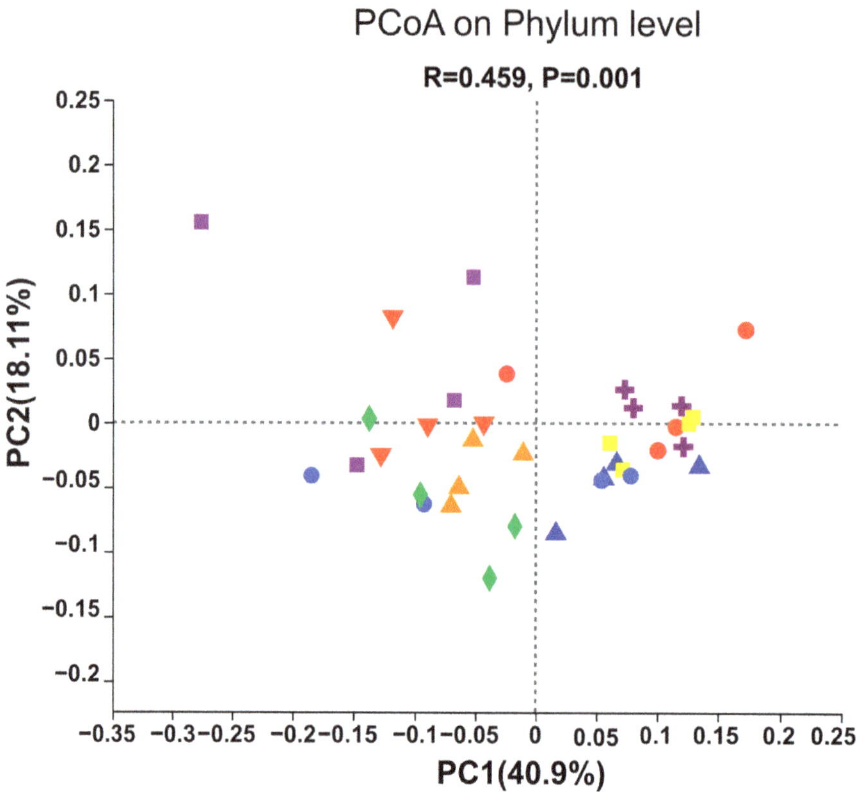

Figure 5. Principal Coordinate Analysis (PCoA) shows soil fungal community structure. C6; continuous cropping of *Coptis chinensis* Franch for 6 years, M2 and M6; maize for 2 and 6 years, P2 and P6; *Polygonum multiflorum* for 2 and 6 years, S2; sweet potato for 2 years, F2; *Fritillaria thunbergia* for 2 years, CB2; cabbage for 2 years, and CK; fallow field.

3.3. *The Correlation Coefficient Analysis of Environmental Factors and Soil Fungal Community*

The link between soil physicochemical parameters, soil enzymatic activities, and fungal phyla relative abundance was evaluated using Spearman's correlation coefficient. The *Mucoromycota* relative abundance showed a negative correlation with soil physicochemical properties, and the correlation coefficient was obtained in pH, TP, AK, TK, and AB. *Glomeromycota* was negatively correlated with TK, TP, OP, and AB. *Chytridiomycota* showed a positive correlation with TK, AK, and AB. *Rozellomycota* phyla relative abundances were positively correlated with TK, AK, TP, OP, and AB (p-values ≤ 0.01, <0.01, <0.01, <0.01, and <0.01, respectively) and not significantly negatively correlated with OM (p-value = 0.385). *Olpidiomycota* phyla had a positively significant correlation with TK, TP, OP, and AB. *Mortierellomycota* phyla are abundant and showed a significant positive correlation with TP, OP, AB, TN, AN, and OM and a negative correlation with TK, AK, and P.H. *Zoopagomycota* phyla showed a positive relationship with TP and OP and a negatively correlated relationship with TK and AB (Figure 6 and Tables S3 and S4).

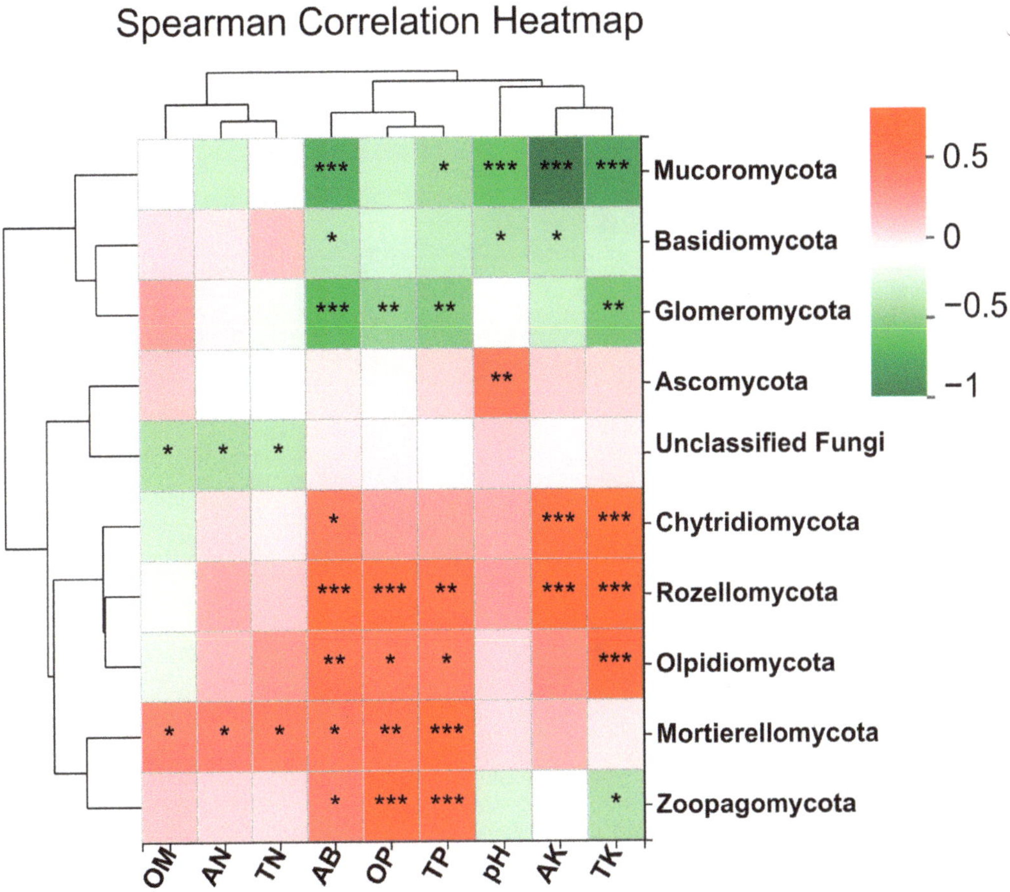

Figure 6. Spearman correlation analysis of soil fungal phyla and soil physicochemical properties under different cropping fields OM; soil organic matter, AN; soil-available nitrogen, TN; total nitrogen, AB; soil-available boron, OP; soil-available phosphorus, TP; total phosphorus, pH; the potential of hydrogen, AK; soil-available potassium, TK; total potassium. Asterisks (*, **, and ***) show significant differences as p-values < 0.05, 0.01, and 0.001, respectively.

Ascomycota phyla abundance was positively correlated with S_SC (p-value < 0.01), and *Basidiomycota* phyla abundance was significantly negatively correlated with S_SC (p-value < 0.01). *Mortierellomycota, Zoopagomycota, and Mucoromycota* were significantly negatively correlated with S_ACP (p-values = 0.01, <0.01, and 0.05, respectively). *Chytridiomycota* was significantly correlated with S_UE (p-value < 0.01), while *Olpidiomycota* was positively correlated with S_SE and had a negative correlation with S_SC (p-values = 0.01 and 0.01, respectively). *Glomeromycota* phyla are significantly positively correlated with S_SC and significantly positively correlated with S_UE (Figure 7 and Table S5).

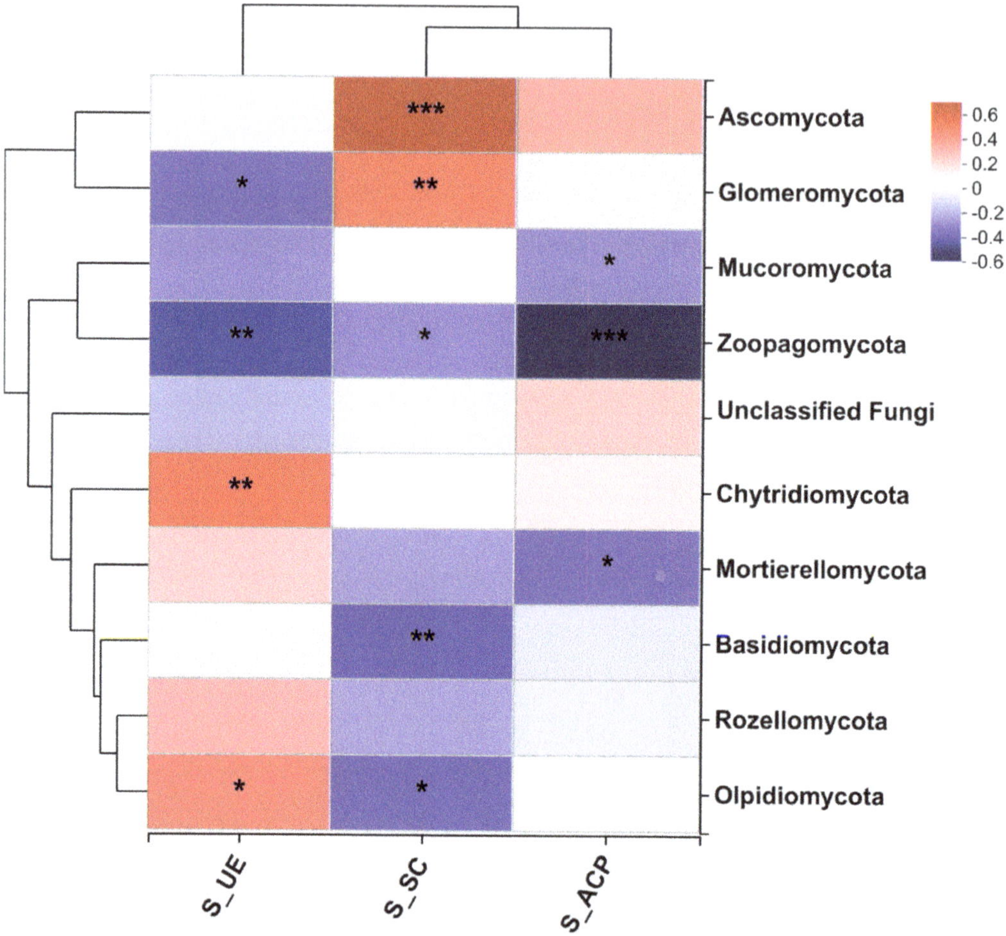

Figure 7. Spearman correlation analysis of soil fungal phyla and soil enzyme activities in different cropping fields S_UE: urease activity; S_SC: sucrase activity; S_ACP: phosphatase activity. Asterisks (*, **, and ***) show significant differences as *p*-values < 0.05, 0.01, and 0.001, respectively.

Different patterns of fungal community structure were found in different soil samples based on the RDA of the most dominant phylum and soil physico-chemical properties. The first two RDA components, which are referred to as RDA1 and RDA2, explain 37.35 and 12.27% of the total variation in the soil's physicochemical characteristics, respectively (Figure 8B), whereas they represent 21.85 and 6.69% (Figure 8A) of the total variance in the soil's enzymatic activities, respectively. We determined the r^2 and *p*-values in order to investigate the importance of environmental influences in the soil on the composition of the fungus community. The total concentration of phosphorus (TP) exhibited the greatest r^2 value among the soil physicochemical values ($r^2 = 0.3507$, *p*-value < 0.01), which indicates that TP had the biggest influence on the fungal community composition. Among the soil physico-chemical properties, TP showed the highest r^2 value. The S_SC soil enzymatic activity exhibited the greatest r^2 value ($r^2 = 0.2885$, *p*-value < 0.01) across all of the soil enzymatic activities.

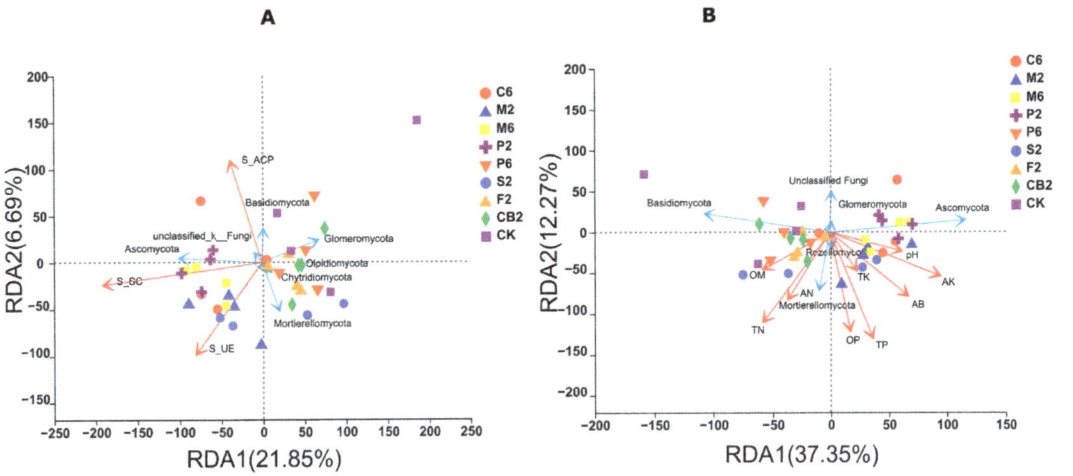

Figure 8. Redundancy analysis (RDA) shows the correlation between soil enzyme activities (**A**), physicochemical properties (**B**), and soil fungal phyla. C6; continuous cropping of *Coptis chinensis* Franch for 6 years, M2 and M6; maize for 2 and 6 years, P2 and P6; *Polygonum multiflorum* for 2 and 6 years, S2; sweet potato for 2 years, F2; *Fritillaria thunbergia* for 2 years, CB2; cabbage for 2 years, and CK; fallow field. OM; soil organic matter, AN; soil-available nitrogen, TN; total nitrogen, AB; soil-available boron, OP; soil-available phosphorus, TP; total phosphorus, pH; the potential of hydrogen, AK; soil-available potassium, TK; total potassium. S_UE; urease activity; S_SC; sucrase activity; S_ACP; phosphatase activity.

3.4. Prediction of Fungal Community Function

FUNGuild identified pathotroph, saprotroph, and symbiotroph major trophic modes and six functional guilds. The unassigned OTUs dominated sequence richness. In assigned OTUs with functions, the animal pathogen was the most dominant functional guild in all soil samples in the identified functional guild and was higher in M6 (22.50%). The lowest abundance was observed in CK (2.22%). The plant pathogen was detected in higher abundances in P6 (11.05%) and lower abundances in CK (1.15%). The Dung Saprotroph functional guild was most abundant in F2 (3.29%) and least abundant in CK (0.43%). C6, compared to other fields, had a lower number of functional guilds following the CK, which meant that cropping systems in *C. chinensis* and other species directly affected the abundance of functional fungi (Figure 9A).

The saprotroph dominated the trophic type in the samples, while the phototroph was the most dominant fungal trophic mode in CB2 (31.11%), followed by M6 (30.62%), and the least abundant was observed in CK (5.51%). Saprotroph trophic type was higher in S2 (48.48%), followed by F2 (42.08%), which was the least abundant in CK (28.48%). Moreover, the somatotrophic trophic mode was the most abundant in CK (30.72%) (Figure 9B).

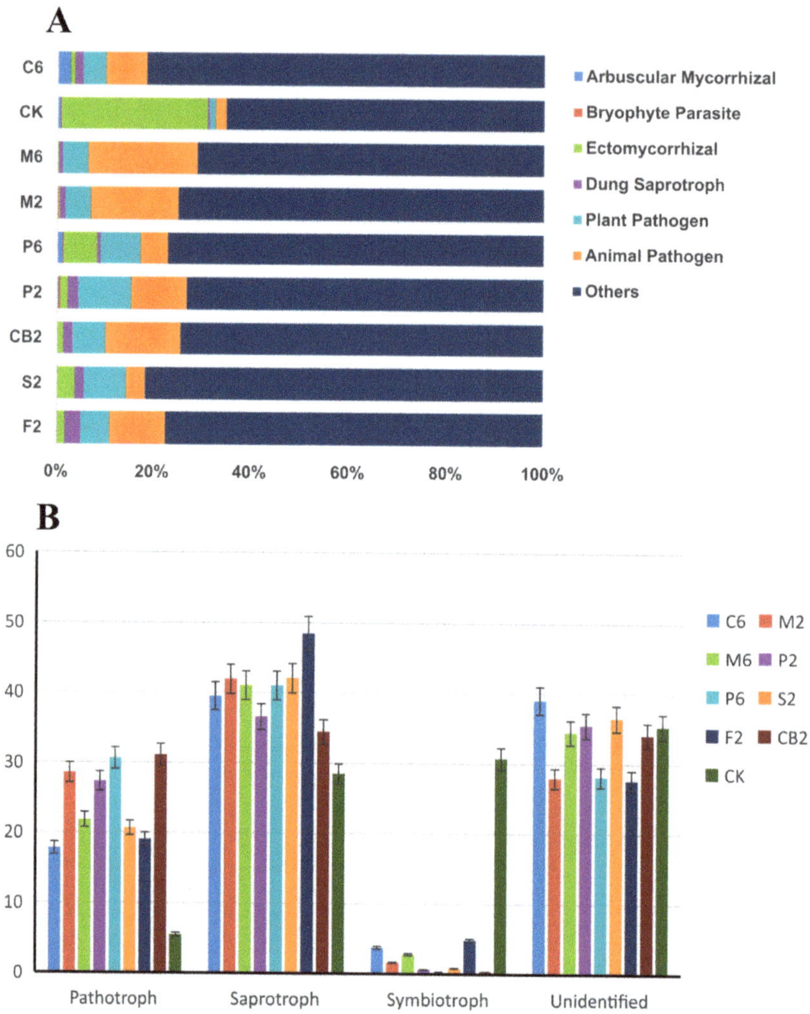

Figure 9. Functional guild (**A**) and trophic mode (**B**) of the fungal community under different rotational cropping patterns C6; continuous cropping of *Coptis chinensis* Franch for 6 years, M2 and M6; maize for 2 and 6 years, P2 and P6; *Polygonum multiflorum* for 2 and 6 years, S2; sweet potato for 2 years, F2; *Fritillaria thunbergia* for 2 years, CB2; cabbage for 2 years, and CK; fallow field.

4. Discussion

Continuous cropping significantly changed the diversity and richness of the fungal community, thereby affecting soil quality, function, and productivity [5,6,23]. Meanwhile, soil microorganisms are essential for plant health, and the microbial population surrounding them is added to the plant's second genome [24–28]. In this study, alpha diversity Sobs and Chao indices estimate species richness, which refers to the total number of different species present in a given area. M2, M6, and F2 plant species have higher alpha diversity indices, suggesting that these species have a greater number of unique species or higher species richness compared to other species mentioned. This could be due to specific ecological characteristics, habitat preferences, or historical factors that have led to a higher diversity of species in these particular plant species. The Shannon index takes into account both species richness and evenness, which refer to the relative abundance of different species.

Higher Shannon index values in M2, P2, P6, and F2 indicate that these species not only have a higher number of species but also a more even distribution of individuals among those species. This suggests that these plant species have a balanced representation of different species, contributing to higher diversity indices. Similarly, Alami et al. (2020) found that soil microbial diversity had almost disappeared after five years of continuous cropping of *C. chinensis*. One-year-old plants had a higher value than five-year-old plants [6]. Similar findings have also been recorded in cucumber and potato, thereby suggesting that plant species and cultivation systems significantly impact the diversity of fungal communities in the rhizosphere. Soil fungus diversity influences soil quality, health, and the soil ecosystem. Root rot diseases are related to a decline in soil fungal diversity [29,30]. A more comprehensive analysis of the specific characteristics, interactions, and environmental factors related to each plant species would be needed to ascertain the exact reasons behind the observed differences in alpha diversity indices.

Continuous cropping and fallow soils had a significant effect on the composition and structure of the fungal population in the field. This study has found that the *Ascomycota* were significantly higher in P2. *Basidiomycota* was higher in CK, indicating that *Basidiomycota* include many important decomposers and mycorrhizal fungi, and their abundance might have been favored in the absence of continuous crop disturbance. The higher abundance of *Mortierellomycota* in S2, which likely had a history of fallow soil, suggests that the accumulation of organic material during the fallow period provided an ideal environment for the proliferation of these fungi. *Rozellomycota* and *Chytridiomycota* are both groups of zoosporic fungi that often inhabit aquatic or wetland environments. The higher presence of these fungal phyla in CB2 indicates that this area might have had high soil moisture content or experienced periods of water saturation, creating suitable conditions for the growth of these aquatic fungi. The dominance of *Glomeromycota* in C6 suggests that continuous cropping practices in this area likely promoted the establishment and colonization of mycorrhizal fungi, as these fungi form symbiotic relationships with plant roots and are commonly found in agricultural soils. *Olpidiomycota* is a recently discovered fungal phylum, and its ecological role is still being investigated. The higher presence of *Olpidiomycota* in CB2 suggests that this particular environment might have provided unique ecological niches or specific conditions that favored the growth and prevalence of these newly identified fungi. Overall, the results indicate that continuous cropping and plant species significantly changed the composition of fungal phyla in the field, which was consistent with the results of previous research [5,25,28,31].

The UniFrac-weighted principal coordinate analysis (PcoA) (Figure 5) indicated a significant disparity in the structural composition of the fungal communities present in the eight fields that alternated between fallow and continuous cropping. The C6 soil samples were grouped together and set apart from the other soil samples in their own little cluster. The distance between each field of continuous cropping was quite far apart, whereas the distance between individual plants of the same species was only somewhat spread out. The findings showed that continuous cropping had a considerable impact on the structure of the fungal population in the rhizosphere; Li et al. also reported findings that were comparable to these [32]. They conducted a 454-pyrosequencing study to get to the conclusion that the makeup of the soil microbial community and the architecture of that community altered dramatically across the three peanut fields due to the distinct monoculture histories. According to Venn diagrams (Figure 3), the number of distinct fungus species altered as one moved from one kind of continuous cropping area to another. According to research that was conducted similarly, the number of distinct fungal OTUs reduced with time in vanilla monoculture [33].

Soil environmental factors such as soil physicochemical properties and soil enzymatic activities closely correlate with certain rhizospheric fungi. The negative correlation between *Mucoromycota* and soil physicochemical properties suggests their ecological preferences and nutrient requirements. *Mucoromycota* prefers a slightly acidic to neutral pH, indicating a decrease in abundance with increasing alkalinity. Lower levels of total phosphorus favor their abundance, indicating their adaptation to phosphorus-limited environments. *Mucoromycota* shows reduced abundance in potassium-rich soils, likely due to lower potassium requirements or competition. Similarly, elevated levels of available boron led to lower *Mucoromycota* abundance. *Glomeromycota* exhibits similar patterns, with reduced abundance in nutrient-rich soils, preferences for phosphorus-limited environments, and lower boron availability. *Chytridiomycota* exhibited a positive correlation with total potassium, soil-available potassium, and soil-available boron, indicating a potential adaptation to environments with higher levels of these nutrients. *Rozellomycota* showed positive correlations with total phosphorus, soil-available potassium, total potassium, soil-available phosphorus, and soil-available boron, suggesting a preference for nutrient-rich soils. *Olpidiomycota* displayed positive correlations with various nutrients such as total potassium, total phosphorus, soil-available phosphorus, and soil-available boron, indicating their reliance on these nutrients for growth. *Mortierellomycota* exhibited a positive correlation with several nutrients, including total phosphorus, soil-available phosphorus, soil-available boron, total nitrogen, soil-available nitrogen, and soil organic matter, while being negatively correlated with total potassium, soil-available potassium, and pH. Lastly, *Zoopagomycota* showed a positive relationship with total phosphorus and soil-available phosphorus while negatively correlating with total potassium and soil-available boron, suggesting specific nutrient preferences within this fungal group. This finding was consistent with previous studies, which showed that the composition and structure of the rhizospheric soil fungal population were significantly influenced by the number of continuous cropping years [34–40].

FUNGuild is a valuable online database for the functions of members within the fungal community. It may be possible to obtain data on the biological processes of fungi through trophic guilds instead of individual taxa [41,42]. The undefined saprotroph dominated the functional guilds (Figure 9B). After the fallow field (CK), the C6 had a less functional guild than other fields, thereby suggesting that continuous cropping of *C. chinensis* directly affected the abundance of functional fungi. OTUs with similar functions had different distributions in different continuous cropping fields and plant species, thereby indicating that the plant species significantly influenced the selection of the fungal community in the rhizosphere.

5. Conclusions

The consistent cultivation of *C. chinensis*, maize, and other plant species had a substantial impact on the diversity, composition, and organization of the fungal community. It was discovered that the fungal communities in the fallow fields were more diverse than those in the continuously cropped fields and that the plant species had a substantial impact on the selection of the fungal community. We have found that the richness and diversity of fungal communities were higher in maize continuous cropping fields; the *Ascomycota* genus is more accumulated in *P. multiflorum* continuous-cropped fields; and after the fallow field, the *C. chinensis* continuous cropping fields had less functional guild than other fields. Our results, therefore, provide a foundation for future agricultural research that aims to boost microbial activity and crop or cash crop output in continuous cropping areas, which are critical for *C. chinensis* and other cash crops.

Supplementary Materials: The following supporting information can be downloaded at: https://www.mdpi.com/article/10.3390/agronomy13071827/s1. Figure S1: Rarefaction curves C6, M2, M6, P2, P6, S2, F2, CB2, and CK; Table S1: The composition of the fungal community at the family level; Table S2: The composition of the fungal community at the phylum level; Table S3: Spearmen correlation analysis of soil fungal phyla and soil physicochemical properties; Table S4: Spearmen correlation p-value for soil physicochemical properties and fungal phyla; Table S5. Spearmen correlation analysis soil physicochemical properties and fungal phyla.

Author Contributions: Conceptualization, Q.P. and X.W.; data curation, M.M.A., S.S. and X.W.; formal analysis, Q.P., M.M.A. and Z.O.; funding acquisition, X.W.; investigation, Q.P.; methodology, Q.P., M.M.A., D.T. and M.J.A.; project administration, X.W.; resources, W.Y., S.S., D.T. and M.J.A.; software, M.M.A., W.Y. and X.W.; supervision, X.W.; validation, S.S. and M.J.A.; visualization, W.Y., Z.O. and X.W.; writing—original draft, Q.P. and M.M.A.; writing—review and editing, Q.P. and M.M.A. All authors have read and agreed to the published version of the manuscript.

Funding: The study was made possible thanks to funding from the National Key R&D Program. A specialized study on the modernization of Traditional Chinese Medicine (TCM), the large-scale production of *Coptis chinensis* Franch, and other Chinese medicinal plants with high value and their involvement in the reduction of poverty (2017YFC1701000).

Informed Consent Statement: The plant that was used in this study was cultivated at Huazhong Agricultural University's experimental base located in Lichuan City, Hubei Province, China, with the cooperation of Zhuanzhuxi Huanglian Professional Cooperative Company. The plants were used with permission from the company and local residents. We confirm that all local, national, or international guidelines and legislation were adhered to in the production of this study.

Data Availability Statement: The datasets generated and/or analyzed during the current study are uploaded to the SRA repository of NCBI with the project number [SUB12266716]. The current status of SRA is under review, and the SRA accession number will be provided as soon as it is accepted by the NCBI. For the availability of data, contact the corresponding author, Xuekui Wang (wang-xuekui@mail.hzau.edu.cn).

Acknowledgments: We appreciate the help of Lichuan City Zhuanzhuxi Huanglian Professional Cooperative Company's Xu Zegang in preparing fields and cooperating with the collection of soil samples. We also thank the Chinese Scholarship Council (CSC) for supporting our studies.

Conflicts of Interest: The authors declare no conflict of interest.

References

1. Yang, T.; Siddique, K.H.M.; Liu, K. Cropping systems in agriculture and their impact on soil health—A review. *Glob. Ecol. Conserv.* **2020**, *23*, e01118. [CrossRef]
2. Xi, H.; Shen, J.; Qu, Z.; Yang, D.; Liu, S.; Nie, X.; Zhu, L. Effects of Long-term Cotton Continuous Cropping on Soil Microbiome. *Sci. Rep.* **2019**, *9*, 18297. [CrossRef]
3. Gao, D.; An, D.; Liu, J.; Shi, J.; Zhou, X.; Wu, F. Wheat cover crop alters soil microbial community and increases cucumber yield under different potassium regimes. *Eur. J. Agron.* **2022**, *139*, 126567. [CrossRef]
4. Li, R.; Liu, Y.; Chu, G. Effects of different cropping patterns on soil enzyme activities and soil microbial community diversity in oasis farmland. *Ying Yong Sheng Tai Xue Bao J. Appl. Ecol.* **2015**, *26*, 490–496.
5. Alami, M.M.; Xue, J.; Ma, Y.; Zhu, D.; Gong, Z.; Shu, S.; Wang, X. Structure, Diversity, and Composition of Bacterial Communities in Rhizospheric Soil of Coptis chinensis Franch under Continuously Cropped Fields. *Diversity* **2020**, *12*, 57. [CrossRef]
6. Alami, M.M.; Xue, J.; Ma, Y.; Zhu, D.; Abbas, A.; Gong, Z.; Wang, X. Structure, Function, Diversity, and Composition of Fungal Communities in Rhizospheric Soil of Coptis chinensis Franch under a Successive Cropping System. *Plants* **2020**, *9*, 244. [CrossRef]
7. Venter, Z.S.; Jacobs, K.; Hawkins, H.-J. The impact of crop rotation on soil microbial diversity: A meta-analysis. *Pedobiologia* **2016**, *59*, 215–223. [CrossRef]
8. Zhang, B.; Li, Y.; Ren, T.; Tian, Z.; Wang, G.; He, X.; Tian, C. Short-term effect of tillage and crop rotation on microbial community structure and enzyme activities of a clay loam soil. *Biol. Fertil. Soils* **2014**, *50*, 1077–1085. [CrossRef]
9. Mu-Chun, Y.; Ting-Ting, X.; Peng-Hui, S.; Jian-Jun, D. Effects of Different Cropping Patterns of Soybean and Maize Seedlings on Soil Enzyme Activities and MBC and MBN. *J. Northeast Agric. Univ. Eng. Ed.* **2012**, *19*, 42–47. [CrossRef]
10. Van Nguyen, S.; Nguyen, P.T.K.; Araki, M.; Perry, R.N.; Ba Tran, L.; Minh Chau, K.; Min, Y.Y.; Toyota, K. Effects of cropping systems and soil amendments on nematode community and its relationship with soil physicochemical properties in a paddy rice field in the Vietnamese Mekong Delta. *Appl. Soil Ecol.* **2020**, *156*, 103683. [CrossRef]
11. Peralta, A.L.; Sun, Y.; McDaniel, M.D.; Lennon, J.T. Crop rotational diversity increases disease suppressive capacity of soil microbiomes. *Ecosphere* **2018**, *9*, e02235. [CrossRef]
12. Xiong, W.; Jousset, A.; Li, R.; Delgado-Baquerizo, M.; Bahram, M.; Logares, R.; Wilden, B.; de Groot, G.A.; Amacker, N.; Kowalchuk, G.A.; et al. A global overview of the trophic structure within microbiomes across ecosystems. *Environ. Int.* **2021**, *151*, 106438. [CrossRef]
13. Shi, P.; Wang, S.-P.; Jia, S.-G.; Gao, Q.; Sun, X.-Q. Effects of three planting patterns on soil microbial community composition. *Chinese J. Plant Ecol.* **2011**, *35*, 965–972. [CrossRef]
14. Xia, Q.; Rufty, T.; Shi, W. Soil microbial diversity and composition: Links to soil texture and associated properties. *Soil Biol. Biochem.* **2020**, *149*, 107953. [CrossRef]

15. Scarlett, K.; Denman, S.; Clark, D.R.; Forster, J.; Vanguelova, E.; Brown, N.; Whitby, C. Relationships between nitrogen cycling microbial community abundance and composition reveal the indirect effect of soil pH on oak decline. *ISME J.* **2021**, *15*, 623–635. [CrossRef]
16. Ding, Y.; Chen, Y.; Lin, Z.; Tuo, Y.; Li, H.; Wang, Y. Differences in Soil Microbial Community Composition Between Suppressive and Root Rot-Conducive in Tobacco Fields. *Curr. Microbiol.* **2021**, *78*, 624–633. [CrossRef]
17. Alami, M.M.; Pang, Q.; Gong, Z.; Yang, T.; Tu, D.; Zhen, O.; Yu, W.; Alami, M.J.; Wang, X. Continuous Cropping Changes the Composition and Diversity of Bacterial Communities: A Meta-Analysis in Nine Different Fields with Different Plant Cultivation. *Agriculture* **2021**, *11*, 1224. [CrossRef]
18. Bolger, A.M.; Lohse, M.; Usadel, B. Trimmomatic: A flexible trimmer for Illumina sequence data. *Bioinformatics* **2014**, *30*, 2114–2120. [CrossRef] [PubMed]
19. Magoč, T.; Salzberg, S.L. FLASH: Fast length adjustment of short reads to improve genome assemblies. *Bioinformatics* **2011**, *27*, 2957–2963. [CrossRef]
20. Caporaso, J.G.; Kuczynski, J.; Stombaugh, J.; Bittinger, K.; Bushman, F.D.; Costello, E.K.; Fierer, N.; Peña, A.G.; Goodrich, J.K.; Gordon, J.I.; et al. QIIME allows analysis of high-throughput community sequencing data. *Nat. Methods* **2010**, *7*, 335–336. [CrossRef]
21. Schloss, P.D. Reintroducing mothur: 10 Years Later. *Appl. Environ. Microbiol.* **2020**, *86*, e02343-19. [CrossRef] [PubMed]
22. Nguyen, N.H.; Song, Z.; Bates, S.T.; Branco, S.; Tedersoo, L.; Menke, J.; Schilling, J.S.; Kennedy, P.G. FUNGuild: An open annotation tool for parsing fungal community datasets by ecological guild. *Fungal Ecol.* **2016**, *20*, 241–248. [CrossRef]
23. Song, X.; Pan, Y.; Li, L.; Wu, X.; Wang, Y. Composition and diversity of rhizosphere fungal community in *Coptis chinensis* Franch. continuous cropping fields. *PLoS ONE* **2018**, *13*, e0193811. [CrossRef]
24. Cardinale, B.J.; Srivastava, D.S.; Emmett Duffy, J.; Wright, J.P.; Downing, A.L.; Sankaran, M.; Jouseau, C. Effects of biodiversity on the functioning of trophic groups and ecosystems. *Nature* **2006**, *443*, 989–992. [CrossRef]
25. de Quadros, P.D.; Zhalnina, K.; Davis-Richardson, A.; Fagen, J.R.; Drew, J.; Bayer, C.; Camargo, F.A.O.; Triplett, E.W. The Effect of Tillage System and Crop Rotation on Soil Microbial Diversity and Composition in a Subtropical Acrisol. *Diversity* **2012**, *4*, 375–395. [CrossRef]
26. Janvier, C.; Villeneuve, F.; Alabouvette, C.; Edel-Hermann, V.; Mateille, T.; Steinberg, C. Soil health through soil disease suppression: Which strategy from descriptors to indicators? *Soil Biol. Biochem.* **2007**, *39*, 1–23. [CrossRef]
27. Enwall, K.; Nyberg, K.; Bertilsson, S.; Cederlund, H.; Stenström, J.; Hallin, S. Long-term impact of fertilization on activity and composition of bacterial communities and metabolic guilds in agricultural soil. *Soil Biol. Biochem.* **2007**, *39*, 106–115. [CrossRef]
28. Wang, R.; Xiao, Y.; Lv, F.; Hu, L.; Wei, L.; Yuan, Z.; Lin, H. Bacterial community structure and functional potential of rhizosphere soils as influenced by nitrogen addition and bacterial wilt disease under continuous sesame cropping. *Appl. Soil Ecol.* **2018**, *125*, 117–127. [CrossRef]
29. Mazzola, M. Assessment and management of soil microbial community structure for disease suppression. *Annu. Rev. Phytopathol.* **2004**, *42*, 35–59. [CrossRef]
30. Kennedy, A.C.; Smith, K.L. Soil microbial diversity and the sustainability of agricultural soils. *Plant Soil* **1995**, *170*, 75–86. [CrossRef]
31. Liu, X.; Zhang, J.; Gu, T.; Zhang, W.; Shen, Q.; Yin, S.; Qiu, H. Microbial Community Diversities and Taxa Abundances in Soils along a Seven-Year Gradient of Potato Monoculture Using High Throughput Pyrosequencing Approach. *PLoS ONE* **2014**, *9*, e86610. [CrossRef]
32. Li, X.-G.; Ding, C.-F.; Zhang, T.-L.; Wang, X.-X. Fungal pathogen accumulation at the expense of plant-beneficial fungi as a consequence of consecutive peanut monoculturing. *Soil Biol. Biochem.* **2014**, *72*, 11–18. [CrossRef]
33. Zhao, J.; Ni, T.; Li, Y.; Xiong, W.; Ran, W.; Shen, B.; Shen, Q.; Zhang, R. Responses of Bacterial Communities in Arable Soils in a Rice-Wheat Cropping System to Different Fertilizer Regimes and Sampling Times. *PLoS ONE* **2014**, *9*, e85301. [CrossRef] [PubMed]
34. Liu, J.; Wu, F.; Yang, Y. Effects of Cinnamic Acid on Bacterial Community Diversity in Rhizosphere Soil of Cucumber Seedlings Under Salt Stress. *Agric. Sci. China* **2010**, *9*, 266–274. [CrossRef]
35. Tan, Y.; Cui, Y.; Li, H.; Kuang, A.; Li, X.; Wei, Y.; Ji, X. Rhizospheric soil and root endogenous fungal diversity and composition in response to continuous Panax notoginseng cropping practices. *Microbiol. Res.* **2017**, *194*, 10–19. [CrossRef] [PubMed]
36. Tayyab, M.; Islam, W.; Lee, C.G.; Pang, Z.; Khalil, F.; Lin, S.; Lin, W.; Zhang, H. Short-term effects of different organic amendments on soil fungal composition. *Sustainability* **2019**, *11*, 198. [CrossRef]
37. Freedman, Z.; Zak, D.R. Soil bacterial communities are shaped by temporal and environmental filtering: Evidence from a long-term chronosequence. *Environ. Microbiol.* **2015**, *17*, 3208–3218. [CrossRef]
38. Xu, H.; Wang, X.; Li, H.; Yao, H.; Su, J.; Zhu, Y. Biochar impacts soil microbial community composition and nitrogen cycling in an acidic soil planted with rape. *Environ. Sci. Technol.* **2014**, *48*, 9391–9399. [CrossRef]
39. Lauber, C.L.; Strickland, M.S.; Bradford, M.A.; Fierer, N. The influence of soil properties on the structure of bacterial and fungal communities across land-use types. *Soil Biol. Biochem.* **2008**, *40*, 2407–2415. [CrossRef]
40. Bai, L.; Sun, H.; Zhang, X.; Cai, B. Next-generation sequencing of root fungal communities in continuous cropping soybean. *Chil. J. Agric. Res.* **2018**, *78*, 528–538. [CrossRef]

41. Treseder, K.K.; Lennon, J.T. Fungal Traits That Drive Ecosystem Dynamics on Land. *Microbiol. Mol. Biol. Rev.* **2015**, *79*, 243–262. [CrossRef] [PubMed]
42. Song, Z.; Kennedy, P.G.; Liew, F.J.; Schilling, J.S. Fungal endophytes as priority colonizers initiating wood decomposition. *Funct. Ecol.* **2017**, *31*, 407–418. [CrossRef]

agronomy

Article

Effect of Root-Knot Nematode Disease on Bacterial Community Structure and Diversity in Peanut Fields

Lijun Wu [1], Yan Ren [1], Xiangsong Zhang [2], Guanghui Chen [1], Chuantang Wang [1], Qi Wu [1], Shuangling Li [1], Fudong Zhan [1], Li Sheng [3], Wenliang Wei [4,*] and Mei Yuan [1,*]

1 Key Laboratory of Peanut Biology, Genitics and Breeding/Ministry of Agriculture and Rural Affairs, Shandong Peanut Research Institute/Shandong Academy of Agricultural Sciences, Qingdao 266100, China; 15863444111@163.com (Y.R.)
2 Linyi Agricultural Technology Extension Center, Linyi 276001, China
3 Qingdao Academy of Agricultural Sciences, Qingdao 266100, China
4 College of Agriculture, Yangtze University, Jingzhou 434000, China
* Correspondence: whwenliang@163.com (W.W.); yuanbeauty@126.com (M.Y.)

Abstract: The root-knot nematode (RKN) disease is a highly destructive soilborne disease that significantly affects peanut yield in Northern China. The composition of the soil microbiome plays a crucial role in plant disease resistance, particularly for soilborne diseases like RKN. However, the relationship between the occurrence of RKN disease and the structure and diversity of bacterial communities in peanut fields remains unclear. To investigate bacterial diversity and the community structure of peanut fields with severe RKN disease, we applied 16S full-length amplicon sequencing based on the third high-throughput sequencing technology. The results indicated no significant differences in soil bacterial α-diversity between resistant and susceptible plants at the same site. However, the Simpson index of resistant plants was higher at the site of peanut-wheat-maize rotation (Ro) than that at the site of peanut continuous cropping (Mo), showing an increase of 21.92%. The dominant phyla identified in the peanut bulk soil included *Proteobacteria*, *Acidobacteria*, *Actinobacteria*, *Planctomycetes*, *Chloroflexi*, *Firmicutes*, and *Bacteroidetes*. Further analysis using LEfSe (Linear discriminant analysis effect size) revealed that *Sulfuricellaceae* at the family level was a biomarker in the bulk soil of susceptible peanut compared to resistant peanut. Additionally, *Singulisphaera* at the genus level was significantly more enriched in the bulk soil of resistant peanut than that of susceptible peanut. Soil properties were found to contribute to the abundance of bacterial operational taxonomic units (OTUs). Available phosphorus (AP), available nitrogen (AN), organic matter (OM), and pH made a positive contribution to the bacterial OTUs, while available potassium (AK) made a negative contribution. The metabolic pathway of novobiocin biosynthesis was only enriched in soil samples from resistant peanut plants. Eleven candidate beneficial bacteria and ten candidate harmful strains were identified in resistant and susceptible peanut, respectively. The identification of these beneficial bacteria provides a resource for potential biocontrol agents that can help improve peanut resistance to RKN disease. Overall, the study demonstrated that severe RKN disease could reduce the abundance and diversity of bacterial communities in peanut bulk soil. The identification of beneficial bacteria associated with resistant peanut offered the possibility for developing biocontrol strategies to enhance peanut resistance to RKN disease.

Keywords: peanut (*Arachis hypogaea* L.); root-knot nematode disease; bulk soil; bacterial community structure

check for updates

Citation: Wu, L.; Ren, Y.; Zhang, X.; Chen, G.; Wang, C.; Wu, Q.; Li, S.; Zhan, F.; Sheng, L.; Wei, W.; et al. Effect of Root-Knot Nematode Disease on Bacterial Community Structure and Diversity in Peanut Fields. *Agronomy* **2023**, *13*, 1803. https://doi.org/10.3390/agronomy13071803

Academic Editor: Jonathan D. Eisenback

Received: 6 May 2023
Revised: 2 July 2023
Accepted: 3 July 2023
Published: 6 July 2023

1. Introduction

Arachis hypogaea L. (peanut or groundnut), commonly known as peanut or groundnut, holds great economic importance as both economic crop and oil crop across the world [1]. It is cultivated worldwide in tropical and subtropical regions. Taxonomically, peanut is considered a legume and is believed to have originated in Central and South America, with cultivation spread to other parts of the world [2]. In many countries, peanut provides a significant nutritious contribution to the diet due to their rich protein, lipid, and fatty acid content [3], and peanut is grown for oil production, peanut butter, confectionary, snacks, and protein extenders globally [4]. Plant-parasitic nematodes are considered one of the major obstacles to the production of peanut crops [5], and one of the most influential nematodes is the root-knot nematode [6].

The root-knot nematode (*Meloidogyne hapla* Chitwood, RKN) can cause significant yield losses in the cultivated peanut and has become an important factor influencing peanut production in Northern China [7]. On the one hand, different peanut varieties have different resistance to RKN. Most peanut cultivars are highly susceptible to RKN [8]. It is well known that the wild diploid peanut relatives showed strong resistance to RKN [9]. But our preliminary tests found that HY9810 was more resistant to RKN than HY20 (Figure 1). They can be used as a good material to study peanut RKN disease. On the other hand, different planting methods have a great influence on RKN. Long-term continuous cropping of peanuts will undoubtedly aggravate the occurrence of RKN disease [10], and the traditional approach tends to use crop rotation to control RKN [11].

Figure 1. Photographs of HY9810 (R) and HY20 (S) at harvest time at the Mo. Site.

Bacterial community and structural changes are closely related to RKN disease. Bacteria are the most abundant and widely distributed microorganisms in soil. They play an important role in maintaining the healthy decomposition of organic matter in the soil, promoting material circulation, and maintaining the balance of the soil ecosystem [12]. Cao et al. observed that RKN infection changed the α-diversity and microbial composition of root microorganisms and drove the transformation of microorganisms [13]. Lu et al. found that the community structure and function of the plant rhizosphere were significantly correlated to the RKN disease [14]. Li et al. demonstrated that community variation and assembly of root endophytic microbiota were significantly affected by RKN [15]. Rani et al. revealed that the bacterial bioagents, namely *B. amyloquefaciens*, *B. megaterium*, *P. fluorescens*, and *P. putida*, showed the potential for controlling RKN [16].

In general, different peanut genotypes and cropping patterns had a great influence on RKN disease. The interaction of bacteria–pathogens is the theoretical basis for improving colonization and controlling the effect of biocontrol bacteria [13]. To explore the relationships between bacterial communities, soil environments, and plant health, bacterial communities were analyzed using the third high-throughput sequencing technology of the 16S full-length rDNA amplicons in peanut samples with different peanut genotypes and cropping patterns. It would provide a theoretical basis for the exploitation and utilization of microbial resources for controlling RKN disease in peanut.

2. Materials and Methods

2.1. Soil Samples Collection and Processing

Two peanut germplasms (HY9810 and HY20) and two planting sites were chosen in this experiment. HY9810 (R-) was an advanced line developed by the disease-resistant breeding team in Shandong Peanut Research Institute (SPRI), which is resistant to RKN (Figure 1). HY20 (S-) was released in 2002 by SPRI, which is susceptible to RKN (Figure 1). One of two planting sites called Mo (35°30′13″ N, 119°11′43″ E) was located at Jiajiagou, Shanzhuan Town, Rizhao City, Shandong Province, and the soil type was sandy loam; another one designated as Ro (36°48′37″ N, 120°30′03″ E) was an experimental base of SPRI, located at Wangcheng Town, Laixi City, Shandong Province, and the soil type was sandy loam. Peanut was planted continuously for seven years at the Mo site which had severe RKN disease. Peanut, wheat, and maize were planted alternatively at the Ro location which had light RKN disease. The two peanut varieties were sown in May 2019. In mid-September, the bulk soils of peanut fields were collected by the five-point sampling method. In each site, the surface soil was removed, and five soil cores at a depth of 5~20 cm near the peanut plants were collected and mixed into one bulk soil. Twelve composite samples were obtained by repeated sampling three times. The samples were sieved through two mm mesh and divided into two groups, with one group stored at room temperature for measurement of soil physicochemical properties, and the other group frozen at −80 °C for DNA extraction of bacteria community. Mo.R and Mo.S stand for soil samples of HY9810 and HY20 from the field of severe RKN disease, respectively. Ro.R and Ro.S were soil samples of HY9810 and HY20 from fields of light RKN disease, respectively.

2.2. 16S rDNA Full-Length Amplification and Sequencing

The soil bacterial diversity and community structure were detected by 16S rDNA full-length sequencing. All the operation processes, including total soil DNA extraction, amplification, library construction, and sequencing, were performed by Novogene (Beijing, China); data analysis was carried out by Gene Denovo (Guangzhou, China). Total genome DNA from soil samples was extracted by the CTAB method. DNA concentration and purity were determined by 1.0% agarose gel and ultraviolet spectrophotometry [17]. The DNA was diluted with sterile water to 1 ng/L, according to the concentration. The specific primers were 16S F (forward primer, 5′-CCTACGGGNGGCWGCAG-3′) and 16S R (reverse primer, 5′-GACTACNVGGGTATCTAATCC-3′) with barcode [18]. The amplified library was sequenced using a PacBio SMRT RS II DNA sequencing platform (Pacific Biosciences, Menlo Park, CA, USA). Low quality was filtered by PacBio circular consensus sequencing technology [19], and the chimera sequences were removed [20] using the UCHIME algorithm [21]. Sequences with ≥97% similarity were assigned to the same OTU (Operational taxonomic unit) by analyzing sequences performed by Uparse software (Uparse v7.0.1001, http://drive5.com/uparse/ accessed on 26 September 2021) [22,23]. The species that were selected to rank top 10 in terms of mean abundance in all samples were visualized using stacked plots, other known species were categorized as others, and unknown species were marked as unclassified with the R project ggplot2 package (version 2.2.1). To identify differences of bacterial communities among the four soil groups, Venn diagrams were plotted with the VennDiagram package [24]. The PCA analysis was performed on the community composition structure of four soil groups at the OTU level to reduce the dimension of the original variables to explore the similarity and differences among groups using the QIIME software package, v. 2 [25]. UPGMA (Unweighted pair-group Method with Arithmetic Mean) is a commonly used clustering analysis method, which mainly refers to the hierarchical clustering analysis method using any distance to evaluate the similarity among the soil groups [25,26]. A species accumulation boxplot was used to investigate the species composition of a sample and predict species abundance in a sample [27]. LEfSe (LDA Effect Size) analysis was used to find biomarkers with statistical differences between groups [28,29], and the Lefse analysis parameter set to the alpha value of factorial Kruskal Wallis test between classes was <0.05, and the threshold value of logarithmic LDA score for distinguishing features was >2.0. Multiple direct gradient regression was used to analyze the correlation between microflora and environmental factors based on soil basic physical and chemical properties and OTU annotation data. The R software language Vegan package, v. 2.6-4, was used for Canonical correlation analyses (CCA), and Pearson, the maximum correlation coefficient between environmental factors and differences in sample community distribution, was used to judge the significance of CCA analysis. Spearman rank correlation was used to study the relationship between environmental factors and bacterial species richness to obtain the correlation and significant p-value between each other. Based on the species and environmental factors, the R language vegan package was used for variance partitioning analysis (VPA) of the contribution (percentage) of each group of environmental factor variables to the species distribution. The relative abundance of screened beneficial and harmful bacteria is displayed using the R language circlize package.

2.3. Determination of Soil Physical and Chemical Properties

Soil basic physicochemical properties of each sample were determined, including alkali-hydrolyzed nitrogen (AN), available phosphorus (AP), available potassium (AK) [30], organic matter (OM) [31], and pH [32]. AN content was determined by the alkali diffusion method [33]. AP content was determined by the molybdenum antimony colorimetric method with a UV-visible spectrophotometer (Shimadzu UV-2700, Kyoto, Japan) [34]. The AK was digested by CH_3COONH_4 and measured by flame atomic absorption spectrophotometry with flame spectrometry (Sherwood M410 Britain, Sherwood scientific Ltd., Cambridge, UK) [33]. OM content was analyzed using dichromate oxidation [31]. Soil pH was measured with a pH meter at a soil to water ratio of 1:2.5 (Meter 3100C, Licor, Lincoln, NE, USA) [33].

2.4. Statistical Analysis

Analysis of variance (ANOVA) with Least-Significant Difference (LSD tests) and Tukey's HSD test were applied to distinguish significant differences between each treatment. All statistics were carried out in IBM-SPSS 22.0 software, and significance was set at $p < 0.05$.

3. Results

3.1. Bacterial Community Diversity of Bulk Soil in Peanuts Field

3.1.1. Sequence Data of 16S Full-Length rDNA

To investigate the bacterial structure and diversity in peanut bulk, we sequenced the 16S full-length amplicon of 12 soil samples. Low-quality reads were corrected or removed, SSR filtered, primers removed with the Cutadapt software v. 4.4, and chimera sequences removed to obtain high-quality reads. Overall, a total of 164,184 raw reads were obtained; 162,706 high-quality clean reads were finally obtained (Table S1), which were used to cluster the analysis of operational taxonomic units (OTUs) with 97% identity (Figure 2).

A total of 3739 OTUs were found in high-quality reads, belonging to 2 domains, 25 phyla, 62 classes, 121 orders, 172 families, 360 genera, and 547 species (File S1). A total of 258 OTUs were common among the 4 groups of soil samples, while 220, 138, 287, and 249 OTUs were unique to Mo.R, Mo.S, Ro.R, and Ro.S, respectively. A total of 53 OTUs were uniquely found in Mo.R and Ro.R, and 61 OTUs were uniquely found in Mo.S and Ro.S (Figure S1).

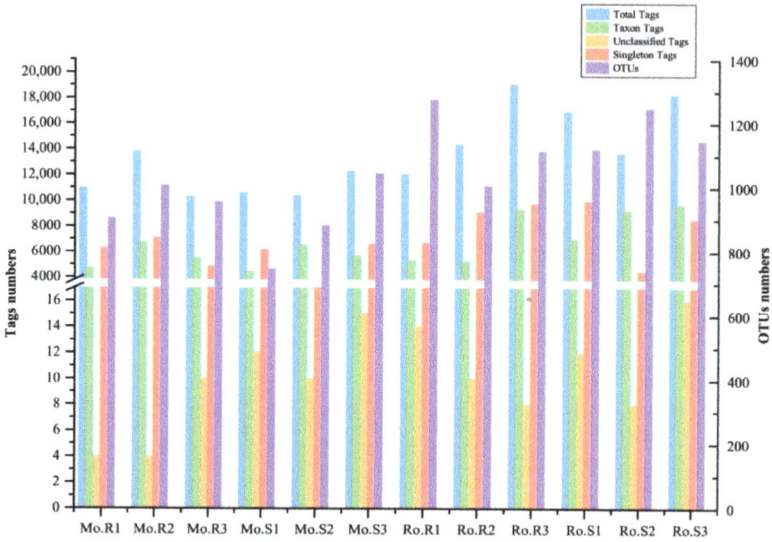

Figure 2. The column chart of preprocessing statistics and quality control of the data.

3.1.2. Alpha Diversity

Through alpha diversity analysis, the community diversity of bacteria was examined by the Simpson and ACE indices. The larger the Simpson implies higher species diversity [35]. The Simpson index analysis showed an extremely significant difference ($p < 0.01$) and ACE indices showed a significant difference ($p < 0.05$) between the Ro site and Mo site; no significant difference in the alpha diversity index was observed between different genotypes ($p > 0.05$) (Table S2). The box plot of α diversity index was drawn and the significance of the difference between every two groups was done by Tukey's HSD test (Figure 3). A comparative analysis of the four groups using the Simpson index revealed that the Mo.R increased by 1.06% compared to Mo.S, the Ro.R increased by 0.15% compared to Ro.S, the Ro.R increased by 1.13% compared to Mo.R, and the Ro.S increased by 2.05% compared to Mo.S. A comparative analysis of the ACE of the four groups revealed that the Mo.R increased by 0.36% compared to Mo.S, the Ro.R increased by −5.61% compared to Ro.S, the Ro.R increased by 21.92% compared to Mo.R, and the Ro.S increased by 29.64% compared to Mo.S (Figure 3 and Table S3).

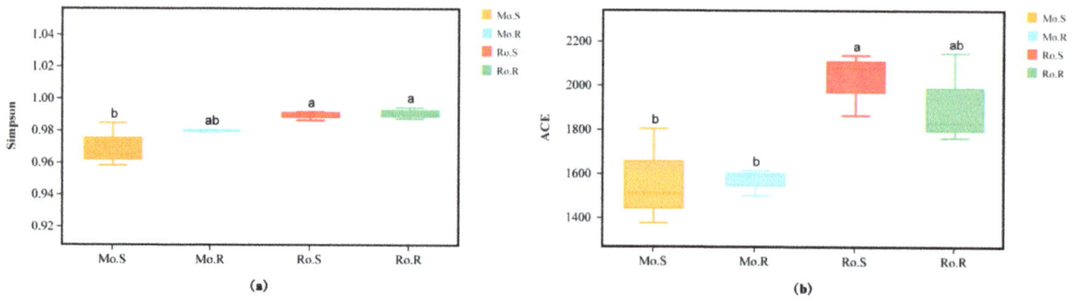

Figure 3. Box plot of alpha diversity in four groups. (**a**) Simpson index. (**b**) ACE Index. Different lowercase letters indicate significant difference at $p < 0.05$.

3.1.3. Beta Diversity

Principal component analysis (PCA) and clustering analysis were used to observe the similarities among the four soil groups. The first two principal components (PC1 and PC2) of PCA explained 59.02% and 18.22% of the total variation, respectively (Figure 4a). Cluster analysis also revealed that soil samples from the same planting sites were classified into a group (Figure 4b).

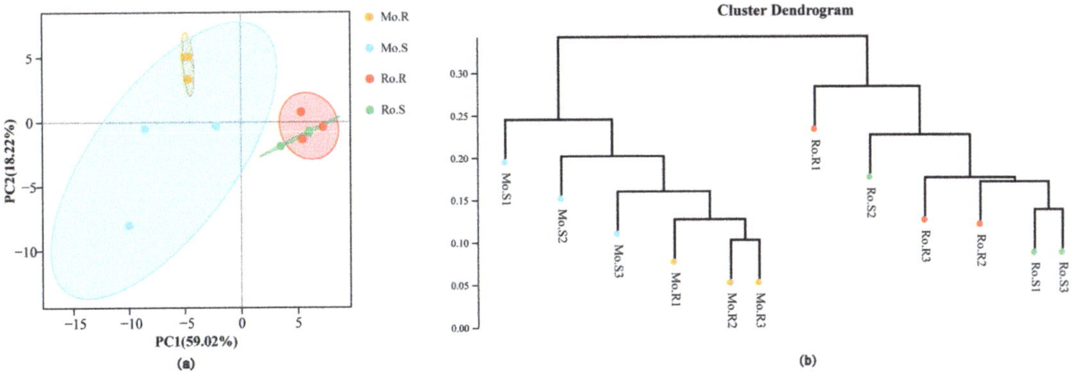

Figure 4. Beta diversity analysis. (**a**) Principal component analysis (PCA) analysis at the OTU level. The same color points belong to the same soil group. (**b**) UPGMA (Unweighted Pair-Group Method with Arithmetic Mean) at the family level.

3.1.4. Bacterial Community Structure of Peanut Bulk Soil Influenced by RKN

In order to further analyze the structure of the bacterial community, the abundance distribution of each group at the levels of phylum, class, order, family, genus, and species was shown according to the results of taxonomic annotation (Figure 5).

All OTUs were classified into 25 phyla, and there were 8 phyla in each group with a relative abundance >1% (Table S4). *Proteobacteria*, *Acidobacteria*, *Actinobacteria*, *Planctomycetes*, *Chloroflexi*, *Firmicutes*, and *Bacteroidetes* were the seven dominant phyla in the peanut bulk soil, accounting for about 90% of all bacterial taxa in each group (Figure 5a). Compared to Mo.S, the relative abundance of *Actinobacteria*, *Planctomycetes*, and *Chloroflexi* increased by 1.12%, 38.22%, and 3.24% in Mo.R, respectively; compared to Ro.S, that increased by 14.84%, 16.96%, and 20.39% in Ro.R, respectively. Conversely, compared to Mo.S, the relative abundance of *Proteobacteria*, *Firmicutes*, *Bacteroidetes*, and *Nitrospirae* decreased by 8.46%, 12.17%, 53.34 and 8.11% in Mo.R, respectively; compared to Ro.S, that decreased by 2.14%, 13.51%, 6.40% and 18.23% in Ro.R, respectively (Figure 5a).

At the class level, most of the bacteria belonged to *Acidobacteria*, *Gammaproteobacteria*, *Alphaproteobacteria*, *Ktedonobacteria*, *Rubrobacteria*, *Betaproteobacteria*, *Planctomycetia*, *Phycisphaerae*, *Acidimicrobiia*, and *Chitinophagia* (Table S5). The most dominant bacterial populations in Mo.R, Mo.S, Ro.R, and Ro.S accounted for 80.84%, 80.98%, 72.49%, and 76.34%, respectively. Compared to Mo.S, the relative abundance of *Ktedonobacteria*, *Planctomycetia*, and *Phycisphaerae* increased by 2.01%, 48.40%, and 15.46% in Mo.R, respectively; compared to Ro.S, that increased by 14.57%, 9.07% and 23.64% in Ro.R, respectively. In contrast, compared to Mo.S, the relative abundance of *Gammaproteobacteria*, *Alphaproteobacteria*, and *Chitinophagia* decreased by 21.33%, 4.67%, and 65.56% in Mo.R, respectively; compared to Ro.S, that decreased by 18.21%, 1.57% and 11.80% in Ro.R, respectively (Figure 5b).

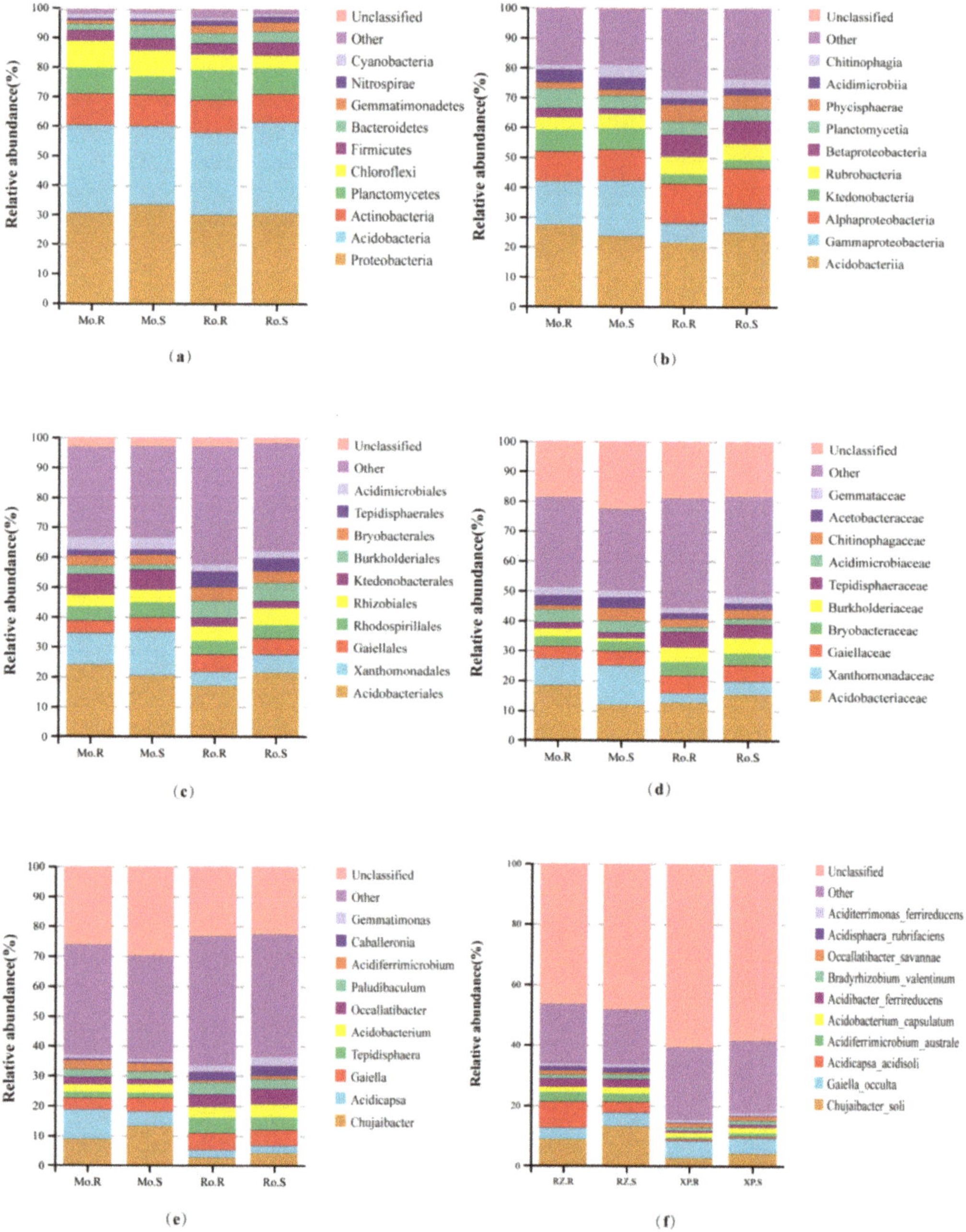

Figure 5. The top 10 species distribution in the average abundance of bacterial communities in peanut bulk soil at the levels of phylum (**a**), class (**b**), order (**c**), family (**d**), genus (**e**), and species (**f**). Other known species are classified as others, and unknown species are marked as unclassified.

Acidobacteriales was the most abundant bacterial order, which accounted for 23.96%, 20.34%, 16.97%, and 21.27% in Mo.R, Mo.S, Ro.R, and Ro.S, respectively (Figure 5c). Compared to Mo.S, the relative abundance of *Ktedonobacterales*, *Bryobacterales*, and *Tepidisphaerales* increased by 1.27%, 1.67%, and 8.79% in Mo.R, respectively; compared to Ro.S, that increased by 20.76%, 18.19% and 21.76% in Ro.R, respectively. While compared to Mo.S, the relative abundance of *Xanthomonadale*, *Rhodospirillales*, and *Rhizobiales* decreased by 27.69%, 8.61%, and 4.41% in Mo.R, respectively; compared to Ro.S, that decreased by 25.41%, 1.84%, and 11.90% in Ro.R, respectively (Figure 5c).

Acidobacteriaceae and *Xanthomonadaceae* were the most abundant families in the four groups collectively (Figure 5d). Compared to Mo.S, the relative abundance of *Bryobacteraceae* and *Tepidisphaeraceae* increased by 1.67% and 8.79% in Mo.R, respectively; compared to Ro.S, that increased by 18.81% and 21.76% in Ro.R, respectively. In contrast, compared to Mo.S, the relative abundance of *Xanthomonadaceae*, *Chitinophagaceae*, and *Acetobacteraceae* decreased by 33.04%, 65.64%, and 4.27% in Mo.R, respectively; compared to Ro.S that decreased by 30.14%, 12.39%, and 2.86% in Ro.R, respectively (Figure 5d).

A thorough investigation at the genus level showed that 360 taxa were classified from the 4 bulk soil communities, whereas most genera were <15%, implying high bacterial diversity in the 4 soil groups (File S2). As shown in Figure 5e, the predominant identifiable genera were *Acidicapsa*, *Chujaibacter*, *Gaiella*, and *Occallatibacter*, which accounted for 9.80%, 13.21%, 5.69%, and 5.02% in Mo.R, Mo.S, Ro.R and Ro.S, respectively. Compared to Mo.S, the relative abundance of *Tepidisphaera* increased by 9.13% in Mo.R; compared to Ro.S, that increased by 20.83% in Ro.R. In contrast, compared to Mo.S, the relative abundance of *Chujaibacter*, *Acidobacterium*, and *Gemmatimonas* decreased by 32.99%, 9.85%, and 2.03% in Mo.R, respectively; compared to Ro.S, that decreased by 35.65%, 19.58% and 30.22% in Ro.R, respectively (Figure 5e and Supplementary File S2).

Chujaibacter soli was the most abundant species in Mo.R and Mo.S, which accounted for 8.85% and 13.21%, respectively; *Gaiella occulta* was the most abundant species in Ro.R and Ro.S, which accounted for 5.65% and 5.17%, respectively (Figure 5f). Compared to Mo.S, the relative abundance of *Acidicapsa acidisoli* increased by 137.21% in Mo.R; compared to Ro.S, that increased by 19.89% in Ro.R. While compared to Mo.S, the relative abundance of *Chujaibacter soli*, *Acidobacterium capsulatum*, *Bradyrhizobium valentinum*, and *Acidisphaera rubrifaciens* decreased by 32.96%, 18.17%, 25.67%, and 13.49% in Mo.R, respectively; compared to Ro.S, that decreased by 35.73%, 23.63%, 18.90%, and 4.86% in Ro.R, respectively (Figure 5f).

3.2. Analysis of the Microbiological Biomarkers in the Peanut Bulk Soil

In order to analyze the biomarkers between different groups, LEfSe (LDA Effect Size) analysis was employed in the four groups of peanut bulk soil. Statistical analysis was performed from the phylum to the genus level in cladograms, and LDA scores of 2 or greater were confirmed by LEfSe between resistant and susceptible peanut (Figures 6 and S2). As can be seen from Figure 6, 17 biomarkers were pointing to susceptible peanut, while *Chujaibacter* and *Xanthomonadaceae* had LDA scores $\geq$ 4.0, and 43 biomarkers pointing to resistant peanut, while *Planctomyceteria*, *Acidobacteria*, *Acidicapsa*, *Acidobacteriaceae*, and *Acidobacteriia* had LDA scores $\geq$ 4.0 at the Mo site; 17 and 20 biomarkers were pointing to susceptible and resistant peanut at the Ro site, respectively.

Sulfuricellaceae at the family level as the specific biomarker was both pointing to susceptible peanut at the same site, which suggested that *Sulfuricellaceae* can be used as biomarkers of the susceptible peanut. *Singulisphaera* at the genus level as the specific biomarkers were both pointing to resistant peanut at the same site, which suggested that *Singulisphaera* can be used as biomarkers of resistant peanut.

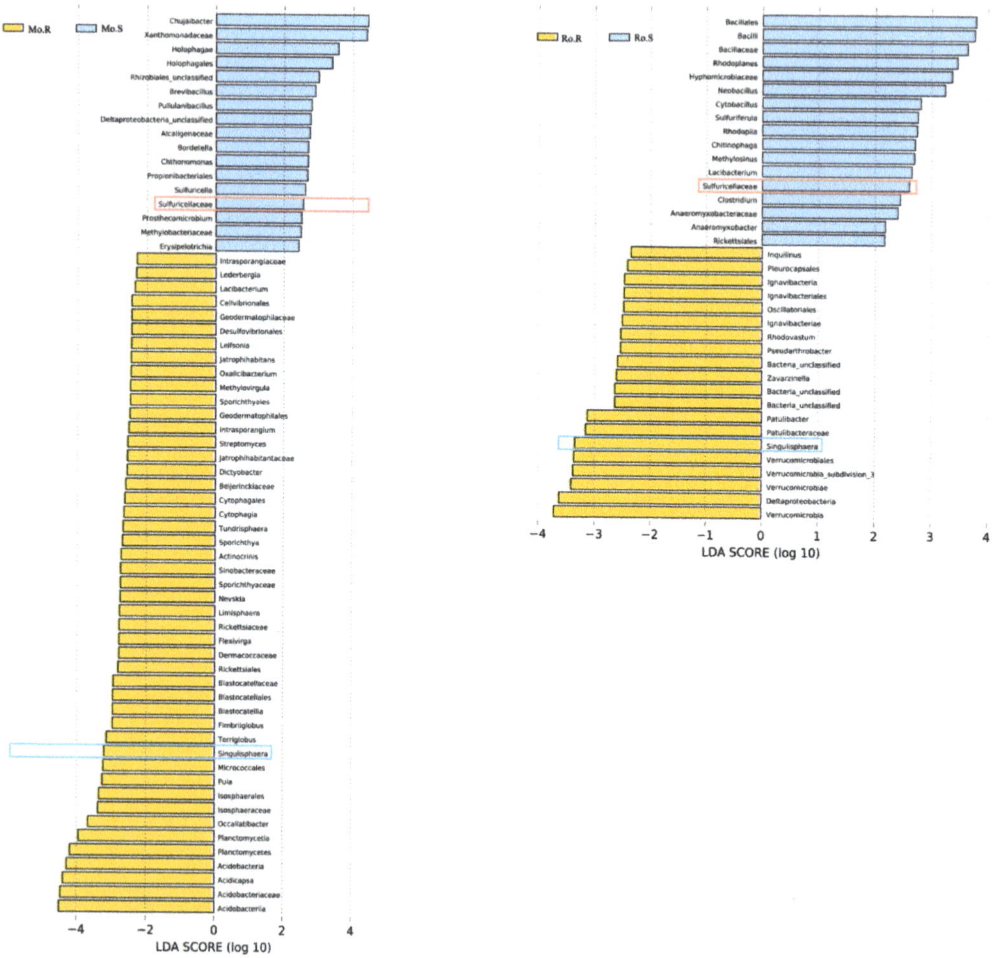

(a) (b)

Figure 6. LEfSe Bar of different abundance between resistant and susceptible peanut at Mo site (**a**) and Ro site (**b**) by linear discriminant analysis (LDA). The yellow horizontal bars are the biomarkers enrichment in bulk soil of RKN-resistant peanut, which had a negative LDA score; the blue horizontal bars are that of RKN susceptible peanut, which had a positive LDA score. The blue frame box is the common biomarker of RKN resistant peanut. The red frame box is the common biomarker of RKN susceptible peanut.

3.3. *Relationship between Bacterial Community Structure and Environment Factors in Peanut Bulk Soil*

Spearman correlation analysis was used to study the relationship between the composition of bacterial community structure and environmental factors. Soil physical and chemical factors were determined, including pH, organic matter (OM), alkali-hydrolyzed nitrogen (AN), available phosphorus (AP), and available potassium (AK) (Table 1). The results showed that the levels of OM, AN, and AP at the Mo site was higher than that at the Ro site, while the pH of the Ro site was higher than that at the Mo site. However, no significant patterns were observed between the soil physicochemical traits and the susceptibility or resistance of the peanut.

Table 1. Environmental chemical characteristics in four groups of peanut bulk soil.

Item	Mo.S	Mo.R	Ro.S	Ro.R
pH	4.5 ± 0.1 a	4.6 ± 0.1 a	4.8 ± 0.2 a	5.1 ± 0.2 a
Organic matter (OM, g/kg)	13.5 ± 0.7 a	14.9 ± 0.7 a	12.1 ± 0.3 a	10.9 ± 0.7 a
Available nitrogen (AN, mg/kg)	82.9 ± 10.8 a	85.5 ± 10.4 a	60.7 ± 5.7 a	54.6 ± 6.3 a
Rapidly available phosphorus (AP, mg/kg)	103.4 ± 8.2 a	121.1 ± 13.3 a	94.3 ± 1.3 a	81.1 ± 6.3 a
Available potassium (AK, mg/kg)	46.1 ± 0.0 a	53.6 ± 14.2 a	61.2 ± 3.3 a	42.3 ± 6.5 a

Note: The same lowercase letters in the table indicate that the differences of each item among different groups are not significant $p > 0.05$.

Multiple direct gradient regression was used to analyze the relationships among sampling points, microflora, and environmental factors, and canonical correspondence analysis (CCA) was constructed. As shown in Figure 7a, the descending order of influences on the distribution of bacterial species in the peanut bulk soil were OM, AP, pH, AN, and AK (the corresponding r2 were 0.8326, 0.8002, 0.6000, 0.7220, and 0.1579, respectively). OM, AP, and AN had extremely significant effects (Pr < 0.01), pH had significant effects (Pr < 0.05), but AK had no significant effects (Pr > 0.05). AP, AN, OM, and pH made a positive contribution to the OTUs of bacterial, while AK made a negative contribution (Figure 7b).

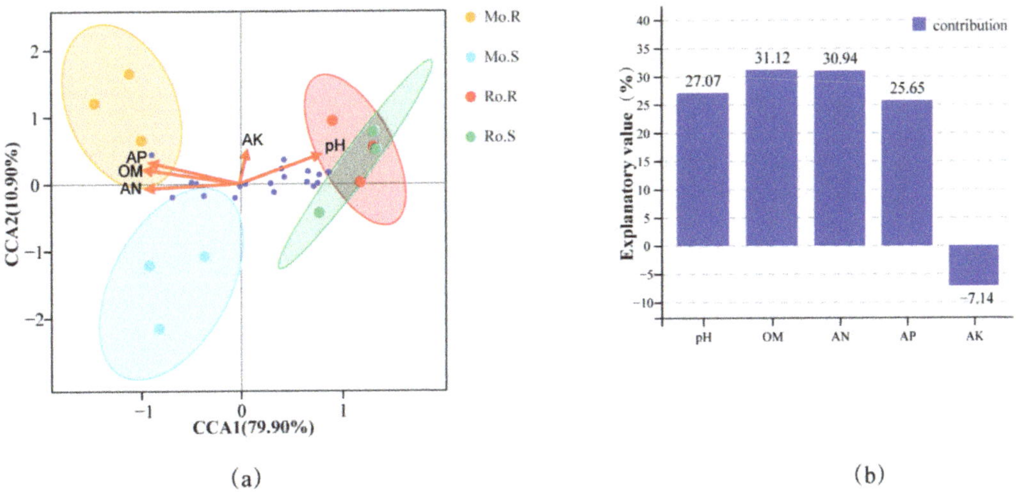

Figure 7. (**a**) Canonical correspondence analysis (CCA) of species information between the peanut bulk soil and environmental factors. Blue spots indicate the top 20 bacterial species. (**b**) Contribution map of environmental factors.

Spearman correlation analysis was performed to study the mutual change relationship between environmental factors and species based on the measured data of soil environment factors and the OTU data of each sample. As seen in Figure 8, OM was the most significant environmental factor in the top 20 of the bulk soil bacteria at specific levels; next were AN and AP, and the last was pH. Moreover, the four environmental factors showed an extremely significant correlation with the abundance of *Acidibacter ferrireducens*, *Acidicapsa acidisoli*, and *Acidiferrimicrobium australe*.

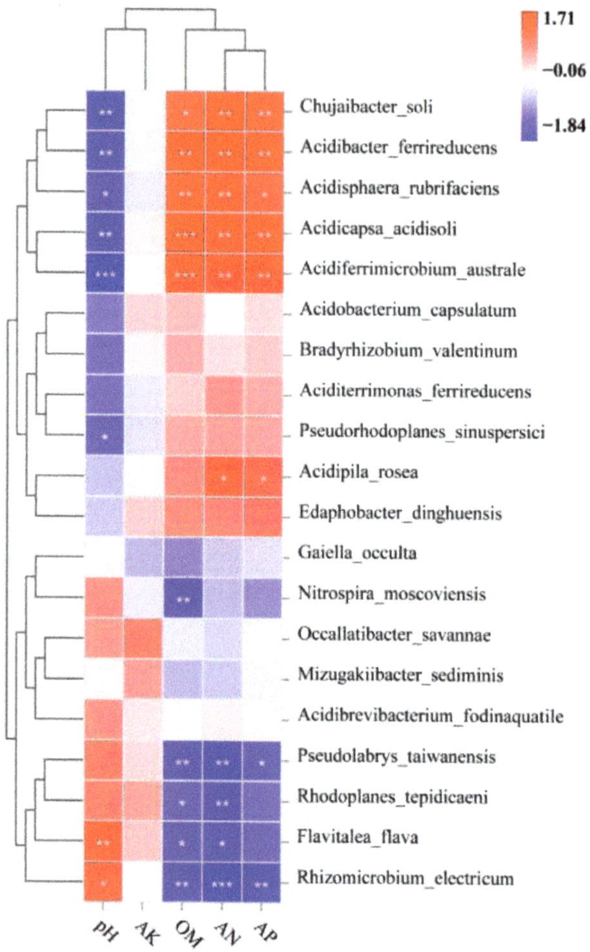

Figure 8. Heat map of Spearman correlation analysis between environmental factors and bacterial species. Colors indicate the strength of the correlation, * indicates a significant correlation with a p-value less than 0.05, ** indicates a highly significant correlation with a p-value less than 0.01; *** indicates p-values less than 0.001.

3.4. Analysis of Beneficial and Harmful Bacteria in the Peanut Bulk Soil

Beneficial bacteria are defined as those that can promote plant growth, help prevent pathogen invasion, and improve plant adaptability to abiotic or biological stresses; they are also called plant growth-promoting rhizobacteria (PGPR) [36,37]. In order to further analyze the bulk soil bacterial diversity of R-cultivars on the basis of relevant literature studies, 11 bacterial species were chosen to draw a Circos diagram. The results showed that the relative abundance of RKN-resistant peanut was higher than that of RKN susceptible peanut (Mo.R-Mo.S $\geq$ 0 and Ro.R-Ro.S $\geq$ 0) (Figure 9a). Among them, the large proportion included *Burkholderia cepacia*, *Jatrophihabitans soli*, *Arthrobacter dokdonellae*, *Rhodanobacter lindaniclasticus*, and *Nitrosospira multiformis*. The relative abundance of *Mucilaginibacter ximonensis* and *Ferruginibacter alkalilentus* in susceptible and resistant plants at the same planting site was a leap from absence to present, although the difference in value was not significant.

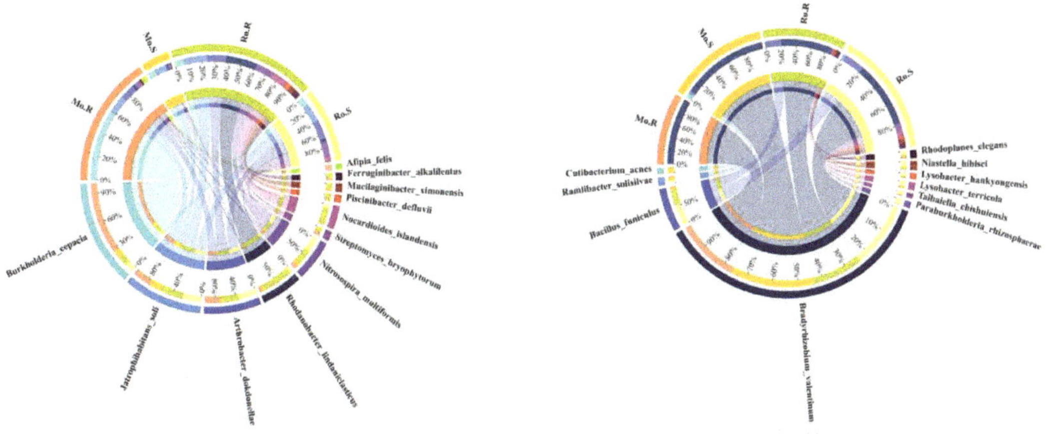

(a) (b)

Figure 9. The Circos map of candidate beneficial and harmful bacteria resistant to RKN in peanut bulk soil; (**a**) beneficial bacteria, (**b**) harmful bacteria.

Harmful bacteria are defined as the opposite of beneficial bacteria, which inhibit plant growth, help pathogens to invade, and reduce plant adaptation to abiotic or biotic stresses [38,39]. To further analyze the bulk soil bacterial diversity of S-cultivars, 10 bacterial species were chosen to draw the Circos diagram, which indicated that the relative abundance of RKN-susceptible peanut was higher than that of RKN-resistant peanut (Mo.R-Mo.S ≤ 0 and Ro.R-Ro.S ≤ 0) (Figure 9b). The large proportions were *Bradyrhizobium valentinum*, *Bacillus funiculus*, *Niastella hibisci*, *Ramlibacter solisilvae*, and *Rhodoplanes elegans*. It is worthy of note that the relative abundance of *Cutibacterium acnes* from resistant to susceptible peanut in the same planting site was a qualitative leap from nothing to something, although there was little difference in value.

3.5. Prediction of Bacterial Functional Potential in the Peanut Bulk Soil

To explore the functional roles of bacteria in peanut bulk soil, PICRUSt2 was used to predict their function based on KEGG metabolic pathways and the relative frequencies of predicted functions. A total of 171 metabolic pathways in KEGG were annotated (File S3). Among the 6 primary pathways, the number of metabolic pathways annotated was at most (118); the next was genetic information processing annotated to 18; the third was cellular processes annotated to 9. In the second level of the metabolic pathway, Xenobiotics biodegradation and metabolism, Carbohydrate metabolism, and the metabolism of terpenoids and polyketides were the 3 most annotated pathways, with 17, 15, and 14 annotated pathways, respectively. Some representative pathways were selected for demonstration and analysis. Seven pathways were both increased in Mo.R compared to Mo.S and Ro.R compared to Ro.S (Figure 10). These seven metabolic pathways mainly focus on the first level of Genetic information processing, Human diseases and metabolism, and the secondary level of Transcription, Translation, Immune diseases, Biosynthesis of other secondary metabolites, and Metabolism of terpenoids and polyketides. Three of the seven metabolic pathways belong to the Metabolism of terpenoids and polyketides. Interestingly, the abundance of novobiocin biosynthesis was over 300 in both resistant peanut, but 0 in both susceptible peanut at the same planting site.

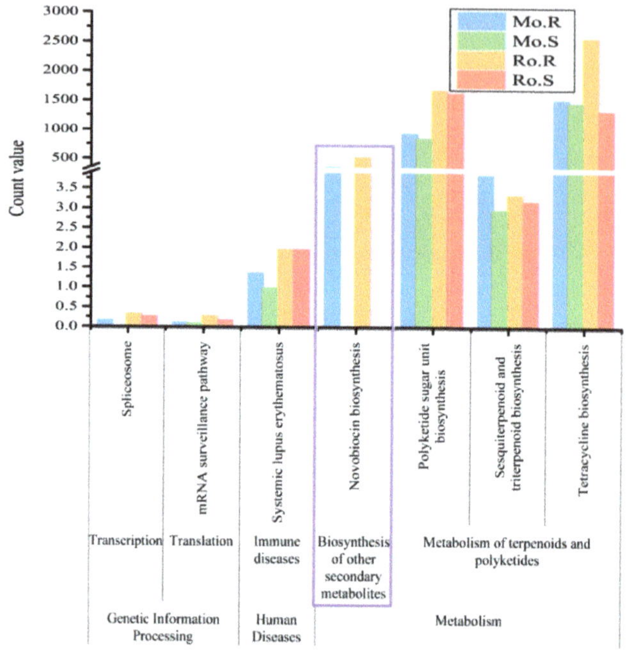

Figure 10. The column chart of KEGG enrichment terms of four groups. Purple frame box is the abundance of novobiocin biosynthesis in four groups.

4. Discussion

4.1. Effect of Severe RKN Disease on the Bacterial Diversity in Peanut Bulk Soil

The structure and function of the root microbiome play an important role in plant immunity and development and are closely related to plant health [40,41]. Some studies have shown that the suppression and outbreak of soilborne diseases are closely related to soil bacterial communities [42,43]. The increase of bacterial quantity and community structure diversity in soil is one of the main reasons to inhibit soilborne plant diseases [44]. Plants regulate their microbiome under biological stresses [45,46]. Our study found that the bacterial diversity of peanut bulk soil in the Ro site was higher than that in the Mo site; and RKN susceptible peanut showed a downward trend compared with RKN resistant peanut at the same planting site, but the difference was not significant (Figure 2). This result is consistent with previous studies that planting area can have a significant impact on bacterial community composition [47,48]. The results of this study supported that the bacterial diversity in bulk soil of RKN susceptible peanut was decreased compared with that of RKN resistant peanut.

We also found that *Acidobacteria*, *Proteobacteria*, *Actinobacteria*, *Planctomycetes*, *Chloroflexi*, *Firmicutes*, and *Bacteroidetes* were the most dominant phyla in peanut fields (Figure 3), which was broadly consistent with previous reports on peanut rhizosphere soil microbiota [22]. It was also found that samples at the same planting site were clustered together by β-diversity analysis (Figure 4). Distinct differences in bacterial communities can be observed between the two different sites, which indicates that cropping pattern or the severity of RKN disease has more influence on a bacterial community than peanut genotypes. It also been found that the abundance of *Actinobacteria*, *Planctomycetes*, and *Chloroflexi* in RKN-resistant peanut bulk soil, and similarly, Cao found that *Actinobacteria* and *Planctomycetes* were also present in healthy tobacco soils compared to RKN-susceptible tobacco soils [49].

4.2. Differential Bacterial Communities between Peanut Bulk Soils of Resistant and Susceptible to RKN

The result showed that *Sulfuricellaceae* at the family level was detectable as a specific biomarker in the bulk soil of RKN susceptible peanut compared to RKN resistant peanut at the same planting site (Figure 6). All members of *Sulfuricellaceae* utilize inorganic sulfur compounds as their energy source and use oxygen or nitrate as terminal electron acceptors for respiration. Its subordinate genus *Sulfuricella* [50,51] is also an indicator of a biomarker in Mo.S compared to Mo.R, and its subordinate genus *Sulfuriferula* is an indicator of a biomarker in Ro.S compared to Ro.R. It seems that the sulfur-like bacteria in the bulk soil of RKN-susceptible peanut are more prominent compared to that of RKN-resistant peanut.

Singulisphaera at the genus level was recognized as the specific biomarker in the bulk soil of RKN-resistant peanut compared to RKN-susceptible peanut at the same planting site (Figure 6). Representatives of this *Singulisphaera* genus are common inhabitants of soils and wetlands [52]. The *Singulisphaera* genus showed remarkable responses to pectin and xylan [53]. Moreover, xylan and pectin are important components of plant cell wall polysaccharides [54], which can prevent invasion and colonization of pathogenic microorganisms [55]. Although *Singulisphaera* seems to be inextricably related to peanut anti-KNF, further study is needed to provide more information on the mechanism.

4.3. The Role of Environment Factors in Bacterial Communities of Peanut Bulk Soil

Soil physicochemical properties are closely related to the bacterial communities of bulk soil [56]. Soil physical and chemical properties have a direct regulatory effect on plant root microenvironment and affect the composition and structure of root bacterial communities [57]. In this study, soil physical and chemical properties displayed an important factor affecting soil bacterial community structure (Figures 7 and 8), which was consistent with previous studies. In addition, we found that AP, AN, OM, and pH made a positive contribution to the OTUs of bacteria, while AK made a negative contribution; pH, OM, AN, and AP all had extremely significant differences along with bacteria *Acidibacter ferrireducens*, *Acidicapsa acidisoli*, and *Acidiferrimicrobium australe* (Figures 7 and 8).

4.4. Bacterial Potential Function in Peanut Bulk Soil

Soil microorganisms are an important part of the farmland ecosystem, which can promote the recycling of matter and energy in soil, especially the recycling and transformation of nutrient elements [58]. In our study, the highest functional enrichment of bacterial communities in peanut bulk soil was the first level of metabolism, and the second level xenobiotic biodegradation and metabolism. Seven metabolic pathways were selected as they were all increased in the resistant peanut bulk soil compared to the susceptible at the same planting site. Notably, we found that the novobiocin biosynthesis pathway differed 300-fold between resistant and susceptible peanut. It is presumed that RKN-resistant peanut could be involved in the biosynthesis of some novobiocin.

4.5. Beneficial and Harmful Bacteria in Peanut Bulk Soil

Beneficial microorganisms in soil can not only promote the transformation of soil organic matter into nutrients that can be absorbed and utilized by plants and improve the soil microecological environment but also produce a variety of bacteriostatic or bactericidal active substances, enhance the resistance of crops to a variety of diseases, and reduce the incidence of soilborne diseases [59]. Nitrifying has a profound effect on the form of mineral nitrogen that plants take up, use, and retain in the soil or lose to the environment [60]. Other beneficial bacteria had similar results, such as *Planctomyces*, *Gemmata*, *Flavisolibacter* [61]. It was found that *Arthrobacter* is a beneficial bacterium in corn fields [62], *Inquilinus* has the function of promoting the growth of ginseng [63], and *Nocardioides* belonging to the phylum actinomycetes can promote the growth of ginseng root [64]. Harmful bacteria, which have the opposite function of beneficial bacteria to plants, also deserve our attention. Literature stated that the relative abundance of beneficial bacteria such as *Sphingomonas*,

Pseudomonas, and *Aspergillus* increased significantly, while the relative abundance of the pathogen *Pythium* decreased significantly after crop rotation [65]. *Rhodoplanes* and *Kaistobacter* were the main bacteria in healthy roots of American ginseng, while *Sphingobium* was the main bacterial group in rotten roots [66]. In soil microorganisms, there are many promoting bacterial populations for plant growth and disease control, which are known as plant growth-promoting rhizobacteria (PGPR). PGPR has a certain biological control effect on soil-pathogenic microorganisms and nonparasitic rhizosphere harmful microorganisms [67]. The relative abundance of beneficial bacteria antagonistic to pathogenic bacteria, such as *Planctomyces*, *Bradyrhizobium*, and *Burkholderia* increased significantly, resulting in a low incidence under pineapple–banana rotation [68]. *Bacillus* bacterial agents can increase the abundance of the beneficial bacteria *Nitrospirae*, *Variovorax*, *Rhodanobacter*, *Nitrosospira*, *Rhodopseudomonas*, and *Mesorhizobium* [69]. It also showed that the family *Pseudomonadaceae* is beneficial to the control of root-knot nematodes [70].

Moreover, some studies have shown that *Bradyrhizobium*, *Rhizobia*, *Burkholderia*, and *Achromobacter* have the potential to biofix nitrogen with cowpea roots [71]. Non-pathogenic *Pseudomonas fluorescens* (WCS417r) and *Moringa oleifera* leaf extracts were effective against wheat aphids [70]. *Stenotrophomonas Maltophilia*, *Serratia Plymuthica*, *Pseudomonas Trivialis*, *P. Fluorescens*, *B. subtilis*, and *Burkholderia cepacia* can produce volatile organic compounds (VOCs) that inhibit the growth of plant pathogenic fungi hyphae [16]. In the research of biological control of plant diseases, the biological control of soilborne diseases has made great achievements. The bacteria used for biocontrol mainly belong to *Trichonderma*, *Streptomyces*, *Gliocladium*, *Bacillus*, *Pseudomonas*, *Agrobacterium*, *Flavobacter*, and *Enterobacter* [72,73].

Referring to the beneficial and harmful bacteria of other plants studied by previous reports, in this study, 11 candidate beneficial bacteria and 10 harmful bacteria were obtained for peanut resistance to RKN (Figure 9). *Burkholderia cepacia* accounted for the largest proportion of beneficial bacteria, and it may benefit from *Burkholderia cepacia* producing volatile organic compounds (VOCs) that inhibit the growth of plant pathogenic fungi hyphae [16]. *Burkholderia cepacia* could be used as a biological agent for peanut resistance to RKN in the future. *Bradyrhizobium* is generally beneficial for plants that can fix nitrogen, and it is also beneficial and antagonistic to pathogenic bacteria in banana rotation [68]. However, in our study, *Bradyrhizobium valentinum* is the largest percentage of potentially harmful bacteria for peanut resistance to RKN. It may be two-sided: it can fix nitrogen and is also pathogenic to peanut resistance to RKN.

5. Conclusions

Our findings strongly indicate that the planting site has more influence on the bacterial community of peanut bulk soil than the peanut genotype. *Singulisphaera* at the genus level was a biomarker in the bulk soil bacteria of RKN-resistant peanut compared to RKN-susceptible peanut in the same planting site and *Sulfuricellaceae* at the family level was detected to be a biomarker in that of RKN susceptible peanut. AP, AN, OM, and pH made a positive contribution to the OTUs of bacteria, while AK made a negative contribution. All pH, OM, AN, and AP had extremely significant differences on *Acidibacter ferrireducens*, *Acidicapsa acidisoli*, and *Acidiferrimicrobium australe*. The function of the novobiocin biosynthesis pathway plays an important role in peanut resistance to RKN. A total of 11 candidate-beneficial and 10 harmful bacteria were obtained for peanut resistance to RKN, and *Burkholderiacepacia*, as a beneficial bacterium against RKN in peanut could be used as a potential bioagent in the future.

These results highlight the significance of planting site, specific bacterial taxa, soil properties, and functional pathways in peanut resistance to RKN. The identification of candidate beneficial bacteria, including *Burkholderia epacia*, suggests the possibility of utilizing them as bioagents in future for RKN management strategies in peanut.

Supplementary Materials: The following supporting information can be downloaded at: https://www.mdpi.com/article/10.3390/agronomy13071803/s1, Figure S1: The Venn diagram of OTUs cluster in four peanut bulk soil groups. Figure S2: Cladogram showing specific phylotypes of peanut bulk soil between resistant or susceptible to RKN in Mo and Ro sites; Table S1: Preprocessing statistics and quality control of the Data. Table S2: ANOVA of Simpon and ACE index. Table S3: The table of Simpson and ACE indices in four groups. Table S4: Distribution and abundance of taxa at the phylum level. Table S5: Distribution and abundance of taxa at the class level; File S1: Details of OTUs. File S2: Details of the genus level of each group of bacterial taxa. File S3: Details of KEGG metabolic pathways the bacteria annotated to.

Author Contributions: Conceptualization, W.W. and M.Y.; methodology, L.W., Y.R., X.Z., F.Z., S.L. and M.Y.; software, L.W.; validation, Q.W. and M.Y.; formal analysis, L.W.; data curation, L.W.; writing—original draft preparation, L.W.; writing—review and editing, L.W., Y.R., G.C., Q.W., C.W., L.S., W.W. and M.Y.; visualization, L.W. and G.C.; supervision, M.Y. All authors have read and agreed to the published version of the manuscript.

Funding: This work was financially supported by the Natural Science Foundation of Shangdong Province, China (ZR2020QC121), the National Natural Science Foundation of China (31471533), the Key Research and Development Project of Shandong Province (2020LZGC001), Agricultural Scientific and Technological Innovation Project of Shandong Academy of Agricultural Sciences (CXGC2023A06, CXGC2021B33) and the Qingdao People's Livelihood Science and Technology Program, China (20-3-4-26-nsh).

Data Availability Statement: Most of the collected data are contained in the tables and figures in the manuscript and Supplementary Materials.

Acknowledgments: The authors thank Novogene Co. (Beijing, China) for data processing and Gene Denovo Co. (Guangzhou, China) for the assistance with data processing and bioinformatics analysis.

Conflicts of Interest: The authors declare that they have no known competing financial interests or personal relationships that could have appeared to influence the work reported in this paper.

Abbreviations

OTU: Operational taxonomic unit; RKN: Root-knot nematode; Mo.R: Bulk soil of HY9810 (R-cultivar) which is resistant to RKN at Mo site; Mo.S: Bulk soil of HY20 (S-cultivar)which is susceptible to RKN at Mo site; Ro.R: Bulk soil of HY9810 (R-cultivar)which is resistant to RKN at Ro site; Ro.S: Bulk soil of HY20 (S-cultivar) which is susceptible to RKN at Ro site; LEfSe: LDA Effect Size; AN: Alkali-hydrolyzed nitrogen; AP: Available phosphorus; AK: Available potassium; OM: organic matter; ANOVA: Analysis of variance; PCA: Principal component analysis; PCoA: Principal coordinate analysis.

References

1. Launio, C.C.; Luis, J.S.; Angeles, Y.B. Factors influencing adoption of selected peanut protection and production technologies in Northern Luzon, Philippines. *Technol. Soc.* **2018**, *55*, 56–62. [CrossRef]
2. Settaluri, V.S.; Kandala, C.V.K.; Puppala, N.; Sundaram, J. Peanuts and their nutritional aspects-a review. *Food Nutr. Sci.* **2012**, *3*, 1644–1650. [CrossRef]
3. Toomer, O.T. Nutritional chemistry of the peanut (*Arachis hypogaea*). *Crit. Rev. Food Sci. Nutr.* **2018**, *58*, 3042–3053. [CrossRef]
4. Toomer, O.T. A comprehensive review of the value-added uses of peanut (*Arachis hypogaea*) skins and by-products. *Crit. Rev. Food Sci. Nutr.* **2020**, *60*, 341–350. [CrossRef]
5. Osman, H.A.; Ameen, H.H.; Mohamed, M.; Elkelany, U.S. Efficacy of integrated microorganisms in controlling root-knot nematode *Meloidogyne javanica* infecting peanut plants under field conditions. *Bull. Natl. Res. Cent.* **2020**, *44*, 134. [CrossRef]
6. Nagy, E.D.; Chu, Y.; Guo, Y.; Khanal, S.; Tang, S.; Li, Y.; Dong, W.B.; Timper, P.; Taylor, C.; Ozias-Akins, P.; et al. Recombination is suppressed in an alien introgression in peanut harboring Rma, a dominant root-knot nematode resistance gene. *Mol. Breed.* **2010**, *26*, 357–370. [CrossRef]
7. Tirumalaraju, S.V.; Jain, M.; Gallo, M. Differential gene expression in roots of nematode-resistant and-susceptible peanut (*Arachis hypogaea*) cultivars in response to early stages of peanut root-knot nematode (*Meloidogyne arenaria*) parasitization. *J. Plant Physiol.* **2011**, *168*, 481–492. [CrossRef]
8. Ballén-Taborda, C.; Chu, Y.; Ozias-Akins, P.; Timper, P.; Jackson, S.A.; Bertioli, D.J.; Leal-Bertioli, S.C. Validation of resistance to root-knot nematode incorporated in peanut from the wild relative *Arachis stenosperma*. *Agron. J.* **2021**, *113*, 2293–2302. [CrossRef]

9. Ballén-Taborda, C.; Chu, Y.; Ozias-Akins, P.; Timper, P.; Holbrook, C.C.; Jackson, S.A.; Bertioli, D.J.; Leal-Bertioli, S.C.M. A new source of root-knot nematode resistance from *Arachis stenosperma* incorporated into allotetraploid peanut (*Arachis hypogaea*). *Sci. Rep.* **2019**, *9*, 17702. [CrossRef] [PubMed]
10. Huang, F.; Deng, X.; Gao, L.; Cai, X.; Yan, D.; Cai, Y.; Chen, X.; Yang, M.; Tong, W.; Yu, L. Effect of marigold (*Tagetes erecta* L.) on soil microbial communities in continuously cropped tobacco fields. *Sci. Rep.* **2022**, *12*, 19632. [CrossRef]
11. Krupinsky, J.M.; Bailey, K.L.; Mcmullen, M.P.; Gossen, B.D.; Turkington, T.K. Managing plant disease risk in diversified cropping systems. *Agron. J.* **2002**, *94*, 198–209. [CrossRef]
12. Ramakrishna, W.; Yadav, R.; Li, K. Plant growth promoting bacteria in agriculture: Two sides of a coin. *Appl. Soil Ecol.* **2019**, *138*, 10–18. [CrossRef]
13. Cao, Y.; Yang, Z.-X.; Yang, D.-M.; Lu, N.; Yu, S.-Z.; Meng, J.-Y.; Chen, X.-J. Tobacco root microbial community composition significantly associated with root-knot nematode infections: Dynamic changes in microbiota and growth stage. *Front. Microbiol.* **2022**, *13*, 807057. [CrossRef]
14. Lu, P.; Shi, H.; Tao, J.; Jin, J.; Wang, S.; Zheng, Q.; Liu, P.; Xiang, B.; Chen, Q.; Xu, Y.; et al. Metagenomic insights into the changes in the rhizosphere microbial community caused by the root-knot nematode *Meloidogyne incognita* in tobacco. *Environ. Res.* **2023**, *216*, 114848. [CrossRef]
15. Li, Y.; Lei, S.; Cheng, Z.; Jin, L.; Zhang, T.; Liang, L.-M.; Cheng, L.; Zhang, Q.; Xu, X.; Lan, C.; et al. Microbiota and functional analyses of nitrogen-fixing bacteria in root-knot nematode parasitism of plants. *Microbiome* **2023**, *11*, 48. [CrossRef] [PubMed]
16. Rani, P.; Singh, M.; Prashad, H.; Sharma, M. Evaluation of bacterial formulations as potential biocontrol agents against the southern root-knot nematode, *Meloidogyne incognita*. *Egypt. J. Biol. Pest Control* **2022**, *32*, 29. [CrossRef]
17. Ullah, A.; Akbar, A.; Luo, Q.; Khan, A.H.; Manghwar, H.; Shaban, M.; Yang, X. Microbiome diversity in cotton rhizosphere under normal and drought conditions. *Microb. Ecol.* **2019**, *77*, 429–439. [CrossRef] [PubMed]
18. Herlemann, D.P.; Labrenz, M.; Jürgens, K.; Bertilsson, S.; Waniek, J.J.; Andersson, A.F. Transitions in bacterial communities along the 2000 km salinity gradient of the Baltic Sea. *ISME J.* **2011**, *5*, 1571–1579. [CrossRef]
19. Bokulich, N.A.; Subramanian, S.; Faith, J.J.; Gevers, D.; Gordon, J.I.; Knight, R.; Mills, D.A.; Caporaso, J.G. Quality-filtering vastly improves diversity estimates from Illumina amplicon sequencing. *Nat. Methods* **2013**, *10*, 57–59. [CrossRef]
20. Haas, B.J.; Gevers, D.; Earl, A.M.; Feldgarden, M.; Ward, D.V.; Giannoukos, G.; Ciulla, D.; Tabbaa, D.; Highlander, S.K.; Sodergren, E.; et al. Chimeric 16S rRNA sequence formation and detection in Sanger and 454-pyrosequenced PCR amplicons. *Genome Res.* **2011**, *21*, 494–504. [CrossRef]
21. Edgar, R.C. MUSCLE: Multiple sequence alignment with high accuracy and high throughput. *Nucleic Acids Res.* **2004**, *32*, 1792–1797. [CrossRef]
22. Dai, L.; Zhang, G.; Yu, Z.; Ding, H.; Xu, Y.; Zhang, Z. Effect of drought stress and developmental stages on microbial community structure and diversity in peanut rhizosphere soil. *Int. J. Mol. Sci.* **2019**, *20*, 2265. [CrossRef] [PubMed]
23. Edgar, R.C. UPARSE: Highly accurate OTU sequences from microbial amplicon reads. *Nat. Methods* **2013**, *10*, 996–998. [CrossRef]
24. Chen, H.; Boutros, P.C. VennDiagram: A package for the generation of highly-customizable Venn and Euler diagrams in R. *BMC Bioinform.* **2011**, *12*, 35. [CrossRef] [PubMed]
25. Jin, S.; Zhao, D.; Cai, C.; Song, D.; Shen, J.; Xu, A.; Qiao, Y.; Ran, Z.; Zheng, Q. Low-dose penicillin exposure in early life decreases Th17 and the susceptibility to DSS colitis in mice through gut microbiota modification. *Sci. Rep.* **2017**, *7*, 43662. [CrossRef]
26. Kim, Y. Bioinformatic and Statistical Analysis of Microbiome Data. In *Statistical Genomics*, 10th ed.; Fridley, B., Wang, X., Eds.; Springer US: New York, NY, USA, 2023; Volume 2629, pp. 183–229.
27. Maughan, H.; Wang, P.W.; Diaz Caballero, J.; Fung, P.; Gong, Y.; Donaldson, S.L.; Yuan, L.; Keshavjee, S.; Zhang, Y.; Yau, Y.C.; et al. Analysis of the cystic fibrosis lung microbiota via serial Illumina sequencing of bacterial 16S rRNA hypervariable regions. *PLoS ONE* **2012**, *7*, e45791. [CrossRef] [PubMed]
28. Segata, N.; Izard, J.; Waldron, L.; Gevers, D.; Miropolsky, L.; Garrett, W.S.; Huttenhower, C. Metagenomic biomarker discovery and explanation. *Genome Biol.* **2011**, *12*, R60. [CrossRef]
29. Zhang, C.; Li, S.; Yang, L.; Huang, P.; Li, W.; Wang, S.; Zhao, G.; Zhang, M.; Pang, X.; Yan, Z.; et al. Structural modulation of gut microbiota in life-long calorie-restricted mice. *Nat. Commun.* **2013**, *4*, 2163. [CrossRef]
30. Gamliel, A.; Austerweil, M.; Kritzman, G. Non-chemical approach to soilborne pest management–organic amendments. *Crop Prot.* **2000**, *19*, 847–853. [CrossRef]
31. Bulluck, L.R.; Brosius, M.; Evanylo, G.K.; Ristaino, J.B. Organic and synthetic fertility amendments influence soil microbial, physical and chemical properties on organic and conventional farms. *Appl. Soil Ecol.* **2002**, *19*, 147–160. [CrossRef]
32. Bakker, M.G.; Glover, J.D.; Mai, J.G.; Kinkel, L.L. Plant community effects on the diversity and pathogen suppressive activity of soil streptomycetes. *Appl. Soil Ecol.* **2010**, *46*, 35–42. [CrossRef]
33. Bao, S.D. *Agricultural and Chemistry Analysis of Soil*, 3rd ed.; Agriculture Press: Beijing, China, 2005; pp. 11–18.
34. Olsen, S.R.; Watanabe, F.S. A method to determine a phosphorus adsorption maximum of soils as measured by the langmuir isotherm. *Soil Sci. Soc. Am. J.* **1957**, *21*, 144–149. [CrossRef]
35. Grice, E.A.; Kong, H.H.; Conlan, S.; Deming, C.B.; Davis, J.; Young, A.C.; Bouffard, G.G.; Blakesley, R.W.; Murray, P.R.; Green, E.D.; et al. Topographical and temporal diversity of the human skin microbiome. *Science* **2009**, *324*, 1190–1192. [CrossRef]
36. Cordero, I.; Balaguer, L.; Rincón, A.; Pueyo, J.J. Inoculation of tomato plants with selected PGPR represents a feasible alternative to chemical fertilization under salt stress. *J. Plant Nutr. Soil Sci.* **2018**, *181*, 694–703. [CrossRef]

37. Wang, J.; Li, R.; Zhang, H.; Wei, G.; Li, Z. Beneficial bacteria activate nutrients and promote wheat growth under conditions of reduced fertilizer application. *BMC Microbiol.* **2020**, *20*, 38. [CrossRef] [PubMed]
38. Guo, L.; Chen, X.; Li, Z.; Wang, M.; Che, Y.; Zhang, L.; Jiang, Z.; Jie, S. Effects of continuous cropping on bacterial community and diversity in rhizosphere soil of industrial hemp: A five-year experiment. *Diversity* **2022**, *14*, 250. [CrossRef]
39. Wang, S.; Cheng, J.; Li, T.; Liao, Y. Response of soil fungal communities to continuous cropping of flue-cured tobacco. *Sci. Rep.* **2020**, *10*, 19911. [CrossRef]
40. Lakshmanan, V.; Selvaraj, G.; Bais, H.P. Functional soil microbiome: Belowground solutions to an aboveground problem. *Plant Physiol.* **2014**, *166*, 689–700. [CrossRef]
41. Zhalnina, K.; Louie, K.B.; Hao, Z.; Mansoori, N.; Da Rocha, U.N.; Shi, S.; Cho, H.; Karaoz, U.; Loque, D.; Bowen, B.P.; et al. Dynamic root exudate chemistry and microbial substrate preferences drive patterns in rhizosphere microbial community assembly. *Nat. Microbiol.* **2018**, *3*, 470–480. [CrossRef]
42. Chen, D.; Wang, X.; Zhang, W.; Zhou, Z.; Ding, C.; Liao, Y.; Li, X. Persistent organic fertilization reinforces soil-borne disease suppressiveness of rhizosphere bacterial community. *Plant Soil.* **2020**, *452*, 313–328. [CrossRef]
43. Shen, Z.; Penton, C.R.; Lv, N.; Xue, C.; Yuan, X.; Ruan, Y.; Li, R.; Shen, Q. Banana Fusarium Wilt Disease Incidence Is Influenced by Shifts of Soil Microbial Communities Under Different Monoculture Spans. *Microb. Ecol.* **2018**, *75*, 739–750. [CrossRef] [PubMed]
44. Chen, L.-J.; Wu, X.-Q.; Xu, Y.; Wang, B.-L.; Liu, S.; Niu, J.-F.; Wang, Z. Microbial diversity and community structure changes in the rhizosphere soils of *Atractylodes lancea* from different planting years. *Plant Signal. Behav.* **2021**, *16*, 1854507. [CrossRef] [PubMed]
45. Sasse, J.; Martinoia, E.; Northen, T. Feed your friends: Do plant exudates shape the root microbiome? *Trends Plant Sci.* **2017**, *23*, 25–41. [CrossRef]
46. Selmar, D.; Kleinwachter, M. Stress enhances the synthesis of secondary plant products: The impact of stress-related over-reduction on the accumulation of natural products. *Plant Cell Physiol.* **2013**, *54*, 817–826. [CrossRef]
47. Njira, K. Soil management practices that improve soil health: Elucidating their implications on biological indicators. *J. Anim. Plant Sci.* **2013**, *18*, 2750–2760.
48. Dorr de Quadros, P.; Zhalnina, K.; Davis-Richardson, A.; Fagen, J.R.; Drew, J.; Bayer, C.; Camargo, F.A.O.; Triplett, E.W. The Effect of Tillage System and Crop Rotation on Soil Microbial Diversity and Composition in a Subtropical Acrisol. *Diversity* **2012**, *4*, 375–395. [CrossRef]
49. Cao, Y. Characteristics of Rhizobacterial Community to Root-Knot Nematode Infected Tobacco and Diversity of Bacterial Flora Associated with *Meloidogyne incongnita*. Ph.D. Thesis, Yunan University, Kunming, China, 2015.
50. Kojima, H.; Kanda, M.; Umezawa, K.; Fukui, M. Sulfurimicrobium lacus gen. nov., sp. nov., a sulfur oxidizer isolated from lake water, and review of the family *Sulfuricellaceae* to show that it is not a later synonym of Gallionellaceae. *Arch. Microbiol.* **2021**, *203*, 317–323. [CrossRef]
51. Watanabe, T.; Kojima, H.; Fukui, M. *Sulfuriferula multivorans* gen. nov., sp. nov., isolated from a freshwater lake, reclassification of '*Thiobacillus plumbophilus*' as *Sulfuriferula plumbophilus* sp. nov., and description of *Sulfuricellaceae* fam. nov. and *Sulfuricellales* ord. nov. *Int. J. Syst. Evol. Microbiol.* **2015**, *65*, 1504–1508. [CrossRef]
52. Kulichevskaya, I.S.; Detkova, E.N.; Bodelier, P.L.E.; Rijpstra, W.I.C.; Sinninghe Damste, J.S.; Dedysh, S.N. *Singulisphaera rosea* sp. nov., a planctomycete from acidic Sphagnum peat, and emended description of the genus Singulisphaera. *Int. J. Syst. Evol. Microbiol.* **2012**, *62*, 118–123. [CrossRef]
53. Costa, O.Y.A.; De Hollander, M.; Pijl, A.; Liu, B.; Kuramae, E.E. Cultivation-independent and cultivation-dependent metagenomes reveal genetic and enzymatic potential of microbial community involved in the degradation of a complex microbial polymer. *Microbiome* **2020**, *8*, 76. [CrossRef]
54. De Vries, R.P.; Kester, H.C.; Poulsen, C.H.; Benen, J.A.; Visser, J. Synergy between enzymes from Aspergillus involved in the degradation of plant cell wall polysaccharides. *Carbohydr. Res.* **2000**, *327*, 401–410. [CrossRef]
55. Huang, G.; Dong, R.; Allen, R.; Davis, E.L.; Baum, T.J.; Hussey, R.S. Developmental expression and molecular analysis of two *Meloidogyne incognita* pectate lyase genes. *Int. J. Parasitol.* **2005**, *35*, 685–692. [CrossRef] [PubMed]
56. Kolb, E.; Legue, V.; Bogeat-Triboulot, M.B. Physical root-soil interactions. *Phys. Biol.* **2017**, *14*, 065004. [CrossRef] [PubMed]
57. Andrew, D.R.; Fitak, R.R.; Munguia-Vega, A.; Racolta, A.; Martinson, V.G.; Dontsova, K. Abiotic factors shape microbial diversity in Sonoran Desert soils. *Appl. Environ. Microbiol.* **2012**, *78*, 7527–7537. [CrossRef] [PubMed]
58. Yang, L.; Chen, Y.M.; He, R.L.; Deng, C.C.; Liu, J.W.; Liu, Y. Responses of soil microbial community structure and function to simulated warming in alpine forest. *J. Appl. Ecol.* **2016**, *27*, 2855–2863.
59. Hayat, R.; Ali, S.; Amara, U.; Khalid, R.; Ahmed, I. Soil beneficial bacteria and their role in plant growth promotion: A review. *Ann. Microbiol.* **2010**, *60*, 579–598. [CrossRef]
60. Subbarao, G.V.; Yoshihashi, T.; Worthington, M.; Nakahara, K.; Ando, Y.; Sahrawat, K.L.; Rao, I.M.; Lata, J.-C.; Kishii, M.; Braun, H.-J. Suppression of soil nitrification by plants. *Plant Sci.* **2015**, *233*, 155–164. [CrossRef]
61. Niu, Q.; Han, Y.; Xu, L. Effects of crop rotation on soil physicochemical properties and bacterial community of foxtail millet rhizosphere soil. *J. Agro-Environ. Sci.* **2018**, *37*, 2802–2809.
62. Jiao, X.-L.; Zhang, X.-S.; Lu, X.-H.; Qin, R.; Bi, Y.-M.; Gao, W.-W. Effects of maize rotation on the physicochemical properties and microbial communities of American ginseng cultivated soil. *Sci. Rep.* **2019**, *9*, 8615. [CrossRef]
63. Jung, H.M.; Lee, J.S.; Bae, H.M.; Yi, T.H.; Kim, S.Y.; Lee, S.T.; Im, W.T. *Inquilinus ginsengisoli* sp. nov., isolated from soil of a ginseng field. *Int. J. Syst. Evol. Microbiol.* **2011**, *61*, 201–204. [CrossRef]

64. Cho, C.H.; Lee, J.S.; An, D.S.; Whon, T.W.; Kim, S.G. *Nocardioides panacisoli* sp. nov., isolated from the soil of a ginseng field. *Int. J. Syst. Evol. Microbiol.* **2010**, *60*, 387–392. [CrossRef] [PubMed]
65. Zhang, X.; Li, K.; Xing, R.; Liu, S.; Chen, X.; Yang, H.; Li, P. miRNA and mRNA expression profiles reveal insight into the chitosan-mediated regulation of plant growth. *J. Agric Food Chem.* **2018**, *66*, 3810–3822. [CrossRef] [PubMed]
66. Jiang, J.; Yu, M.; Hou, R.; Li, L.; Ren, X.; Jiao, C.; Yang, L.; Xu, H. Changes in the soil microbial community are associated with the occurrence of *Panax quinquefolius* L. root rot diseases. *Plant Soil.* **2019**, *438*, 143–156. [CrossRef]
67. Vessey, J.K. Plant growth promoting rhizobacteria as biofertilizers. *Plant Soil.* **2003**, *255*, 571–586. [CrossRef]
68. Wang, B.; Li, R.; Ruan, Y.; Ou, Y.; Zhao, Y.; Shen, Q. Pineapple–banana rotation reduced the amount of *Fusarium oxysporum* more than maize–banana rotation mainly through modulating fungal communities. *Soil Biol. Biochem.* **2015**, *86*, 77–86. [CrossRef]
69. Zuo, W.; Song, B.; Shi, Y.; Zupanic, A.; Guo, S.; Huang, H.; Jiang, L.; Yu, Y. Using *Bacillus thuringiensis* HM-311@hydroxyapatite@ biochar beads to remediate Pb and Cd contaminated farmland soil. *Chemosphere* **2022**, *307*, 135797. [CrossRef]
70. Basumatary, B.; Das, D.; Choudhury, B.; Dutta, P.; Bhattacharyya, A. Isolation and characterization of endophytic bacteria from tomato foliage and their in vitro efficacy against root-knot nematodes. *J. Nematol.* **2021**, *53*, 1–16. [CrossRef]
71. Azarias Guimaraes, A.; Duque Jaramillo, P.M.; Simao Abrahao Nobrega, R.; Florentino, L.A.; Barroso Silva, K.; De Souza Moreira, F.M. Genetic and symbiotic diversity of nitrogen-fixing bacteria isolated from agricultural soils in the western Amazon by using cowpea as the trap plant. *Appl. Environ. Microbiol.* **2012**, *78*, 6726–6733. [CrossRef]
72. Harris, A.R.; Schisler, D.A.; Ryder, M.H.; Adkins, P.G. Bacteria suppress damping-off caused by Pythium ultimumvar. Sporangiiferum, and promote growth, in bedding plants. *Soil Biol. Biochem.* **1994**, *26*, 1431–1437. [CrossRef]
73. Lee, S.-W.; Cooksey Donald, A. Genes Expressed in Pseudomonas putidaduring Colonization of a Plant-Pathogenic Fungus. *Appl. Environ. Microbiol.* **2000**, *66*, 2764–2772. [CrossRef]

agronomy

Article

The Nitrogen Cycling Key Functional Genes and Related Microbial Bacterial Community α−Diversity Is Determined by Crop Rotation Plans in the Loess Plateau

Rui Liu [1], Yang Liu [2], Yuan Gao [2], Fazhu Zhao [2] and Jun Wang [1,2,*]

[1] State Key Laboratory of Soil Erosion and Dryland Farming on the Loess Plateau, Institute of Soil and Water Conservation, Chinese Academy of Sciences and Ministry of Water Resources, Yangling 712100, China; liurui216@mails.ucas.ac.cn

[2] Shaanxi Key Laboratory of Earth Surface System and Environmental Carrying Capacity, College of Urban and Environmental Science, Northwest University, Xi'an 710127, China

* Correspondence: wangj@nwu.edu.cn; Tel.: +86-29-8830-8783

Abstract: Soil nitrogen cycling microbial communities and functional gene α−diversity indicate soil nitrogen cycling ecological functions and potentials. Crop rotation plans affect soil nitrogen fractions and these indicators. We sequenced soil samples from four crop rotation plans (fallow, winter wheat monoculture, pea-winter wheat-winter wheat-millet rotation, and corn-wheat-wheat-millet rotation) in a long-term field experiment. We examined how microbial communities and functional gene α−diversity changed with soil nitrogen fractions and how nitrogen fractions regulated them. Planting crops increased the abundance and richness of nitrogen cycling key functional genes and bacterial communities compared with fallow. The abundance and richness correlated positively with nitrogen fractions, while Shannon index did not. The abundance increased with soil total nitrogen (STN) and potential nitrogen mineralization (PNM), while Shannon index showed that nitrogen cycling key functional genes increased and then decreased with increasing STN and PON. Introducing legumes into the rotation improved the α−diversity of nitrogen cycling key functional genes. These results can guide sustainable agriculture in the Loess Plateau and clarify the relationship between nitrogen fractions and nitrogen cycling key functional genes.

Keywords: nitrogen cycling genes; nitrogen fractions; microbial community; crop rotation; loess Plateau

Citation: Liu, R.; Liu, Y.; Gao, Y.; Zhao, F.; Wang, J. The Nitrogen Cycling Key Functional Genes and Related Microbial Bacterial Community α−Diversity Is Determined by Crop Rotation Plans in the Loess Plateau. *Agronomy* **2023**, *13*, 1769. https://doi.org/10.3390/agronomy13071769

Academic Editors: Yong-Xin Liu and Peng Yu

Received: 23 May 2023
Revised: 19 June 2023
Accepted: 21 June 2023
Published: 29 June 2023

1. Introduction

The abundance of functional genes is an important indicator for evaluating ecosystem functionality [1]. Nitrogen cycling functional genes, including amoA, amoB, and amoC for nitrification, nirS, nirK, and nosZ for denitrification, and ureC for ammonification, have been applied to gauge the nitrogen transformation capacity in agricultural soils at varying stages [2–4]. Several studies have demonstrated that agricultural management practices alter functional gene abundance, for instance, reducing tillage or implementing no-till practices, which increase the abundance of amoA, nirK, and nosZ [5–7]. Additionally, the application of organic fertilizers can boost the abundance of ureC, amoA, nirS, nirK, and nxrB in nitrogen cycling [8,9]. In recent years, functional gene diversity has also evolved as an essential factor for assessing the functional potential of soil microbial communities in ecosystems [10,11]. Researchers have explored the functional gene diversity in different ecosystems, such as grasslands, forests, and agricultural lands, to obtain knowledge on the functional potential of soil microbial communities [12–14]. Moreover, Chukwuneme et al. [15] used metagenomic sequencing to investigate the impact of different agricultural management practices on the diversity of carbon cycle genes in the corn rhizosphere soils, while a meta-analysis by You et al. [16] revealed that nitrogen addition affects the diversity of crucial functional genes linked to N_2O emission in agricultural ecosystems. Hence, exploring the abundance and diversity of functional genes is essential in understanding the ecological functions and potential of soil microbial communities.

The Loess Plateau is facing severe soil nutrient deficiency due to long-term soil erosion. To address this issue, crop rotation has been widely used to enhance the soil nitrogen pool [17,18]. Fu et al. [19] conducted research indicating that long-term diversified crop rotation can effectively improve soil nitrogen fractions and sustain dryland planting regimes. The corn-legume-wheat rotation, for example, can increase soil particulate organic nitrogen and mineralized nitrogen [20], while rice-rape rotation can increase soil particulate organic nitrogen and microbial nitrogen [21]. Additionally, crop rotation can improve soil structure, the living environment for microorganisms, and increase microbial diversity [22]. Additionally, long-term crop rotation can increase crop diversity, and the input of diversified crop residues and litter can result in changes to the substrate available to soil microbes, leading to an increase in microbial $\alpha-$diversity [23–25]. Nonetheless, studies suggest that long-term crop rotation may lead to the continuous accumulation of high-C/N crop residues in the soil [26], which could result in a decrease in microbial diversity [27]. Therefore, introducing crops with different C/N ratios, such as legumes or non-legumes, into long-term crop rotation plans may have varying effects on microbial $\alpha-$diversity.

Changes in soil microbial communities can result in alterations in functional genes involved in nitrogen cycling [28,29]. Several studies have demonstrated that crop rotation can regulate the abundance of denitrification genes (nirK, nosZ, and nirS), thereby influencing N_2O emissions [30,31]. However, soil nitrogen fractions, as important factors reflecting soil productivity and nitrogen status [32], can also affect changes in nitrogen cycling functional genes [4,33], but there are few studies in this direction. In crop rotation plan, different crops have varied litter properties that can impact soil nutrient content [22]. Legumes, for instance, release nutrient-rich, juicy, and protein- and sugar-rich residues, which result in faster nutrient release than the more stable organic matter derived from the fibrous plant residues of crops such as corn and cereals [34]. Therefore, further research is necessary to assess the impact of nitrogen fractions in different long-term crop rotation plans on functional gene $\alpha-$diversity.

This study aims to investigate the impact of different planting plans on soil nitrogen fractions, key functional genes and related microbial communities' $\alpha-$diversity through 36 years of field experiments. It is hypothesized that: (1) crop rotation plans can increase the abundance of key functional genes for nitrogen cycling and the richness of related microbial communities; (2) long-term crop residue from rotation plans may result in differences in nitrogen cycling key functional genes and related microbial communities' diversity due to varying C/N ratios. The objectives of this study are to elucidate the effects of different crop rotation plans on key functional genes and related microbial communities involved in nitrogen cycling and to explore the regulatory modes of soil nitrogen fractions on key functional genes $\alpha-$diversity.

2. Materials and Methods

2.1. Experimental Site and Treatments

A long-term field experiment was established in September 1984 at the Changwu Agroecological Station (35°12′ N, 107°44′ E) in Changwu County, Shaanxi Province, China. The study area is representative of a typical rainfed farming system in the Loess Plateau of China. The experimental site has a continental monsoon climate, with a mean annual temperature of 9.1 °C and a frost-free period of 194 d. The long-term (1984–2020) average annual precipitation is 580 mm, half of which occurs from July to September. The soil is a Heilutu silt loam (Calcarid Regosol according to the FAO classification system or Ultisol according to the U.S. soil taxonomy), with 45 g kg^{-1} sand, 656 g kg^{-1} silt, 309 g kg^{-1} clay, 8.4 pH, 105 g kg^{-1} $CaCO_3$, 10.5 g kg^{-1} organic C, 1.0 g kg^{-1} total N, and 1.4 Mg m^{-3} bulk density at the 0–30 cm depth at the beginning of the experiment.

There were four cropping designs for the study, including (1) fallow (F); (2) winter wheat monoculture (W); (3) 1 year of pea, 1 year of winter wheat, and 1 year of winter wheat and millet (PWWM); (4) 1 year of corn, 1 year of winter wheat, and 1 year of winter wheat and millet (CWWM). A randomized block design was used in this experiment with three replications. Each plot had 10.3 m by 6.5 m size separated by 0.5 m strip, and each block was separated by 1 m strips. Crops were planted by hand under conventional tillage using animal-drawn (first 16 yr) and hand tractor-drawn (second 18 yr) plows to a depth of 10 cm. Crops were planted at 20 cm row spacing, except for corn which was planted at 70 cm spacing. Plant populations were 2.23, 0.60 and 0.04 million plants ha^{-1} for winter wheat, pea and corn, respectively, and the seeding rate was 28 kg ha^{-1} for millet, respectively. At planting, chemical fertilizers were broadcasted to winter wheat, corn and millet using urea (46% N) and monoammonium phosphate (11% N, 23% P) at rates of 120 kg N ha^{-1} and 20 kg P ha^{-1}. Pea received N and P from monoammonium phosphate at 10 kg N ha^{-1} and 20 kg P ha^{-1}, respectively. Because of the high soil potassium content, no potassium fertilizer was applied. Weed management was carried out by hand before, during, and after crop growth. Pesticides were applied as needed to control pests.

2.2. Soil Samplings

In September 2020, samples were collected after peas and corn were harvested and before wheat was planted. Soil samples were collected from a depth of 15 cm from three places in central rows of each plot using a hand probe (5 cm inside diameter) in September 2020. A separate undisturbed soil core (5 cm inside diameter) was collected simultaneously at 0–15 from each plot for the bulk density. The collected soil samples were immediately stored in sterile plastic bags, placed in iceboxes, and brought back to the laboratory immediately. Then, all the samples were sifted through a 2 mm mesh, crop residues, root materials, and stones were removed. and were thoroughly homogenized to be further divided into three parts: one part was air-dried to analyze the nitrogen fractions, another part was stored at −4 °C for determination of MBN (microbial biomass nitrogen), while the rest of the samples were stored at −80 °C until DNA extraction and metagenomic sequencing.

2.3. Analyses of Soil Nitrogen Fractions

The soil total nitrogen (STN) concentration was determined by the combustion method using a high induction furnace N analyzer (Euro Vector EA3000, Manzoni, Italy) [35]. The particulate organic nitrogen (PON) was determined with the sodium hexametate separation method [36]. The potential nitrogen mineralization (PNM) concentration was determined by the incubation method modified by Haney et al. [37]. The NH_4^+-N and NO_3^--N concentrations in the extract were determined using the modified Griess-Ilosvay method with an autoanalyzer (Lachat Instruments, Loveland, CO) [35]. The soil water content (SWC) and bulk density (BD) were measured from the gravimetric weight of the core before and after oven drying at 105 °C for 24 h. The other container with moist soil was subsequently used for determining MBN concentration using the modified fumigation-incubation method for airdried soils [38].

2.4. DNA Extraction, Sequencing, and Data Processing

Total genomic DNA was extracted from 0.5 g soil samples using the E.Z.N.A.® Soil DNA Kit (Omega Bio-tek, Norcross, GA, USA) according to manufacturer's instructions. Concentration and purity of extracted DNA was determined using TBS-380 and NanoDrop2000, respectively. DNA extract quality was checked on 1% agarose gel. There were three replicates for each soil sample to obtain sufficient DNA for shotgun metagenome sequencing. The metagenome was sequenced in an Illumina HiSeq 2000 platform (Personal, Shanghai, China) to generate 150 bp paired-end reads at a greater sequencing depth. Reads aligned to the human and vegetation genome were removed, and the lengths were trimmed using Sickle. Sequence data associated with this project have been deposited in the NCBI Short Read Archive database (Accession Number: PRJNA985043; https://www.ncbi.nlm.nih.gov/sra/PRJNA985043 accessed on 4 June 2022).

2.5. Metagenomic Analysis

To enhance the reliability and quality of subsequent analysis, we removed adapter sequences, and discarded those quality-trimmed reads of less than 50 bp in length or containing N (ambiguous) bases [39]. Megahit software (https://hku-bal.github.io/megabox/, accessed on 15 April 2021) was used to obtain high-quality reads [40], MetaGeneMark (http://exon.gatech.edu/GeneMark/metagenome, accessed on 20 April 2021) was used to predict genes in contigs longer than 200 bp [41]. Gene abundance for each sample was the trans per million values [TPM: (Reads Number/Gene Length)_Relative] × 1,000,000) [42].

According to the results of the KEGG database search, functional annotation and taxonomic assignment of each sample were carried out for further analysis using eggNOG-mapper v2, a tool based on precomputed orthology assignments. Based on a previous study, 17 N-cycling functional genes were defined as nitrogen cycling genes, associated with the processes of Nitrification, Denitrification, Dissimilatory Nitrate Reduction (DNR), Assimilatory Nitrate Reduction (ANR), Ammoniation, and Assimilation [43]. Detailed information on N-cycling functional genes is listed in Table S1.

2.6. Screening Key N-Cycling Functional Genes

We screened key N-cycling functional genes using network analysis and random forests [44]. Firstly, the 17 N-cycling functional genes were screened based on the KEGG databases. To explore the relationships among N-cycling functional genes, Spearman correlation analyses were performed using the "Hmisc" and "igraph" packages [45], genes with a relative abundance higher than 0.1% were kept for network construction. A valid co-occurrence was considered a statistically robust correlation ($|r| > 0.6$, $p < 0.05$). The network was visualized by Gephi platform [46]. We combined the overall samples and used the network analyses to select 11 genes with respect to hub nodes in network, including amoA, amoC, napA, nirK, nirS, norB, nasA, nasB, nirA, ureC and gdh (Supplementary Materials Figure S1A) according to degree (degree > 0). Secondly, we used random forests to disentangle the contributions of N-cycling functional genes to variations in different nitrogen fractions (Supplementary Materials Figure S1B) using the 'randomforest' package [47]. Finally, on the basis of the results of co-occurrence network and random forest analyses, 6 genes, including amoC, nirS, norB, nasA, nasB, gdh, were selected as the key N-cycling functional genes in our study.

2.7. Statistical Analyses

All statistical analyses were conducted in the R environment (v4.0.5; http://www.r-project.org/, accessed on 30 April 2021). One-way analysis of variance (ANOVA) was performed based on the 'stats' package to assess the effect of different crop rotation plans on nitrogen fractions content, abundance and the Shannon index of N-cycling genes, and abundance and Shannon index of microbial functional community at the 0.05 level of significance. Diversity of Shannon index and richness index (Chao1, ACE) were calculated using "phyloseq" package [48]. Nonlinear regression and spearman correlation analyses were performed using the "survival" and "basicTrendline" packages, respectively [49]. Heatmaps were used to illustrate the Z-score-normalized relative abundance of N-cycling functional genes using the "pheatmap" package [50]. Other graphs were drawn using the "ggplot2" package [51].

3. Results

3.1. Soil Nitrogen Fractions

Different crop rotation plans have been found to affect the nitrogen fractions in soil (Figure 1). Compared with F, other cropping plans significantly increased the content of total nitrogen (STN) and potential mineralizable nitrogen (PNM) in the soil. Compared with W, CWWM significantly increased STN by 23.9%, while PNM decreased by 12.7% ($p < 0.05$); Compared with W, PWWM or CWWM significantly increased the content of microbial nitrogen (MBN) by 47.4% and 60.3%, respectively ($p < 0.05$). Compared with

PWWM, CWWM increased STN by 28.6% and particulate organic nitrogen (PON) by 88.2% ($p < 0.05$). In addition, the content of ammonium nitrogen and nitrate nitrogen in the soil did not show significant differences in different crop rotation plans.

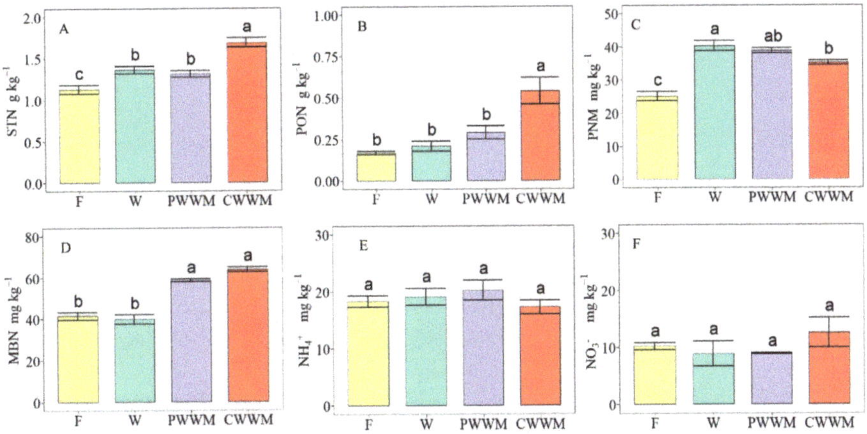

Figure 1. Effects of different crop rotation plans on soil nitrogen fractions. Soil N fractions including (**A**) STN, soil total nitrogen; (**B**) PON, particulate organic nitrogen; (**C**) PNM, potential nitrogen mineralization; (**D**) MBN, microbial biomass N; (**E**) NH_4^+, ammonium nitrogen; (**F**) NO_3^-, nitrate nitrogen. Mean values ± standard deviation of nitrogen fractions. Different small letters indicate significant differences among crop rotation plans ($p < 0.05$, ANOVA, Tukey HSD).

3.2. *Alpha Diversity of Key Function Genes in Nitrogen Cycling*

There were significant differences in the abundance and α−diversity of key functional genes under different crop rotation plans in our investigation. The abundance of key functional genes varied significantly in different crop rotation plans. The abundance of amoC and nasA in F was significantly lower than in other treatments. Moreover, the abundance of norB and nasB in F was significantly lower than in PWWM. PWWM had a significantly higher abundance of nasB than CWWM. In addition, there was no significant difference in the abundance of nirS and gdh among different crop rotation plans (Table 1). Furthermore, compared with W, PWWM and CWWM significantly increased the abundance of amoA and nirK (Table S1). The abundance and α−diversity of key functional genes also differed under different crop rotation plans. The total abundance of key functional genes in F was significantly lower than in other treatments (Figure 2A). There was no significant difference in the total abundance of key functional genes among PWWM, CWWM and W. The Shannon of F was significantly lower than that of PWWM (Figure 2B); the Shannon of W and PWWM was significantly higher than that of CWWM; there was no significant difference in Shannon between PWWM and W.

Table 1. Effects of different crop rotation plans on the abundance of nitrogen cycle key genes values.

Processes	Nitrification	Denitrification		ANR (Assimilatory Nitrogen Reduction)		Assimilation
Crop Rotation Planss	amoC	norB	nirS	nasA	nasB	gdh
	($NH_4^+ \to NH_2OH$)	($NO \to N_2O$)	($NO_2^- \to NO$)	($NO_3^- \to NO_2^-$)	($NO_3^- \to NO_2^-$)	($NH_4^+ \to$ Org)
F	124.67 ± 7.33 b	714.67 ± 18.56 b	57.33 ± 8.11 a	1149.33 ± 53.07 b	99.33 ± 10.73 bc	1720.67 ± 70.70 a
W	184.00 ± 7.57 a	753.33 ± 20.80 ab	39.33 ± 11.79 a	1351.33 ± 28.99 a	127.33 ± 5.81 ab	1906.67 ± 75.10 a
PWWM	174.00 ± 9.17 a	858.67 ± 20.34 a	44.67 ± 2.40 a	1362.67 ± 35.63 a	154.67 ± 10.48 a	1960.67 ± 16.18 a
CWWM	170.00 ± 2.00 a	788.00 ± 69.41 ab	56.00 ± 2.00 a	1473.33 ± 34.57 a	88.00 ± 7.21 c	2004.67 ± 134.39 a

Note: F, fallow; W, wheat monoculture; PWWM, pea-wheat-wheat-millet rotation; CWWM, corn-wheat-wheat-millet rotation. Values within the same column followed by different letters indicate significant differences ($p < 0.05$, ANOVA, Tukey HSD). are mean ± standard deviation ($n = 3$).

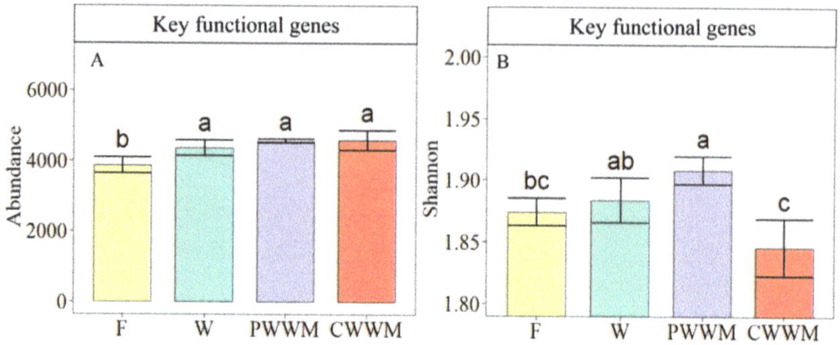

Figure 2. Values of α−diversity of N-cycling key functional genes in different crop rotation plans. (**A**) the abundance of key N-cycling functional genes; (**B**) the Shannon of key N-cycling functional genes. Mean values ± standard deviation of the abundance and the Shannon of nitrogen cycle genes. Different small letters indicate significant differences among crop rotation plans ($p < 0.05$, ANOVA, Tukey HSD).

3.3. *Alpha Diversity of Microbial Communities Related to Key Functional Genes*

In species related to key functional genes in the nitrogen cycle at the phylum level, the relative abundance of *Proteobacteria* in CWWM was higher than that in W, while the relative abundance of *Actinobacteria* was lower in CWWM than in PWWM (Figure 3; Table S3). There were no significant changes in the relative abundance of other species. However, there was no significant difference in the relative abundance of microorganisms between CWWM and PWWM at class, order, family, and genus levels (Figure S2). In this study, a total of 546 microbial species related to key functional genes in the nitrogen cycle were detected, and there were differences in bacteria α−diversity among different crop rotation plans. Compared with the W, the Chao1 index of the rotation plans was significantly increased, and the ACE index of CWWM was significantly increased compared with the F. However, there was no significant difference in the richness index (Chao1, ACE) between PWWM and CWWM. In addition, there was no significant difference in the Shannon index of microbial communities among different crop rotation plans (Table 2).

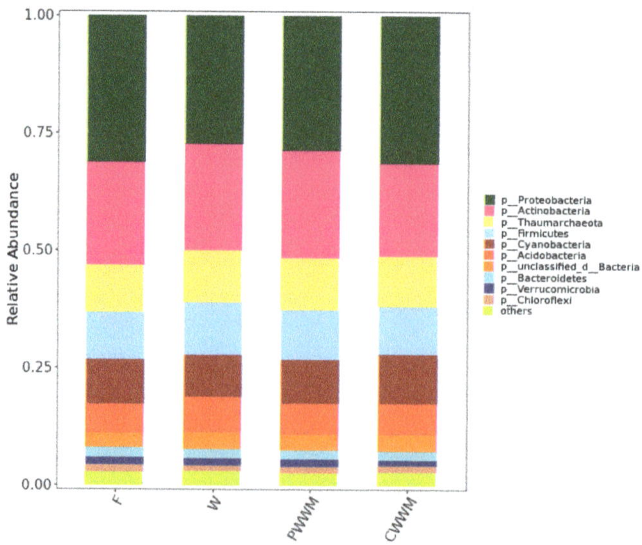

Figure 3. Relative abundance of the phyla of species related to key functional genes.

Table 2. Effects of different crop rotation plans on α−diversity of microbial community. Values are mean ± standard deviation (n = 3).

Kingdom	Bacteria			Archaea		
Crop Rotation Planss	Shannon	Chao1	ACE	Shannon	Chao1	ACE
F	6.94 ± 0.04 a	297.33 ± 3.53 bc	7.47 ± 0.12 b	0.94 ± 0.08 a	3.67 ± 0.67 a	0.91 ± 0.09 a
W	6.87 ± 0.08 a	292.33 ± 8.51 c	7.79 ± 0.13 a	1.08 ± 0.03 a	4.67 ± 0.33 a	1.06 ± 0.03 a
PWWM	6.89 ± 0.01 a	313.33 ± 2.60 ab	7.78 ± 0.06 ab	1.00 ± 0.04 a	3.67 ± 0.33 a	0.89 ± 0.05 a
CWWM	6.97 ± 0.05 a	325.67 ± 4.33 a	7.95 ± 0.02 a	0.98 ± 0.01 a	4.00 ± 0.58 a	0.97 ± 0.08 a

Note: F, fallow; W, wheat monoculture; PWWM, pea-wheat-wheat-millet rotation; CWWM, corn-wheat-wheat-millet rotation. Values within the same column followed by different letters indicate significant differences ($p < 0.05$, ANOVA, Tukey HSD). Bold values indicate significant differences.

3.4. The Regulatory Modes of Soil Nitrogen Fractions on Key Functional Genes and Microbial Communities

As the key functional genes that affect the changes of nitrogen fractions, their abundance changes must be positively correlated with nitrogen fractions, which was shown by Spearman correlation analysis and confirmed by regression models in this study (Table 3; Figure 4A,B). However, Spearman correlation analysis revealed that The Chao1 index of the bacteria community was significantly positively correlated with PON and MBN. Furthermore, nitrogen fractions had no significant effect on the Shannon index of the key functional genes and microbial community. In addition, the results of the regression model also showed that the Shannon index of the key functional genes increased first and then decreased with the increase in STN and PON, and higher levels of STN and PON led to a decrease in the Shannon index of the key functional genes.

Table 3. Spearman correlation coefficients between the N fractions and α−diversity of microbial community.

	Key Functional Genes		Bacteria			Archaea		
Crop Rotation Plans	Abundance	Shannon	Shannon	Chao1	ACE	Shannon	Chao1	ACE
STN	0.720 **	−0.399	−0.05	0.482	0.557	0.02	0.177	0.234
PON	0.698 *	−0.407	0.053	0.583 *	0.455	−0.03	0.155	0.2
PNM	0.273	0.476	−0.27	0.1	0.564	0.531	0.298	0.288
MBN	0.706 *	−0.238	0.146	0.809 **	0.52	−0.249	−0.209	−0.231
NH_4^+	0.259	0.42	−0.402	−0.212	−0.006	−0.039	−0.003	−0.164
NO_3^-	−0.399	−0.448	0.375	0.365	0.1	−0.225	−0.303	−0.284

Note: ** Correlation is significant at $p < 0.01$; * Correlation is significant at $p < 0.05$.

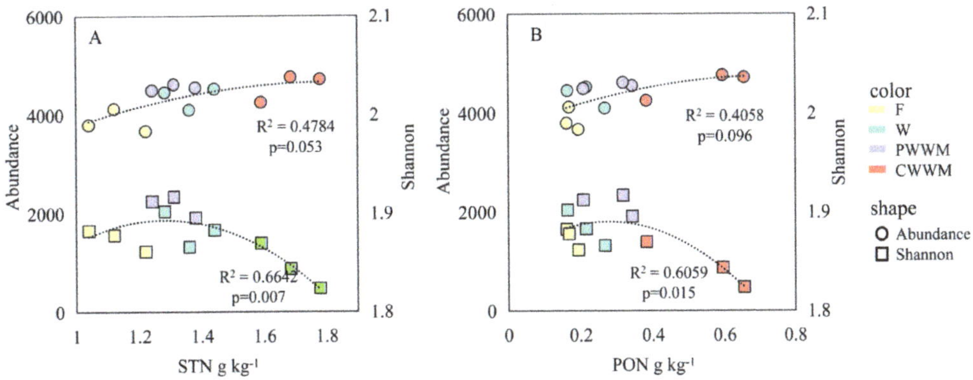

Figure 4. Relationships between α−diversity of key functional genes and nitrogen fractions. STN, soil total nitrogen; PON, particulate organic nitrogen. (**A**) Relationships between α−diversity of key functional genes and STN; (**B**) Relationships between α−diversity of key functional genes and PON; The nonlinear fitting method is quadratic fitting. The fitting degree (R^2) of the curve is provided.

4. Discussion

4.1. Variations in Soil Nitrogen Fractions under Different Crop Rotation Plans

Soil nitrogen (N) plays a central role in soil quality and biogeochemical cycles [52], and is available in various chemical fractions, including soil total N (STN), particulate organic matter (PON), potential nitrogen mineralization (PNM), microbial biomass N (MBN), ammonium N (NH_4^+-N), and nitrate N (NO_3^--N) [53]. Different nitrogen fractions could better reflect changes in soil quality and nitrogen supply potential that alter nutrient dynamics in agroecosystems [54]. Therefore, it is essential to study the transformations of nitrogen fractions under different agricultural management measures for assessing changes in soil quality and function [55].

The results showed that corn-based crop rotation significantly increased STN and PON compared to winter wheat monoculture, while pea-based crop rotation had no significant effect. However, both corn-based and pea-based crop rotations increased PNM compared to winter wheat monoculture. The higher STN and PON under corn—based crop rotation may be attributed to the quality and quantity of crop residue. Corn produces more biomass than leguminous crops, which can provide more external nitrogen input to the soil [56,57]. Moreover, corn residue has a high C/N ratio, which can slow down its decomposition rate and increase soil nitrogen accumulation [58]. On the other hand, leguminous crops have a lower C/N ratio, which can increase soil mineralization rate and PNM. This is consistent with previous studies that reported higher PNM after legume rotation than after corn rotation [59].

The increase in PNM under both corn-based and pea-based crop rotations may also be related to the symbiotic nitrogen fixation by leguminous crops. Legumes can enhance soil nitrogen fixation by establishing a symbiotic relationship with rhizobia bacteria. This can increase the availability of nitrogen for subsequent crops and improve soil fertility. Previous studies have also reported that rotation with legumes can increase PNM compared to monocropping [60]. The results of this study also indicated that crop rotation plans can increase microbial biomass nitrogen (MBN) compared to monocropping. This may be due to the increased diversity and quantity of crop residues and root exudates under rotation plans, which can provide more substrates for microbial growth and activity [61]. This finding is in agreement with other studies that reported higher MBN under rotation plans than under monocropping [62,63].

The findings of this study suggest that corn-based crop rotation can improve soil nitrogen status and fertility more than pea-based crop rotation or winter wheat monoculture. However, this study has some limitations that need to be addressed. First, the study was conducted in a single site with a specific soil type and climate condition, which may limit the generalizability of the results. Second, the study only measured soil nitrogen parameters at one time point after harvest, which may not reflect the seasonal dynamics of soil nitrogen processes.

4.2. Effects of Different Crop Rotation Plans on Alpha Diversity of Key Functional Genes

Environmental factors can effectively regulate the abundance of soil nitrogen cycling genes [64]. In this study, compared with F, the total abundance of key functional genes in other treatments was significantly increased, which may be related to the increased organic matter input and PNM by planting crops (Figure 1C); moreover, crop roots can also improve soil bulk density (BD) and provide a more diversified living space for microbial communities, thereby increasing microbial richness [65]. Brandan et al. [66] reported a significant negative correlation between soil BD and the abundance of nitrogen cycling functional gene in a long-term rotation study, so the higher BD in F may also be one of the reasons for its lower key functional gene abundance than other treatments (Table S2). In addition, soil pH and soil water content (SWC) are also the main factors affecting the abundance of soil nitrogen cycling genes [67,68]. However, due to the factors such as climate, tillage and rainfall in the Loess Plateau, there were no significant differences in pH, BD and SWC between monoculture and two rotation plans, which was also an important reason for the lack of significant difference in the total abundance of key functional genes among these three treatments.

Soil functional gene diversity reflects the potential of soil microorganisms to perform various metabolic activities and respond to environmental changes, which is important for maintaining soil health and function [69]. In this study, we found that CWWM significantly reduced the Shannon index of key functional genes compared with W and PWWM, indicating that introducing maize into rotation reduced the functional potential of nitrogen cycling in the ecosystem, while introducing legumes into rotation had more positive effects on the functional potential of nitrogen cycling in the ecosystem. Soil functional gene diversity can also influence the stability and resilience of soil ecosystems, as it can buffer against disturbances and enhance functional redundancy [70]. Especially in the Loess Plateau region, introducing legumes into rotation can more effectively improve the risk resistance of soil ecosystems. We speculate that the main reasons for the difference in key gene diversity between the two different rotation systems are as follows: first, legume crops can form symbiotic relationships with specific bacteria (i.e., *Actinobacteria*) during their growth, resulting in increased abundance of specific genes (i.e., nasB) [71]; second, legume crop residues have a lower C/N, which makes them easier to decompose and provide nutrients for microorganisms faster [72]; meanwhile, in practical agricultural management, the nitrogen application rate for legume crops is only 1/6 of that for maize crops, and crop residue amount is also lower than that for maize crops, resulting in soil nitrogen deficiency, leading to a competitive relationship between microorganisms and crops, with microorganisms having an advantage [73,74]. After long-term introduction of maize into rotation, although it can increase yield and economic benefits, the risk of reducing key gene diversity should be considered. Hence, future studies should explore the combination of ecological agriculture measures, such as organic fertilization or introduction of legume into rotation, with introduction of maize into rotation to enhance the ecosystem sustainability.

4.3. *Effects of Different Crop Rotation Plans on Alpha Diversity of Microbial Communities*

The quality and quantity of crop residues may be the main driving factor affecting the composition of microbial communities [75]. Among the microbial communities related to key functional genes, the relatively high abundance of *Proteobacteria* and *Actinobacteria* changes in different crop rotation plans. The relative abundance of *Proteobacteria* in crop rotation is higher than that in single cropping, which is similar to the law of STN content. Li et al. and Wang et al. [76,77] also confirmed that the relative abundance of *Proteobacteria* is related to organic matter content because *Proteobacteria* can decompose recalcitrant carbon sources into small intermediate molecules to provide nutrition for other microorganisms [78]. Therefore, the relative abundance of *Proteobacteria* in CWWM is significantly higher than that in monoculture. However, the relative abundance of *Actinobacteria* in CWWM is significantly lower than that in PWWM and W. The refractory corn residue makes it difficult for *Actinobacteria* to grow in the soil, as *Actinobacteria* mainly use available carbon sources for growth [79]. This can also explain why there is no difference in STN and PON content between PWWM and W.

The $\alpha-$diversity of microbial communities plays an important role in ecosystem processes [80]. Numerous studies have demonstrated that crop rotation plans can effectively increase the $\alpha-$diversity of soil microbial communities [81,82]. This is because both legume-wheat and corn-wheat rotations can provide diverse substrates for microbial utilization through diverse plant residues and root secretion, respectively; Moreover, the growth of different crops can alter soil nutrient conditions and soil environment, which can directly or indirectly enhance the diversity and abundance of soil microbial communities [83–86]. In this study, crop rotation improved the richness of microbial communities but did not change the Shannon index of microbial communities. This may be due to the dominant role of long-term fertilizer inputs and mechanical cultivation in regulating soil biodiversity in farmland [87,88]. Additionally, the sampling period in this study was after crop harvest, and the farmland soils were in a fallow state, which may have led to non-significant changes in the Shannon index of microbial communities.

4.4. Effects of Nitrogen Fractions on the $\alpha-$Diversity of Key Functional Genes and Associated Microbial Communities

The richness of microbial communities is closely related to the quantity and quality of their available resources [89,90], with soil nitrogen content being a limiting factor for microbial growth, and soil microbial community richness increasing as soil nitrogen accumulated [91,92]. In the crop rotation plan, the input of different crop residues can increase the availability of substrates, thus increasing microbial community richness [93]. This study also showed a significant positive correlation between microbial community richness and STN, PON (Table 3). However, while exogenous input of crop residues may change soil microbial community richness, different species can have the same ecological function in an ecosystem [94], and so changes in microbial communities may not necessarily result in the corresponding changes in functional genes.

We found a significant positive correlation between the abundance of key functional genes and nitrogen fractions, but the Shannon index showed an initial increase followed by a decrease with an increase in STN and PON, which is consistent with our hypothesis. Compared with F, different agricultural cropping plans lead to an increase in soil organic matter, which will provide a food source for microorganisms participating in nutrient cycling during mineralization [95]. Different crop rotation plans change the microbial community and also regulate nitrogen cycling functional genes [96]. Therefore, in this study, the abundance of key functional genes was positively correlated with STN and PNM (Table 3). However, the change in the Shannon index of the key functional genes may be related to the nitrogen environment and crop residue quality (Figure 4). When the soil STN or PON content is low, the Shannon index of key functional genes in nitrogen cycling increases with an increase in STN, as diversified crop residue inputs in crop rotation plans will provide more ecological niches for microorganisms, thereby increasing microbial diversity [97,98]. However, when the STN or PON content is high, the Shannon index of key functional genes in nitrogen cycling decreases. Similar findings have been reported, such as a significant increase in soil organic matter and total nitrogen content in long-term organic management but a decrease in soil microbial diversity [99,100]. This may be because higher soil nitrogen provides abundant resources for microorganisms, allowing microorganisms with lower resource utilization efficiency to occupy the main ecological niche [76]. This may lead to a reduction in the diversity of key functional genes. In addition, the excessive input of high C/N ratio crop residues into the soil in the CWWM causes a disturbance in the soil C/N, affecting the metabolic activities of microorganisms towards carbon sources [93]. Therefore, although introducing corn into the wheat planting can effectively increase the soil nitrogen fractions, long-term crop residue input will have a negative impact on the potential of nitrogen cycling function in soil on the Loess Plateau.

5. Conclusions

Based on a 36-year field experiment in the Loess Plateau, the results showed that compared with wheat monoculture, maize rotation contributed to the accumulation of soil STN and PON, while legume rotation effectively maintained the soil STN and PON contents. Crop rotation did not change the Shannon index of microbial communities in the Loess Plateau but increased the microbial biomass nitrogen and also increased the Chao1 of bacterial communities related to key functional genes involved in nitrogen cycling. As different crop rotation plans have different effects on STN and PON content, introducing peas or corn into crop rotation could have completely different regulatory effects on the Shannon index of nitrogen cycling key functional genes. Specifically, compared with wheat monoculture, the corn-based crop rotation reduced the Shannon index of key functional genes involved in the nitrogen cycle; however, the pea-based crop rotation significantly increased the Shannon index of these genes compared to the corn-based crop rotation.

Therefore, in order to enhance the ecological potential of soil nitrogen cycling and sustain agricultural development in the Loess Plateau region, it is recommended to introduce leguminous crops into the crop rotation plan.

Supplementary Materials: The following supporting information can be downloaded at: https://www.mdpi.com/article/10.3390/agronomy13071769/s1, Table S1. Effects of different planting regimes on the abundance of nitrogen-cycling functional genes ($n = 3$). Values are mean ± standard deviation. Table S2. Effects of different planting regimes on pH, BD and SWC ($n = 3$). Values are mean ± standard deviation. Table S3. Effects of distinct planting regimes on the relative abundance of species related to key functional genes ($n = 3$). Values are mean ± standard deviation. Figure S1. Key functional genes associated with N-cycling processes in different planting regimes. (A) Co-occurrence network of N-cycling functional genes. The circle size represents the degree centrality. (B) Random forest analyses show the significant predictors of functional genes for different nitrogen fractions. MES (%) means the percentage of increase of mean square error. STN, soil total nitrogen; PON, particulate organic nitrogen; PNM, potential nitrogen mineralization; MBN, microbial biomass N; NH4+, ammonium nitrogen; NO3−, nitrate nitrogen; PON/STN, ratio of PON to STN; PNM/STN, ratio of PNM to STN; MBN/STN, ratio of MBN to STN. Figure S2. Relative abundance of the other level classification of species related to key functional genes.

Author Contributions: Conceptualization, J.W. and R.L.; methodology, R.L. and F.Z.; software, R.L. and Y.L.; validation, R.L., Y.L. and J.W.; formal analysis, R.L. and Y.L.; investigation, Y.G.; resources, Y.G. and R.L.; data curation, Y.G. and R.L.; writing—original draft preparation, R.L. and F.Z.; writing—review and editing, R.L. and Y.L.; visualization, R.L.; supervision, J.W.; project administration, J.W.; funding acquisition, J.W. All authors have read and agreed to the published version of the manuscript.

Funding: This research was funded by the National Natural Science Foundation of China, grant number 42277322, and the Chinese Academy of Sciences "Light of West China", and Shaanxi Agricultural Science & Technology Innovation-Driven Project (NYKJ-2021-XA-005, NYKY-2022-XA-004).

Data Availability Statement: All data generated or analyzed during this study are included in this published article.

Acknowledgments: The authors thank ChengJie Ren for his help in the determination of soil Metagenomic sequencing of this research.

Conflicts of Interest: The authors declare no conflict of interest.

References

1. Hallin, S.; Jones, C.M.; Schloter, M.; Philippot, L. Relationship between N-cycling communities and ecosystem functioning in a 50-year-old fertilization experiment. *ISME J.* **2009**, *3*, 597–605. [CrossRef] [PubMed]
2. Gaby, J.C.; Buckley, D.H. A comprehensive aligned nifH gene database: A multipurpose tool for studies of nitrogen-fixing bacteria. *Database* **2014**, *2014*, bau001. [CrossRef] [PubMed]
3. Avrahami, S.; Conrad, R. Cold-temperate climate: A factor for selection of ammonia oxidizers in upland soil? *Can. J. Microbiol.* **2005**, *51*, 709–714. [CrossRef] [PubMed]
4. Zhang, X.; Liu, W.; Schloter, M.; Zhang, G.; Chen, Q.; Huang, J.; Li, L.; Philippot, L.; Chen, T.; He, Z. Response of the abundance of key soil microbial nitrogen-cycling genes to multi-factorial global changes. *PLoS ONE* **2013**, *8*, e76500. [CrossRef] [PubMed]
5. Zhang, L.; Lv, J. Land-use change from cropland to plantations affects the abundance of nitrogen cycle-related microorganisms and genes in the Loess Plateau of China. *Appl. Soil Ecol.* **2021**, *161*, 103873. [CrossRef]
6. Tatti, E.; Goyer, C.; Burton, D.L.; Wertz, S.; Zebarth, B.J.; Chantigny, M.; Filion, M. Tillage management and seasonal effects on denitrifier community abundance, gene expression and structure over winter. *Microb. Ecol.* **2015**, *70*, 795–808. [CrossRef]
7. Badagliacca, G.; Benítez, E.; Amato, G.; Badalucco, L.; Giambalvo, D.; Laudicina, V.A.; Ruisi, P. Long-term no-tillage application increases soil organic carbon, nitrous oxide emissions and faba bean (*Vicia faba* L.) yields under rain-fed Mediterranean conditions. *Sci. Total Environ.* **2018**, *639*, 350–359. [CrossRef]
8. Ouyang, Y.; Reeve, J.R.; Norton, J.M. Soil enzyme activities and abundance of microbial functional genes involved in nitrogen transformations in an organic farming system. *Biol. Fertil. Soils* **2018**, *54*, 437–450. [CrossRef]
9. Sun, R.; Guo, X.; Wang, D.; Chu, H. Effects of long-term application of chemical and organic fertilizers on the abundance of microbial communities involved in the nitrogen cycle. *Appl. Soil Ecol.* **2015**, *95*, 171–178. [CrossRef]
10. He, Z.; Deng, Y.; Van Nostrand, J.D.; Zhou, J.; Luo, Y.; Xia, B.; Wang, A.; Yuan, T. GeoChip 3.0 as a high-throughput tool for analyzing microbial community composition, structure and functional activity. *ISME J.* **2010**, *4*, 1167–1179. [CrossRef]

11. Philippot, L.; Spor, A.; Hénault, C.; Bru, D.; Bizouard, F.; Jones, C.M.; Sarr, A.; Maron, P.A. Loss in microbial diversity affects nitrogen cycling in soil. *ISME J.* **2013**, *7*, 1609–1619. [CrossRef] [PubMed]
12. Yang, Y.; Gao, Y.; Wang, S.; Xu, D.; Yu, H.; Wu, L.; Lin, Q.; Hu, Y.; Li, X. The microbial gene diversity along an elevation gradient of the Tibetan grassland. *ISME J.* **2014**, *8*, 430–440. [CrossRef] [PubMed]
13. Wang, F.; Liang, Y.; Jiang, Y.; Yang, Y.; Xue, K.; Xiong, J.; Zhou, J.; Sun, B. Planting increases the abundance and structure complexity of soil core functional genes relevant to carbon and nitrogen cycling. *Sci. Rep.* **2015**, *5*, 14345. [CrossRef] [PubMed]
14. Ren, T.; Yu, X.; Liao, J.; Liang, Y.; Zhang, W.; Wang, F.; Sun, B. Application of biogas slurry rather than biochar increases soil microbial functional gene signal intensity and diversity in a poplar plantation. *Soil Biol. Biochem.* **2020**, *146*, 107825. [CrossRef]
15. Chukwuneme, C.F.; Ayangbenro, A.S.; Babalola, O.O. Metagenomic analyses of plant growth-promoting and carbon-cycling genes in maize rhizosphere soils with distinct land-use and management histories. *Genes* **2021**, *12*, 1431. [CrossRef]
16. You, L.; Ros, G.H.; Chen, Y.; Liu, X.; Zhang, X.; Wang, J.; Liang, Y.; Sun, B. Global meta-analysis of terrestrial nitrous oxide emissions and associated functional genes under nitrogen addition. *Soil Biol. Biochem.* **2022**, *165*, 108523. [CrossRef]
17. Liu, W.; Wang, J.; Sun, L.; Zhang, X.; Liang, Y.; Sun, B. Sustainability evaluation of soybean-corn rotation systems in the Loess Plateau region of Shaanxi, China. *J. Clean. Prod.* **2019**, *210*, 1229–1237. [CrossRef]
18. Jiang, J.; Guo, S.; Zhang, Y.; Liu, W.; Liang, Y.; Sun, B. Changes in temperature sensitivity of soil respiration in the phases of a three-year crop rotation system. *Soil Tillage Res.* **2015**, *150*, 139–146. [CrossRef]
19. Fu, X.; Wang, J.; Sainju, U.M.; Zhang, X.; Liang, Y.; Sun, B. Soil nitrogen fractions under long-term crop rotations in the Loess Plateau of China. *Soil Tillage Res.* **2019**, *186*, 42–51. [CrossRef]
20. Munera-Echeverri, J.L.; Martinsen, V.; Strand, L.T.; Cornelissen, G.; Mulder, J. Effect of conservation farming and biochar addition on soil organic carbon quality, nitrogen mineralization, and crop productivity in a light textured Acrisol in the sub-humid tropics. *PLoS ONE* **2020**, *15*, e0228717. [CrossRef]
21. Bu, R.; Lu, J.; Ren, T.; Liang, Y.; Zhang, W.; Wang, F.; Sun, B. Particulate organic matter affects soil nitrogen mineralization under two crop rotation systems. *PLoS ONE* **2015**, *10*, e0143835. [CrossRef] [PubMed]
22. Li, N.; Li, X.; Li, S.; Wang, F.; Zhang, X.; Liang, Y.; Sun, B. Impacts of Rotation-Fallow Practices on Bacterial Community Structure in Paddy Fields. *Microbiol. Spectr.* **2022**, *10*, e00227-22. [CrossRef] [PubMed]
23. Tiemann, L.K.; Grandy, A.S.; Atkinson, E.E.; Marin-Spiotta, E.; McDaniel, M.D. Crop rotational diversity enhances belowground communities and functions in an agroecosystem. *Ecol. Lett.* **2015**, *18*, 761–771. [CrossRef]
24. Zhang, X.; Johnston, E.R.; Wang, Y.; Wang, J.; Yuan, T.; He, Z. Distinct drivers of core and accessory components of soil microbial community functional diversity under environmental changes. *Msystems* **2019**, *4*, e00374-19. [CrossRef] [PubMed]
25. Zhou, Q.; Zhang, P.; Wang, Z.; Liu, W.; Liang, Y.; Sun, B. Winter crop rotation intensification to increase rice yield, soil carbon, and microbial diversity. *Heliyon* **2023**, *9*, e12903. [CrossRef]
26. Hu, G.; Liu, X.; He, H.; Liu, W.; Liang, Y.; Sun, B. Multi-seasonal nitrogen recoveries from crop residue in soil and crop in a temperate agro-ecosystem. *PLoS ONE* **2015**, *10*, e0133437. [CrossRef]
27. Maron, P.A.; Sarr, A.; Kaisermann, A.; Lévêque, J.; Mathieu, O.; Guigue, J.; Karimi, B.; Bernard, L.; Dequiedt, S.; Terrat, S.; et al. High microbial diversity promotes soil ecosystem functioning. *Appl. Environ. Microbiol.* **2018**, *84*, e02738-17. [CrossRef]
28. Jetten, M.S.M. The microbial nitrogen cycle. *Environ. Microbiol.* **2008**, *10*, 2903–2909. [CrossRef]
29. Griffiths, B.S.; Philippot, L. Insights into the resistance and resilience of the soil microbial community. *FEMS Microbiol. Rev.* **2013**, *37*, 112–129. [CrossRef]
30. Maul, J.E.; Cavigelli, M.A.; Vinyard, B.; Buyer, J.S. Cropping system history and crop rotation phase drive the abundance of soil denitrification genes nirK.; nirS and nosZ in conventional and organic grain agroecosystems. *Agric. Ecosyst. Environ.* **2019**, *273*, 95–106. [CrossRef]
31. Linton, N.F.; Machado, P.V.F.; Deen, B.; Wagner-Riddle, C. Long-term diverse rotation alters nitrogen cycling bacterial groups and nitrous oxide emissions after nitrogen fertilization. *Soil Biol. Biochem.* **2020**, *149*, 107917. [CrossRef]
32. Zhong, Y.; Yan, W.; Shangguan, Z. Soil carbon and nitrogen fractions in the soil profile and their response to long-term nitrogen fertilization in a wheat field. *Catena* **2015**, *135*, 38–46. [CrossRef]
33. Lindsay, E.A.; Colloff, M.J.; Gibb, N.L.; Wakelin, S.A. The abundance of microbial functional genes in grassy woodlands is influenced more by soil nutrient enrichment than by recent weed invasion or livestock exclusion. *Appl. Environ. Microbiol.* **2010**, *76*, 5547–5555. [CrossRef]
34. Sarrantonio, M.; Gallandt, E. The role of cover crops in North American cropping systems. *J. Crop Prod.* **2003**, *8*, 53–74. [CrossRef]
35. Wang, J.; Fu, X.; Ghimire, R.; Zhang, X.; Liang, Y.; Sun, B. Responses of soil bacterial community and enzyme activity to organic matter components under long-term fertilization on the loess plateau of China. *Appl. Soil Ecol.* **2021**, *166*, 103992. [CrossRef]
36. Cambardella, C.A.; Elliott, E.T. Particulate soil organic-matter changes across a grassland cultivation sequence. *Soil Sci. Soc. Am. J.* **1992**, *56*, 777–783. [CrossRef]
37. Haney, R.L.; Franzluebbers, A.J.; Porter, E.B.; Hons, F.M.; Zuberer, D.A. Soil carbon and nitrogen mineralization: Influence of drying temperature. *Soil Sci. Soc. Am. J.* **2004**, *68*, 489–492. [CrossRef]
38. Franzluebbers, A.J.; Haney, R.L.; Hons, F.M.; Zuberer, D.A. Determination of microbial biomass and nitrogen mineralization following rewetting of dried soil. *Soil Sci. Soc. Am. J.* **1996**, *60*, 1133–1139. [CrossRef]
39. Zhang, X.; Johnston, E.R.; Barberán, A.; Wang, J.; Yuan, T.; He, Z. Decreased plant productivity resulting from plant group removal experiment constrains soil microbial functional diversity. *Glob. Chang. Biol.* **2017**, *23*, 4318–4332. [CrossRef]

40. Li, D.; Liu, C.M.; Luo, R.; Sadakane, K.; Lam, T.W. MEGAHIT: An ultra-fast single-node solution for large and complex metagenomics assembly via succinct de Bruijn graph. *Bioinformatics* **2015**, *31*, 1674–1676. [CrossRef]
41. Hyatt, D.; LoCascio, P.F.; Hauser, L.J.; Uberbacher, E.C. Gene and translation initiation site prediction in metagenomic sequences. *Bioinformatics* **2012**, *28*, 2223–2230. [CrossRef]
42. Qin, J.; Li, Y.; Cai, Z.; Li, S.; Zhu, J.; Zhang, F.; Liang, S.; Zhang, W.; Guan, Y.; Shen, D.; et al. A metagenome-wide association study of gut microbiota in type 2 diabetes. *Nature* **2012**, *490*, 55–60. [CrossRef] [PubMed]
43. Wu, X.; Peng, J.; Liu, P.; Wang, F.; Zhang, X.; Liang, Y.; Sun, B. Metagenomic insights into nitrogen and phosphorus cycling at the soil aggregate scale driven by organic material amendments. *Sci. Total Environ.* **2021**, *785*, 147329. [CrossRef] [PubMed]
44. Wang, J.; He, L.; Xu, X.; Zhang, X.; Liang, Y.; Sun, B. Linkage between microbial functional genes and net N mineralisation in forest soils along an elevational gradient. *Eur. J. Soil Sci.* **2022**, *73*, e13276. [CrossRef]
45. Csardi, G.; Nepusz, T. The igraph software package for complex network research. *Interj. Complex Syst.* **2006**, *1695*, 1–9.
46. Bastian, M.; Heymann, S.; Jacomy, M. Gephi: An open source software for exploring and manipulating networks. *Proc. Int. AAAI Conf. Web Soc. Media* **2009**, *3*, 361–362. [CrossRef]
47. Liaw, A.; Wiener, M. Classification and regression by randomForest. *R News* **2002**, *2*, 18–22.
48. McMurdie, P.J.; Holmes, S. phyloseq: An R package for reproducible interactive analysis and graphics of microbiome census data. *PLoS ONE* **2013**, *8*, e61217. [CrossRef] [PubMed]
49. Cheng, H.; Jin, B.; Luo, K.; Pei, J.; Zhang, X.; Zhang, Y.; Tang, J.; Yang, Q.; Sun, G. Vegetation response to goats grazing intensity in semiarid hilly Grassland of the loess Plateau, Lanzhou, China. *Sustainability* **2021**, *13*, 3569. [CrossRef]
50. Kolde, R. Pheatmap: Pretty Heatmaps, Version 1.0.12; CRAN. 2015. Available online: https://cran.r-project.org/web/packages/pheatmap/index.html (accessed on 4 June 2022).
51. Gómezrubio, V. ggplot2—Elegant Graphics for Data Analysis. *J. Stat. Softw.* **2017**, *77*, 1–3.
52. Xu, S.; Wang, M.; Shi, X.; Liang, Y.; Wang, Z.; Zhang, X.; Liang, C.; Zhang, Z. Integrating hyperspectral imaging with machine learning techniques for the high-resolution mapping of soil nitrogen fractions in soil profiles. *Sci. Total Environ.* **2021**, *754*, 142135. [CrossRef]
53. Fu, X.; Wang, J.; Sainju, U.M.; Zhang, X.; Liang, Y.; Sun, B. Soil microbial community and carbon and nitrogen fractions responses to mulching under winter wheat. *Appl. Soil Ecol.* **2019**, *139*, 64–68. [CrossRef]
54. Wang, J.; Sainju, U.M. Soil carbon and nitrogen fractions and crop yields affected by residue placement and crop types. *PLoS ONE* **2014**, *9*, e105039. [CrossRef] [PubMed]
55. Bai, Z.; Caspari, T.; Gonzalez, M.R.; Müller, A.; Reidsma, P.; König, H.J.; Qin, P.; Gemmer, M.; Oosterveer, P.; Ewert, F. Effects of agricultural management practices on soil quality: A review of long-term experiments for Europe and China. *Agric. Ecosyst. Environ.* **2018**, *265*, 1–7. [CrossRef]
56. Sainju, U.M.; Lenssen, A.W.; Caesar-TonThat, T.; Waddell, J.T.; Hatfield, P.G.; Scholljegerdes, E.J.; Lartey, R.T. Dryland soil nitrogen cycling influenced by tillage, crop rotation, and cultural practice. *Nutr. Cycl. Agroecosyst.* **2012**, *93*, 309–322. [CrossRef]
57. Mazzilli, S.R.; Kemanian, A.R.; Ernst, O.R.; Jackson, R.B.; Piñeiro, G. Greater humification of belowground than aboveground biomass carbon into particulate soil organic matter in no-till corn and soybean crops. *Soil Biol. Biochem.* **2015**, *85*, 22–30. [CrossRef]
58. Mazzilli, S.R.; Kemanian, A.R.; Ernst, O.R.; Jackson, R.B.; Piñeiro, G. Priming of soil organic carbon decomposition induced by corn compared to soybean crops. *Soil Biol. Biochem.* **2014**, *75*, 273–281. [CrossRef]
59. Paśmionka, I.B.; Bulski, K.; Boligłowa, E. The Participation of Microbiota in the Transformation of Nitrogen Compounds in the Soil—A Review. *Agronomy* **2021**, *11*, 977. [CrossRef]
60. Akley, E.K.; Rice, C.W.; Adotey, N.; Schapaugh, W.T.; Ciampitti, I.A. Residual Bradyrhizobium inoculation effects on soybean performance and selected soil health parameters. *Agron. J.* **2022**, *114*, 1627–1641. [CrossRef]
61. Benitez, M.S.; Osborne, S.L.; Lehman, R.M. Previous crop and rotation history effects on maize seedling health and associated rhizosphere microbiome. *Sci. Rep.* **2017**, *7*, 15709. [CrossRef]
62. McDaniel, M.D.; Grandy, A.S. Soil microbial biomass and function are altered by 12 years of crop rotation. *Soil* **2016**, *2*, 583–599. [CrossRef]
63. Su, Y.; Hu, Y.; Zi, H.; Liang, Y.; Wang, Z.; Zhang, X.; Liang, C.; Zhang, Z. Contrasting assembly mechanisms and drivers of soil rare and abundant bacterial communities in 22-year continuous and non-continuous cropping systems. *Sci. Rep.* **2022**, *12*, 3264. [CrossRef] [PubMed]
64. Munroe, J.W.; McCormick, I.; Deen, W.; Scott-Dupree, C.; Dunfield, K.E. Effects of 30 years of crop rotation and tillage on bacterial and archaeal ammonia oxidizers. *J. Environ. Qual.* **2016**, *45*, 940–948. [CrossRef] [PubMed]
65. Brankatschk, G.; Finkbeiner, M. Crop rotations and crop residues are relevant parameters for agricultural carbon footprints. *Agron. Sustain. Dev.* **2017**, *37*, 1–14. [CrossRef]
66. Brandan, C.P.; Meyer, A.; Meriles, J.M.; Conforto, C.; Figoni, G.; Erijman, L.; Wall, L.G. Relationships between soil physicochemical properties and nitrogen fixing, nitrifying and denitrifying under varying land-use practices in the northwest region of Argentina. *Soil Water Res.* **2019**, *14*, 1–9. [CrossRef]
67. Behnke, G.D.; Kim, N.; Zabaloy, M.C.; Schipanski, M.E.; Fonte, S.J. Soil microbial indicators within rotations and tillage systems. *Microorganisms* **2021**, *9*, 1244. [CrossRef]
68. Strickland, M.S.; Rousk, J. Considering fungal: Bacterial dominance in soils-methods, controls, and ecosystem implications. *Soil Biol. Biochem.* **2010**, *42*, 1385–1395. [CrossRef]

69. Paula, F.S.; Rodrigues, J.L.M.; Zhou, J.; Wu, L.; Mueller, R.C.; Mirza, B.S.; Bohannan, B.J.M.; Nüsslein, K.; Deng, Y.; Tiedje, J.M.; et al. Land use change alters functional gene diversity, composition and abundance in Amazon forest soil microbial communities. *Mol. Ecol.* **2014**, *23*, 2988–2999. [CrossRef]
70. Tilman, D.; Isbell, F.; Cowles, J.M. Biodiversity and ecosystem functioning. *Annu. Rev. Ecol. Evol. Syst.* **2014**, *45*, 471–493. [CrossRef]
71. Greenlon, A.; Chang, P.L.; Damtew, Z.M.; Wu, C.H.; Fong, S.S.; Rodrigue, S.; Tiedje, J.M.; Cole, J.R.; Konstantinidis, K.T.; Zhou, J. Global-level population genomics reveals differential effects of geography and phylogeny on horizontal gene transfer in soil bacteria. *Proc. Natl. Acad. Sci. USA* **2019**, *116*, 15200–15209. [CrossRef]
72. Nguyen, T.T.; Cavagnaro, T.R.; Ngo, H.T.T.; Marschner, P. Soil respiration, microbial biomass and nutrient availability in soil amended with high and low C/N residue–Influence of interval between residue additions. *Soil Biol. Biochem.* **2016**, *95*, 189–197. [CrossRef]
73. Hodge, A.; Fitter, A.H. Substantial nitrogen acquisition by arbuscular mycorrhizal fungi from organic material has implications for N cycling. *Proc. Natl. Acad. Sci. USA* **2010**, *107*, 13754–13759. [CrossRef] [PubMed]
74. Wang, X.X.; Wang, X.; Sun, Y.; Dong, Y.Y.; Wang, F.Y.; Gao, Y.N.; Feng, G.L. Arbuscular mycorrhizal fungi negatively affect nitrogen acquisition and grain yield of maize in a N deficient soil. *Front. Microbiol.* **2018**, *9*, 418. [CrossRef] [PubMed]
75. Fierer, N.; Bradford, M.A.; Jackson, R.B. Toward an ecological classification of soil bacteria. *Ecology* **2007**, *88*, 1354–1364. [CrossRef] [PubMed]
76. Li, R.; Khafipour, E.; Krause, D.O.; Entz, M.H.; de Kievit, T.R.; Fernando, W.G.D. Pyrosequencing reveals the influence of organic and conventional farming systems on bacterial communities. *PLoS ONE* **2012**, *7*, e51897. [CrossRef]
77. Wang, J.; Xue, C.; Song, Y.; Wang, L.; Huang, Q.; Shen, Q. Wheat and rice growth stages and fertilization regimes alter soil bacterial community structure, but not diversity. *Front. Microbiol.* **2016**, *7*, 1207. [CrossRef]
78. Campbell, B.J.; Polson, S.W.; Hanson, T.E.; Mack, M.C.; Schuur, E.A.G. The effect of nutrient deposition on bacterial communities in Arctic tundra soil. *Environ. Microbiol.* **2010**, *12*, 1842–1854. [CrossRef]
79. Zeng, J.; Liu, X.; Song, L.; Lin, X.; Zhang, H.; Shen, C.; Chu, H. Nitrogen fertilization directly affects soil bacterial diversity and indirectly affects bacterial community composition. *Soil Biol. Biochem.* **2016**, *92*, 41–49. [CrossRef]
80. Jeanne, T.; Parent, S.É.; Hogue, R. Using a soil bacterial species balance index to estimate potato crop productivity. *PLoS ONE* **2019**, *14*, e0214089. [CrossRef]
81. Potter, T.S.; Vereecke, L.; Lankau, R.A.; Smith, A.B.; Williams, M.A. Long-term management drives divergence in soil microbial biomass, richness, and composition among upper Midwest, USA cropping systems. *Agric. Ecosyst. Environ.* **2022**, *325*, 107718. [CrossRef]
82. Li, M.; Guo, J.; Ren, T.; Wang, J.; Zhang, J.; Zhang, F.; Chen, X. Crop rotation history constrains soil biodiversity and multifunctionality relationships. *Agric. Ecosyst. Environ.* **2021**, *319*, 107550. [CrossRef]
83. D'Acunto, L.; Andrade, J.F.; Poggio, S.L.; Giménez, D.O.; Arolfo, R.V.; Montico, S.; Zabaloy, M.C. Diversifying crop rotation increased metabolic soil diversity and activity of the microbial community. *Agric. Ecosyst. Environ.* **2018**, *257*, 159–164. [CrossRef]
84. Liu, Q.; Zhao, Y.; Li, T.; Wang, Y.; Liang, C.; Zhou, J. Changes in soil microbial biomass, diversity, and activity with crop rotation in cropping systems: A global synthesis. *Appl. Soil Ecol.* **2023**, *186*, 104815. [CrossRef]
85. Venter, Z.S.; Jacobs, K.; Hawkins, H.J. The impact of crop rotation on soil microbial diversity: A meta-analysis. *Pedobiologia* **2016**, *59*, 215–223. [CrossRef]
86. Liu, Z.; Liu, J.; Yu, Z.; Yao, H.; Liu, X.; Liang, C.; Zhao, B. Long-term continuous cropping of soybean is comparable to crop rotation in mediating microbial abundance, diversity and community composition. *Soil Tillage Res.* **2020**, *197*, 104503. [CrossRef]
87. Culman, S.W.; Young-Mathews, A.; Hollander, A.D.; Ferris, H.; Sánchez-Moreno, S.; O'Geen, A.T.; Jackson, L.E. Biodiversity is associated with indicators of soil ecosystem functions over a landscape gradient of agricultural intensification. *Landsc. Ecol.* **2010**, *25*, 1333–1348. [CrossRef]
88. Kozjek, K.; Kundel, D.; Kushwaha, S.K.; Kastelec, D.; Vodnik, D. Long-term agricultural management impacts arbuscular mycorrhizal fungi more than hort-term experimental drought. *Appl. Soil Ecol.* **2021**, *168*, 104140. [CrossRef]
89. Bell, T.; Newman, J.A.; Silverman, B.W.; Turner, S.L.; Lilley, A.K. The contribution of species richness and composition to bacterial services. *Nature* **2005**, *436*, 1157–1160. [CrossRef]
90. Saleem, M.; Fetzer, I.; Harms, H.; Chatzinotas, A. Trophic complexity in aqueous systems: Bacterial species richness and protistan predation regulate dissolved organic carbon and dissolved total nitrogen removal. *Proc. R. Soc. B Biol. Sci.* **2016**, *283*, 20152724. [CrossRef]
91. Sarathchandra, S.U.; Ghani, A.; Yeates, G.W.; Burch, G.; Cox, N.R. Effect of nitrogen and phosphate fertilisers on microbial and nematode diversity in pasture soils. *Soil Biol. Biochem.* **2001**, *33*, 953–964. [CrossRef]
92. Callaway, J.C.; Sullivan, G.; Zedler, J.B. Species-rich plantings increase biomass and nitrogen accumulation in a wetland restoration experiment. *Ecol. Appl.* **2003**, *13*, 1626–1639. [CrossRef]
93. Cai, L.; Guo, Z.; Zhang, J.; Wang, Y.; Zhang, Y.; Zhang, X.; Liang, W. No tillage and residue mulching method on bacterial community diversity regulation in a black soil region of Northeastern China. *PLoS ONE* **2021**, *16*, e0256970. [CrossRef] [PubMed]
94. Louca, S.; Polz, M.F.; Mazel, F.; Albright, M.B.N.; Huber, J.A.; O'Connor, M.I.; Ackermann, M.; Hahn, A.S.; Srivastava, D.S.; Crowe, S.A.; et al. Function and functional redundancy in microbial systems. *Nat. Ecol. Evol.* **2018**, *2*, 936–943. [CrossRef]

95. Veres, Z.; Kotroczó, Z.; Fekete, I.; Tóth, J.A.; Lajtha, K.; Townsend, K.; Tóthmérész, B. Soil extracellular enzyme activities are sensitive indicators of detrital inputs and carbon availability. *Appl. Soil Ecol.* **2015**, *92*, 18–23. [CrossRef]
96. Cai, F.; Luo, P.; Yang, J.; Wang, Y.; Liang, W.; Zhang, X.; Zhang, Y. Effect of long-term fertilization on ammonia-oxidizing microorganisms and nitrification in brown soil of northeast China. *Front. Microbiol.* **2021**, *11*, 622454. [CrossRef]
97. Xue, K.; Wu, L.; Deng, Y.; He, Z.; Van Nostrand, J.; Robertson, G.P.; Schmidt, T.M.; Zhou, J. Functional gene differences in soil microbial communities from conventional, low-input, and organic farmlands. *Appl. Environ. Microbiol.* **2013**, *79*, 1284–1292. [CrossRef]
98. Shu, X.; Zou, Y.; Shaw, L.J.; Jiang, Y.; Xu, M.; Kuzyakov, Y. Applying cover crop residues as diverse mixtures increases initial microbial assimilation of crop residue-derived carbon. *Eur. J. Soil Sci.* **2022**, *73*, e13232. [CrossRef]
99. Tian, W.; Wang, L.; Li, Y.; Wang, Y.; Zhang, X.; Liang, W. Responses of microbial activity, abundance, and community in wheat soil after three years of heavy fertilization with manure-based compost and inorganic nitrogen. *Agric. Ecosyst. Environ.* **2015**, *213*, 219–227. [CrossRef]
100. Hartmann, M.; Frey, B.; Mayer, J.; Mäder, P.; Widmer, F. Distinct soil microbial diversity under long-term organic and conventional farming. *ISME J.* **2015**, *9*, 1177–1194. [CrossRef]

agronomy

Article

High-Throughput Sequencing Reveals the Effect of the South Root-Knot Nematode on Cucumber Rhizosphere Soil Microbial Community

Fan Yang [1,†], Huayan Jiang [1,2,†], Shen Liang [1,†], Gaozheng Chang [1], Kai Ma [1], Lili Niu [1], Guoquan Mi [1], Yanling Tang [1], Baoming Tian [1,2,*] and Xuanjie Shi [1,*]

[1] Institute of Horticulture, Henan Academy of Agricultural Sciences, Graduate T&R Base of Zhengzhou University, Zhengzhou 450002, China
[2] Henan International Joint Laboratory of Crop Gene Resources and Improvement, School of Agricultural Sciences, Zhengzhou University, Zhengzhou 450001, China
* Correspondence: tianbm@zzu.edu.cn (B.T.); shixuanjie@hnagri.org.cn (X.S.)
† These authors contributed equally to this work.

Abstract: Due to long-term cultivation in greenhouses, cucumbers are susceptible to root-knot nematode (RKN), resulting in reduced yield and quality. The objective of this study was to investigate the effect of RKN on the rhizosphere microbial community of cucumber. Understanding the composition of rhizosphere bacterial and fungal communities and the possible interaction between microorganisms and RKN is expected to provide a reference for the eco-friendly control of *M. incognita* in the future. Three different groups were selected for sampling based on the RKN incidence and root galling scale (NHR, 0%, no root galling; NR, 5–15%, root galling scale 1–2; NS, 60–75%, root galling scale 4–5). Soil properties were determined to evaluate the effect of *M. incognita* on rhizosphere soil. High-throughput sequencing was used to examine the bacterial and fungal communities in rhizosphere soil. The results showed that the contents of soil nutrients and enzyme activities were significantly lower in the NS than in the NHR. The alpha diversity showed that *M. incognita* had a greater effect on rhizosphere soil bacteria than on fungi. In beta diversity, there were significant differences among the three groups by PCoA (p = 0.001). Furthermore, bacteria and fungi with significant differences in relative abundance were screened at the genus level for a correlation analysis with soil factors, and a correlation analysis between the bacteria and fungi was performed to study their relationships. A redundancy analysis (RDA) of rhizosphere microorganisms and soil properties showed a negative correlation between nematode contamination levels and soil nutrient content. Finally, we predicted the interaction among RKN, soil factors, and the rhizosphere microbial community, which provided evidence for the prevention of RKN via microecological regulation in the future.

Keywords: *Cucumis sativus*; root-knot nematode; soil properties; high-throughput sequencing; rhizosphere microbial community

Citation: Yang, F.; Jiang, H.; Liang, S.; Chang, G.; Ma, K.; Niu, L.; Mi, G.; Tang, Y.; Tian, B.; Shi, X. High-Throughput Sequencing Reveals the Effect of the South Root-Knot Nematode on Cucumber Rhizosphere Soil Microbial Community. *Agronomy* **2023**, *13*, 1726. https://doi.org/10.3390/agronomy13071726

Academic Editors: Yong-Xin Liu and Peng Yu

Received: 25 May 2023
Revised: 23 June 2023
Accepted: 26 June 2023
Published: 27 June 2023

1. Introduction

The cucumber (*Cucumis sativus* L.) is widely cultivated and provides a high number of vitamins and minerals and a high amount of fiber and roughage [1]. China plays an essential role in the cultivation and consumption of cucumbers. In terms of planting methods, greenhouse planting accounts for a considerable proportion, and it is still increasing year by year [2,3]. However, greenhouses with a single crop and a high multiple cropping index provide a suitable place for a variety of cucumber diseases and pests, among which, root-knot nematode (RKN) disease is the major disease affecting the yield and quality of cucumber [4]. RKN (*Meloidogyne arenaria*, *M. enterolobii*, *M. incognita*, and *M. javanica*) has been reported to cause large economic losses [5]. Among them, the southern root-knot nematode (SRKN, *M. incognita*) is particularly harmful, and it is known to be parasitic in cucumbers, tomatoes, watermelons, and many other plants [6]. Nematodes invade plant root cells, absorb nutrients, and suppress plant immune systems [7]. They will generate a large number of galls, which affects the absorption of nutrients by the host, causing a significant decrease in quality and yield [8].

Until now, conventional chemical nematicides have been used to control nematode diseases, whereas the massive use of nematicides has caused problems in food safety and environmental pollution and has ultimately affected human health and the sustainable development of agriculture. Therefore, it is very important and urgent to develop an environmentally friendly nematode control strategy to ensure food safety and crop production. By isolating plant endophytes and soil microorganisms, the development of biocontrol methods utilizing beneficial antagonists as a promising option has received increasing attention. In recent years, researchers have isolated many microorganisms from plant tissue or soil that are able to suppress different plant pathogens. For example, *Bacillus subtilis* YB-15 was isolated from wheat roots and soil and has been shown to be effective in reducing the growth of *Fusarium graminearum* [9]; *Flavobacterium* TRM1 isolated from rhizosphere soil can effectively inhibit *Ralstonia solanacearum* [10]; and *Pseudomonas aeruginosa* NXHG29 isolated from soil can also reduce the occurrence of tobacco bacterial wilt caused by *R. solanacearum* [11]. An increasing number of rhizosphere bacterial, fungal, and endophytic bacteria have been demonstrated to successfully reduce plant disease incidence [12]. In addition, some studies have confirmed that the interaction among various microorganisms may also be an important condition for plants to enhance their resistance to diseases [13]. Recent studies have shown that plant microbiota can confer a broad range of immune functions to plant hosts and that this process is strongly dependent on the interaction between soil nutrient status and the plant immune system [ss,sss]. The root microbial-mediated innate immune stimulation of plants can develop resistance to various pathogens. In *Arabidopsis*, the transcription factor MYB72 plays a key role in the regulation of induced systemic resistance triggered by *Trichoderma* spp. fungi and the bacterium *Pseudomonas simiae* [14,15]. Interestingly, MYB72 is also involved in the *Arabidopsis* response to iron deficiency [16], suggesting a direct interaction between nutritional stress and immunity. The direct suppression of pathogens by members of the microbiota in the plant root system has been reported several times [17,18]. Modalities include the secretion of antimicrobial compounds [19], hyperparasitism [20], and competition for resources such as nutrients or space to suppress pathogenic microorganisms [21].

Consequently, to develop biological agents for the biological control of *M. incognita*, it is essential to comprehend the microbial composition of the cucumber rhizosphere infected with *M. incognita*. However, research in this field is relatively limited. Based on the evidence obtained thus far, we speculated that the rhizosphere microbial community structure of *M. incognita*-infected cucumbers would be significantly changed. Moreover, RKN infection may have various influences on cucumber rhizosphere bacteria and fungi, to different degrees. To validate our hypotheses, different plots of rhizosphere soil were collected to explore the effect of *M. incognita* on the rhizosphere microbial community structure. Our findings will contribute to a better understanding of how *M. incognita* affects microbial communities in cucumber rhizosphere soils and provide a reference for the biological control of *M. incognita*.

2. Materials and Methods

2.1. Study Area and Experimental Description

Cucumber cultivar BoJie616 was planted in greenhouse facilities of Beiwang village in Luoyang, Henan province, China (34°66′ N, 112°52′ E), and the rhizosphere soil was obtained from different greenhouses. To study the effects of *M. incognita* on the cucumber rhizosphere microbial community, we selected greenhouses with different degrees of RKN incidence for sampling. The specific groups are as follows: NHR (no disease, no root galling), NR (incidence rate 5%–15%, root galling scale 1–2), and NS (incidence rate

60%–75%, root galling scale 4–5). The root galling scale was determined at the time of collection of the rhizosphere soil. Briefly, 10–15 cucumber plants were taken randomly from each plot to determine the degree of root galling, which was graded according to the method of Taylor and Sasser (1978) [22]. Consistent field management was applied during the experiment.

2.2. Soil Sampling

All cucumber seedlings in this study were factory planted to the three-leaf stage and then uniformly transplanted to different greenhouses for standardized management. Sampling was carried out in late May of 2022, which was approximately the 80th day after transplanting the cucumber seedlings to different greenhouses. Each greenhouse was approximately 8 m in width and 65 m in length, and the distance between adjacent greenhouses was 2 m. Based on the RKN incidence and root galling scale, the three adjacent greenhouses were divided into NHR, NR, and NS, and the rhizosphere soil was collected. Rhizosphere soil was sampled as described in a previous study [23]. Briefly, the roots were first carefully dug out from the soil, and large pieces of soil around the root system were removed. After gently shaking off the soil particles attached to the root surface and removing impurities such as fallen leaves and roots with a 1-mm sieve, the soil was finally collected into 15 mL centrifuge tubes. Soil tightly adhered to the root surface was referred to as rhizospheric soil. For each greenhouse, the soil samples were taken in an S-type sampling trajectory, and 5 soil samples were collected and pooled to make a composite sample. All soil samples were rapidly frozen in liquid nitrogen for 1 h and then stored in a −80 °C refrigerator for DNA extraction and detection of soil properties.

2.3. DNA Extraction, PCR Amplification, and Sequencing

Rhizosphere soil pellets were removed from the −80 °C freezer, and 0.2 g was transferred into 96-well DNA extraction plates. Microbial genomic DNA was extracted from 18 samples using the E.Z.N.A.® soil DNA Kit (Omega Bio-Tek, Norcross, GA, USA) according to the manufacturer's instructions. Finally, a NanoDrop 2000 UV-vis spectrophotometer (Thermo Fisher Scientific, Wilmington, DE, USA) was used to detect the concentration and purity of DNA using standard methods. Bacterial 16S rRNA gene amplicons were amplified with universal primers 338F (5′-ACTCCTACGGGAGGCAGCAG-3′) and 806R (5′-GGACTACHVGGGTWTCTAAT-3′), which anneal to the hypervariable region V3-V4 of the bacterial 16S rRNA gene. For identification of fungi, the internal transcribed spacer (ITS) region of nuclear rDNA was amplified using the primer pair ITS1F (5′-CTTGGTCATTTAGAGGAAGTAA-3′) and ITS2R (5′-GCTGCGTTCTTCATCGATGC-3′) using an ABI GeneAmp® 9700 PCR thermocycler (Applied Biosystems, Foster City, CA, USA). PCR amplification was performed with default parameters. PCR was performed in triplicate. The PCR product was extracted from a 2% agarose gel, purified using an AxyPrep DNA Gel Extraction Kit (Axygen Biosciences, Union City, CA, USA) according to the manufacturer's instructions, and quantified using a Quantus™ Fluorometer (Promega, Madison, WI, USA). Then, the 16S rRNA gene and the ITS region of fungal ribosomal DNA were sequenced on the NovaSeq PE250 platform (Illumina, San Diego, CA, USA) with six repetitions in each soil plot according to standard protocols.

2.4. Bioinformatics Analysis

Paired-end reads were assigned to each sample based on their unique barcode and truncated by cutting off the barcode and primer sequence. Paired-end reads were merged using FLASH [24] for 16S rRNA gene sequencing and PEAR (v0.9.6) [25] for ITS sequencing. To produce high-quality clean tags, raw reads were filtered under specific conditions using fqtrim (v0.94). Vsearch software (v2.3.4) was used to filter chimeric sequences [26]. DADA2 [27] was used to dereplicate features and obtain the feature sequence and feature table. Alpha and beta diversity were calculated using QIIME2 [28].

2.5. Soil Chemical Properties and Enzymatic Activity Assay

Twelve soil properties were measured according to previously reported methods. To determine soil available nitrogen (AN), alkaline hydrolysis diffusion was used [29]. Soil available phosphorus (AP) was quantified using molybdenum-antimony anti-spectrophotometry [30]. Available copper (ACu), available iron (AFe) and exchangeable calcium (ECa) were measured using atomic absorption spectrophotometry [31]. Electrical conductivity (EC) was determined using a conductivity monitor [32]. Soil organic matter (OM) was determined through oxidization [33]. Microbial biomass carbon (MBC) was determined using the chloroform fumigation extraction method [34]. Soil dehydrogenase (S-DHA) was assessed using a colorimetric procedure [35]. Soil sucrase activity (S-SR) was assayed using the 3,5-dinitrosalicylic acid colorimetric method [36]. Soil urease activity (S-UR) was assayed using the indophenol blue colorimetric method [37]. Soil acid phosphatase (S-ACP) was assayed using the standard method [38]. Soil properties were measured with three repetitions in each soil plot. The above experimental technology service was provided by Norminkoda Biotechnology Co., Ltd. (Wuhan, China).

2.6. Statistical Analyses

The IBM SPSS software program (IBM Corporation, V.20.0, New York, NY, USA) and R software (version 3.5.2) were used for statistical analysis. One-way analysis of variance (ANOVA) with Tukey's post hoc test was performed for comparisons among multiple groups. Relative abundance differences at the phylum and genus levels between two groups were analyzed using the Wilcoxon rank sum test. All statistical tests performed in this study were considered significant and extremely significant at $p < 0.05$ and $p < 0.01$.

3. Results

3.1. Changes in Rhizosphere Soil Properties of Cucumber Infected with *M. incognita*

To evaluate the effects of *M. incognita* on the chemical properties and enzymatic activities of rhizosphere soil, we measured 12 soil properties. Our results revealed significant differences among the properties. AN, AP, and S-ACP exhibited consistent trends, with significantly higher levels observed in NR than in NHR, while NS showed significantly lower levels than NHR. The remaining nine properties showed a significant decrease in both NR and NS compared with NHR (Figure 1). These results indicated a significant effect of RKNs on cucumber rhizosphere soil properties.

3.2. Diversities of Rhizosphere Bacterial and Fungal Communities

After data filtering and removing low-quality reads for the bacteria and fungi, the filtered reads ranged from 73,068 to 80,495 and 68,048 to 104,095, respectively, and the average reads were 76,632 and 86,007, respectively. The Q30 of 16S rRNA and ITS sequencing was above 91.2% and 90.66%, respectively (Table S1). The above data indicated that the quality of the sequencing data was reliable. High-quality paired-end reads were then connected to tags and clustered into 11,381 and 1922 amplicon sequence variants (ASVs) in bacteria and fungi, respectively. Among the identified ASVs of the bacteria and fungi, 5027 and 294 were common among the three groups, respectively (Figure 2A,D).

To comprehensively evaluate the alpha diversity of rhizosphere bacteria and fungi, we used Chao1 and Observed_features to represent the richness, the Shannon, and the Simpson indices to indicate the diversity, the Pielou evenness index, and the Goods coverage (Table S2). The Observed_features and Simpson of the bacteria were significantly different among the three groups with p values = 0.009 and 0.004, respectively. The Observed_features and Simpson of the fungi showed no significant difference among the three groups, with p values of 0.13 and 0.135, respectively. The Goods coverage showed no significant difference in either the bacteria or the fungi, with p values of 0.067 and 0.126, respectively (Figure 2B,E). The beta diversity revealed the distinct clustering of the three soil groups by PCoA (Figure 2C,F). In brief, the ASV abundance of the NS was separated from that of the NHR and NS. However, NHR and NS had a small overlap, indicating that the microbial composition of NHR and NR was closer, which was consistent in the fungi and the bacteria. The ANOSIM analysis showed the different compositions of the three soil groups (R = 0.673, p = 0.001 for bacteria; R = 0.451, p = 0.001 for fungi).

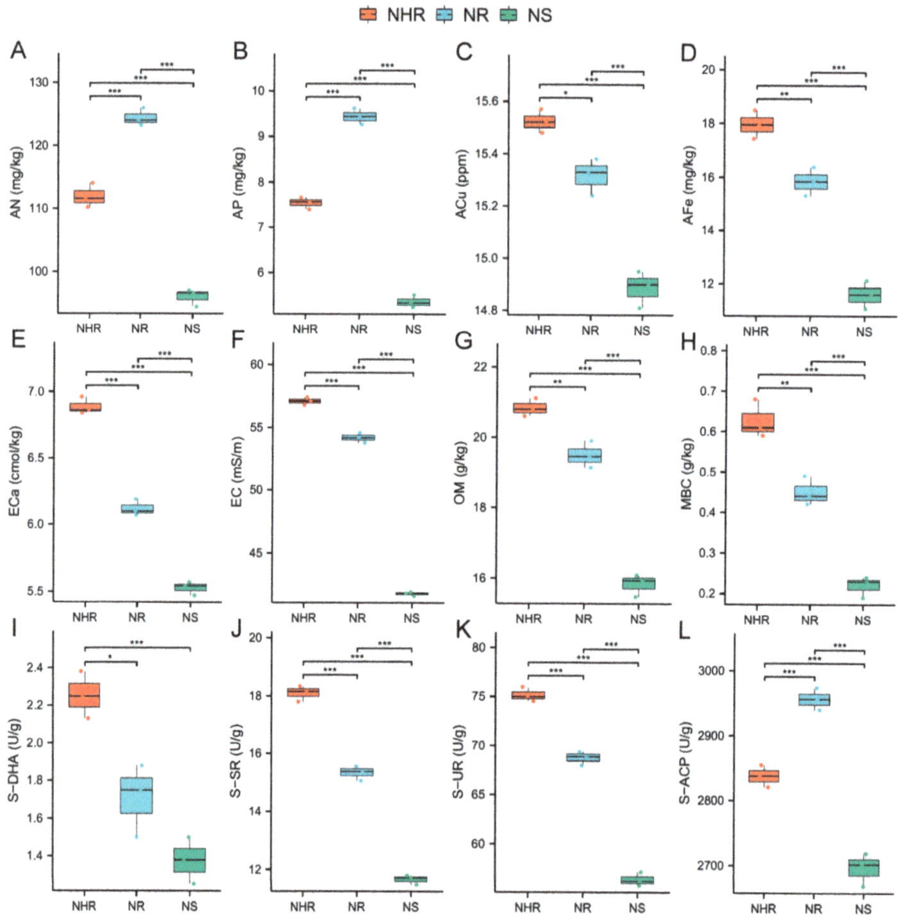

Figure 1. Effect of *M. incognita* on rhizosphere soil chemical properties and enzymatic activities. (**A**) AN, available nitrogen; (**B**) AP, available phosphorus; (**C**) ACu, available copper; (**D**) AFe, available iron; (**E**) ECa, exchangeable calcium; (**F**) EC, electrical conductivity; (**G**) OM, organic matter; (**H**) MBC, microbial biomass carbon; (**I**) S-DHA, soil dehydrogenase; (**J**) S-SR, soil sucrase activity; (**K**) S-UR, soil urease activity; (**L**) S-ACP, soil acid phosphatase. The symbol "*" indicates significant differences ($p < 0.05$, n = 3) among three groups, based on one-way ANOVA followed by the LSD test. * $p < 0.05$, ** $p < 0.01$, *** $p < 0.001$.

3.3. Characterization of Bacterial and Fungal Communities at the Phylum and Genus Levels

To investigate the influence of microbial communities by *M. incognita*, the relative abundance of soil bacteria and fungi at the phylum and genus levels was compared (Figure 3). In bacterial communities, the dominant phyla were *Proteobacteria Actinobacteriota*, *Acidobacteriota*, *Chloroflexi*, *Gemmatimonadota*, and *Myxococcota* (Figure 3A). The dominant genera were *Vicinamibacteraceae*, *A4b*, *MND1*, *Rokubacteriales*, and *RB41* (Figure 3B). In fungal communities, the dominant phyla were *Ascomycota*, *Mortierellomycota*, *Basidiomycota*, and *Rozellomycota* (Figure 3C). The dominant genera were *Chaetomium*, *Mortierella*, *Acremonium*, *Scedosporium*, *Alternaria*, and *Nigrospora* (Figure 3D).

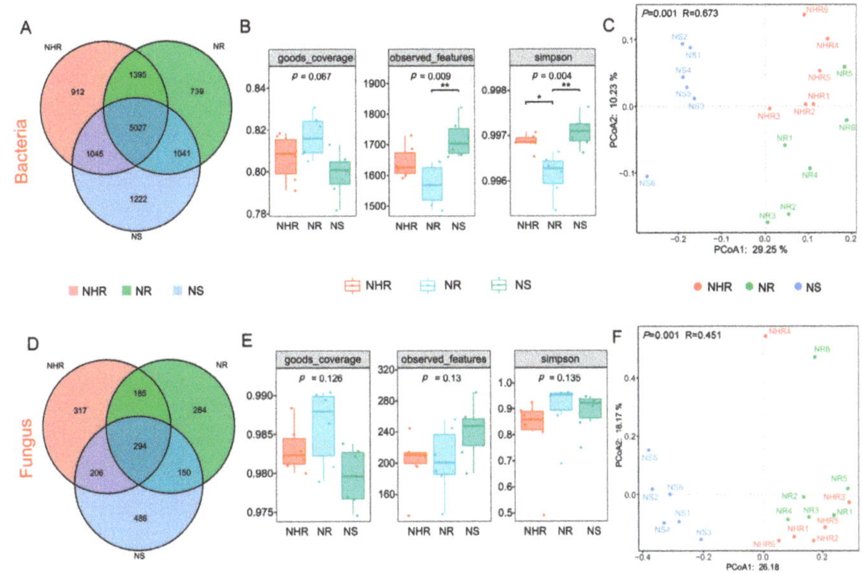

Figure 2. The alpha and beta diversity of rhizospheric bacterial and fungal communities. Venn diagram of bacterial (**A**) and fungal (**D**) ASVs identified in the three groups. Statistical analysis of differences in alpha diversity indicators (Goods coverage, Observed features, and Simpson index) of bacteria (**B**) and fungi (**E**) based on the Kruskal–Wallis rank sum test with Dunn's test (* $p < 0.05$; ** $p < 0.01$). Principal coordinates analysis (PCoA) shows the beta diversity of bacteria (**C**) and fungi (**F**) based on Bray–Curtis distances. The p value represents the analysis of ANOSIM.

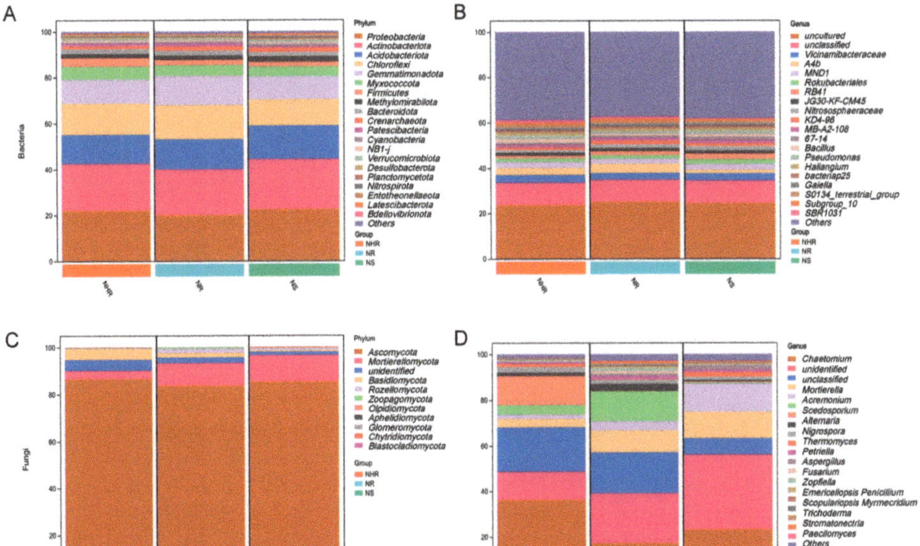

Figure 3. The composition of microorganism communities at the phylum and genus levels (relative abundance top 20). Phylum (**A**) and genus (**B**) levels of bacteria. Phylum (**C**) and genus (**D**) levels of fungi.

To further confirm the differences in the microorganism communities at the genus level, we used the Kruskal-Wallis rank sum test to select microbial communities with significant differences in rhizosphere soils. A fold change ≥ 2 and $p < 0.05$ were set as the thresholds for significantly differential abundance. At the bacterial genus level, the number of significantly different genera in "NR vs. NHR" and "NS vs. NHR" was 50 and 141, respectively, of which 23 were shared (Figure 4A). A clustering heatmap was drawn to demonstrate their relative abundance levels (Figure 4B). Similarly, at the level of fungal genera, a total of four genera were screened. The clustering heatmap is shown in Figure 4C,D.

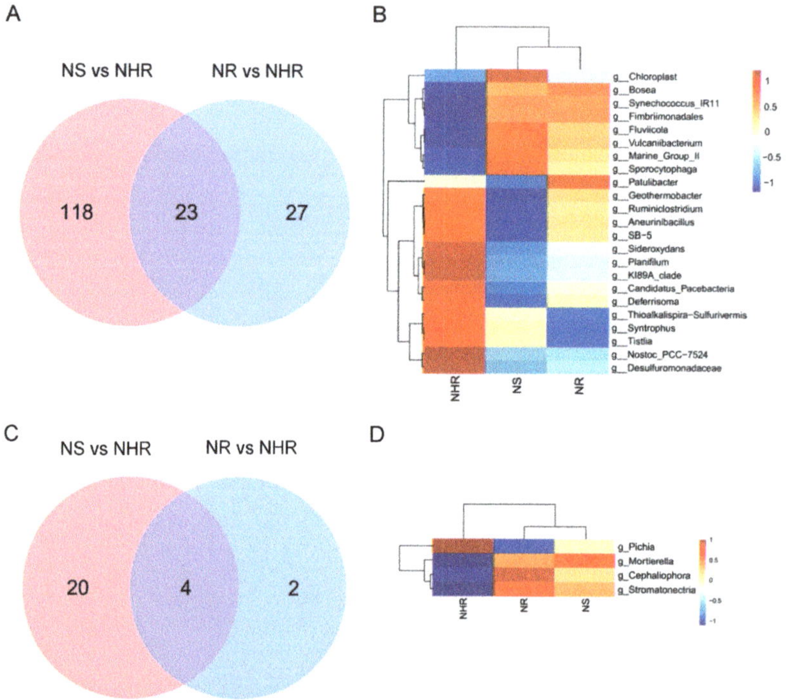

Figure 4. Differential analysis of rhizosphere soil microorganisms at the genus level. Venn diagrams (**A**) and clustering heatmap (**B**) of bacteria with significant differences at the genus level in "NR vs. NHR" and "NS vs. NHR". Venn diagrams (**C**) and clustering heatmap (**D**) of fungi with significant differences at the genus level in "NR vs. NHR" and "NS vs. NHR".

3.4. Potential Relationships among Microbial Community Composition and Soil Properties

We performed a Spearman correlation analysis to investigate the relationship between microbial community changes and soil properties (Figure 5). At the bacterial genus level, the relative abundance of seven genera was positively correlated with the soil factor, and eleven genera were negatively correlated (Figure 5A). At the fungal genus level, the abundance of three genera was positively correlated with the soil factors, while *Pichia* showed a lower level of negative correlation with the soil factors (Figure 5B). Additionally, we performed a correlation analysis of the relative abundance between the bacteria and the fungi at the genus level, and the results showed that there was a very significant positive correlation between *Stromatonectria* in fungi and *Synechococcus_IR11* and *Vulcaniibacterium* in bacteria and a very significant negative correlation with *Desulfuromonadaceae*. Another fungus, *Mortierella*, had a very significant positive correlation with the bacteria *Marine_Group_II* and a very significant negative correlation with *Ruminiclostridium* (Figure 5C). A redundancy analysis (RDA) was conducted to further study the relationship between the community composition of microorganisms and soil factors. The detrended correspondence analysis (DCA1) values of bacteria and fungi were 2.07474 and 0.71634, respectively, both less than 3.0, indicating that the data may be linearly distributed, so we chose to use RDA rather than CCA as the method of analysis. The first and second axes in the bacterial and fungal communities of the RDA model explained 50.95% and 14.97% and 55.87% and 34.55% of the variation, respectively (Figure 5D,E).

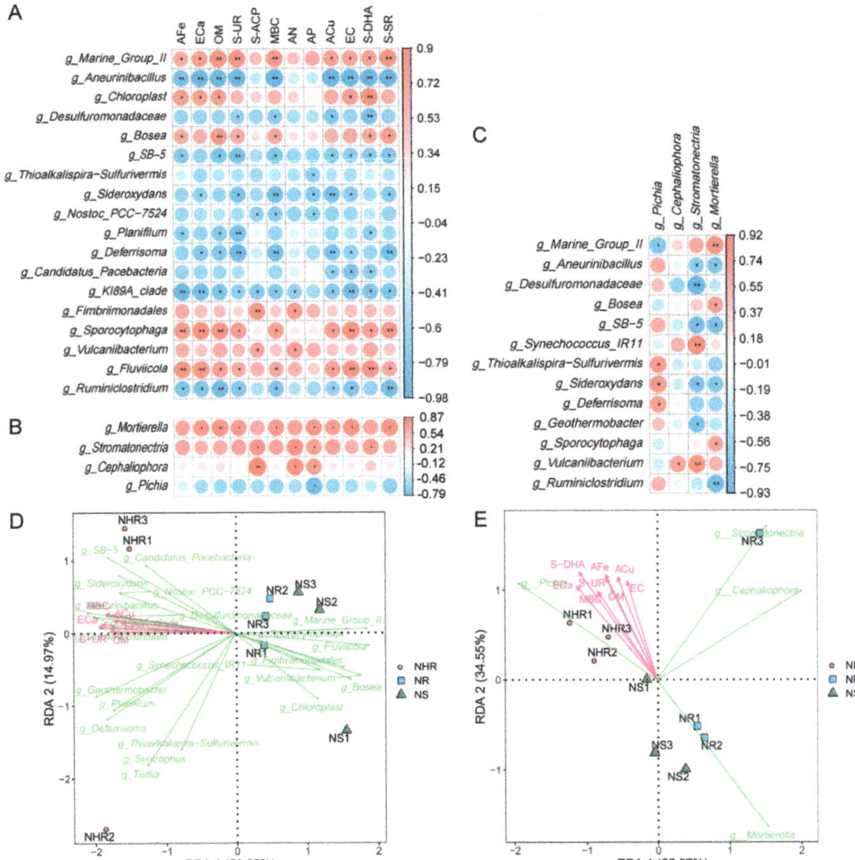

Figure 5. Relationships among microbial community composition and soil properties. Correlation heatmap between the key rhizosphere soil bacterial community (**A**) and fungal community (**B**) at the genus level and soil properties of rhizosphere soil. (**C**) Correlation analysis between the key rhizosphere soil bacterial community and fungal community. Redundancy analysis (RDA) of the relationship between soil bacterial (**D**) and fungal (**E**) community structures and soil factors in different soil groups. In the correlation analysis, the legend shows the correlation coefficient value, red represents the positive correlation and blue represents the negative correlation; color depth indicates the strength of the correlation; * $p < 0.05$, ** $p < 0.01$.

3.5. The Effect of Rhizosphere Microorganism Interactions on *M. incognita*

Based on the interaction between soil factors and microbial communities and the correlation between microorganisms, we speculated that soil properties, rhizosphere bacteria, and fungi may have direct effects on the occurrence and development of *M. incognita* in the cucumber rhizosphere. Then, the difference analysis of *Stromatonectria* and *Mortierella* screened in RDA and the correlation analysis, as well as *Desulfuromonadaceae*, *Synechococcus_IR11*,

Vulcaniibacterium, *Ruminiclostridium*, and *Marine_Group_II*, were performed (Figure 6A). A schematic diagram of soil factors and the interactions of rhizosphere microorganisms was drawn to reveal the possible mechanisms of interroot microbial resistance to *M. incognita* (Figure 6B). The results showed that the relative abundance of *Stromatonectria* and *Mortierella* in fungal genera was significantly lower in NHR than in NR and NS. In bacteria, the abundance of *Desulfuromonadaceae* and *Ruminiclostridium* in the NHR was significantly higher than that in the NR and NS. The abundance of *Synechococcus_IR11*, *Vulcaniibacterium*, and *Marine_Group_II* in the NR and NS was significantly higher than that in the NHR.

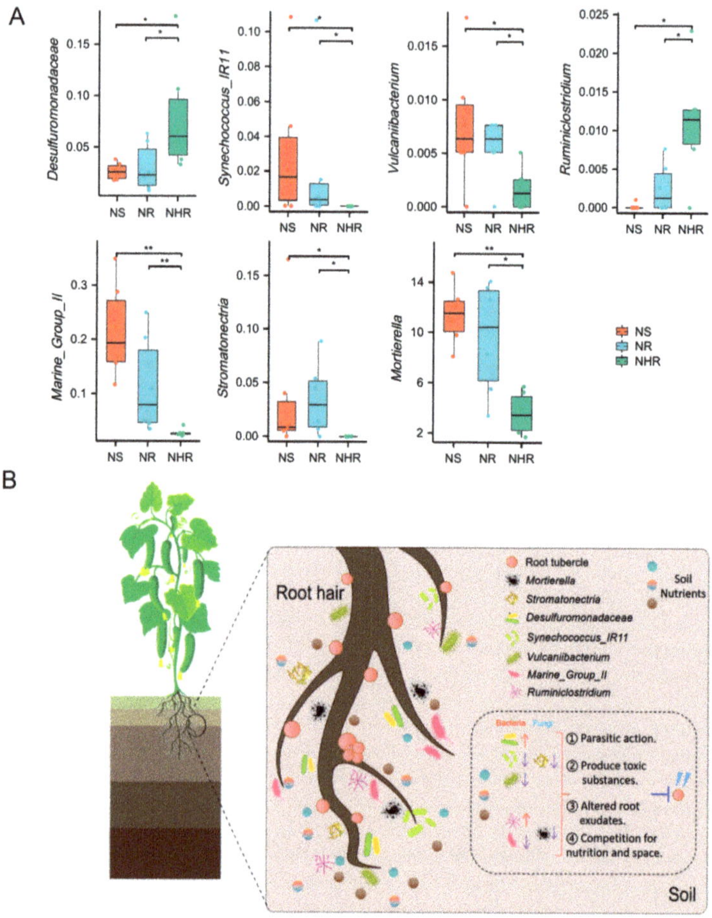

Figure 6. A predictive model for the synergistic effect of rhizosphere microorganisms combined with soil factors against *M. incognita*. (**A**) Boxplot of the relative abundance of microorganisms. Differences between groups were analyzed using the Wilcoxon rank sum test. (n = 6. * $p < 0.05$; ** $p < 0.01$). (**B**) Schematic diagram of the model for synergistic suppression of RKNs by soil factors and rhizosphere microorganisms. Red arrows indicate elevated microbial abundance and blue arrows indicate decreased microbial abundance.

4. Discussion

Cucumber is a typical cold-sensitive plant that is usually grown in greenhouses. However, cucumbers grown in greenhouses are generally planted continuously for many years, leading to the accumulation of pathogenic microorganisms and nematodes, which can easily cause soil-borne diseases. For decades, RKN infestations in crops have been controlled through the application of chemical nematicides. However, the excessive use of chemical nematicides has a huge negative impact on the environment. Biological control is usually considered a much more cost-effective alternative with less impact on the environment compared with both chemical and physical methods [39]. Previous studies have also demonstrated that microorganisms play an important role in soil properties and plant disease resistance. Among them, *Streptomyces platensis* could inhibit the growth of the spores of *Plasmodiophora brassicae* [40]. In addition, single or multiple rhizosphere bacteria in the soil are separated and made into biological control agents, which can control RKN. For example, a mix of *Bacillus amyloliquefaciens* and *B. subtilis* strains can reduce the amount of *M. incognita* in the soil [41]. Therefore, it is of great value to study the composition of microbial communities in rhizosphere soil and the changes in the composition of bacteria and fungi with different pathogenicity degrees for the future biological control of *M. incognita*. In recent years, with the promotion of high-throughput sequencing technology, researchers have carried out a large number of studies on the diversity of soil microbial communities, which can provide an in-depth understanding of the composition of soil and rhizosphere microbial communities [42].

4.1. Changes in Rhizosphere Soil Properties of Cucumber Infected with M. incognita

Soil properties have an important influence on plant growth. Generally, the richness of microorganisms is in direct proportion to the nutrients of the soil. [43]. This study showed that different incidences of *M. incognita* could significantly reduce organic matter, microbial biomass carbon, soil fertility (AFe, ACu, ECa, OM, EC), and enzyme activity (S-SR, S-UR, S-DHA) (Figure 1). Another study also showed that the available phosphorus content in the soil was negatively correlated with the incidence of Fusarium wilt [44]. It may be that a higher content of nutrients such as OM, nitrogen, and phosphorus in the soil can promote plant growth and improve plant disease resistance. Another important factor in the soil is the activity of various soil enzymes, which is also crucial for the growth and development of plants [45]. As the incidence of RKN increased, the activities of soil sucrase, urease, and dehydrogenase were significantly reduced compared with NHR, suggesting that soil enzyme activities may enhance the protection of plant cells; however, it was unfavorable for *M. incognita* to affect plant nitrogen utilization and carbon metabolism.

4.2. Rhizosphere Soil Microbial Diversity and Community Compositions

Maintaining healthy and stable soil microbial diversity has a certain inhibitory effect on pathogens. The perturbation of the rhizosphere microbial community composition is a significant contributor to soil-borne diseases. The multifunctionality of ecosystems in soil is determined by soil biodiversity and microbial composition, which have a suppressive impact on plant pathogens [46,47]. In general, a more diverse microbial community composition is associated with a greater capacity to impede pathogens [48]. A study has reported that there is a significant correlation between the abundance of rhizosphere microbial communities and the ability of plants to suppress pathogens. It was shown that there is a close correlation between higher bacterial and fungal diversity and a lower incidence of wilt disease in tomato rhizosphere soils [49]. Similar to previous research, in terms of bacterial community composition, *Proteobacteria*, *Actinobacteria*, and *Acidobacteria* were the dominant phyla [50]. In addition, several other dominant bacterial phyla, such as *Acidobacteria* and *Bacteroidetes*, have also been reported to play important roles in element utilization and the decomposition of organic matter [51]. At the genus level, the results showed that the major microbial groups were *Vicinamibacteraceae*, *Rokubacteriales*, *Nitrososphaeraceae*, *Bacillus*, and *Pseudomonas*. *Rokubacteriales* were previously reported to be nitrogen-efficient bacteria that affect the nitrogen reduction process in rhizosphere soil and accelerate the loss of nitrogen sources from the rhizosphere soil, thus reducing the nitrogen metabolic pathways of rhizosphere microorganisms and plants [52]. The results were consistent with the constant decrease in available nitrogen in the NHR that we observed (Figure 1).

Pseudomonas are one of the most abundant bacterial genera in the soil and inter-rhizosphere and play an important role in promoting plant health. It has been demonstrated that *Pseudomonas putida* Sneb821 can stimulate resistance to SRKNs in tomato. *Pseudomonas putida* Sneb821 was able to suppress the expression of Sly-miR482d, which in turn up-regulated the expression of its target gene NBS-LRR, promoted H2O2 accumulation in tomato roots, and regulated SOD and POD enzyme activities, thereby regulating the immune response of tomato to SRKN infestation [53]. Potato scab and blight are two major diseases that can cause severe crop losses. They are caused by the bacterium *Streptomyces scabies* and an oomycete known as *Phytophthora infestans*, respectively. Studies have confirmed that Pseudomonas can produce hydrogen cyanide (HCN) and cyclic lipopeptide (CLP), key compounds that inhibit Streptomyces scab and the disease-causing blight of potato, thereby suppressing disease development [54]. Moreover, we also observed a significant decrease in the relative abundance of Bacillus among rhizobacteria as the incidence of RKN increased. *Bacillus* spp. can directly inhibit RKN disease and promote other bacteria to inhibit the occurrence of plant diseases [55]. Another previous study showed that *Bacillus* spp. can effectively inhibit various potential plant pathogens, including soil bacterial pathogens, soil fungal pathogens, soil viruses, and nematodes [56]. Fungal communities are simpler in composition than bacterial communities. The main fungal phyla are *Ascomycota*, *Morphomycota*, and *Basidiomycota*. Similar results were obtained by other investigators [57]. At the fungal genus level, Trichoderma is an important biocontrol fungus that not only fights and controls ecologically important plant pathogenic fungi, viruses, and bacteria that are widely present, but also effectively controls nematodes, especially RKNs. Fungal species of the genus Trichoderma are considered as biocontrol agents against plant-associated fungal pathogens and can suppress root-knot nematodes through competitive action as well as by producing specific metabolites [58,59].

4.3. Mechanisms of Interactions between Rhizosphere Microorganisms Resisting M. incognita

Furthermore, root-associated microbial communities may play an important role in modulating root-soil nutrient–nematode interactions [60]. There are several species of bacteria and fungi that are closely associated with nematodes, and some have been proven to have harmful effects on them [61]. Some fungi in the soil can secrete toxins or lytic enzymes to kill or inhibit nematodes, such as the toxin produced by *Pleurotus ostreatus*, which is fudecenedioic acid [62]. In addition, *Paecilomyces lilacinus* and *Trichoderma strictum* can control RKN through parasitism [63,64]. Earlier research has demonstrated that some *Mortierella* spp. may generate antibiotics, and numerous isolates have been widely used as prospective antagonistic agents against various pathogens [65]. It has also been discovered that *Acremonium strictum* could reduce *M. incognita* populations in tomato roots, indicating that some of the identified microbial taxa may be useful for biocontrol strategies [64]. Previous studies focused on bacteria or fungi in the soil, and there has been more research on bacteria than fungi [18]; hence, fungal communities have sometimes been neglected. In the correlation analysis of bacteria and fungi in this study, some key bacteria and fungi had significant positive or negative associations, indicating that they may have synergistic or antagonistic effects against RKN (Figure 5C). As reported in previous studies, *Mortierella* spp. can suppress *Fusarium oxysporum* in the soil to reduce vanilla Fusarium wilt disease [66]. In addition, the latest results also prove that symbiotic bacteria in fungi can help to protect them from nematode attack, suggesting that the interaction with bacteria can inhibit the development of nematodes [67]. In response to *M. incognita*, rhizosphere bacteria and fungi may use the following four pathways: (1) causing parasitic effects on nematodes; (2) using toxic and harmful substances to nematodes; (3) changes in root exudates to affect nematode development; and (4) competition with nematodes for nutrition and space position. There may also be other pathways, such as promoting plant growth and inducing host resistance (Figure 6B).

Supplementary Materials: The following supporting information can be downloaded at: https://www.mdpi.com/article/10.3390/agronomy13071726/s1. Supplementary Table S1: Overview of bacterial and fungal diversity sequencing data; Table S2: Bacteria and fungi alpha diversity index table.

Author Contributions: Conceptualization, B.T. and X.S.; Software, H.J.; Validation, S.L., G.C. and K.M.; Formal Analysis, L.N.; Investigation, F.Y. and Y.T.; Resources, F.Y.; Data Curation, G.M.; Writing—Original Draft Preparation, F.Y.; Writing—Review and Editing, B.T. and X.S.; Visualization, F.Y. All authors have read and agreed to the published version of the manuscript.

Funding: This research was funded by the Scientific and Technological Breakthrough Foundation of Henan Province Project (212102110426, 222102110004) and the Technology System of Watermelon Industry in Henan Province (HARS-22-10-G1).

Data Availability Statement: All raw sequence data have been made available in the NCBI Sequence Read Archive (SRA) database under the bioproject accession numbers PRJNA907508 and PRJNA907474.

Conflicts of Interest: The authors declare no conflict of interest.

References

1. Huang, S.W.; Li, R.Q.; Zhang, Z.H.; Li, L.; Gu, X.F.; Fan, W.; Lucas, W.J.; Wang, X.W.; Xie, B.Y.; Ni, P.X.; et al. The genome of the cucumber *Cucumis sativus* L. *Nat. Genet.* **2009**, *41*, 1275–1281. [CrossRef] [PubMed]
2. Zhang, J.; Yang, J.; Yang, Y.; Luo, J.; Zheng, X.; Wen, C.; Xu, Y. Transcription Factor CsWIN1 Regulates Pericarp Wax Biosynthesis in Cucumber Grafted on Pumpkin. *Front. Plant Sci.* **2019**, *10*, 1564. [CrossRef] [PubMed]
3. Zhang, X.; Zhang, Y.; Xu, C.; Liu, K.; Bi, H.; Ai, X. H_2O_2 Functions as a Downstream Signal of IAA to Mediate H2S-Induced Chilling Tolerance in Cucumber. *Int. J. Mol. Sci.* **2021**, *22*, 12910. [CrossRef] [PubMed]
4. Gine, A.; Lopez-gomez, M.; Vela, M.D.; Ornat, C.; Talavera, M.; Verdejo-lucas, S.; Sorribas, F.J. Thermal requirements and population dynamics of root-knot nematodes on cucumber and yield losses under protected cultivation. *Plant Pathol.* **2014**, *63*, 1446–1453. [CrossRef]
5. Castagnone-Sereno, P. Genetic variability and adaptive evolution in parthenogenetic root-knot nematodes. *Heredity* **2006**, *96*, 282–289. [CrossRef]
6. Li, J.; Zou, C.G.; Xu, J.P.; Ji, X.L.; Niu, X.M.; Yang, J.K.; Huang, X.W.; Zhang, K.Q. Molecular mechanisms of nematode-nematophagous microbe interactions: Basis for biological control of plant-parasitic nematodes. *Annu. Rev. Phytopathol.* **2015**, *53*, 67–95. [CrossRef]
7. Siddique, S.; Grundler, F.M. Parasitic nematodes manipulate plant development to establish feeding sites. *Curr. Opin. Microbiol.* **2018**, *46*, 102–108. [CrossRef]
8. Kayani, M.Z.; Mukhtar, T.; Hussain, M.A. Effects of southern root knot nematode population densities and plant age on growth and yield parameters of cucumber. *Crop Prot.* **2017**, *92*, 207–212. [CrossRef]
9. Xu, W.; Yang, Q.; Xie, X.; Goodwin, P.H.; Deng, X.; Zhang, J.; Sun, R.; Wang, Q.; Xia, M.; Wu, C.; et al. Genomic and Phenotypic Insights into the Potential of *Bacillus subtilis* YB-15 Isolated from Rhizosphere to Biocontrol against Crown Rot and Promote Growth of Wheat. *Biology* **2022**, *11*, 778. [CrossRef]
10. Kwak, M.J.; Kong, H.G.; Choi, K.; Kwon, S.K.; Song, J.Y.; Lee, J.; Lee, P.A.; Choi, S.Y.; Seo, M.; Lee, H.J.; et al. Rhizosphere microbiome structure alters to enable wilt resistance in tomato. *Nat. Biotechnol.* **2018**, *36*, 1100. [CrossRef]
11. Ma, L.; Zhang, H.Y.; Zhou, X.K.; Yang, C.G.; Zheng, S.C.; Duo, J.L.; Mo, M.H. Biological control tobacco bacterial wilt and black shank and root colonization by bioorganic fertilizer containing bacterium *Pseudomonas aeruginosa* NXHG29. *Appl. Soil Ecol.* **2018**, *129*, 136–144. [CrossRef]
12. Kloeppe, J.W.; Rodríguez-Kábana, R.; Zehnder, A.W.; Murphy, J.F.; Sikora, E.; Fernández, C. Plant root-bacterial interactions in biological control of soilborne diseases and potential extension to systemic and foliar diseases. *Aust. Plant Pathol.* **1999**, *28*, 21–26. [CrossRef]
13. Miliute, I.; Buzaite, O.; Baniulis, D.; Stanys, V. Bacterial endophytes in agricultural crops and their role in stress tolerance: A review. *Zemdirbyste* **2015**, *102*, 465–478. [CrossRef]
14. Segarra, G.; Van der Ent, S.; Trillas, I.; Pieterse, C.M.J. MYB72, a node of convergence in induced systemic resistance triggered by a fungal and a bacterial beneficial microbe. *Plant Biol.* **2009**, *11*, 90–96. [CrossRef] [PubMed]
15. Van der Ent, S.; Verhagen, B.W.; Van Doorn, R.; Bakker, D.; Verlaan, M.G.; Pel, M.J.; Joosten, R.G.; Proveniers, M.C.; Van Loon, L.C.; Ton, J.; et al. MYB72 is required in early signaling steps of rhizobacteria-induced systemic resistance in arabidopsis. *Plant Physiol* **2008**, *146*, 1293–1304. [CrossRef] [PubMed]
16. Palmer, C.M.; Hindt, M.N.; Schmidt, H.; Clemens, S.; Guerinot, M.L. MYB10 and MYB72 are required for growth under iron-limiting conditions. *PLoS Genet.* **2013**, *9*, e1003953. [CrossRef]
17. Mendes, R.; Kruijt, M.; de Bruijn, I.; Dekkers, E.; van der Voort, M.; Schneider, J.H.; Piceno, Y.M.; DeSantis, T.Z.; Andersen, G.L.; Bakker, P.A.; et al. Deciphering the rhizosphere microbiome for disease-suppressive bacteria. *Science* **2011**, *332*, 1097–1100. [CrossRef]

18. Cha, J.Y.; Han, S.; Hong, H.J.; Cho, H.; Kim, D.; Kwon, Y.; Kwon, S.K.; Crüsemann, M.; Bok Lee, Y.; Kim, J.F.; et al. Microbial and biochemical basis of a Fusarium wilt-suppressive soil. *ISME J.* **2016**, *10*, 119–129. [CrossRef]
19. Chen, Y.; Wang, J.; Yang, N.; Wen, Z.Y.; Sun, X.P.; Chai, Y.R.; Ma, Z.H. Wheat microbiome bacteria can reduce virulence of a plant pathogenic fungus by altering histone acetylation. *Nat. Commun.* **2018**, *9*, 3429. [CrossRef]
20. Parratt, S.R.; Laine, A.L. Pathogen dynamics under both bottom-up host resistance and top-down hyperparasite attack. *J. Appl. Ecol.* **2018**, *55*, 2976–2985. [CrossRef]
21. Zelezniak, A.; Andrejev, S.; Ponomarova, O.; Mende, D.R.; Bork, P.; Patil, K.R. Metabolic dependencies drive species co-occurrence in diverse microbial communities. *Proc. Natl. Acad. Sci. USA* **2015**, *112*, 6449–6454. [CrossRef]
22. Taylor, A.L.; Sasser, J.N. *Biology, Identification and Control of Root-Knot Nematodes (Meloidogyne Species)*; North Carolina State University Graphics: Raleigh, NC, USA, 1978; Volume 111.
23. Lyu, J.; Jin, L.; Jin, N.; Xie, J.; Xiao, X.; Hu, L.; Tang, Z.; Wu, Y.; Niu, L.; Yu, J. Effects of different vegetable rotations on fungal community structure in continuous tomato cropping matrix in greenhouse. *Front. Microbiol.* **2020**, *11*, 829. [CrossRef] [PubMed]
24. Magoč, T.; Salzberg, S.L. FLASH: Fast length adjustment of short reads to improve genome assemblies. *Bioinformatics* **2011**, *27*, 2957–2963. [CrossRef]
25. Zhang, J.J.; Kobert, K.; Flouri, T.; Stamatakis, A. PEAR: A fast and accurate Illumina Paired-End reAd mergeR. *Bioinformatics* **2014**, *30*, 614–620. [CrossRef] [PubMed]
26. Rognes, T.; Flouri, T.; Nichols, B.; Quince, C.; Mahé, F. VSEARCH: A versatile open source tool for metagenomics. *PeerJ* **2016**, *18*, e2584. [CrossRef] [PubMed]
27. Callahan, B.J.; McMurdie, P.J.; Rosen, M.J.; Han, A.W.; Johnson, A.J.; Holmes, S.P. DADA2: High-resolution sample inference from Illumina amplicon data. *Nat. Methods* **2016**, *13*, 581–583. [CrossRef] [PubMed]
28. Bolyen, E.; Rideout, J.R.; Dillon, M.R.; Bokulich, N.A.; Abnet, C.C.; Al-Ghalith, G.A.; Alexander, H.; Alm, E.J.; Arumugam, M.; Asnicar, F.; et al. Reproducible, interactive, scalable and extensible microbiome data science using QIIME 2. *Nat. Biotechnol.* **2019**, *37*, 852–857. [CrossRef]
29. Song, X.; Pan, Y.; Li, L.; Wu, X.; Wang, Y. Composition and diversity of rhizosphere fungal community in *Coptis chinensis* Franch. continuous cropping fields. *PLoS ONE* **2018**, *13*, e0193811. [CrossRef]
30. Kim, J.; Kreller, C.R.; Greenberg, M.M. Preparation and analysis of oligonucleotides containing the c4′-oxidized abasic site and related mechanistic probes. *J. Org. Chem.* **2005**, *70*, 8122–8129. [CrossRef]
31. Xu, Y.X.; Du, A.P.; Wang, Z.C.; Zhu, W.K.; Li, C.; Wu, L.C. Effects of different rotation periods of *Eucalyptus plantations* on soil physiochemical properties, enzyme activities, microbial biomass and microbial community structure and diversity. *For. Ecol. Manag.* **2020**, *456*, 117683. [CrossRef]
32. Shi, Y.; Li, Y.; Xiang, X.; Sun, R.; Yang, T.; He, D.; Zhang, K.; Ni, Y.; Zhu, Y.G.; Adams, J.M.; et al. Spatial scale affects the relative role of stochasticity versus determinism in soil bacterial communities in wheat fields across the North China Plain. *Microbiome* **2018**, *6*, 27. [CrossRef] [PubMed]
33. Yu, Y.Y.; Xu, J.D.; Huang, T.X.; Zhong, J.; Yu, H.; Qiu, J.P.; Guo, J.H. Combination of beneficial bacteria improves blueberry production and soil quality. *Food Sci. Nutr.* **2020**, *8*, 5776–5784. [CrossRef]
34. Vance, E.D.; Brookes, P.C.; Jenkinson, D.S. An extraction method for measuring soil microbial biomass C. *Soil Biol. Biochem.* **1987**, *19*, 703–707. [CrossRef]
35. Klein, D.A.; Loh, T.C.; Goulding, R.L. A rapid procedure to evaluate dehydrogenase activity of soils low in organic matter. *Soil Biol. Biochem.* **1971**, *3*, 385–387. [CrossRef]
36. Yang, R.; Xia, X.; Wang, J.; Zhu, L.; Wang, J.; Ahmad, Z.; Yang, L.; Mao, S.; Chen, Y. Dose and time-dependent response of single and combined artificial contamination of sulfamethazine and copper on soil enzymatic activities. *Chemosphere* **2020**, *250*, 126161. [CrossRef] [PubMed]
37. Kandeler, E.; Gerber, H. Short-term assay of soil urease activity using colorimetric determination of ammonium. *Biol. Fertil. Soils* **1988**, *6*, 68–72. [CrossRef]
38. Tabatabai, M.A.; Bremner, J.M. Use of p-nitrophenylphosphate for assay of soil phosphatase activity. *Soil Biol. Biochem.* **1969**, *1*, 301–307. [CrossRef]
39. Fravel, D.R. Commercialization and implementation of biocontrol. *Annu. Rev. Phytopathol.* **2005**, *43*, 337–359. [CrossRef]
40. Shakeel, Q.; Lyu, A.; Zhang, J.; Wu, M.; Chen, S.; Chen, W.; Li, G.; Long, Y. Optimization of the cultural medium and conditions for production of antifungal substances by *Streptomyces platensis* 3-10 and evaluation of its efficacy in suppression of clubroot disease (*Plasmodiophora brassicae*) of oilseed rape. *Biol. Control* **2016**, *101*, 59–68. [CrossRef]
41. Burkett-Cadena, M.; Kokalis-Burelle, N.; Lawrence, K.S.; Santen, E.V.; Kloepper, J.W. Suppressiveness of root-knot nematodes mediated by rhizobacteria. *Biol. Control* **2008**, *47*, 55–59. [CrossRef]
42. Liu, C.M.; Yang, Z.F.; He, P.F.; Munir, S.; Wu, Y.X.; Ho, H.; He, Y.Q. Deciphering the bacterial and fungal communities in clubroot-affected cabbage rhizosphere treated with *Bacillus subtilis* XF-1. *Agric. Ecosyst. Environ.* **2018**, *256*, 12–22. [CrossRef]
43. Bünemann, E.K.; Bongiorno, G.; Bai, Z.G.; Creamer, R.E.; De Deyn, G.; De Goede, R.; Fleskens, L.; Geissen, V.; Kuyper, T.W.; Mäder, P.; et al. Soil quality-A critical review. *Soil Biol. Biochem.* **2018**, *120*, 105–125. [CrossRef]
44. Shen, Z.Z.; Ruan, Y.Z.; Xue, C.; Zhong, S.T.; Li, R.; Shen, Q.R. Soils naturally suppressive to banana *Fusarium* wilt disease harbor unique bacterial communities. *Plant Soil* **2015**, *393*, 21.e33. [CrossRef]

45. Acosta-Martinez, V.; Cano, A.; Johnson, J. Simultaneous determination of multiple soil enzyme activities for soil health-biogeochemical indices. *Appl. Soil Ecol.* **2018**, *126*, 121–128. [CrossRef]
46. van Elsas, J.D.; Chiurazzi, M.; Mallon, C.A.; Elhottovā, D.; Krištůfek, V.; Salles, J.F. Microbial diversity determines the invasion of soil by a bacterial pathogen. *Proc. Nat. Acad. Sci. USA* **2012**, *109*, 1159–1164. [CrossRef]
47. Cameron, W.; Franz, B.S.; Franco, W.; Van der Heijden, M.G.A. Soil biodiversity and soil community composition determine ecosystem multifunctionality. *Proc. Natl. Acad. Sci. USA* **2014**, *111*, 5266–5270.
48. Raaijmakers, J.M.; Paulitz, T.C.; Steinberg, C.; Alabouvette, C.; Moënne-Loccoz, Y. The rhizosphere: A playground and battlefield for soilborne pathogens and beneficial microorganisms. *Plant Soil Fertil. Sci. China* **2009**, *321*, 341–361. [CrossRef]
49. Zhou, X.; Wang, J.; Liu, F.; Liang, J.; Zhao, P.; Tsui, C.K.M.; Cai, L. Cross-kingdom synthetic microbiota supports tomato suppression of Fusarium wilt disease. *Nat. Commun.* **2022**, *13*, 7890. [CrossRef] [PubMed]
50. Delgado-Baquerizo, M.; Oliverio, A.M.; Brewer, T.E.; Benavent-González, A.; Eldridge, D.J.; Bardgett, R.D.; Maestre, F.T.; Singh, B.K.; Fierer, N. A global atlas of the dominant bacteria found in soil. *Science* **2018**, *359*, 320–325. [CrossRef]
51. Eichorst, S.A.; Trojan, D.; Roux, S.; Herbold, C.; Rattei, T.; Woebken, D. Genomic insights into the *Acidobacteria* reveal strategies for their success in terrestrial environments. *Environ. Microbiol.* **2018**, *20*, 1041–1063. [CrossRef]
52. Becraft, E.D.; Woyke, T.; Jarett, J.; Ivanova, N.; Godoy-Vitorino, F.; Poulton, N.; Brown, J.M.; Brown, J.; Lau, M.C.Y.; Onstott, T.; et al. Genomic Giants among the Uncultured Bacterial Phyla. *Front. Microbiol.* **2017**, *8*, 2264. [CrossRef]
53. Yang, F.; Ding, L.; Zhao, D.; Fan, H.; Zhu, X.; Wang, Y.; Liu, X.; Duan, Y.; Chen, L. Identification and Functional Analysis of Tomato MicroRNAs in the Biocontrol Bacterium Pseudomonas putida Induced Plant Resistance to Meloidogyne incognita. *Phytopathology* **2022**, *112*, 2372–2382. [CrossRef]
54. Pacheco-Moreno, A.; Stefanato, F.L.; Ford, J.J.; Trippel, C.; Uszkoreit, S.; Ferrafiat, L.; Grenga, L.; Dickens, R.; Kelly, N.; Kingdon, A.D.; et al. Pan-genome analysis identifies intersecting roles for Pseudomonas specialized metabolites in potato pathogen inhibition. *Elife* **2021**, *10*, e71900. [CrossRef] [PubMed]
55. Son, S.H.; Khan, Z.; Kim, S.G.; Kim, Y.H. Plant growth-promoting rhizobacteria, *Paenibacillus polymyxa* and *Paenibacillus lentimorbus* suppress disease complex caused by root-knot nematode and *fusarium* wilt fungus. *J. Appl. Microbiol.* **2009**, *107*, 524–532. [CrossRef] [PubMed]
56. Kloepper, J.W.; Ryu, C.M.; Zhang, S. Induced Systemic Resistance and Promotion of Plant Growth by *Bacillus* spp. *Phytopathology* **2004**, *94*, 1259–1266. [CrossRef] [PubMed]
57. Bahram, M.; Netherway, T.; Hildebrand, F.; Pritsch, K.; Drenkhan, R.; Loit, K.; Anslan, S.; Bork, P.; Tedersoo, L. Plant nutrient-acquisition strategies drive topsoil microbiome structure and function. *New Phytol.* **2020**, *227*, 1189–1199. [CrossRef]
58. Oyekanmi, E.O.; Coyne, D.L.; Fagade, O.E.; Osonubi, O. Improving root-knot nematode management on two soybean genotypes through the application of *Bradyrhizobium japonicum*, *Trichoderma pseudokoningii* and *Glomus mosseae* in full factorial combinations. *Crop Prot.* **2007**, *26*, 1006–1012. [CrossRef]
59. Dunlop, R.W.; Simon, A.; Sivasithampara, K.; Ghisalberti, E.L. An antibiotic from Trichoderma Koningii active against soilborne plant pathogens. *J. Nat. Prod.* **1989**, *52*, 67–74. [CrossRef]
60. Toju, H.; Tanaka, Y. Consortia of anti-nematode fungi and bacteria in the rhizosphere of soybean plants attacked by root-knot nematodes. *R Soc. Open Sci.* **2019**, *6*, 181693. [CrossRef]
61. Zhou, D.M.; Feng, H.; Schuelke, T.; De Santiago, A.; Zhang, Q.M.; Zhang, J.F.; Luo, C.P.; Wei, L.H. Rhizosphere microbiomes from root knot nematode Non-infested plants suppress nematode Infection. *Microb. Ecol.* **2019**, *78*, 470–481. [CrossRef]
62. Kwok, O.C.; Plattner, R.; Weisleder, D.; Wicklow, D.T. A nematicidal toxin from *Pleurotus ostreatus* NRRL 3526. *J. Chem. Ecol.* **1992**, *18*, 127–136. [CrossRef] [PubMed]
63. Anastasiadis, I.A.; Giannakou, I.O.; Prophetou-Athanasiadou, D.A.; Gowen, S.R. The combined effect of the application of a biocontrol agent *Paecilomyces lilacinus*, with various practices for the control of root-knot nematodes. *Crop Prot.* **2008**, *27*, 352–361. [CrossRef]
64. Goswami, J.; Pandey, R.K.; Tewari, J.P.; Goswami, B.K. Management of root knot nematode on tomato through application of fungal antagonists, *Acremonium strictum* and *Trichoderma harzianum*. *J. Environ. Sci. Health* **2008**, *43*, 237–240. [CrossRef] [PubMed]
65. Tagawa, M.; Tamaki, H.; Manome, A.; Koyama, O.; Kamagata, Y. Isolation and characterization of antagonistic fungi against potato scab pathogens from potato field soils. *FEMS Microbiol. Lett.* **2010**, *305*, 136–142. [CrossRef]
66. Xiong, W.; Li, R.; Ren, Y.; Liu, C.; Zhao, Q.Y.; Wu, H.S.; Jousset, A.; Shen, Q.R. Distinct roles for soil fungal and bacterial communities associated with the suppression of vanilla *Fusarium* wilt disease. *Soil Biol. Biochem.* **2017**, *107*, 198–207. [CrossRef]
67. Büttner, H.; Niehs, S.P.; Vandelannoote, K.; Cseresnyés, Z.; Dose, B.; Richter, I.; Gerst, R.; Figge, M.T.; Stinear, T.P.; Pidot, S.J.; et al. Bacterial endosymbionts protect beneficial soil fungus from nematode attack. *Proc. Natl. Acad. Sci. USA* **2021**, *37*, e2110669118. [CrossRef] [PubMed]

agronomy

MDPI

Article

Long-Term Chemical Fertilization Drove Beneficial Bacteria for Rice Soil to Move from Bulk Soil to the Rhizosphere

Jian Xiao [1,2,3,†], Jianglin Zhang [2,3,†], Yajie Gao [2,3], Yanhong Lu [2,3], Xue Xie [2,3,4], Changyu Fang [2,3,5], Yulin Liao [1,2,3,*] and Jun Nie [1,2,3,4,*]

1 Longping Branch, College of Biology, Hunan University, Changsha 410125, China; xiaojian1997@hnu.edu.cn
2 Hunan Soil and Fertilizer Research Institute, Changsha 410125, China; zhangjianglin@hunaas.cn (J.Z.); gyj06107048@163.com (Y.G.); luyanhong6376432@163.com (Y.L.); xiexue@hunau.edu.cn (X.X.); 18684663642@163.com (C.F.)
3 Scientific Observing and Experimental Station of Arable Land Conservation (Hunan), Ministry of Agriculture, Changsha 410125, China
4 College of Resource, Hunan Agricultural University, Changsha 410128, China
5 College of Agronomy, Hunan Agricultural University, Changsha 410128, China
* Correspondence: yulliao2006@126.com (Y.L.); neijun197@163.com (J.N.)
† These authors contributed equally to this work.

Abstract: Overuse of chemical fertilizer (CF) causes damage to soil and the environment. To reveal the process of the response of crop rhizospheric and bulk soil fertility and the bacterial community to long-term CF conditions, CF application and nonfertilization (CK, control) treatments were used in a long-term (12-year) fertilization experiment. Long-term CF application significantly increased the soil organic matter, total nitrogen, and available phosphorus contents ($p < 0.05$), increased the available nitrogen (AN) and potassium (AK) contents to varying degrees, and decreased the soil pH in both rice rhizospheric soil and bulk soil. In addition, the bacterial Shannon and Ace indices in rice rhizospheric soil under the CF treatment were all higher than those under the control (CK) treatment, and the bulk soil bacteria showed the opposite trend. The LEfSe results showed that *unidentified_Gammaproteobacteria* and *Geobacter* (genera) were significantly enriched in the rhizospheric and bulk soil of rice under the CK treatment, respectively. Gemmatimonadetes (phylum) and Nitrospirae (phylum) + *Thiobacillus* (genus) were significantly enriched in the rice rhizospheric and bulk soil under the CF treatment. Only AK and AN had strong positive correlations with soil bacteria. Long-term CF application accelerated the migration of soil bacteria from the bulk soil to the rhizosphere, thus improving soil fertility and nutrient cycling.

Keywords: paddy soil; fertilization; rhizosphere; bulk soil; bacterial community

check for updates

Citation: Xiao, J.; Zhang, J.; Gao, Y.; Lu, Y.; Xie, X.; Fang, C.; Liao, Y.; Nie, J. Long-Term Chemical Fertilization Drove Beneficial Bacteria for Rice Soil to Move from Bulk Soil to the Rhizosphere. *Agronomy* **2023**, *13*, 1645. https://doi.org/10.3390/agronomy13061645

Academic Editors: Yong-Xin Liu and Peng Yu

Received: 4 June 2023
Revised: 16 June 2023
Accepted: 18 June 2023
Published: 19 June 2023

1. Introduction

Rice (*Oryza sativa* L.) plays a crucial role in agricultural fields worldwide as one of the main food crops [1]. China has the second largest cultivation area and is the largest producer in the world [2]. Yuan et al. [3] and Zhao et al. [4] suggested that the distribution of grain production will change with increasing population and the improvement of people's living standards in the future. However, the increase in crop yield is overdependent on the use of chemical fertilizers (CFs) in the process of food production in China [5]. To date, in conventional agricultural fertilizer management, agricultural producers manage fields with high CF inputs to maintain soil productivity and increase crop yields [2,6]. The amounts of nutrients required by crops are often far less than the amount of CF (i.e., the amount of conventional CF input) [6,7]. Potential harms, such as soil acidification [8], declines in soil organic matter (SOM) and soil fertility [9], and dramatic reductions in soil biodiversity [10], can be caused by excess nutrients in the soil.

Soil microbes play an important role in maintaining soil fertility and health by decomposing litter and recycling nutrients [11]. Wu et al. [10] found that soil microbial community structures can be affected by changing soil properties through tilling, irrigating, fertilizing, etc. For example, fertilization directly changes soil nutrients and then affects the structure and diversity of soil microbial communities [12].

Previous studies have reported that microbial populations associated with the nitrogen cycle are positively or negatively affected by long-term chemical or mineral fertilizer application [13]. Thus, Wu et al. [10] reported that soil quality, soil fertility, and the soil microbial community can be negatively affected by repeatedly overusing CFs. Sun et al. [14] reported that soil pH was significantly reduced, soil bacterial diversity was decreased, and soil bacterial community composition was significantly changed after long-term CF application. However, Geisseler and Scow [15] reported that in comparison with a nonfertilized treatment, fertilizer application increased soil microbial biomass by 15.1%.

In the rhizosphere, the organic compounds secreted by plants enrich diversified microbial communities to provide beneficial nutrients for plants [16–18]. Edwards et al. [19] exhaustively characterized the root-associated microbiome of rice to support a multistep model for the assembly of a root microbiome from soil. Previous studies reported that the compositions of soil microbial communities in rhizospheric soil were significantly different from those in bulk soil [20,21]. In the rhizosphere of plants, their roots coordinate development and interact with rhizospheric microorganisms [22]. The structure, composition, and functioning of plant-associated rhizospheric microbiota can be shaped by free-living soil microorganisms. Beneficial effects, including nitrogen fixation [23], organic phosphate mineralization [24], production of plant growth regulators or phytohormones [25,26], production of siderophores [27], stress tolerance [28], and biological control [29,30], are facilitated by the rhizospheric microbiome. There are differences in the microbial metabolism and ecological processes between rhizospheric soil and bulk soil [18] because bulk soil is far away from plants' living areas and is not affected by plants.

Fertilizers that are applied to soil have an aftereffect and leave lasting residuals on crops. Compared with short-term tests, long-term positioning tests can be used to objectively characterize the effects of different management measures on the physical, chemical, and biological properties of soil. Many studies have reported the effects of soil quality, soil fertility, the soil microbial community, and crop growth indicators on yield with long-term CF application. However, few studies [31,32] have investigated the effects of the bacterial community structure in rhizospheric and bulk soils after long-term CF application. For this reason, in the present study, the effects of long-term (12-year) CF application on bacterial community structures in rhizospheric and bulk soils of rice were investigated through a 12-year field experiment. The main objective of the study was to provide a theoretical basis for revealing the migration mechanism of rice soil microorganisms under long-term chemical fertilizer application.

2. Materials and Methods

2.1. Location and Experimental Design with Rice

The continuous outdoor experiment began in 2008 in Sanxianhu Village (GPS coordinates 29°13′03″ N, 112°28′53″ E), Nan County, Yiyang City, Hunan Province, China, which is located in a humid subtropical monsoon zone. The altitude, annual average temperature, and precipitation were approximately 28.8 m, 16.6 °C, and 1237.7 mm, respectively. At the experimental location, the soil was a typical sandy purple clay soil developed from Dongting Lake sediment with 7.8% sand, 72.6% silt, and 19.6% clay [33]. Before transplanting rice, the soil chemical properties of the initial plow-layer soil (0–20 cm) were as follows: soil pH, 8.18 ± 0.12 (n = 3); soil organic matter (SOM), 31.85 ± 0.48 g kg^{-1} (n = 3); total nitrogen (TN), 3.04 ± 0.06 g kg^{-1} (n = 3); alkaline hydrolyzable nitrogen (AN), 169.9 ± 2.51 mg kg^{-1} (n = 3); available P (AP), 15.4 ± 0.65 mg kg^{-1} (n = 3); available K (AK), 81.9 ± 1.30 mg kg^{-1} (n = 3).

The conventional fertilizers used were urea (46% N, Xinlianxin, Xinxiang, China), calcium superphosphate (12% P_2O_5, Shuoling, Huaihua, China), and potassium chloride (60% K_2O, Lingkelongzi, Changsha, China). The treatment and control were as follows:

(i) Nonfertilization was used as the control (CK) and (ii) the recommended conventional chemical fertilizer application for local paddy fields, as used by local farmers, was used as the treatment (CF, 150 kg ha^{-1} urea, 75 kg ha^{-1} P_2O_5, and 120 kg ha^{-1} K_2O) [34]. The area of each test plot was 20 m^2 (4 m × 5 m), with three replicates; they were randomly arranged, and the transplanting spacing of the rice was 0.20 m × 0.20 m. Around each plot, polyethylene films were draped over ridges (0.3 m wide and 0.15 m above the ground) to prevent the transfer of water and nutrients between plots. Early rice (Yuanzao No. 1 (Hunan Academy of Agricultural Sciences, China) was planted from 2008 to 2010, and Xiangzaoxian No. 25 (Hunan Academy of Agricultural Sciences, China) was planted after 2010) was planted in middle or late April and harvested in early July. Since 2008, Huanghuazhan (Hunan Academy of Agricultural Sciences, China) has been planted as late rice in mid- or late July and harvested in early November. We ensured that other identical field management practices (irrigation, weeding, etc., other than fertilization) were consistent between the CF treatment and the control.

2.2. Soil Sampling

In mid-July 2019, bulk and rhizospheric soil samples were collected from the 6 plots. Debris (grass roots, litter, stones, etc.) and surface soil (0–1 cm) were removed before soil sampling. Five rice plants with the same growth potential were selected from each plot. The unvegetated soil cores (5 cm in diameter) adjacent to the rice plants (i.e., bulk soil) were sampled at a depth of 0–20 cm. The loosely adhered soil was shaken off by violently shaking the rice roots, and the soil attached to the roots was called rhizospheric soil. The roots were forcibly shaken with sterile forceps to separate the roots from the soil, and then five soil samples from each plot were mixed together to generate a composite sample. Finally, all of the soil separated from the roots was sieved by using a 2 mm mesh sieve and put into a sterile zip-lock bag [35,36].

2.3. Analysis of Soil Chemical Properties and Soil Bacterial Diversities

A pH meter (FE20K, Mettler Toledo, Switzerland) was used to measure the soil pH; the soil organic matter (SOM) and total nitrogen (TN) contents were determined with the wet digestion (120 °C, 2 h) and Kjeldahl methods, respectively [37]; the soil available nitrogen (AN), phosphorus (AP), and potassium (AK) contents were determined by using microdiffusion after the alkaline hydrolysis method, the Olsen method, and flame photometry, respectively [37].

The extraction, PCR amplification, and sequencing of the total genomic DNA of the soil samples were completed according to previous protocols [36]. The primer pair 515F (5′-GTGCCAGCMGCCGCGGTAA-3′) and 806R (5′-GGACTACHVGGGTWTCTAAT-3′) was used to amplify the soil bacterial 16S rRNA gene in the V4 hypervariable region [38,39]. A library was constructed by using the TruSeq® DNA PCR-Free Sample Preparation Kit (Illumina, San Diego, CA, USA), and Qubit and Q-PCR were used to quantify the constructed library. Finally, NovaSeq6000 was used for sequencing, and 250 bp paired-end reads were generated. The raw reads of the soil bacteria were deposited in the NCBI Sequence Read Archive (SRA) database (Accession Number: PRJNA955403).

The UPARSE software (Version 7.0.1001, http://drive5.com/uparse/, accessed on 2 April 2023) was used to perform the sequence analysis. Sequences with ≥97% similarity were assigned to the same operational taxonomic units (OTUs). A representative sequence for each OTU was screened for further annotation. The SILVA database (Release 132, http://www.arb-silva.de/, accessed on 2 April 2023) was used to annotate the taxonomic information. All alpha diversity indices were calculated with QIIME (Version 1.7.0) and displayed with the R software (Version 3.3.1). Whether the sequencing findings accurately reflected the microorganisms in the sample was shown by the index of coverage [39]. The higher the diversity index (Shannon index, http://www.mothur.org/wiki/Shannon, accessed on 2 April 2023) and richness indices (Chao1 index, http://www.mothur.org/wiki/Chao, accessed on 2 April 2023, and Ace index, http://www.mothur.org/wiki/

Ace, accessed on 2 April 2023) of a community, the higher the diversity and richness of that community [40]. Graphics of the bacterial community compositions and Venn diagrams were generated with the R (Version 3.3.1) software. The "vegan" package was used, and graphs for a heatmap analysis of the bacterial communities were created by using the R language (Version 3.3.1) tool. The "WGCNA" package, "stat" packages, and "ggplot2" package were used for weighted UniFrace Principal Coordinate Analysis (PCoA) in the R software (Version 3.3.1). Unweighted Pair–Group Method with Arithmetic Means (UPGMA) Clustering was performed as a type of hierarchical clustering method to interpret the distance matrix by using the average linkage and was conducted with the QIIME software (Version 1.7.0). The MUSCLE software (Version 3.8.31) was used to conduct multiple sequence alignment and construct phylogenetic relationships. LEfSe (http://huttenhower.sph.harvard.edu/galaxy/root?tool_id=lefse_upload, accessed on 2 April 2023) was used to perform linear discriminant analysis (LDA) on samples according to different grouping conditions based on the taxonomic composition to identify clusters that had a significant differential ($p < 0.05$, LDA $\geq$ 3.0) impact on sample delineation. FAPROTAX [38] was employed to predict the bacterial functions involved in processes of the biochemical cycle as an artificially constructed functional classification database. The "pheatmap" package was run with the R (Version 3.3.1) software to conduct a heatmap analysis of the Spearman correlation between the soil chemical properties and soil bacteria. The data processing and microbial diversity analysis were the same as those in previous studies [39,40].

2.4. Statistical Analyses

The results are shown as the mean ± standard deviation (SD) values calculated in Microsoft Excel 2019. One-way analysis of variance (ANOVA), a two-tailed Duncan test, and multiple comparisons ($p < 0.05$) were used to evaluate significant differences. IBM SPSS Statistics (Version 26.0, IMB SPSS Inc., Chicago, IL, USA) and Origin 2021 (OriginLab Corporation, Northampton, MA, USA) were used to analyze and plot the experimental data, respectively.

3. Results

3.1. Soil Chemical Properties

In comparison with the CK treatment, the soil pH in the CF treatment was similar in both the rhizospheric and bulk soil, and there were no significant differences ($p > 0.05$). In contrast, the opposite trends were shown in the AN and AK contents. Furthermore, the SOM, TN, and AP concentrations in both the rhizospheric and bulk soil were significantly increased by the CF treatment ($p < 0.05$) (Table 1).

Table 1. Soil chemical properties in rice fields after long-term chemical fertilization.

Treatment	pH	SOM (g kg^{-1})	TN (g kg^{-1})	AN (mg kg^{-1})	AP (mg kg^{-1})	AK (mg kg^{-1})
RCK	7.89 ± 0.06 a	29.67 ± 0.60 bc	2.73 ± 0.15 bc	167.23 ± 7.68 ab	13.90 ± 0.41 b	78.49 ± 3.33 ab
RCF	7.77 ± 0.12 a	32.50 ± 0.61 a	3.13 ± 0.19 a	176.97 ± 9.64 a	18.97 ± 0.82 a	82.73 ± 11.64 a
BCK	7.75 ± 0.09 a	28.37 ± 0.85 c	2.65 ± 0.24 c	144.87 ± 4.47 b	13.65 ± 0.19 b	67.97 ± 1.12 b
BCF	7.73 ± 0.08 a	31.23 ± 2.57 ab	3.03 ± 0.05 ab	155.63 ± 20.72 ab	18.91 ± 0.76 a	71.54 ± 1.06 ab

All data values indicate the mean ± SD (n = 3). Different lowercase letters in the same column indicate significant differences ($p < 0.05$). Prefix R, rhizospheric soil; Prefix B, bulk soil. The same applies below in all tables and figures.

3.2. Soil Bacterial Alpha and Beta Diversities

The soil bacterial OTU numbers rose as a function of the number of samples according to a boxplot of species accumulation (Figure 1a). With the saturation of the OTU numbers, the curve became asymptotically stable, and fewer new OTUs were added to each soil sample, showing sufficient sequencing depth to adequately characterize the rice soil bacterial compositions (Figure 1b) with a good coverage of >99.4%, and the data were reliable (Figure 1c). Moreover, the results showed that the rhizospheric and bulk soil bacterial

Shannon (describing bacterial diversity) and Ace (describing bacterial richness) indices in the paddy field were not significantly changed by the CF treatment compared with the CK treatment (Figure 1d,e). However, we found, quite interestingly, that the bacterial Shannon and Ace indices in the rice rhizospheric soil under the CF treatment were all higher than those under the CK treatment, and the bulk soil bacteria showed the opposite trend (Figure 1d,e). The results of the PCoA showed that the soil bacterial compositions in each group were clustered separately, and they were quite similar (Figure 1f). All of the above results suggest that the bacterial diversity and richness in the rice rhizospheric soil were increased, and the bacterial diversity and richness in the rice bulk soil were decreased by the long-term application of CF.

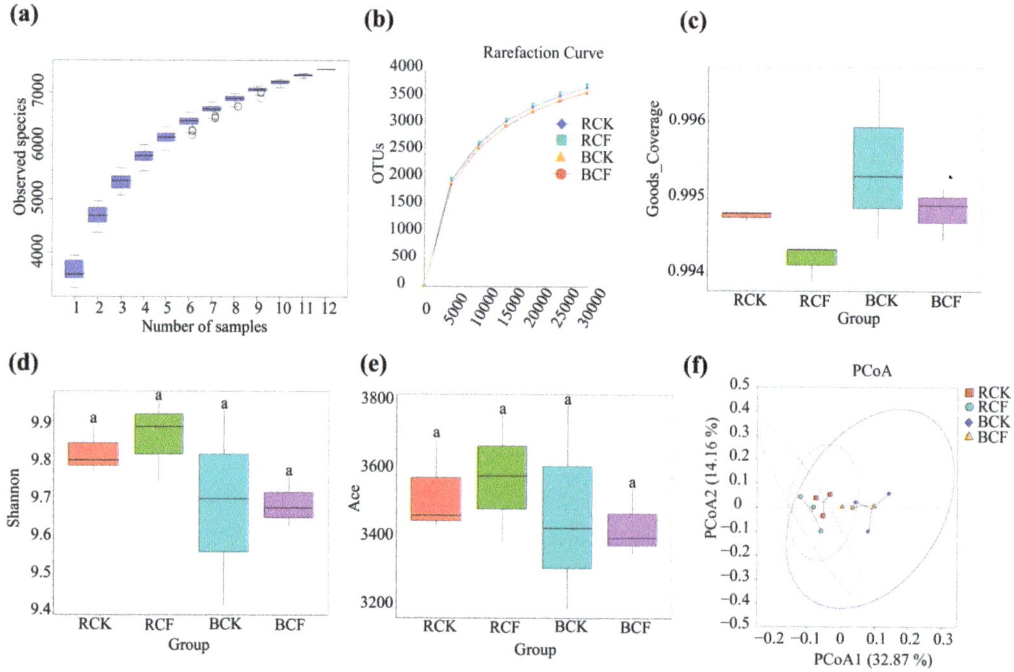

Figure 1. Soil bacterial alpha and beta diversities at the OTU level after long-term chemical fertilization. (**a**) Species accumulation boxplot for soil bacteria. (**b**) The rarefaction curve for soil bacteria. (**c**) Good coverage of soil bacteria. (**d**) Shannon index for soil bacteria. (**e**) Ace index for soil bacteria. (**f**) Soil bacterial PCoA. Different color blocks indicate different groups. Different lowercase letters indicate significant differences ($p < 0.05$).

3.3. UPGMA Cluster Analysis and Species Phylogenetic Tree

The results of UPGMA showed that the cluster of the RCF and RCK treatments was a branch and that of the BCK and BCF treatments was another branch, indicating that there were differences in the bacterial community compositions, relative abundance, and phylogenetic relationships between the rhizospheric and bulk soil of rice under long-term lack of fertilization (CK) and long-term fertilization (CF) (Figure 2a). In both rice rhizospheric soil and bulk soil, there was little difference in the soil bacterial community compositions, relative abundance, and phylogenetic relationships between the CK and CF treatments (Figure 2a).

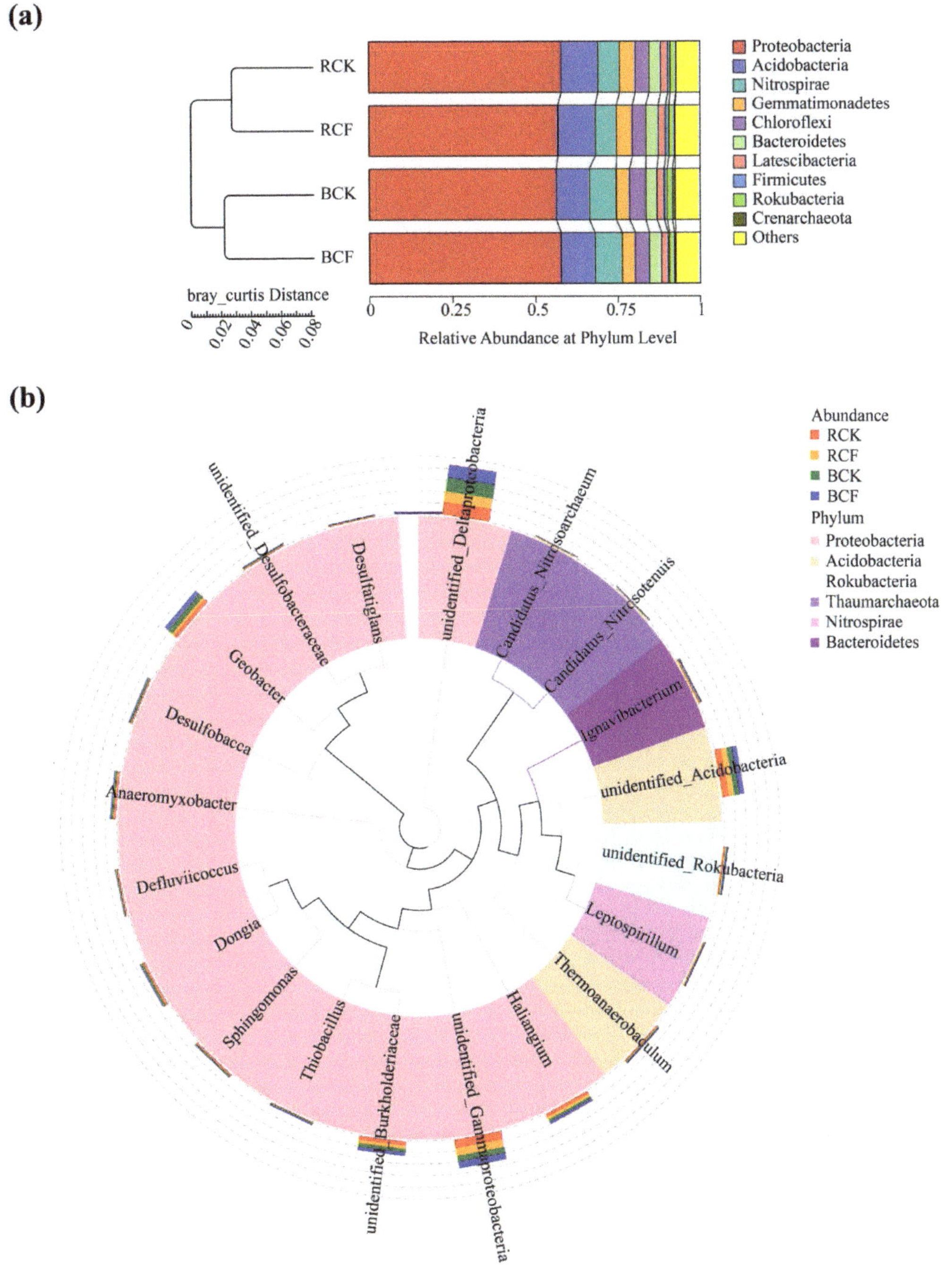

Figure 2. Bray–Curtis UPGMA clustering tree (**a**) and species evolutionary tree (**b**). The colors of the branches and fan-shaped sections represent their corresponding phyla, and the stacked columns on the outside of the fan ring represent the abundance distribution information of different species in the genus.

The representative sequences of the top 20 most abundant soil bacterial genera were obtained by using a multisequence comparison to further study the species' phylogenetic relationships at the genus level (Figure 2b). The top 20 bacterial genera in the soil in terms of relative abundance mainly belonged to Proteobacteria, Acidobacteria, Rokubacteria, Thaumarchaeota, Nitrospirae, and Bacteroidetes. The soil bacteria in the rhizospheric soil and bulk soil samples were mainly distributed in the following six genera: *unidentified_Deltaproteobacteria, unidentified_Gammaproteobacteria, unidentified_Acidobacteria, unidentified_Burkholderiaceae, Geobacter*, and *Haliangium*. *Unidentified_Deltaproteobacteria* and *unidentified_Gammaproteobacteria*, which were relatively abundant, both belonged to the branch of Proteobacteria (Figure 2b).

3.4. Soil Bacterial Composition

In the rhizospheric soil of rice, the common dominant (i.e., relative abundance percentages > 1.00%) bacterial phyla were Proteobacteria, Acidobacteria, Nitrospirae, Gemmatimonadetes, Chloroflexi, Bacteroidetes, Latescibacteria, and Rokubacteria in both the CK (averages of 57.88%, 11.22%, 6.59%, 4.84%, 4.24%, 3.45%, 2.16%, and 1.67%, respectively) and the CF (averages of 56.86%, 11.43%, 6.41%, 4.90%, 4.09%, 3.81%, 2.18%, and 1.62%, respectively) treatments (Figure 3a). In addition, Firmicutes was a unique dominant bacterial phylum in the CF treatment. In the bulk soil of rice, the dominant bacterial phyla were Proteobacteria, Acidobacteria, Nitrospirae, Chloroflexi, Gemmatimonadetes, Bacteroidetes, Latescibacteria, and Rokubacteria in both the CK (averages of 56.35%, 10.04%, 8.15%, 4.96%, 4.14%, 3.41%, 2.17%, and 1.60%, respectively) and the CF (averages of 57.78%, 10.32%, 8.27%, 4.22%, 4.06%, 3.74%, 1.98%, and 1.46%, respectively) treatments (Figure 3a).

At the genus level, Haliangium, Defluviicoccus, unidentified_Acidobacteria, Sphingomonas, Dongia, unidentified_Gammaproteobacteria, and unidentified_Burkholderiaceae were enriched in the rhizospheric soil of rice. Geobacter, Desulfobacca, Sulfurifustis, unidentified_Deltaproteobacteria, and Leptospirillum were enriched in the bulk soil of rice. Candidatus_Nitrosotenuis and Candidatus_Nitrosoarchaeum were enriched in the rhizospheric soil of rice under the CF treatment. Thiobacillus, Ignavibacterium, and Thermoanaerobaculum were enriched in the bulk soil of rice under the CF treatment (Figure 3b).

The numbers of soil bacteria at the OTU level in the RCK, RCF, BCK, and BCF treatments were 3719, 3775, 3781, and 3739 OTUs, respectively. Furthermore, the numbers of unique soil bacteria in the RCK, RCF, BCK, and BCF treatments were 338, 394, 400, and 358 OTUs, respectively (Figure 3c).

3.5. Soil Bacterial Communities with Statistically Significant Differences

The effect size (LEfSe) in linear discriminant analysis (LDA) was calculated to identify high-dimensional biomarkers (from phylum to species) with significantly different soil bacterial abundances in the rhizospheric and bulk soils of rice between the CK and CF treatments (Figure 4). Cladograms were used to show different groups (Figure 4a), and LDA scores of three or greater were confirmed with the LEfSe tool (Figure 4b). In total, 18 abundant soil bacterial clades exhibited significant differences ($p < 0.05$, LDA ≥ 3.0) (Figure 4b). The LEfSe results showed that *unidentified_Gammaproteobacteria* (genus) was significantly enriched in the rhizospheric soil of rice under the CK treatment, *Geobacter* (genus) was significantly enriched in the bulk soil of rice under the CK treatment, Gemmatimonadetes (phylum) was significantly enriched in the rhizospheric soil of rice under the CF treatment, and Nitrospirae (phylum) and *Thiobacillus* (genus) were significantly enriched in the bulk soil of rice under the CF treatment.

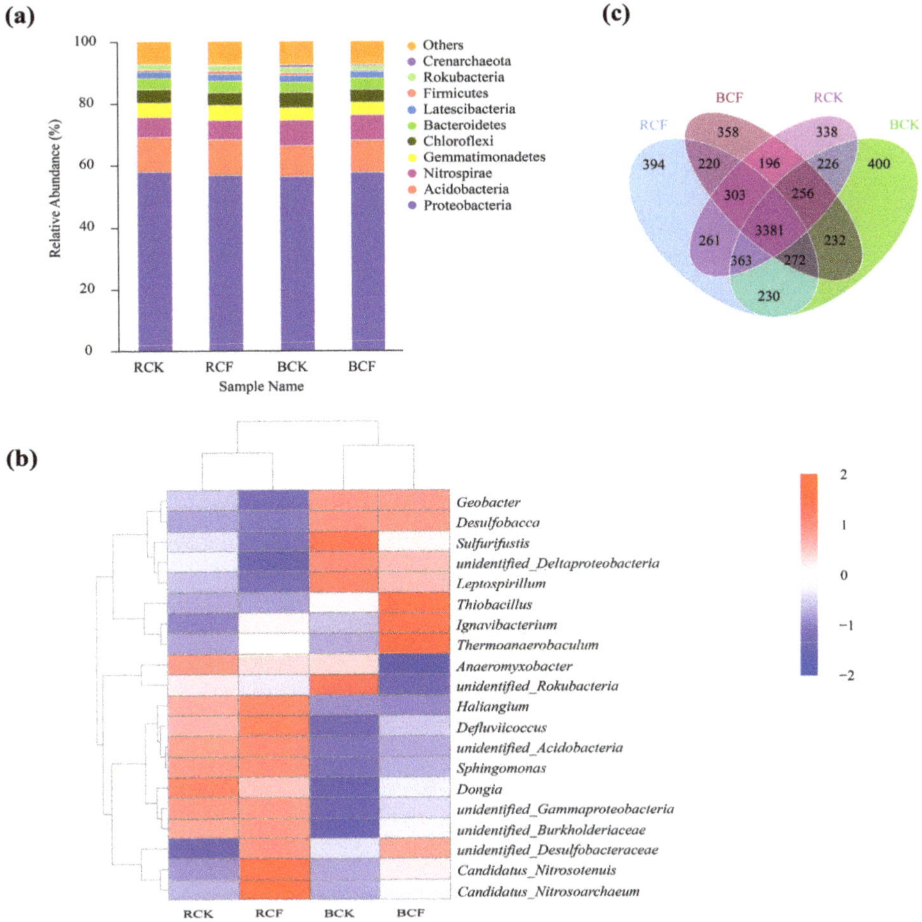

Figure 3. Soil bacterial composition of rice after long-term chemical fertilization. (**a**) Relative abundance of the top ten soil bacterial phyla. (**b**) Heatmap of the top 20 most abundant soil bacterial genera based on relative abundances. (**c**) Venn analyses of soil bacteria at the OTU level.

3.6. Functional Predictions of the Soil Bacterial Communities with FAPROTAX

The proportions of soil bacterial abundances related to human_pathogens_pneumonia, aerobic_chemoheterotrophy, animal_parasites_or_symbionts, human_pathogens_all, chemoheterotrophy, and predatory_or_exoparasitic functions were higher in the rice rhizospheric soil than in the rice bulk soil. In addition, the guilds related to iron respiration and nitrification of soil bacteria were enriched in the bulk soil compared to the rhizospheric soil. Furthermore, in comparison with the CK treatment, the guilds related to respiration_of_sulfur_compounds and sulfate_respiration in soil bacteria were enriched in the rhizospheric and bulk soil of rice under the CF treatment (Figure 5).

3.7. Soil Chemical Factors Structuring Soil Bacterial Communities after Long-Term Fertilization

Spearman correlation heatmaps were used to show the relationships between the top 10 soil bacterial phyla (Figure 6a) and genera (Figure 6b) and the soil chemical factors. First, Acidobacteria and Gemmatimonadetes showed a significant positive correlation with the AN (r = 0.79 and r = 0.66) and AK (r = 0.62 and r = 0.74) contents, respectively. Second, Nitrospirae and Crenarchaeota showed significant negative correlations with the

AN (r = −0.68 and r = −0.71) and AK (r = −0.72 and r = −0.66) contents, respectively. Furthermore, Chloroflexi showed a significant negative correlation with the SOM (r = −0.67), TN (r = −0.67), and AN (r = −0.61) contents. Finally, Latescibacteria showed a significant positive correlation with the AN (r = 0.62) content (Figure 6a). Moreover, only the AN and AK contents were correlated with the top 10 bacterial genera. The AK content was significantly positively correlated with *Haliangium* (r = 0.75), *unidentified_Acidobacteria* (r = 0.60), and *unidentified_Gammaproteobacteria* (r = 0.78) and significantly negatively correlated with *Desulfobacca* (r = −0.80) and *Thiobacillus* (r = −0.81). In addition to the same results as those for the AK content in the Spearman correlation heatmap, AP content was positively correlated with *unidentified_Burkholderiaceae* (r = 0.62) (Figure 6b).

(a)

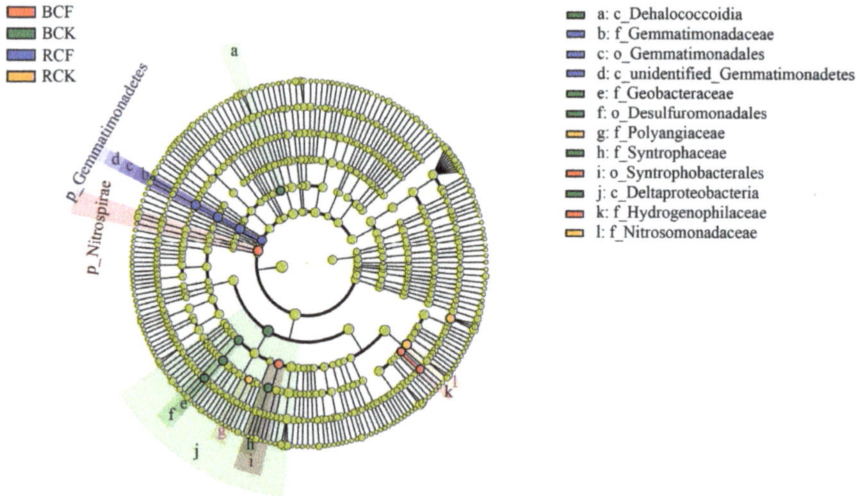

(b)

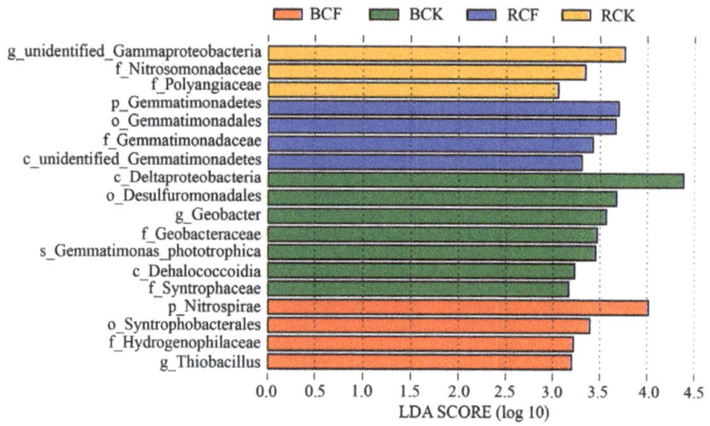

Figure 4. Soil bacterial cladograms (**a**) and biomarkers with LDA scores ≥ 3.0 (**b**) after long-term chemical fertilization. Different levels are indicated by different prefixes (p, phylum; c, class; o, order; f, family; g, genus; s, species).

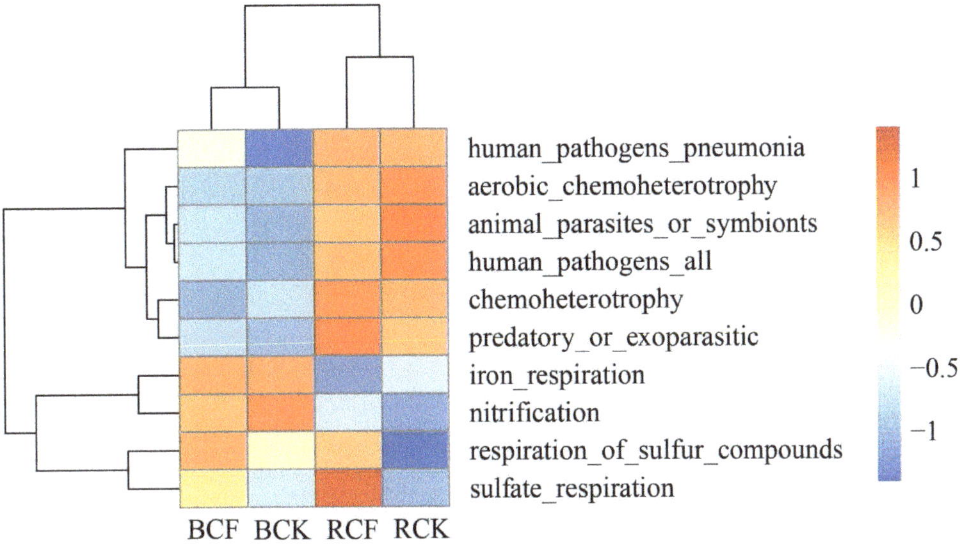

Figure 5. Heatmap of functional predictions of the soil bacterial communities. The X and Y axes are the treatment groups and soil bacterial functional predictions, respectively, and each lattice represents a Spearman correlation coefficient between a treatment group and a bacterial functional prediction. Red represents a positive correlation, whereas blue represents a negative correlation.

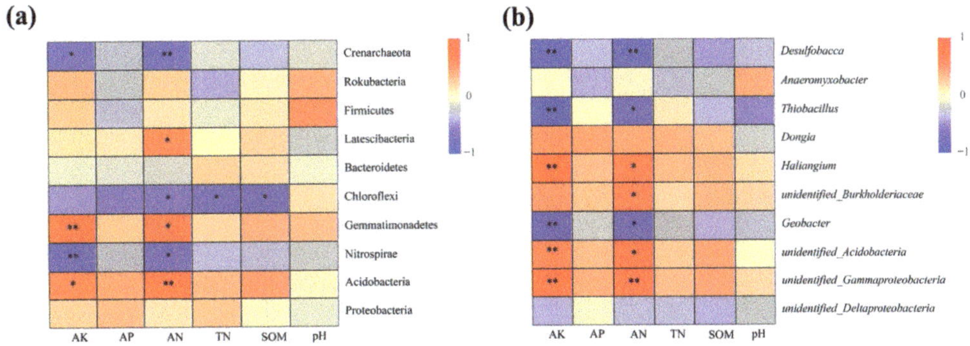

Figure 6. Correlation heatmap of soil physicochemical properties and the top 10 soil bacterial phyla (**a**) and genera (**b**). The X and Y axes are the soil physicochemical factors (pH, soil pH value; SOM, soil organic matter; TN, soil total nitrogen; AN, soil alkaline hydrolysable nitrogen; AP, soil available phosphorus; AK, soil available potassium) and soil bacterial phyla or genera, respectively, and each lattice represents a Spearman correlation coefficient between a soil physicochemical factor and a bacterial phylum or genus. Red represents a positive correlation, whereas blue represents a negative correlation (* $0.01 < p \leq 0.05$, ** $0.001 < p \leq 0.01$).

4. Discussion

4.1. Effects on Soil Chemical Properties after Long-Term Chemical Fertilization

In the present study, the results showed that some soil chemical properties were significantly changed by the long-term application of chemical fertilizers, which was similar to the findings of previous studies [10,41–43]. Zhong et al. [44] and Dan et al. [43] reported that the SOC and TN contents (as organic substrates) were significantly increased by the long-term application of synthetic chemical fertilizers, which was consistent with our results in both the rhizospheric and bulk soils (Table 1). Pinggera et al. [45] found that regulating the C/N ratio by changing the N input was beneficial for the decomposition of organic matter by microorganisms, the release of organic nutrients, and the increase in SOM content, which may explain the increased SOM in the rhizospheric soil and bulk soil with the CF treatment in this study. Jin et al. [6] also verified this explanation by changing the N input content. On the other hand, we found that the SOM content in the rhizospheric soil was higher than that in the bulk soil, which was similar to the findings of a previous study [46]. This result was interpreted to mean that root exudates promoted soil organic carbon decomposition [46,47]. Previous studies found that the soil pH was significantly reduced after long-term fertilization in calcareous soil [48], Hapli-Udic Cambisol [49], and acidic soil [50]. In our study, long-term fertilization also reduced soil pH (in both the rhizospheric and bulk soils) in the typical sandy purple clay soil, and there were no significant differences, which may have been caused by differences in soil types and N inputs. Wu et al. [10] also reported that the soil pH was not significantly changed by the application of chemical fertilizers.

4.2. *Effects on Soil Bacterial Diversity after Long-Term Chemical Fertilization*

Previous studies reported that conventional fertilization reduced the soil bacterial richness in the plow layer (0–20 cm depth) of wheat [51] and maize [44]. Tang et al. [52] also discovered that CF alone reduced the soil microbial richness and Shannon indices in the plow layer of rice. In this study, the bacterial Shannon (used to measure diversity) and Ace (used to measure richness) indices in rice bulk soil with the CF treatment were all lower—but not significantly—than those with the CK treatment (Figure 1d,e); thus, these results were consistent with those of the aforementioned studies. On the other hand, we also found that the bacterial diversity and richness in the rhizospheric soil were higher than those in the bulk soil. Tang et al. [46] also reported the same results, explaining that root exudates enriched microorganisms and accelerated microbial activities and that extracellular enzyme activities were higher in the rhizosphere [53]; the soil quality and soil fertility indices were different [54].

4.3. *Effects on Soil Bacterial Communities after Long-Term Chemical Fertilization*

The ecological consistency of microbial communities can be revealed by elucidating the soil microbial taxa at the phylum level [6,55]. In our study, Proteobacteria was the most dominant phylum, and the relative abundances of Acidobacteria and Bacteroidetes increased in both the rhizospheric and bulk soils after long-term (12-year) application of chemical fertilizers according to high-throughput sequencing, which was consistent with the findings of a previous study by Wang et al. [42]. The possible reason is that Acidobacteria are acidophiles, and the long-term fertilization reduced the pH of the rhizospheric soil and bulk soil, which was not conducive to growth. Tang et al. [46] found that Acidobacteria were the main C source and thrived in both rhizospheric and bulk soils. Hungria et al. [56] also suggested that the enrichment of Acidobacteria in the plant rhizosphere may be due to their ability to utilize root exudates. In this study, we also found that Acidobacteria were the dominant bacteria in the rice rhizospheric and bulk soils. However, the abundance of Acidobacteria in the rhizospheric soil was higher than that in the bulk soil of rice, regardless of whether the CF treatment or the CK treatment was used. Wu et al. [57], Bulgarelli et al. [58], and Tang et al. [46] all suggested that the enhanced Acidobacteria in roots could be generally adapted to the plant rhizosphere as saprophytic bacteria.

Bacteroidetes are copiotrophes that rapidly proliferate in eutrophic environments [59] and mineralized organic matter [60], and they are mainly distributed in the rhizosphere [61]. In this study, the proportions of the abundance of Bacteroidetes between different treatments were in the following order (from large to small): RCF > RCK > BCF > BCK, which was consistent with the findings of previous studies [61].

Long-term fertilization results in the enrichment of specific microorganisms that can use these nutrients efficiently [42]. According to the results of the LEfSe analyses, 18 abundant soil bacterial taxa in the two treatments were distinctly different. *Geobacter* is an obligate anaerobic N-fixing bacterium that is commonly observed in flooded conditions or anoxic environments in deep soils and plays an important role in nitrate reduction and iron restoration [62–64]. This explained why *Geobacter* responded to the nitrogen requirements of rice growth by reducing nitrite in the nutrient-poor environment. Nitrospirae can oxygenate nitrite into nitrate under aerobic conditions and play an important role in the nitrogen cycle [65]. Nitrate production leads to pH reduction in the soil. *Thiobacillus* is a beneficial soil bacterium that plays an important role in the soil sulfur cycle. Sulfuric acid is produced after the oxidation of sulfur by *Thiobacillus*, and the soil pH is reduced [66]. The significant enrichment of Nitrospirae and *Thiobacillus* led to nitrate and sulfate production in the rice bulk soil with the CF treatment, which might have caused the pH of the bulk soil to decrease under the CF treatment [66].

5. Conclusions

Overall, in this study, we demonstrated that long-term CF had different effects on rhizospheric and bulk soil fertility, soil bacterial diversity, and the structure and function of soil bacterial communities. After long-term fertilization, rhizospheric soil fertility indices (such as pH, SOM, TN, AN, AP, AK) and the soil bacterial diversity and richness were higher than those in the bulk soil. Based on our study, long-term fertilization can improve the fertility of rhizospheric and bulk soils, increase the rhizospheric soil bacterial diversity and richness, and decrease the bulk soil bacterial diversity and richness. Only AK and AN had significant correlations with soil bacteria. These results provide useful insights for studying nutrient accumulation, turnover, and microbial migration in rhizospheric and bulk soils under long-term fertilization strategies.

Author Contributions: Conceptualization, Y.L. (Yanhong Lu), Y.L. (Yulin Liao) and J.N.; methodology, J.X., X.X. and C.F.; software, J.Z.; validation, J.Z., Y.G. and J.X.; formal analysis, X.X. and C.F.; investigation, J.X.; resources, Y.L. (Yulin Liao); data curation, Y.L. (Yanhong Lu); writing—original draft preparation, J.X.; writing—review and editing, J.X.; J.Z., Y.L. (Yulin Liao) and J.N.; visualization, X.X.; supervision, Y.L. (Yanhong Lu); project administration, Y.L. (Yulin Liao); funding acquisition, Y.L. (Yulin Liao) and J.N. All authors have read and agreed to the published version of the manuscript.

Funding: This work was financially supported by the National Key Research and Development Program of China (2021YFD1700200), the National Natural Science Foundation of China (32202607), the earmarked fund for CARS—Green manure (CARS-22), and the Joint Funds of the National Natural Science Foundation of China Natural (U19A2046).

Data Availability Statement: Raw reads obtained in the current study have been deposited in the NCBI Sequence Read Archive (SRA) database (Accession Number: PRJNA955403).

Conflicts of Interest: The authors declare no conflict of interest.

References

1. Liu, K.; Chen, Y.; Li, S.; Wang, W.; Zhang, W.; Zhang, H.; Gu, J.; Yang, J.; Liu, L. Differing responses of root morphology and physiology to nitrogen application rates and their relationships with grain yield in rice. *Crop J.* **2023**, *11*, 618–627. [CrossRef]
2. Yang, M.; Xu, X.; Li, Z.; Meng, Y.; Yang, X.; Song, X.; Yang, G.; Xu, S.; Zhu, Q.; Xue, H. Remote sensing prescription for rice nitrogen fertilizer recommendation based on improved NFOA model. *Agronomy* **2022**, *12*, 1804. [CrossRef]
3. Yuan, S.; Linquist, B.A.; Wilson, L.T.; Cassman, K.G.; Stuart, A.M.; Pede, V.; Miro, B.; Saito, K.; Agustiani, N.; Aristya, V.E.; et al. Sustainable intensification for a larger global rice bowl. *Nat. Commun.* **2021**, *12*, 7163. [CrossRef] [PubMed]
4. Zhao, W.; Chou, J.; Li, J.; Xu, Y.; Li, Y.; Hao, Y. Impacts of extreme climate events on future rice yields in global major rice-producing regions. *Int. J. Environ. Res. Public Health* **2022**, *19*, 4437. [CrossRef] [PubMed]
5. Chen, Q.; Tian, Y.; Yao, X.; Cao, W.; Zhu, Y. Comparison of five nitrogen dressing methods to optimize rice growth. *Plant Prod. Sci.* **2014**, *17*, 66–80. [CrossRef]
6. Jin, N.; Jin, L.; Wang, S.; Li, J.; Liu, F.; Liu, Z.; Luo, S.; Wu, Y.; Lyu, J.; Yu, J. Reduced chemical fertilizer combined with bio-organic fertilizer affects the soil microbial community and yield and quality of lettuce. *Front. Microbiol.* **2022**, *13*, 863325. [CrossRef]
7. Zhu, J.; Peng, H.; Ji, X.; Li, C.; Li, S. Effects of reduced inorganic fertilization and rice straw recovery on soil enzyme activities and bacterial community in double-rice paddy soils. *Eur. J. Soil Biol.* **2019**, *94*, 103116. [CrossRef]
8. Sun, R.; Wang, F.; Hu, C.; Liu, B. Metagenomics reveals taxon-specific responses of the nitrogen-cycling microbial community to long-term nitrogen fertilization. *Soil Biol. Biochem.* **2021**, *156*, 108214. [CrossRef]

9. Wang, J.; Li, R.; Zhang, H.; Wei, G.; Li, Z. Beneficial bacteria activate nutrients and promote wheat growth under conditions of reduced fertilizer application. *BMC Microbiol.* **2020**, *20*, 38. [CrossRef]
10. Wu, L.; Jiang, Y.; Zhao, F.; He, X.; Liu, H.; Yu, K. Increased organic fertilizer application and reduced chemical fertilizer application affect the soil properties and bacterial communities of grape rhizosphere soil. *Sci. Rep.* **2020**, *10*, 9568. [CrossRef]
11. Liao, H.; Zhang, Y.; Zuo, Q.; Du, B.; Chen, W.; Wei, D.; Huang, Q. Contrasting responses of bacterial and fungal communities to aggregate-size fractions and long-term fertilizations in soils of northeastern China. *Sci. Total. Environ.* **2018**, *635*, 784–792. [CrossRef] [PubMed]
12. Bell, C.W.; Asao, S.; Calderon, F.; Wolk, B.; Wallenstein, M.D. Plant nitrogen uptake drives rhizosphere bacterial community assembly during plant growth. *Soil Biol. Biochem.* **2015**, *85*, 170–182. [CrossRef]
13. Singh, J.S.; Gupta, V.K. Soil microbial biomass: A key soil driver in management of ecosystem functioning. *Sci. Total. Environ.* **2018**, *634*, 497–500. [CrossRef]
14. Sun, M.; Xiao, T.; Ning, Z.; Xiao, E.; Sun, W. Microbial community analysis in rice paddy soils irrigated by acid mine drainage contaminated water. *Appl. Microbiol. Biotechnol.* **2015**, *99*, 2911–2922. [CrossRef] [PubMed]
15. Geisseler, D.; Scow, K.M. Long-term effects of mineral fertilizers on soil microorganisms—A review. *Soil. Biol. Biochem.* **2014**, *75*, 54–63. [CrossRef]
16. Fan, K.; Weisenhorn, P.; Gilbert, J.A.; Shi, Y.; Bai, Y.; Chu, H. Soil pH correlates with the co-occurrence and assemblage process of diazotrophic communities in rhizosphere and bulk soils of wheat fields. *Soil Biol. Biochem.* **2018**, *121*, 185–192. [CrossRef]
17. Ren, C.; Zhou, Z.; Guo, Y.; Yang, G.; Zhao, F.; Wei, G.; Han, X.; Feng, L.; Feng, Y.; Ren, G. Contrasting patterns of microbial community and enzyme activity between rhizosphere and bulk soil along an elevation gradient. *Catena* **2021**, *196*, 104921. [CrossRef]
18. Kang, H.; Xue, Y.; Yan, C.; Lu, S.; Yang, H.; Zhu, J.; Fu, Z.; Wang, D. Contrasting patterns of microbial nutrient limitations between rhizosphere and bulk soil during stump sprout restoration in a clear-cut oak forest. *Forest Ecol. Manag.* **2022**, *515*, 120241. [CrossRef]
19. Edwards, J.; Johnson, C.; Santos-Medellin, C.; Lurie, E.; Podishetty, N.K.; Bhatnagar, S.; Eisen, J.A.; Sundaresan, V. Structure, variation, and assembly of the root-associated microbiomes of rice. *Proc. Natl. Acad. Sci. USA* **2015**, *112*, E911–E920. [CrossRef]
20. Fan, K.; Cardona, C.; Li, Y.; Shi, Y.; Xiang, X.; Shen, C.; Wang, H.; Gilbert, J.A.; Chu, H. Rhizosphere-associated bacterial network structure and spatial distribution differ significantly from bulk soil in wheat crop fields. *Soil Biol. Biochem.* **2017**, *113*, 275–284. [CrossRef]
21. Zhang, K.; Adams, J.M.; Shi, Y.; Yang, T.; Sun, R.; He, D.; Ni, Y.; Chu, H. Environment and geographic distance differ in relative importance for determining fungal community of rhizosphere and bulk soil. *Environ. Microbiol.* **2017**, *19*, 3649–3659. [CrossRef] [PubMed]
22. Kuzyakov, Y.; Razavi, B.S. Rhizosphere size and shape: Temporal dynamics and spatial stationarity. *Soil Biol. Biochem.* **2019**, *135*, 343–360. [CrossRef]
23. Lin, L.; Li, Z.; Hu, C.; Zhang, X.; Chang, S.; Yang, L.; An, Q. Plant growth-promoting nitrogen-fixing Enterobacteria are in association with sugarcane plants growing in Guangxi, China. *Microbes Environ.* **2012**, *27*, 391–398. [CrossRef] [PubMed]
24. Zaheer, A.; Malik, A.; Sher, A.; Qaisrani, M.M.; Mehmood, A.; Khan, S.U.; Ashraf, M.U.; Mirza, Z.; Karim, S.; Rasool, M. Isolation, characterization, and effect of phosphate-zinc-solubilizing bacterial strains on chickpea (*Cicer arietinum* L.) growth. *Saudi J. Biol. Sci.* **2019**, *26*, 1061–1067. [CrossRef]
25. Matsuda, R.; Handayani, M.L.; Sasaki, H.; Takechi, K.; Takano, H.; Takio, S. Production of indoleacetic acid by strains of the epiphytic bacteria *Neptunomonas* spp. isolated from the red alga *Pyropia yezoensis* and the seagrass *Zostera marina*. *Arch Microbiol.* **2018**, *200*, 255–265. [CrossRef]
26. Cappellari, L.D.R.; Santoro, M.V.; Schmidt, A.; Gershenzon, J.; Banchio, E. Induction of essential oil production in *Mentha* x *piperita* by plant growth promoting bacteria was correlated with an increase in jasmonate and salicylate levels and a higher density of glandular trichomes. *Plant Physiol. Bioch.* **2019**, *141*, 142–153. [CrossRef]
27. Sulochana, M.B.; Jayachandra, S.Y.; Kumar, S.A.; Parameshwar, A.B.; Reddy, K.M.; Dayanand, A. Siderophore as a potential plant growth-promoting agent produced by *Pseudomonas aeruginosa* JAS-25. *Appl. Biochem. Biotech.* **2014**, *174*, 297–308. [CrossRef]
28. Banerjee, S.; van der Heijden, M.G.A. Soil microbiomes and one health. *Nat. Rev. Microbiol.* **2022**, *21*, 6–20. [CrossRef]
29. Sun, G.; Yao, T.; Feng, C.; Chen, L.; Li, J.; Wang, L. Identification and biocontrol potential of antagonistic bacteria strains against *Sclerotinia sclerotiorum* and their growth-promoting effects on *Brassica napus*. *Biol. Control.* **2017**, *104*, 35–43. [CrossRef]
30. Etesami, H.; Emami, S.; Alikhani, H.A. Potassium solubilizing bacteria (KSB): Mechanisms, promotion of plant growth, and future prospects a review. *J. Soil Sci. Plant Nut.* **2017**, *17*, 897–911. [CrossRef]
31. Ran, J.; Liu, X.; Hui, X.; Ma, Q.; Liu, J. Differentiating bacterial community responses to long-term phosphorus fertilization in wheat bulk and rhizosphere soils on the Loess Plateau. *Appl. Soil Ecol.* **2021**, *166*, 104090. [CrossRef]
32. Wu, Q.; Chen, D.; Zhou, W.; Zhang, X.; Ao, J. Long-term fertilization has different impacts on bacterial communities and phosphorus forms in sugarcane rhizosphere and bulk soils under low-P stress. *Front. Plant Sci.* **2022**, *13*, 1019042. [CrossRef]
33. Zhou, X.; Lu, Y.H.; Liao, Y.L.; Zhu, Q.D.; Cheng, H.D.; Nie, X.; Cao, W.D.; Nie, J. Substitution of chemical fertilizer by Chinese milk vetch improves the sustainability of yield and accumulation of soil organic carbon in a double-rice cropping system. *J. Integr. Agr.* **2019**, *18*, 2381–2392. [CrossRef]

34. Zhang, J.; Nie, J.; Cao, W.; Gao, Y.; Lu, Y.; Liao, Y. Long-term green manuring to substitute partial chemical fertilizer simultaneously improving crop productivity and soil quality in a double-rice cropping system. *Eur. J. Agron.* **2023**, *142*, 126641. [CrossRef]
35. Lu, S.; Song, H.; Guan, C.; Lepo, J.E.; Wu, Z.; He, X.; Zhang, Z. Long-term rice-rice-rape rotation optimizes 1,2-benzenediol concentration in rhizosphere soil and improves nitrogen-use efficiency and rice growth. *Plant Soil.* **2019**, *445*, 23–37. [CrossRef]
36. Tang, Z.; Zhang, L.; He, N.; Liu, Z.; Ma, Z.; Fu, L.; Wang, H.; Wang, C.; Sui, G.; Zheng, W. Influence of planting methods and organic amendments on rice yield and bacterial communities in the rhizosphere soil. *Front. Microbiol.* **2022**, *13*, 918986. [CrossRef]
37. Lu, R.K. *Soil Argrochemistry Analysis Protocoes*; China Agriculture Science Press: Beijing, China, 1999. (In Chinese)
38. Xiao, J.; Chen, S.Y.; Liang, T.; Yang, S.D.; Tan, H.W. Response of endophytic bacteria in sugarcane roots to different slow-release fertilizers with dicyandiamide (DCD) and humic acid (HA) applications. *Environ. Technol. Innov.* **2023**, in press. [CrossRef]
39. Zhao, Z.; Ma, Y.; Feng, T.; Kong, X.; Wang, Z.; Zheng, W.; Zhai, B. Assembly processes of abundant and rare microbial communities in orchard soil under a cover crop at different periods. *Geoderma* **2022**, *406*, 115543. [CrossRef]
40. Xiao, J.; Chen, S.; Sun, Y.; Wu, S.; Liang, W.; Yang, S. Effects of mechanical weeding on soil fertility and microbial community structure in star anise (*Illicium verum* Hook.f.) plantations. *PLoS ONE* **2022**, *17*, e0266949. [CrossRef] [PubMed]
41. Wang, J.; Zhang, L.; Lu, Q.; Raza, W.; Huang, Q.; Shen, Q. Ammonia oxidizer abundance in paddy soil profile with different fertilizer regimes. *Appl. Soil Ecol.* **2014**, *84*, 38–44. [CrossRef]
42. Wang, J.; Song, Y.; Ma, T.; Raza, W.; Li, J.; Howland, J.G.; Huang, Q.; Shen, Q. Impacts of inorganic and organic fertilization treatments on bacterial and fungal communities in a paddy soil. *Appl. Soil Ecol.* **2017**, *112*, 42–50. [CrossRef]
43. Dan, X.; He, M.; Meng, L.; He, X.; Wang, X.; Chen, S.; Cai, Z.; Zhang, J.; Zhu, B.; Müller, C. Strong rhizosphere priming effects on N dynamics in soils with higher soil N supply capacity: The 'Matthew effect' in plant-soil systems. *Soil Biol. Biochem.* **2023**, *178*, 108949. [CrossRef]
44. Zhong, W.; Gu, T.; Wang, W.; Zhang, B.; Lin, X.; Huang, Q.; Shen, Q. The effects of mineral fertilizer and organic manure on soil microbial community and diversity. *Plant Soil* **2010**, *326*, 511–522. [CrossRef]
45. Pinggera, J.; Geisseler, D.; Piepho, H.P.; Joergensen, R.G.; Ludwig, B. Effect of substrate quality on the N uptake routes of soil microorganisms in different soil depths. *Pedobiologia* **2015**, *58*, 211–218. [CrossRef]
46. Tang, H.; Li, C.; Wen, L.; Li, W.; Shi, L.; Cheng, K.; Xiao, X. Microbial carbon source utilization in rice rhizosphere and non-rhizosphere soils in a 34-year fertilized paddy field. *J. Basic Microbiol.* **2020**, *60*, 1004–1013. [CrossRef]
47. Kuzyakov, Y. Review: Factors affecting rhizosphere priming effects. *J. Plant Nut. Soil Sci.* **2002**, *165*, 382–396. [CrossRef]
48. Zhang, Y.; Zhang, S.; Wang, R.; Cai, J.; Zhang, Y.; Li, H.; Huang, S.; Jiang, Y. Impacts of fertilization practices on pH and the pH buffering capacity of calcareous soil. *Soil Sci. Plant Nutr.* **2016**, *62*, 432–439. [CrossRef]
49. Ge, S.; Zhu, Z.; Jiang, Y. Long-term impact of fertilization on soil pH and fertility in an apple production system. *J. Soil Sci. Plant Nut.* **2018**, *18*, 282–293. [CrossRef]
50. Liu, K.L.; Han, T.F.; Huang, J.; Asad, S.; Li, D.M.; Yu, X.C.; Huang, Q.H.; Ye, H.C.; Hu, H.W.; Hu, Z.H.; et al. Links between potassium of soil aggregates and pH levels in acidic soils under long-term fertilization regimes. *Soil Till. Res.* **2020**, *197*, 104480. [CrossRef]
51. Shen, J.P.; Zhang, L.M.; Guo, J.F.; Ray, J.L.; He, J.Z. Impact of long-term fertilization practices on the abundance and composition of soil bacterial communities in Northeast China. *Appl. Soil Ecol.* **2010**, *46*, 119–124. [CrossRef]
52. Tang, H.M.; Xu, Y.L.; Xiao, X.P.; Li, C.; Li, W.Y.; Cheng, K.K.; Pan, X.C.; Sun, G. Impacts of long-term fertilization on the soil microbial communities in double-cropped paddy fields. *J. Agric. Sci.* **2018**, *156*, 857–864. [CrossRef]
53. Ai, C.; Liang, G.; Sun, J.; Wang, X.; Zhou, W. Responses of extracellular enzyme activities and microbial community in both the rhizosphere and bulksoil to long-term fertilization practices in a Fluvo-aquic soil. *Geoderma* **2012**, *173–174*, 330–338. [CrossRef]
54. Aciego Pietri, J.C.; Brookes, P.C. Substrate inputs and pH as factors controlling microbial biomass, activity and community structure in an arable soil. *Soil Biol. Biochem.* **2009**, *41*, 1396–1405. [CrossRef]
55. Guo, J.; Liu, W.; Zhu, C.; Luo, G.; Kong, Y.; Ling, N.; Wang, M.; Dai, J.; Shen, Q.; Guo, S. Bacterial rather than fungal community composition is associated with microbial activities and nutrient-use efficiencies in a paddy soil with short-term organic amendments. *Plant Soil* **2018**, *424*, 335–349. [CrossRef]
56. Hungria, M.; Franchini, J.C.; Brandão-Junior, O.; Kaschuk, G.; Souza, R.A. Soil microbial activity and crop sustainability in a long-term experiment with three soil-tillage and two crop-rotation systems. *Appl. Soil Ecol.* **2009**, *42*, 288–296. [CrossRef]
57. Wu, M.N.; Qin, H.L.; Chen, Z.; Wu, J.S.; Wei, W.X. Effect of long-term fertilization on bacterial composition in rice paddy soil. *Biol. Fert. Soils* **2011**, *47*, 397–405. [CrossRef]
58. Bulgarelli, D.; Rott, M.; Schlaeppi, K.; van Themaat, E.V.L.; Ahmadinejad, N.; Assenza, F.; Rauf, P.; Huettel, B.; Reinhardt, R.; Schmelzer, E.; et al. Revealing structure and assembly cues for *Arabidopsis* root-inhabiting bacterial microbiota. *Nature* **2012**, *488*, 91–95. [CrossRef]
59. Fierer, N.; Bradford, M.A.; Jackson, R.B. Toward an ecological classification of soil bacteria. *Ecology* **2007**, *88*, 1354–1364. [CrossRef] [PubMed]
60. Guo, W.; Andersen, M.N.; Qi, X.B.; Li, P.; Li, Z.Y.; Fan, X.Y.; Zhou, Y. Effects of reclaimed water irrigation and nitrogen fertilization on the chemical properties and microbial community of soil. *J. Integr. Agric.* **2017**, *16*, 679–690. [CrossRef]
61. Li, H.; Yang, S.; Semenov, M.V.; Yao, F.; Ye, J.; Bu, R.; Ma, R.; Lin, J.; Kurganova, I.; Wang, X.; et al. Temperature sensitivity of SOM decomposition is linked with a K-selected microbial community. *Global Change Biol.* **2021**, *27*, 2763–2779. [CrossRef] [PubMed]

62. Lucia, F.; Ana, F.S. Strong shift in the diazotrophic endophytic bacterial community inhabiting rice (*Oryza sativa*) plants after flooding. *FEMS Microbiol Ecol.* **2015**, *91*, five104. [CrossRef]
63. Lin, Y.B.; Ye, Y.M.; Wu, C.F.; Hu, Y.M.; Shi, H.K. Changes in microbial community structure under land consolidation in paddy soils: A case study in eastern China. *Ecol. Eng.* **2020**, *145*, 105696. [CrossRef]
64. Wang, H.; Li, X.; Li, X.; Li, F.; Su, Z.; Zhang, H. Community composition and co-occurrence patterns of diazotrophs along a soil profile in paddy fields of three soil types in China. *Microb. Ecol.* **2021**, *82*, 961–970. [CrossRef]
65. Tu, Y.; Li, H.X.; Jiang, L.; Dong, K.; Wang, D.Q. Bacterial communities structure and diversity in rhizosphere of different plants from Huixian wetland, Guangxi. *Ecol. Environ. Sci.* **2019**, *28*, 252–261. (In Chinese) [CrossRef]
66. Beheshti Ale Agha, A.; Kahrizi, D.; Ahmadvand, A.; Bashiri, H.; Fakhri, R. Identification of *Thiobacillus* bacteria in agricultural soil in Iran using the 16S rRNA gene. *Mol. Biol. Rep.* **2018**, *45*, 1723–1731. [CrossRef] [PubMed]

agronomy

MDPI

Article

Variations in Methanogenic and Methanotrophic Communities Resulted in Different Methane Emissions from Paddy Soil Applied with Two Types of Manure

Beibei Zhou [1,2], Ruirui Chen [2], Shuang Peng [1,2], Jianwei Zhang [2], Xiangui Lin [2] and Yiming Wang [2,*]

1 School of Environment and Ecology, Jiangsu Open University, Nanjing 210017, China; zhou_bae@163.com (B.Z.)
2 State Key Laboratory of Soil and Sustainable Agriculture, Institute of Soil Science, Chinese Academy of Sciences, Nanjing 210008, China
* Correspondence: ymwang@issas.ac.cn

Abstract: Organic manure application is crucial for the maintenance and improvement of soil fertility. However, it inevitably results in increased paddy CH_4 emissions, restricting the use of organic manure in the rice fields. In the present study, two kinds of manures, rapidly composted manure (RCM) and non-composted manure (NCM), were investigated through a 19-week greenhouse experiment, during which the dynamics of CH_4 emission, soil parameters (DOC, acetate, NH_4^+, NO_3^-, and SO_4^{2-}), and communities of methanogens and methanotrophs were simultaneously measured. The results showed that NCM significantly enhanced CH_4 emission, while RCM decreased CH_4 emission by 65.03%; there was no significant difference with the manure-free treatment. In order to well understand the methanogenic process, the seasonal CH_4 flux was divided into two periods, namely Stage 1 (before drainage) and Stage 2 (after drainage), on the basis of CH_4 emission intensity. The different CH_4 production abilities among the three treatments could contribute to the varied CH_4 emissions at Stage 1. The much higher soil DOC concentrations were observed in the manure-amended soils (NCM- and RCM-treatments), which could correspondingly lead to the relative higher CH_4 emissions compared to the control during Stage 1. Furthermore, the increased methanogenic abundance and the shifted methanogenic archaeal community characterized by the functionally stimulated growth of *Methanosarcina* genus were observed in the NCM-treated soils, which could consequently result in a higher CH_4 emission from the NCM treatment relative to the RCM treatment. As for Stage 2, apart from the significant decrease in soil DOC, the increased contents of soil NO_3^- and SO_4^{2-}, especially with the RCM-treated soils, were also detected following the drainage, which might retard CH_4 production. The lower CH_4 emission at Stage 2 could also be attributed to the vigorous aerobic CH_4 oxidations, especially in the RCM-treated soils. As a support, the amount of methanotrophs revealed an increasing trend during the late rice growth period, as did the predominance of the methylotrophy of *Methylophilaceae* species, which showed robust co-occurrence with methanotrophs, inferring interspecies cooperation in methane oxidation.

Keywords: manure; CH_4 emission; methanogenic community; methanotrophs; qPCR

Citation: Zhou, B.; Chen, R.; Peng, S.; Zhang, J.; Lin, X.; Wang, Y. Variations in Methanogenic and Methanotrophic Communities Resulted in Different Methane Emissions from Paddy Soil Applied with Two Types of Manure. *Agronomy* **2023**, *13*, 1268. https://doi.org/10.3390/agronomy13051268

Academic Editors: Yong-Xin Liu and Peng Yu

Received: 9 March 2023
Revised: 18 April 2023
Accepted: 27 April 2023
Published: 28 April 2023

1. Introduction

CH_4 is the second-most potent greenhouse gas in the atmosphere, with its concentration increasing steadily from 722 ppb (1750) to 1845 ppb (2016) [1]. Rice paddies are recognized as a major source of CH_4, which accounts for nearly 11% of the total CH_4 emissions [2,3]. China is one of the most important rice-producing countries in the world, accounting for 16% and 28% of the world's rice area and rice production [4]. Whereas, the growing demand for rice to feed the ever-increasing global population challenges the present management strategies in rice cultivation [5]. To guarantee sustainable increases in rice yield, soil fertility must be maintained. Nowadays, the importance of organic agriculture has been confirmed by more and more research, especially in combination with inorganic fertilization. Incorporation of organic materials into the crop fields can help to enhance plant growth and prevent soil desertification by improving soil structure and fertility by increasing soil carbon (C) content as well as soil microbial diversity and activity [6–8]. However, there are concerns that organic amendments (OAs) can significantly intensify paddy CH_4 emissions with more C materials introduced [9–11].

The total emission of CH_4 is the balance between its production and oxidation. In paddy fields, during the anaerobic degradation of soil organic matter, sequential reduction of soil NO_3^-, Mn (IV), Fe (III), and SO_4^{2-} happened, and then methanogenesis became the exclusive process [12]. Methanogenic archaea (methanogens) are the sole CH_4 producers, which produce CH_4 by metabolizing either acetate or H_2/CO_2 that were accumulated from the microbial fermentation of organic matters, mainly polysaccharides, in rice fields [13,14]. At present, all known methanogens fall into the phylum *Euryarchaeota*, where they form seven distinct orders: *Methanobacteriales, Methanosarcinales, Methanomicrobiales, Methanococcales, Methanopyrales, Methanocellales*, and *Methanoplasmatales* [15,16].

The oxidation of CH_4 conducted by methanotrophs plays an important role in controlling paddy CH_4 emissions. It is estimated that almost up to 90% of the CH_4 produced in wetlands can be oxidized before reaching the atmosphere [17,18]. Methanotrophs generally inhabit the rhizosphere and surface soil [19] and utilize CH_4 as their sole carbon and energy sources, including two major groups, namely Type I and Type II methanotrophs. Type I methanotrophs, which belong to Gamma-Proteobacteria, are split into two more groups (Type Ia and Type Ib) and are housed within the family Methylococcaceae, which has fifteen genus: *Methylococcus, Methylocaldum, Methylomonas, Methylobacter, Methylomicrobium, Methylosarcina, Methylohalobius, Methylosoma, Methylothermus, Methylomarinum, Methylovulum, Methylogaea, Methylosphaera, Crenothrix*, and *Clonothrix*. The first two genera are also referred to as Type Ib methanotrophs. The Type II methanotrophs belonging to Alpha-Proteobacteria are grouped in two families, Methylocystaceae and Beijerinckiaceae, with *Methylocystis, Methylosinus, Methylocella, Methylocapsa*, and *Methyloferula* as the member genus [20]. The CH_4 oxidation process involves an enzyme called methane monooxygenase (MMO), which plays an important role in the initial enzymatic reaction of the aerobic oxidation of CH_4. Generally, the highly conserved *pmoA* gene that encodes the membrane-bound subunit of MMO is widely used as a phylogenetic marker of methanotrophs in ecological studies [19].

Strategies to reduce CH_4 production or enhance CH_4 oxidation can help retard the increased paddy CH_4 emission due to the introduction of OAs, and a large number of relevant experiments have been conducted. The mitigative potential of electron acceptor supplementation has been well confirmed in field and laboratory studies with ferrihydrite, ammonium sulfate, gypsum, and phosphogympsum amended [21–23]. In those cases, it is predicted that the methanogens are thermodynamically outcompeted for common substrates like H_2 or acetate by ferric iron-reducing bacteria (FRB) or sulfate-reducing bacteria (SRB). Additionally, a few field studies have also demonstrated the importance of the proper pretreatment of the incorporated OAs in effectively mitigating paddy CH_4 emissions. As revealed, the incorporation of composted crop straw and livestock manures can significantly decrease paddy CH_4 emissions compared to untreated manures, with reduction rates ranging between 50% and 90% [9,24,25]. But unfortunately, most of the research did not pay attention to the responses of methanogens and methanotrophs, which is of great significance to better understanding the microbial mechanisms underlying those measures in paddy CH_4 emission control.

In this study, a new type of rapidly composted manure (RCM) was applied to paddy fields. Basically, the manure is prepared in a closed rolling reactor with a catalyst rich in sulfate and nitrogen and an ammonia fixator and treated at 120 °C for just three hours, similar to biochar production. This processing technique realizes the rapid disposal of awkward manures, potentially advancing the sustainable development of the livestock and poultry industries and the. In our previous study, the decreased CH_4 emission of

RCM relative to non-composted manure (NCM) was demonstrated [26]. However, how the methanogens and methanotrophs, both in populations and community structures, as well as the soil parameters, changed along with the cultivation of rice is still unclear. Therefore, a new pot experiment was conducted in a greenhouse, including three treatments: the NCM treatment, the RCM treatment, and the control. The main objectives of this study are: (1) to confirm the mitigation of seasonal CH_4 emission from paddy soils treated with RCM relative to NCM; (2) to determine the shifts of methanogens and methanotrophs along rice cultivation, both including population size and community structure, as well as the variations of corresponding soil parameters; and (3) to eventually explicate the underlying mechanisms that contributed to the significantly lower CH_4 emission of the RCM-treated soils in detail.

2. Materials and Methods

2.1. Greenhouse Experiment

The experiment was set up with three replicates and conducted in the greenhouse of the Institute of Soil Science, Chinese Academy of Sciences, during 19 June 2014 to 27 October 2014. The soil used was collected from the Changshu Agroecological Experiment Station (123°38′ E, 31°33′ N), Chinese Academy of Sciences, in Jiangsu Province, eastern China. The soil was air dried, ground, sieved through a 5-mm mesh screen, and then thoroughly mixed. Approximately 7 kg of dried soil was put into a quadrate pot (20 × 24 × 20 cm), with a water groove acting as the chamber base in this experiment. The soil was classified as Gletic-Stagnic Anthrosols. The relevant soil properties were as follows: 20.7 g/kg soil organic C (SOC), 2.41 g/kg total N, 1.67 g/kg total P, 27.02 g/kg total K, and pH 7.30.

In this study, two types of cattle manures were used as basal fertilizer, which were collected from WoLvBao Organic Agriculture (Suqian, China). The raw manure (Non-composted manure, NCM) contains total organic C (TOC), 332.75 g/kg; total N, 17.26 g/kg; total P, 12.66 g/kg; total K, 12.78 g/kg; SO_4^{2-}, 10.43 g/kg; and pH, 7.80. The RCM manure had a pH of 6.67 and contained 335.68 g/kg TOC, 32.65 g/kg total N, 10.19 g/kg total P, 11.5 g/kg total K, and 61.18 g/kg SO_4^{2-}.

Manures were applied to the pots a week before transplantation, considering the slow release of nutrients in organic fertilizers. Then, twelve 30-day-old seedlings (*Oryza sativa* L.) from four hills were transplanted into each pot 7 days after flooding (7 d). For the whole rice cultivation period, apart from the cattle manures applied at a loading rate of 1% (dry weight) of the total pot soil at the beginning of the experiment, no more chemical fertilizer was amended. During rice growth, all the pots were permanently flooded, with the water level maintained at approximately 3 cm, except for a drainage carried out from 28 July (39 d) to 3 August (45 d). The rice was harvested 126 days after transplantation.

2.2. Methane Emission

CH_4 flux was measured after flooding. A bottom-open PVC chamber (20 × 24 × 80 cm) equipped with a battery-driven fan, a thermometer, and a sampling septum was used for gas sampling. Water was added to the groove to seal the chamber. 20 mL gas samples were collected at 0, 5, 10, 20, and 30 min with a syringe and injected into pre-evacuated 18 mL vials for CH_4 analysis (Agilent 7890, Palo Alto, CA, USA). The gas sampling interval was 3–7 days, and the sampling time was fixed between 8:00 and 10:00 am. The CH_4 emission rate was calculated by a linear regression of CH_4 concentration over time.

The CH_4 emission rate was calculated as follows: $F = \rho \times V/A \times (dc/dt) \times 273/T$ [9]. Where F is the emission rate; ρ is the CH_4 density (0.714 kg/m^3); V is the chamber volume; A is the surface area enclosed by the chamber; dc/dt is the CH_4 increase rate; and T is the air temperature of the chamber. Cumulative CH_4 emissions were calculated by linear interpolation of daily CH_4 emissions during the monitoring period.

2.3. Soil Sampling and DNA Extraction

Soil was sampled with a core sampler on days 7 d (26 June), 21 d (1 July), 39 d (28 July), 90 d (16 September), and 133 d (29 October), five times in total. The third and fourth samplings corresponded to the maximum rice tillering stage and the grain filling stage, respectively. Genomic DNA was extracted from 0.5 g of moist soil using the FastDNA® SPIN Kit for soil (MP Biomedicals, Santa Ana, CA, USA) according to the manufacturer's protocol. The extracted DNA was dissolved in 100 μL of TE buffer, quantified by a NanoDrop ND-1000 spectrophotometer (Thermo Scientific, Waltham, MA, USA), and stored at −20 °C until use.

2.4. Chemical Analysis of Soil Samples

The collected soil samples were slightly dried and sieved through a 2-mm mesh screen. Soil dissolved organic carbon (DOC) and inorganic nitrogen (NH_4^+ and NO_3^-) were determined according to the methods described by Li et al. [27]. Briefly, fresh moist soil (10 g) was shaken using a reciprocating shaker for 30 min with double-distilled water (20 mL), and the extracts were then centrifuged at 4000 rpm for 20 min and filtered through the 0.45 μm membrane. The DOC in the extract was analyzed by Multi N/C 3000 total organic carbon (TOC) analyzer (Analytik Jena AG, Jena, Germany). The extracts were also analyzed by ion chromatography Dionex ICS-1100 (Thermo Scientific, Waltham, MA, USA) for the determination of SO_4^{2-} [28]. In addition, acetate in the filtrate was analyzed using an Agilent 7890A GC with a Headspace Sampler Agilent 7697A (Agilent Technologies, Palo Alto, CA, USA). Ten grams of soil were suspended in 40 mL of KCl (2M) and shaken for 1 h. After filtration, the NH_4^+ and NO_3^- from the supernatant were analyzed using an autoanalyzer (Skalar Scan++, Skalar, Breda, The Netherlands).

2.5. Real-Time Quantitative PCR

For each soil, qPCR was used to determine the abundance of the methanogenic archaeal 16S rRNA (primer set 1106F (TTWAGTCAGGCAACGAGC)/1378R (TGTGCAAGGAGCAGGGAC)), methanotrophic *pmoA* gene (primer set A189 (GGNGACTGGGACTTCTGG)/mb661 (CCG-GMGCAACGTCYTTACC)) [29], and sulfate-reducing bacterial *dsr*AB (primer set DSR1-F+ (ACSCACTGGAAGCACGGCGG)/DSR-R (GTGGMRCCGTGCAKRTTGG) [30] genes. The target gene copy numbers were quantified by a C1000™ Thermal Cycler equipped with a CFX96™ Real-Time system (Bio-Rad, Hercules, CA, USA). A 20 μL reaction mixture containing 10 μL of SYBR Premix Ex Taq™ (TaKaRa, Kyoto, Japan), primer sets (0.5 μM each), and a 1.0 μL template containing approximately 10 ng of DNA was set up. The negative control was run with water as the template instead of soil DNA extract. Plasmids carrying the target gene were extracted and purified. Then, the plasmid DNA content was quantified by NanoDrop and diluted 10-fold to generate a 7-point standard curve. The specificity of the target gene amplification was confirmed by melting curve analysis, and the expected sizes of the PCR amplicons were checked by agarose gel electrophoresis. Real-time PCR was performed in triplicate, and amplification efficiencies of 97–102% were obtained with R^2 values of 0.990–0.998. The final number of target genes was obtained by calibrating the extracted total DNA concentrations and soil water content.

2.6. PCR and Illumina Miseq Pyrosequencing

For the soil samples, the above-mentioned primer sets 1106F/1378R and 519F (CAGCM GCCGCGGTAATWC)/907R (CCGTCAATTCMTTTRAGTTT) were used to amplify the methanogenic archaeal and bacterial 16S rRNA gene fragments of approximately 280 bp and 400 bp, respectively, for sequencing on the Illumina Miseq pyrosequencing platform. We chose soils sampled on 7 d, 39 d and 133 d for bacterial 16S rRNA gene analysis. The oligonucleotide sequences included a 5-bp barcode fused to the forward primer as follows: barcode + forward primer. PCR was carried out in 50-μL reaction mixtures with the following components: 4 μL (initial 2.5 mM each) of deoxynucleoside triphosphates, 2 μL (initial 10 μM each) of forward and reverse primers, 2 U of Taq DNA polymerase with 0.4 μL

(TaKaRa, Kyoto, Japan), and 1 μL of template containing approximately 50 ng of genomic community DNA. For the methanogenic archaeal 16S rRNA gene, the PCR conditions were as follows: 35 cycles of denaturation at 95 °C for 45 s, annealing at 55 °C for 45 s, and extension at 72 °C for 60 s, with a final extension at 72 °C for 10 min. As for the bacterial 16S rRNA gene, 35 cycles (95 °C for 45 s, 56 °C for 45 s, and 72 °C for 60 s) were performed with a final extension at 72 °C for 7 min. The bar-coded PCR products from all samples were normalized to equimolar amounts and purified before pyrosequencing.

2.7. Processing of the Pyrosequencing Data

The methanogenic archaeal and bacterial 16S rRNA gene data were processed using the Quantitative Insights Into Microbial Ecology (QIIME) 1.4.0-dev pipeline [31]. Briefly, raw data were quality filtered (>25 quality score and 200 bp in length), denoised, and then binned into OTUs at a 97% similarity level. The sequence with the highest abundance was selected from each OTU as the representative sequence of that OTU. Taxonomic analysis was conducted by comparing it with the Silva 104 database. PyNAST and QIIME were used to align representative sequences and remove chimeric sequences, respectively. A total of 721,250 methanogenic sequences and 398,775 bacterial sequences were obtained, respectively. The raw sequences were deposited in the NCBI SRA with an accession number of SRP388055.

2.8. Statistical Analysis

The statistical analysis was performed using SPSS v 13.0 (SPSS Inc., Chicago, IL, USA). The data were expressed as the means ± standard deviation (SD). The letters above the error bars in the figure indicate significant differences between treatments. The mean separation was assessed by Tukey's test. Differences at $p < 0.05$ were considered statistically significant. The data were visualized by ImageGP [32].

A partial least squares path model (PLS-PM) was performed to infer the potential direct and indirect effects of NCM and RCM amendment, soil properties, microbial abundance, methanogenic community, and methanotrophic community on total CH_4 emissions at five sampling time points. Soil properties are latent variables measured by concentrations of NO_3^-, NH_4^+, DOC, acetate, and SO_4^{2-}; microbial abundances are latent variables reflected by abundances of three functional microbial groups: methanogens, SRB, and methanotrophs; and microbial communities are the first PCA axis of methanogenic and methanotrophic communities. The quality of the PLS-PM is evaluated by the goodness of fit (GOF) index, which >0.7 means a good overall prediction performance of the model, and R^2 values represent the variance of dependent variables explained by the inner model [33].

3. Results

3.1. CH_4 Emission

Similar CH_4 flux patterns were found from three treatments, with the amplitudes varying (Figure 1a). Generally, no obvious CH_4 emission was observed at the beginning (before 17 d) of the experiment, followed by steadily increased emissions, with the emission peaks appearing on 33 d Thereafter, the drainage carried out between 39 d and 45 d resulted in significantly decreased CH_4 emission rates for all three treatments. In addition, pulse CH_4 emissions were monitored during the drainage. The average CH_4 fluxes before drainage were approximately 9.73, 6.30, and 3.55 mg m^{-2} h^{-1} for NCM-, RCM-, and control treatments, respectively. After drainage, slight variations of the CH_4 emission rates were measured, and the values ranged between 0.09 and 3.57 mg m^{-2} h^{-1}.

In order to have a clear vision of the CH_4 fluxes with three different treatments, the whole rice growing season was divided into two parts, namely Stage 1 (before drainage) and Stage 2 (after drainage). In general, vigorous CH_4 emissions were observed in Stage 1, accounting for 52.6%, 96.8%, and 32.9% of the total CH_4 emissions of the NCM-, RCM-, and control treatments, respectively. In comparison with the control, NCM and RCM applications significantly increased the total CH_4 emission of Stage 1, with an increase of

279.4% and 120.8%, respectively (Table S1). However, for Stage 1, approximately 34.3–55.4% of CH_4 emissions were mitigated with the RCM treatment compared to the NCM treatment. On the contrary, the cumulative CH_4 emission of three treatments during Stage 2 decreased in the following order: NCM < Ctrl < RCM. The CH_4 emission with the RCM-treated soils was weak, and the total CH_4 emission was reduced by 90.1% relative to that of the control (Table S1). The reduced CH_4 emissions from the RCM-treated soils compared to the NCM treatment at Stage 1 and Stage 2, respectively, accounted for 21.9% and 46.7% of the total emitted CH_4 of the NCM treatment, indicating the main mitigation occurred at Stage 2 for RCM.

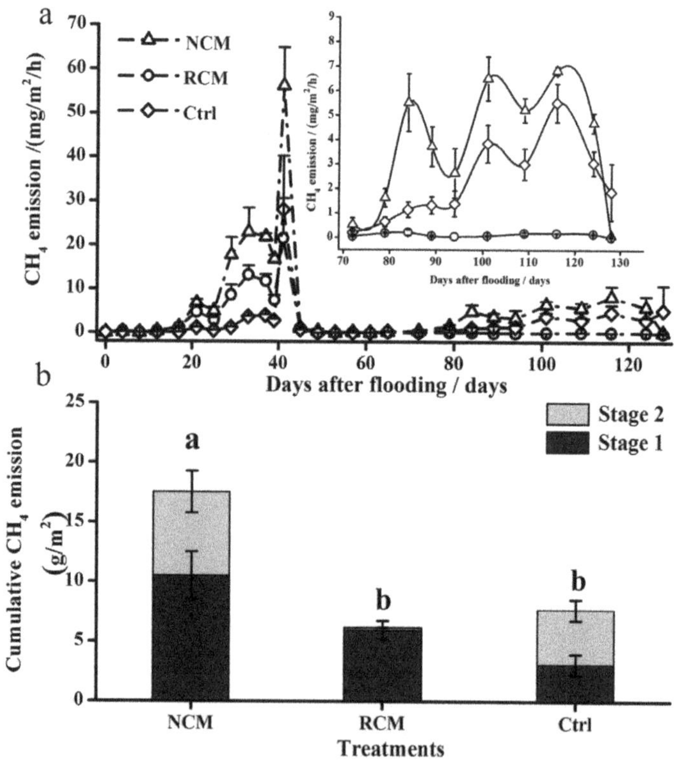

Figure 1. Seasonal variations of CH_4 flux and the cumulative CH_4 emissions of different treatments. (**a**) Seasonal variations of CH_4 flux. The zoom-in view for the late rice cultivation period was displayed on the right. (**b**) Cumulative CH_4 emissions. Different letters above the error bars indicate significant differences ($p < 0.05$).

3.2. Soil Chemical Characteristics

Regardless of treatments, soil DOC represented an accumulation during the first three weeks when soils were subjected to flooding (Figure 2a). After the maximum recorded at the time point of 21 d, the DOC concentrations began to decrease gradually. Applications of NCM and RCM significantly increased soil DOC concentration during rice cultivation relative to the control ($p < 0.05$), except for a minimum detected from the RCM-treated soils on 90 d. When harvested (133 d), relatively similar DOC concentrations were observed among the three treatments, but at a low level.

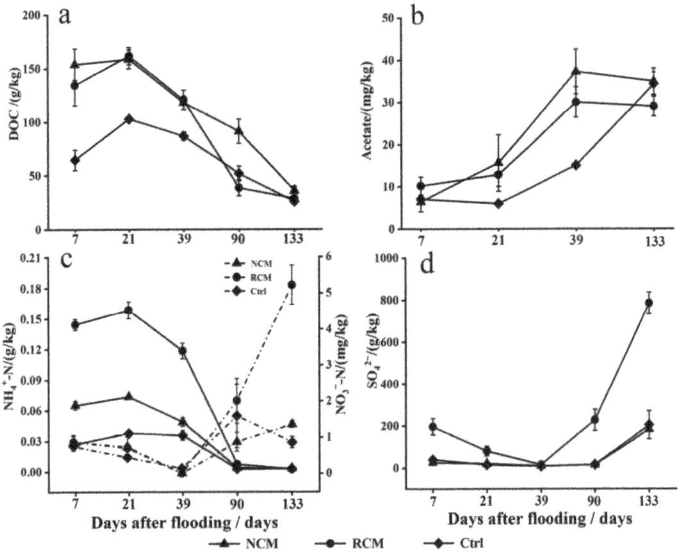

Figure 2. The dynamics of soil dissolved organic carbon (DOC) (**a**), acetate (**b**), NH_4^+, and NO_3^-. The full line and the dotted line refer to NH_4^+ and NO_3^-, respectively. (**c**) and SO_4^{2-} (**d**) concentrations.

Variations in the concentrations of soil acetate were also monitored over time. In contrast to the dynamics of DOC in Figure 2a, generally, the acetate concentration gradually increased with rice growth for these three treatments (Figure 2b). For the control, apart from a slight decline from 7 to 21 d, the soil acetate contents gradually increased, which has been more apparent lately. Consistent with the case of DOC, NCM and RCM applications also led to significantly increased soil acetate concentrations, and the acetate concentrations from the NCM treatment seemed higher than those from the RCM treatment ($p > 0.05$).

Concentrations of soil nitrate (NO_3^-) and ammonium nitrogen (NH_4^+) during rice growth changed in opposite ways (Figure 2c). Generally, the soil NH_4^+ concentrations showed a decreasing trend and remained at low levels during the late rice growth period. NCM and RCM amendments significantly increased soil NH_4^+ concentrations ($p < 0.05$) compared to those of the control for the first three sampling dates. Moreover, the NH_4^+ concentration with the RCM treatment was approximately twice that with the NCM treatment. There was no significant difference observed in the soil NO_3^- concentration among the three treatments before drainage ($p > 0.05$), and the values were always low. After drainage, the soil NO_3^- concentrations began to increase, which was remarkably apparent with the RCM-treated soils and much higher than those from the other two treatments ($p < 0.05$).

The SO_4^{2-} concentrations from these three treatments varied in similar patterns to those of NO_3^- (Figure 2d). Before drainage, the SO_4^{2-} concentration gradually decreased and then apparently started to increase, following the reflooding of soils. Both the NCM treatment and the control contained a similar amount of SO_4^{2-}, which was much lower than that of the RCM treatment ($p < 0.05$). Relatively large amplitudes of the soil SO_4^{2-} content variations were observed with the RCM-treated soils.

3.3. Abundance of Functional Microbial Groups

The abundance of the methanogens from soils treated with NCM ranged between 0.95 and 1.54 × 10^9 /g and was consistently higher than that from RCM treated soils (0.65–1.20 × 10^9 /g) and the control (0.51–1.27 × 10^9 /g), during the cultivation of rice (Figure 3a). However, no significant difference was found between the RCM treatment and the control in the abundance of soil methanogens ($p > 0.05$), except for an apparent increase in the control at the beginning of the rice tillering stage (21 d) ($p < 0.05$).

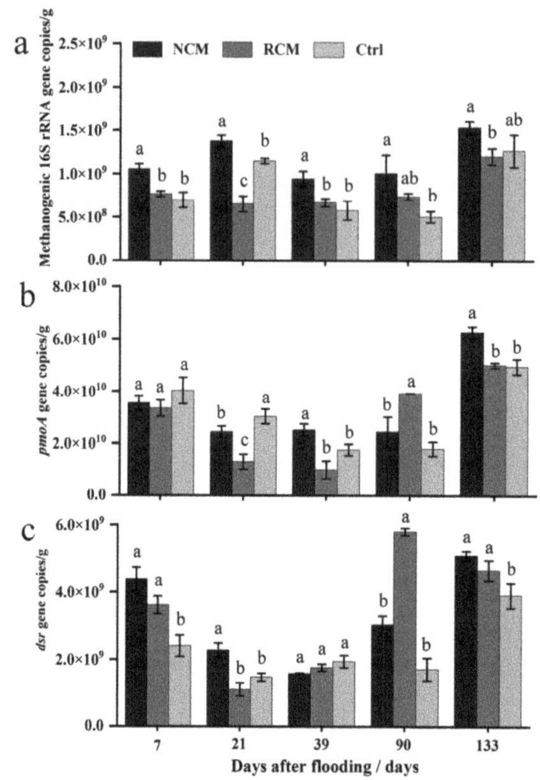

Figure 3. Copy numbers of the methanogenic archaeal 16S rRNA genes (**a**), the methanotrophic *pmoA* genes (**b**), and the *dsr* gene (**c**) of paddy soils under different treatments along rice cultivation. Significant differences are indicated by different letters above the error bars ($p < 0.05$).

Variations of the number of methanotrophs from all treatments were similar, with the abundance gradually decreasing before drainage while obviously increasing from 90 to 133 d (Figure 3b). It is noteworthy that, at the initial stage (7 d), relatively similar abundances of the methanotrophs were found with all three treatments, and then the methanotrophic numbers decreased, with the RCM treatment containing the lowest methanotrophic abundance on 21 and 39 d. However, the number of methanotrophs in the soils treated with RCM emerged to be the largest on 90 d. At maturation, the methanotrophic abundances of the three treatments reached peaks, with the largest number observed in the NCM-treated soils.

The population size of sulfate-reducing bacteria (SRB) varied in similar ways as the changes in the methanotrophs. As flooding continued, the SRB abundance showed a decreasing trend until the drainage was carried out at 39 d (Figure 3c). Then, after drainage, the SRB abundance began to increase again and reached its maximum at the harvest. Similarly, an apparent increase in the number of SRB was observed over 90 d in soils treated with the RCM.

3.4. Community Compositions of Methanogens and Methanotrophs

Pyrosequencing revealed that the soil methanogenic archaeal community of all treatments during the whole rice growing period was dominated by genera of *Methanosarcina* (12.2–45.4%), *Methanocella* (14.4–29.4%), *Methanosaeta* (12.2–26.9%), and *Methanobacterium* (11.7–21.0%), whose relative abundances could represent serious shifts with the cultivation of rice, as illustrated in Figure 4a. *Methanosarcina* was the most predominant methanogenic group for soils amended with NCM, while the proportions largely varied, ranging between 23.0 and 45.4%. In contrast, the RCM-treated soils were dominated by *Methanocella* (24.3–29.4%), with the percentage varying slightly. The methanogenic species in the control seemed evenly distributed, with almost no dominant methanogenic species observed. In addition, the relative abundance of *Methanosarcina* from the NCM treatment and the control showed a similar decreasing trend with the growth of rice, which was more apparent in the NCM treatment and not found in the RCM treatment. Besides, the *Methanosaeta* seemed to increase at the late rice growing stage, while the *Methanobacterium* decreased within all three treatments.

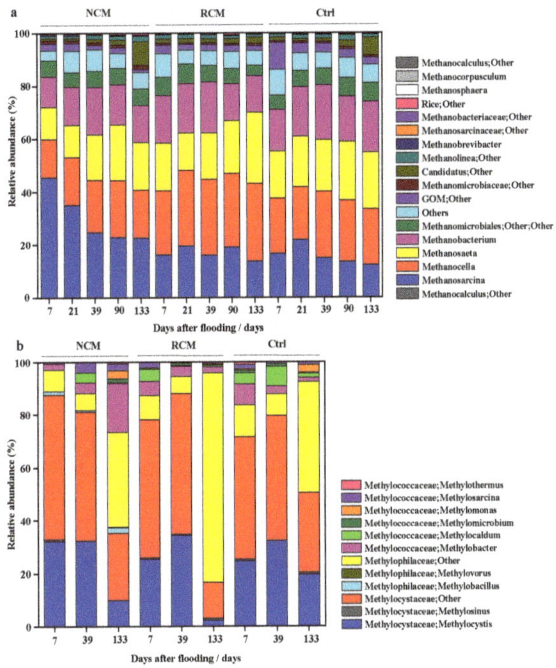

Figure 4. A 100% stacked column chart of the relative abundances of the dominant methanogenic (**a**) and methanotrophic (**b**) genera derived from 16S rRNA genes of paddy soils under different treatments along rice cultivation. The relative abundances of the methylotrophs of Methylophilaceae were also displayed, as they cooperated with methanotrophs in methane oxidation.

The total bacterial sequences assigned to each family were classified at genus level, and sequences belonging to methanotrophs were selected. The relative abundance of each methanotrophic genera is illustrated in Figure 4b. It was observed that the composition of methanotrophs varied over time, especially at the rice maturation stage. At the beginning of the experiment, the methanotrophic community structures of all treatments seemed relatively similar; only the genus less dominant differed. Type II methanotrophs, especially *Methylocystis* and other undefined genus from the family of Methylocystaceae, overwhelmingly dominated the paddy methanotrophic community before drainage, accounting for approximately 60.3–73.7% of the total methanotrophs. However, the structure of the methanotrophic community significantly changed at the late growing stage of rice ($p < 0.05$), as shown in Figure 4b on 133 d. The dominance of Type II methanotrophs decreased, ranging from 19.9% to 48.8% among the three treatments. Interestingly, data from 133 d revealed that methylotrophs of the family Methylophilaceae emerged to be the most abundant group, with the proportions increasing from the initial 7.4%, 6.9%, and 8.4%, respectively, to 34.1%, 73.5%, and 37.0% for the treatments of NCM, RCM, and control. It deserved to be mentioned that, in the mature stage, the methylotrophs of the soils treated with RCM seemed to greatly grow and ultimately dominate the methanotrophs.

3.5. Shifts in the Communities of Methanogens and Methanotrophs

The variations in the paddy methanogenic archaeal and methanotrophic communities of the different samples along rice cultivation were statistically evaluated using a principal component analysis (PCA) of the weighted pairwise UniFrac community distances, which are displayed in Figure 5. PCA revealed that the methanogenic communities from the NCM-treated soils apparently shifted with the rice cultivation, which seemed much different from those of the other treatments (Figure 5a). The samples from the RCM treatment and the control were grouped together and dissimilated from samples of the NCM treatment, which could indicate the much similar methanogenic communities between the RCM treatment and the control. Moreover, the methanogenic communities of the RCM treatment and the control slightly shifted over time, especially for those of the RCM treatment.

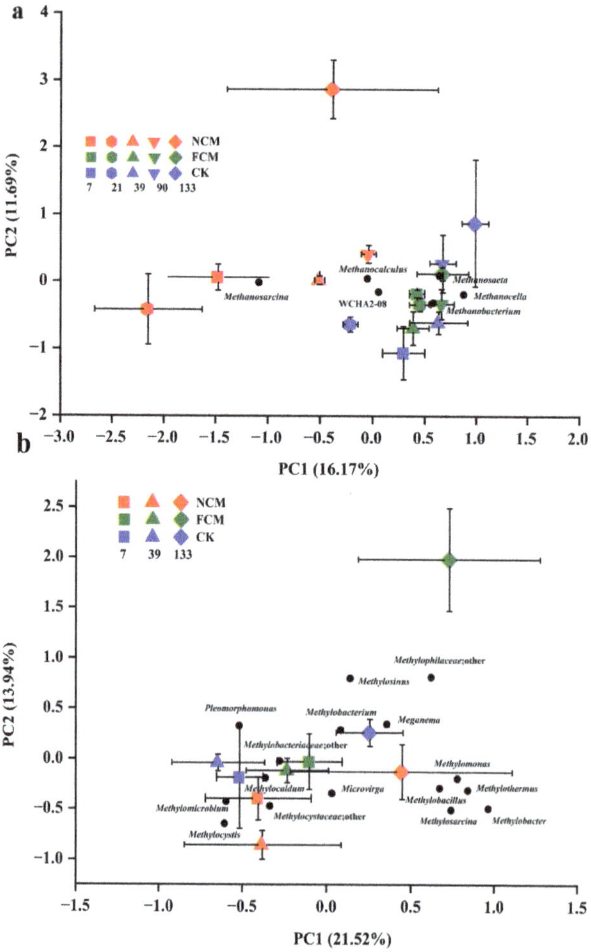

Figure 5. The community compositional structures of the methanogens (**a**) and methanotrophs (**b**) in the paddy soils as indicated by principal component analysis (PCA) of the weighted pairwise UniFrac community distances between different treatments along rice cultivation.

As shown in Figure 5b, the soil samples from the first two sampling dates were gathered for all three treatments, suggesting similar and less varied methanotrophic communities in the three treatments. However, all soil samples on 133 d separated from the rest of the samples, indicating the methanotrophic communities largely changed, which was more apparent for the RCM-treated soils.

3.6. Potential Direct and Indirect Effects of NCM and RCM Amendments on Total CH_4 Emission

PLS-PM indicated that, for five sampling time points, all predictor variables explained 91.3%, 90.5%, 89.4%, 95.7%, and 93.5% of variations in total CH_4 emission, respectively. Both NCM and RCM amendments showed significant effects on soil properties, except for 7 d, when just RCM application significantly affected soil SO_4^{2-} concentration (Figure 6a). Amendment of NCM exerted significant impacts on soil methanogenic communities for 7 d, 21 d, 39 d and 90 d, while the effects of RCM amendment were non-significant for all time.

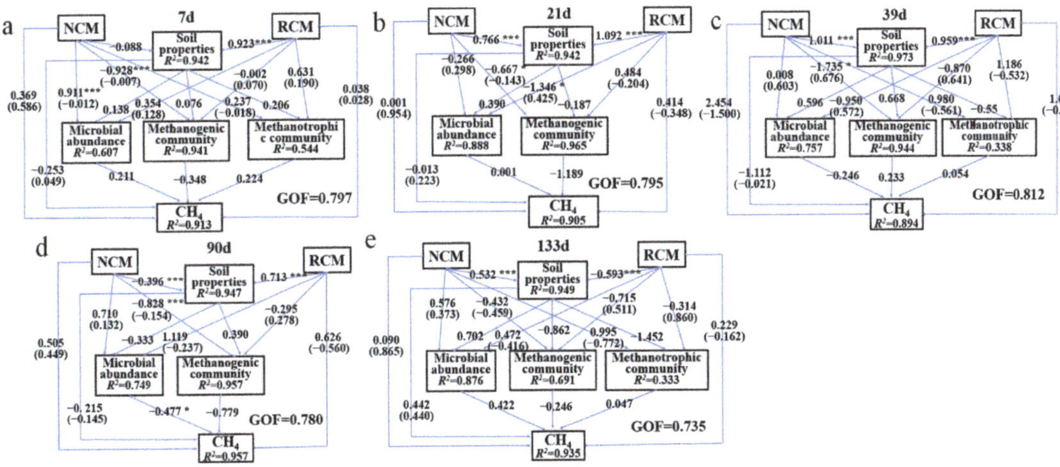

Figure 6. The partial least squares path models (PLS-PM) illustrate the direct and indirect effects of NCM and RCM, soil properties, microbial abundance, methanogenic community, and methanotrophic community (except 21 d and 90 d) on total CH_4 emissions at 7 d (**a**), 21 d (**b**), 39 d (**c**), 90 d (**d**), and 133 d (**e**), respectively. Soil properties are latent variables measured by SO_4^{2-} concentrations for 7 d (**a**); NO_3^-, NH_4^+, DOC, acetate, and SO_4^{2-} concentrations for 21 d (**b**) and 133 d (**e**); NO_3^-, DOC and acetate concentrations for 39 d (**c**); and NO_3^-, NH_4^+, DOC, and SO_4^{2-} concentrations for 90 d (**d**). Microbial abundances are latent variables reflected by the abundances of three functional microbial groups: methanogens, SRB, and methanotrophs, with the abundances of methanotrophs and SRB excluded for 7 d and 39 d, respectively. The main soil properties and microbial abundances are selected through PLS-PM analysis, and microbial communities are the first PCA axis of methanogenic and methanotrophic communities. The numbers on the arrowed lines indicate the normalized direct (above) and indirect (below) path coefficients. R^2 below the latent variables represents the variations of dependent variables explained by the inner model. The GOF index represents the goodness of fit. Asterisks represent significant effects. *, $p < 0.05$; ***, $p < 0.001$.

4. Discussion

4.1. RCM Application Significantly Decreases Paddy CH_4 Emission

Livestock manures contain high amounts of labile organic C and nutrients, which are regarded as valuable organic fertilizers in agriculture, favoring crop growth as well as improving soil structure and fertility [34,35]. As anything has two sides, the high C content of the livestock manure could otherwise inevitably lead to vigorous paddy CH_4 emissions when applied to the rice fields [11,36], triggering concerns about global warming. However, the substantial CH_4 emission resulting from the incorporation of the raw manure could be mitigated when that is properly pretreated. In the present study, a new rapid composting method was utilized to handle the raw cattle manures, which can promise the rapid disposal of vast amounts of ever-growing waste generated from intensive livestock and poultry operations. CH_4 emission fluxes with all three treatments were simultaneously measured during the whole rice cultivation, revealing that the RCM-treated rice soils emitted almost the same amount of CH_4 in comparison with that from the control, whereas approximately double CH_4 emission was detected with the NCM-treated rice soils (Figure 1a). Therefore, referring to the above-mentioned results, we can conclude that RCM seems to be more superior relative to raw manure as an organic amendment, which can effectively enhance rice growth, increase grain production (Table S2), and emit less CH_4. Numerous studies have also shown that composting manure, irrespective of whether it is straw or livestock manure, can effectively reduce CH_4 emissions compared to raw manures when used in paddies [9,24,26]. Field experiments conducted by Kim et al. [24] reveal that the application of composted manures reduces CH_4 emissions by up to 50% compared to air-dried manures without pretreatment.

4.2. Insights into the CH_4 Emission Patterns of Soils Treated with NCM and RCM

In the present study, organic manures from NCM and RCM were amended as basal fertilizers without other fertilizers being applied during the whole rice growth period. Before transplantation, all the pots were subjected to flooding for one week, considering the slow release of nutrients contained in the organic manures. As shown in Figure 1a, the CH_4 emissions from all three treatments during the early 20 days after flooding were relatively lower, which could be partially attributed to the weak CH_4 transport of rice seedlings as the rice aerenchyma tissues were not well developed then [37]. It is estimated that approximately 90% of the emitted paddy CH_4 is released to the atmosphere through plant-mediated transport, with only 2–3% percolating through flooded water [38]. In addition, the relatively low temperature during the early experiment could also limit CH_4 production. Then, CH_4 emission rates significantly increased, with the appearance of three CH_4 flux peaks, despite those being much less conspicuous in the control (Figure 1a). It is reported that the occurrences of such CH_4 emission peaks correspond to maxima in air temperature [39]. Whereas, the relatively higher CH_4 emissions observed in Stage 1 with the NCM- and RCM-treated soils could be ascribed to the extra application of organic manures. It is noteworthy that relatively lower CH_4 emissions were detected with the RCM-treated soils compared to those from the NCM-treated soils, and the possible underlying reasons will be interpreted below.

After drainage, the intensity of paddy CH_4 emissions significantly decreased (Figure 1a), with emission rates consistently maintaining low levels, which was extremely apparent in the RCM treatments. Moreover, the CH_4 fluxes between the NCM and Ctrl treatments were very similar, as shown in the zoomed-in view shown on the right. The almost comparative CH_4 fluxes between the NCM and Ctrl treatments could indicate the consumption of the labile C contained in the NCM and the CH_4 production during late rice cultivation, which mostly originated from degradations of exudates and autolysis products of roots. Studies have shown that the presence of rice plants promoted the emission of CH_4 at the rice mature stage [40,41], indicating the important influence of rice roots on CH_4 production in the late growth stage of rice. Whereas, the situation was strange with the RCM-treated soils, whose rice biomasses were 68.7% and 35.4% higher than those of the Ctrl and NCM treatments (data not shown), respectively, which means relatively higher root exudates, but there was almost no CH_4 emission detected with the RCM treatment. Thus, we speculated that either reduced CH_4 production or increased CH_4 oxidation conducted by methanotrophs could take place in the RCM-treated soils.

4.3. The Lower CH_4 Production with the RCM Treatment at Rice Growth Stage 1 Led to the Lower CH_4 Emissions Relative to the NCM Treatment

DOC is the most labile part of soil organic C, which is an important C source and easily accessible for soil microorganisms. A large amount of research has demonstrated that CH_4 emission rates positively correlate with DOC concentrations in soil [24,42]. Soil DOC is the major C source for methanogens and is considered an important factor affecting paddy CH_4 emissions. The dynamics of soil DOC concentrations during rice cultivation are depicted in Figure 2a. Amendment of NCM and RCM significantly increased soil DOC concentration, except in the case of the RCM treatment at 90 d, ultimately resulting in higher emissions of CH_4 by increasing CH_4 production relative to Ctrl.

Additionally, the lower CH_4 emission from the RCM-treated soils relative to that from the NCM-treated soils seemed strange, considering the similar DOC concentrations between these two treatments. However, this paradoxical phenomenon points toward the significance of the functionally diverse microorganisms that are involved in the production and oxidation of paddy CH_4. First, methanogens are the sole producers of paddy CH_4, broadly classified into two groups: hydrogenotrophic and acetoclastic methanogens, which can obtain energy from the reduction of CO_2 with H_2 and the cleavage of acetate, respectively [14]. Organic manure amendments introduce more C into soils, which can provide sufficient C substrates for methanogens, potentially influencing both the abundance and composition of the methanogens and finally having an impact on paddy CH_4 emissions. A field study conducted by Feng et al. [29] reveals that decreased soil DOC content, resulting from elevated ground-level O_3, reduces both the abundance and diversity of the dominant methanogen of *Methanosaeta*, consequently contributing to the reduced CH_4 emission in the plots under elevated ground-level O_3. In the present study, the amendment of NCM seemed to significantly stimulate the growth of methanogens, whereas no obvious enhancement was observed in the methanogenic archaeal population size of the RCM-treated soils over the whole rice growth stage (Figure 3a). Moreover, pyrosequencing results further revealed that NCM application also resulted in apparent shifts in the paddy methanogenic composition, selectively enhancing the growth of the methanogens of *Methanosarcina*; whereas, the same variation was not found in both the RCM and Ctrl treatments (Figure 4a). It is well known that *Methanosarcina* methanogens are metabolically versatile, using both CO_2 and acetate as C sources, and can rapidly grow to become the dominant genera [43]. It is estimated that acetoclastic methanogenesis contributes more than 67% of the paddy CH_4 emissions. Thus, as corresponding results show, the lower relative abundance of *Methanosarcina* in RCM-treated soils may consistently lead to lower CH_4 emissions. Second, the methanotrophic numbers of the RCM-treated soils appeared lower than those of the NCM-treated soils for Stage 1, indicating the lower CH_4 oxidation ability of the RCM-treated soils. Additionally, PCA results also displayed a similar community structure of methanotrophs between all three treatments (Figure 5b). Therefore, we can conclude that, for Stage 1, the relatively lower CH_4 production ability led to the lower CH_4 emission of the RCM treatment compared to that of the NCM treatment.

4.4. The Enhanced CH_4 Oxidation with the RCM Treatment at Rice Growth Stage 2 Contributed to the Lower CH_4 Emission

After drainage, dramatically decreased CH_4 fluxes were observed in all three treatments, and almost no CH_4 emission was detected in the RCM-treated soils. Usually, soil drainage always results in changed soil conditions, primarily increased soil Eh, re-oxidation of soil reductants [19]. In the present study, at the late rice growth stage, the concentrations of NO_3^- and SO_4^{2-} showed apparent increasing trends, which were especially intensive in the RCM-treated soils (Figure 2c,d). The increased NO_3^- and SO_4^{2-} contents could partially be responsible for the extremely low CH_4 emissions of RCM treatment. On one hand, the high content of NO_3^- and SO_4^{2-} could retard the decrease of soil Eh, which could inhibit the CH_4 production of RCM-treated soils as the reduced conditions needed have not been formed [21]. On the other hand, the abundant inorganic acceptors such as

NO_3^- and SO_4^{2-} may activate the corresponding nitrate- and sulfate-reducing bacteria, effectively competing C substrates with methanogens [44]. Coincidentally, as support, both an apparent drop in DOC concentration (Figure 2a) and a significant increase in *dsr* gene copies (Figure 3c) were observed at 90 d with the RCM-treated soils.

Furthermore, the apparent CH_4 emission reduction of RCM treatment at rice growth Stage 2 might be related to the predominance of the methylotrophs belonging to the family *Methylophilaceae* (Figure 4b). More recently, a CH_4 consumption metabolic mode linked to alternative electron donors, such as nitrate/nitrite-dependent anaerobic/microaerobic bacterial methane oxidation in freshwater environments, was proposed [45]. A number of studies have already revealed the cooperation between the *Methylococcaceae* and the *Methylophilaceae* in response to both CH_4 and NO_3^- stimuli, implying the importance of such syntrophism in CH_4 consumption [46,47]. Krause et al. [48] deduced the cross-feeding mechanism and verified methanol as the dominant carbon and energy source the methanotroph provides to support the growth of the nonmethanotrophs. Likewise, in the present study, after drainage, the soil NO_3^- from all three treatments showed obvious accumulations and was obvious in the RCM-treated soils (Figure 2c), coordinately providing conditions for the symbiotic growth of *Methylophilaceae*. This, to some extent, could explain the relatively lower CH_4 emission from the RCM treatment at Stage 2.

5. Conclusions

NCM is a valuable organic fertilizer that contains rich, labile organic C and nutrients. However, the application of NCM to rice soils significantly enhanced CH_4 emission by both promoting the growth of methanogens and shifting the methanogenic archaeal composition, primarily stimulating the growth of *Methanosarcina*, during whole rice cultivation. On the contrary, the incorporation of RCM did not increase CH_4 emissions, which were almost the same as those of the control. Unlike the NCM-treated soils, neither the stimulation of the total methanogenic population size nor the specific methanogenic species, such as *Methanosarcina*, was detected with the RCM-treated soils during the rice growth, which primarily resulted in the low CH_4 production at Stage 1. In addition, the extremely low CH_4 emission from the RCM-treated soils during the late rice growth period (Stage 2) could partially be attributed to aerobic/anaerobic oxidation of CH_4, as the methylotrophy of *Methylophilaceae* species appeared dominant. As a summary, the statistical results of contrasting NCM and RCM, respectively, with control at five sampling time points based on a *t*-test are presented in Table S1.

Supplementary Materials: The following supporting information can be downloaded at: https://www.mdpi.com/article/10.3390/agronomy13051268/s1, Table S1: Statistic results of contrasting NCM and RCM, respectively, with control on five sampling days based on *t*-test.; Table S2: Effects of different treatments on rice biomass.

Author Contributions: Data curation, B.Z. and J.Z.; Funding acquisition, Y.W.; Investigation, B.Z. and S.P.; Methodology, R.C.; Supervision, X.L.; Writing—original draft, B.Z.; Writing—review and editing, Y.W. All authors have read and agreed to the published version of the manuscript.

Funding: This work was supported by the Major Science and Technology Projects of the Inner Mongolia Autonomous Region (NMKJXM202009), CAS Key Technology Talent Program and Project 333 of Jiangsu Province.

Data Availability Statement: Rawreads are deposited in theNCBI under BioProject IDPRJNA861115 in the fastq format (https://www.ncbi.nlm.nih.gov/bioproject/PRJNA861115).

Conflicts of Interest: The authors declare no conflict of interest.

References

1. Kharitonov, S.; Semenov, M.; Sabrekov, A.; Kotsyurbenko, O.; Zhelezova, A.; Schegolkova, N. Microbial Communities in Methane Cycle: Modern Molecular Methods Gain Insights into Their Global Ecology. *Environments* **2021**, *8*, 16. [CrossRef]
2. Qin, Y.; Liu, S.; Guo, Y.; Liu, Q.; Zou, J. Methane and nitrous oxide emissions from organic and conventional rice cropping systems in Southeast China. *Biol. Fertil. Soils* **2010**, *46*, 825–834. [CrossRef]
3. Wang, Z.; Zhang, X.; Liu, L.; Wang, S.; Zhao, L.; Wu, X.; Zhang, W.; Huang, X. Estimates of methane emissions from Chinese rice fields using the DNDC model. *Agric. For. Meteorol.* **2021**, *303*, 108368. [CrossRef]
4. Hou, H.; Yang, S.; Wang, F.; Li, D.; Xu, J. Controlled irrigation mitigates the annual integrative global warming potential of methane and nitrous oxide from the rice-winter wheat rotation systems in Southeast China. *Ecol. Eng.* **2016**, *86*, 239–246. [CrossRef]
5. Lennartz, B.; Horn, R.; Duttmann, R.; Gerke, H.; Tippkötter, R.; Eickhorst, T.; Janssen, I.; Janssen, M.; Rüth, B.; Sander, T. Ecological safe management of terraced rice paddy landscapes. *Soil Tillage Res.* **2009**, *102*, 179–192. [CrossRef]
6. Thangarajan, R.; Bolan, N.S.; Tian, G.L.; Naidu, R.; Kunhikrishnan, A. Role of organic amendment application on greenhouse gas emission from soil. *Sci. Total Environ.* **2013**, *465*, 72–96. [CrossRef]
7. Feng, Y.; Chen, R.; Hu, J.; Zhao, F.; Wang, J.; Chu, H.; Zhang, J.; Dolfing, J.; Lin, X. Bacillus asahii comes to the fore in organic manure fertilized alkaline soils. *Soil Biol. Biochem.* **2015**, *81*, 186–194. [CrossRef]
8. Sun, R.; Zhang, X.; Guo, X.; Wang, D.; Chu, H. Bacterial diversity in soils subjected to long-term chemical fertilization can be more stably maintained with the addition of livestock manure than wheat straw. *Soil Biol. Biochem.* **2015**, *88*, 9–18. [CrossRef]
9. Chen, R.; Lin, X.; Wang, Y.; Hu, J. Mitigating methane emissions from irrigated paddy fields by application of aerobically composted livestock manures in eastern China. *Soil Use Manag.* **2011**, *27*, 103–109. [CrossRef]
10. Lee, C.H.; Kim, S.Y.; Villamil, M.B.; Pramanik, P.; Hong, C.O.; Kim, P.J. Different response of silicate fertilizer having electron acceptors on methane emission in rice paddy soil under green manuring. *Biol. Fertil. Soils* **2012**, *48*, 435–442. [CrossRef]
11. Kim, S.Y.; Pramanik, P.; Bodelier, P.L.E.; Kim, P.J. Cattle Manure Enhances Methanogens Diversity and Methane Emissions Compared to Swine Manure under Rice Paddy. *PLoS ONE* **2014**, *9*, 113593. [CrossRef]
12. Yao, H.; Conrad, R. Effect of temperature on reduction of iron and production of carbon dioxide and methane in anoxic wetland rice soils. *Biol. Fertil. Soils* **2000**, *32*, 135–141. [CrossRef]
13. Kruger, M.; Frenzel, P.; Kemnitz, D.; Conrad, R. Activity, structure and dynamics of the methanogenic archaeal community in a flooded Italian rice field. *FEMS Microbiol. Ecol.* **2005**, *51*, 323–331. [CrossRef]
14. Conrad, R. Methane production in soil environments—Anaerobic biogeochemistry and microbial life between flooding and desiccation. *Microorganisms* **2020**, *8*, 881. [CrossRef]
15. Angel, R.; Claus, P.; Conrad, R. Methanogenic archaea are globally ubiquitous in aerated soils and become active under wet anoxic conditions. *ISME J.* **2012**, *6*, 847–862. [CrossRef]
16. Paul, K.; Nonoh, J.O.; Mikulski, L.; Brune, A. "Methanoplasmatales", Thermoplasmatales-related archaea in termite guts and other environments, are the seventh order of methanogens. *Appl. Environ. Microb.* **2012**, *78*, 8245–8253. [CrossRef]
17. Ma, K.; Lu, Y. Regulation of microbial methane production and oxidation by intermittent drainage in rice field soil. *FEMS Microbiol. Ecol.* **2011**, *75*, 446–456. [CrossRef]
18. Chistoserdova, L. Methylotrophy in a lake: From metagenomics to single-organism physiology. *Appl. Environ. Microb.* **2011**, *77*, 4705–4711. [CrossRef]
19. Wang, J.; Chen, Z.; Ma, Y.; Sun, L.; Xiong, Z.; Huang, Q.; Sheng, Q. Methane and nitrous oxide emissions as affected by organic-inorganic mixed fertilizer from a rice paddy in southeast China. *J. Soil. Sediments* **2013**, *13*, 1408–1417. [CrossRef]
20. Deng, Y.; Che, R.; Wu, Y.; Wang, Y.; Cui, X. A review of the physiological and ecological characteristics of methanotrophs and methanotrophic community diversity in the natural wetlands. *Acta Ecol. Sin.* **2015**, *35*, 4579–4591.
21. Linquist, B.A.; Adviento-Borbe, M.A.; Pittelkow, C.M.; Kessel, C.V.; Groenigen, K.J.V. Fertilizer management practices and greenhouse gas emissions from rice systems: A quantitative review and analysis. *Field Crop. Res.* **2012**, *135*, 10–21. [CrossRef]
22. Zhou, S.; Xu, J.; Yang, G.; Zhuang, L. Methanogenesis affected by the co-occurrence of iron(III) oxides and humic substances. *FEMS Microbiol. Ecol.* **2014**, *88*, 107–120. [CrossRef] [PubMed]
23. Luo, D.; Meng, X.; Zheng, N.; Li, Y.; Yao, H.; Chapman, S.J. The anaerobic oxidation of methane in paddy soil by ferric iron and nitrate, and the microbial communities involved. *Sci. Total Environ.* **2021**, *788*, 147773. [CrossRef] [PubMed]
24. Kim, S.Y.; Pramanik, P.; Gutierrez, J.; Hwang, H.Y.; Kim, P.J. Comparison of methane emission characteristics in air-dried and composted cattle manure amended paddy soil during rice cultivation. *Agric. Ecosyst. Environ.* **2014**, *197*, 60–67. [CrossRef]
25. Jeong, S.T.; Kim, G.W.; Hwang, H.Y.; Kim, P.J.; Kim, S.Y. Beneficial effect of compost utilization on reducing greenhouse gas emissions in a rice cultivation system through the overall management chain. *Sci. Total Environ.* **2018**, *613*, 115–122. [CrossRef]
26. Zhou, B.; Wang, Y.; Feng, Y.; Lin, X. The application of rapidly composted manure decreases paddy CH_4 emission by adversely influencing methanogenic archaeal community: A greenhouse study. *J. Soil. Sediments* **2016**, *16*, 1889–1990. [CrossRef]
27. Li, Z.; Han, C.; Han, F. Organic C and N mineralization as affected by dissolved organic matter in paddy soils of subtropical China. *Geoderma* **2010**, *157*, 206–213. [CrossRef]
28. Lueders, T.; Friedrich, M.W. Effects of amendment with ferrihydrite and gypsum on the structure and activity of methanogenic populations in rice field soil. *Appl. Environ. Microb.* **2002**, *68*, 2484–2494. [CrossRef]

29. Feng, Y.; Xu, Y.; Yu, Y.; Xie, Z.; Lin, X. Mechanisms of biochar decreasing methane emission from Chinese paddy soils. *Soil Biol. Biochem.* **2012**, *46*, 80–88. [CrossRef]
30. Kondo, R.; Nedwell, D.B.; Purdy, K.J.; Silva, S.Q. Detection and enumeration of sulphate-reducing bacteria in estuarine sediments by competitive PCR. *Geomicrobiol. J.* **2004**, *21*, 145–157. [CrossRef]
31. Caporaso, J.G.; Kuczynski, J.; Stombaugh, J.; Bittinger, K.; Knight, R. QIIME allows analysis of high-throughput community sequencing data. *Nat. Methods* **2010**, *7*, 335–336. [CrossRef]
32. Chen, T.; Liu, Y.; Huang, L. ImageGP: An easy-to-use data visualization web server for scientific researchers. *iMeta* **2022**, *1*, e5. [CrossRef]
33. Tian, S.; Zhu, B.; Yin, R.; Wang, M.; Jiang, Y.; Zhang, C.; Li, D.; Chen, X.; Kardol, P.; Liu, M. Organic fertilization promotes crop productivity through changes in soil aggregation. *Soil Biol. Biochem.* **2022**, *165*, 108533. [CrossRef]
34. Bhunia, S.; Bhowmik, A.; Mallick, R.; Mukherjee, J. Agronomic efficiency of animal-derived organic fertilizers and their effects on biology and fertility of soil: A review. *Agronomy* **2021**, *11*, 823. [CrossRef]
35. Shi, A.; Nunan, N.; Figueira, J.; Herrmann, A.M.; Wetterlind, J. Long-term ley and manure managements have consistent effects on microbial functional profiles and organic C groups across soils from a latitudinal gradient. *Agron. Sustain. Dev.* **2022**, *42*, 107. [CrossRef]
36. Nguyen, B.T.; Trinh, N.N.; Bach, Q.V. Methane emissions and associated microbial activities from paddy salt-affected soil as influenced by biochar and cow manure addition. *Appl. Soil Ecol.* **2020**, *152*, 103531. [CrossRef]
37. Ding, H.; Jiang, Y.; Cao, C. Deep rice root systems reduce methane emissions in rice paddies. *Plant Soil* **2021**, *468*, 337–352. [CrossRef]
38. Bhattacharyya, P.; Dash, P.K.; Swain, C.K.; Padhy, S.R.; Roy, K.S.; Neogi, S.; Berliner, J.; Adak, T.; Pokhare, S.S.; Baig, M.J.; et al. Mechanism of plant mediated methane emission in tropical lowland rice. *Sci. Total Environ.* **2019**, *651*, 84–92. [CrossRef]
39. Watanabe, A.; Yamada, H.; Kimura, M. Effects of shifting growth stage and regulating temperature on seasonal variation of CH_4 emission from rice. *Glob. Biogeochem. Cycles* **2001**, *15*, 729–739. [CrossRef]
40. Jia, Z.; Cai, Z.; Xu, H.; Li, X. Effect of rice plants on CH4 production, transport, oxidation and emission in rice paddy soil. *Plant Soil* **2001**, *230*, 211–221. [CrossRef]
41. Zou, J.; Huang, Y.; Zong, L.; Zheng, X.; Wang, S. A field study on CO_2, CH_4 and N_2O emissions from rice paddy and impact factors. *Acta Sci. Circum.* **2003**, *23*, 758–764.
42. Lou, Y.S.; Inubushi, K.; Mizuno, T.; Hasegawa, T.; Lin, Y.; Sakai, H.; Cheng, W.; Kobayashi, K. CH_4 emission with differences in atmospheric CO_2 enrichment and rice cultivars in a Japanese paddy soil. *Glob. Chang. Biol.* **2008**, *14*, 2678–2687. [CrossRef]
43. Dong, D.; Li, J.; Ying, S.; Wu, J.; Han, X.; Teng, Y.; Zhou, M.; Ren, Y.; Jiang, P. Mitigation of methane emission in a rice paddy field amended with biochar-based slow-release fertilizer. *Sci. Total Environ.* **2021**, *792*, 148460. [CrossRef] [PubMed]
44. Andrews, S.E.; Schultz, R.; Frey, S.D.; Bouchard, V.; Varner, R.; Ducey, M.J. Plant community structure mediates potential methane production and potential iron reduction in wetland mesocosms. *Ecosphere* **2013**, *4*, 44. [CrossRef]
45. Wu, M.L.; Ettwig, K.F.; Jetten, M.S.; Strous, M.; Keltjens, J.T.; Niftrik, L. A new intra-aerobic metabolism in the nitrite-dependent anaerobic methane-oxidizing bacterium Candidatus '*Methylomirabilis oxyfera*'. *Biochem. Soc. Trans.* **2011**, *39*, 243–248. [CrossRef]
46. Beck, D.A.C.; Kalyuzhnaya, M.G.; Malfatti, S.; Tringe, S.G.; del Rio, T.G.; Ivanova, N.; Lidstrom, M.E.; Chistoserdova, L. A metagenomic insight into freshwater methane-utilizing communities and evidence for cooperation between the Methylococcaceae and the Methylophilaceae. *PeerJ* **2013**, *1*, 23. [CrossRef]
47. Chistoserdova, L. Methylotrophs in natural habitats: Current insights through metagenomics. *Appl. Environ. Microb.* **2015**, *99*, 5763–5779. [CrossRef]
48. Krause, S.M.B.; Johnson, T.; Karunaratne, Y.S.; Fu, Y.; Lidstrom, M. Lanthanide-dependent cross-feeding of methane-derived carbon is linked by microbial community interactions. *Proc. Natl. Acad. Sci. USA* **2017**, *114*, 358–363. [CrossRef]

agronomy

Article

Changes in Soil Rhizobia Diversity and Their Effects on the Symbiotic Efficiency of Soybean Intercropped with Maize

Zeyu Cheng [1,†], Lingbo Meng [2,†], Tengjiao Yin [1], Ying Li [1], Yuhang Zhang [1] and Shumin Li [1,*]

1 Resource and Environmental College, Northeast Agricultural University, Harbin 150030, China
2 School of Geography and Tourism, Harbin University, Harbin 150086, China
* Correspondence: shuminli@neau.edu.cn
† These authors contributed to this work equally.

Abstract: It has been established that maize/soybean intercropping can improve nitrogen use efficiency. However, few studies have addressed how maize/soybean intercropping affects nitrogen-fixing bacterial diversity and N fixation efficiency of intercropped soybean. In this study, nitrogen-fixing bacterial communities, N fixation efficiency, and their relationships with soil properties under three nitrogen fertilization application rates (N0 0 kg/ha, N1 40 kg/ha, N2 80 kg/ha) were explored through field experiments. Nitrogen fixation and nitrogen-fixing bacteria diversity were assessed using ^{15}N natural abundance, Illumina high-throughput sequencing, and *nifH* (nitrogen fixation) gene copies quantification in the rhizosphere soil of intercropped soybean. The results showed that nitrogen application rates significantly decreased the nitrogen-fixing bacteria diversity, nitrogen fixation efficiency, and *nifH* gene copies in the rhizosphere soil. Nitrogen fixation efficiency, nodule number, and dry weight of intercropped soybean were highest in the N0 treatment, and nitrogen fixation was the highest in the N1 treatment. The nitrogen-fixing efficiency in N0, N1, and N2 treatments increased by 69%, 59%, and 42% and the nodule number of soybean was 10%, 22%, and 21%, respectively, compared with monocultures. The soybean nitrogen-fixing bacteria diversity in intercropping under N0 and N1 treatments significantly increased compared with monocultures. There was a significant positive correlation between soil *nifH* gene copies and N fixation efficiency and a negative correlation with soil available nitrogen. *Bradyrhizobium* abundance in soybean rhizosphere soil decreased significantly with the increase in nitrogen application rates and was significantly correlated with soil AN (available nitrogen) and pH content in the soybean rhizosphere. These results help us to understand the mechanisms by which nitrogen use efficiency was improved, and nitrogen fertilizer could be reduced in legume/Gramineae intercropping, which is important to improve the sustainability of agricultural production.

Keywords: nitrogen-fixing efficiency; nitrogen-fixing bacteria; intercropped soybean; ^{15}N natural abundance; *Bradyrhizobium*

Citation: Cheng, Z.; Meng, L.; Yin, T.; Li, Y.; Zhang, Y.; Li, S. Changes in Soil Rhizobia Diversity and Their Effects on the Symbiotic Efficiency of Soybean Intercropped with Maize. *Agronomy* **2023**, *13*, 997. https://doi.org/10.3390/agronomy13040997

Academic Editors: Yong-Xin Liu and Peng Yu

Received: 28 February 2023
Revised: 24 March 2023
Accepted: 26 March 2023
Published: 28 March 2023

1. Introduction

Intercropping systems are among the important cultural practices to further increase yield in China's agriculture. Selecting the appropriate crop intercropped, yield [1], and nutrient use efficiency [2] have more significant advantages than the corresponding monocropping. Many legumes/gramineous intercropping combinations, such as wheat/soybean [3], cowpea/corn [4], lentil/barley, etc., are common beneficial combination systems. Maize/soybean intercropping is a frequently used intercropping planting pattern and has been found to have certain advantages in nutrient utilization, significantly increasing the nitrogen equivalent ratio [5]. At the same time, maize/soybean intercropping can not only significantly promote rhizosphere microbial abundance but also affect rhizosphere microbial diversity [6,7]. Some studies showed that Gramineae could absorb nitrogen fixed by intercropped legumes and, at the same time, secrete root exudates to influence legumes' N fixation [8–10]. Except that, in the legume/cereal intercropping system, due to the large nitrogen amount absorbed by the cereal crops, soil mineral nitrogen is maintained at a relatively low level, reducing the nitrogen-induced inhibition of legume nitrogen fixation, which can promote legume nitrogen fixation from the air to satisfy their own growth [11]. In addition, due to the ecological niche differences [12], the two intercropping crops have different plant heights and alternate planting rows, which reduces the competitive effect. Moreover, the different root depths of the two crops lead to growth at different soil levels, reducing their competition. These would also lead to differences in nitrogen fixation efficiency in intercropping compared to monoculture. Studies have shown that intercropping significantly increased crop yield and promoted nodulation and nitrogen fixation of legumes [13]. The nitrogen application levels and interspecific interaction in intercropping systems affected the complementary utilization of nutrients, growth of nodules, and nitrogen fixation of legumes. Therefore, the advantages of intercropping partially depend on below-ground interspecific plant interactions [14], which include interspecific facilitation, such as the complementary utilization of N resources [10] and niche differences [15].

In general, legumes fix air nitrogen by forming a symbiotic system between nitrogen-fixing soil microorganisms and legume roots. Plants can only absorb free nitrogen in the atmosphere under the action of nitrogen-fixing microorganisms. Nitrogen-fixing bacteria are among the bacteria involved in nitrogen fixing. Rhizosphere nitrogen bacterial diversity and structure will influence the nitrogen fixation of legumes in intercropping. Different planting cultivation would affect soil microbial community structure, microbial quantity, and microbial activity [16]. Therefore, it is necessary to study the changes of nitrogen-fixing bacteria in the rhizosphere to verify the mechanism of high N fixation efficiency in the maize/soybean intercropping system. It has been shown that the intercropping of legumes and grasses could significantly increase the abundance of rhizosphere soil microorganisms [17,18]. Nitrogen-fixing bacteria regulate the nitrogen-fixing process through the 70K Da nitrogen-fixing enzyme encoded by the *nifH* (nitrogen fixation) gene [19]. Therefore, exploring the changes in the *nifH* gene number in the rhizosphere soil is helpful in clarifying the influence on legume fixation efficiency and development.

Nitrogen-fixing bacteria are important in the soil nitrogen cycle [20] and are affected by soil nitrogen levels. The change in Azotobacter diversity is beneficial to plant growth [21]. A certain amount of nitrogen fertilizer would improve the nitrogen-fixing efficiency of nitrogen-fixing bacteria in soybean [22]. Nitrogen application levels have an important effect on rhizosphere microorganisms and community growth and development changes. Some studies have found that soil available nitrogen has a significant effect on bacterial community composition [23], and short-term fertilization also changes the community composition of rhizobia [24]. Fertilization increased the abundance of nitrogen-fixing bacteria species and was beneficial to increase the number and variety of nitrogen-fixing bacterial genes in the black soil area of Northeast China [25]. Evaluation of rhizosphere *nifH* gene diversity in two sorghum (*Sorghum Moench*) varieties suggested that nitrogen application was the main factor affecting the nitrogen-fixing bacteria community structure in sorghum [26]. However, long-term fertilization would greatly inhibit bacterial nitrogen fixation; other studies showed that the increase in nitrogen fertilizer inhibited the expression of the *nifH* gene [5]. Nitrogen-rich soil inhibited soybean symbiosis and nitrogen fixation, which resulted in decreased soybean nodule quantity and size and decreased nitrogen fixation activity [27]. Planting methods and soil nutrients are important factors determining rhizosphere microbial diversity and community composition [28]. However, there is a limited number of studies about the community structure of nitrogen-fixing bacteria and *nifH* gene abundance in the maize/soybean intercropping under different N application rates.

Therefore, in the present experiment, three nitrogen application levels were evaluated in the maize/soybean intercropping system. The ^{15}N natural abundance method, high-throughput sequencing technology, and real-time fluorescent quantitative PCR were used to study the community structure of nitrogen-fixing bacteria and *nifH* gene abundance in the soybean rhizosphere. The aims of this study were: (i) to quantify the nitrogen fixation efficiency of intercropped soybean as affected by N application rates. (ii) to evaluate the changes of nitrogen-fixing bacteria community diversity and abundance in the rhizosphere of intercropped soybean; (iii) to reveal the relationship between *nifH* gene abundance and nitrogen fixation efficiency of intercropped soybean. These results will further explain why N utilization efficiency is improved in the maize/soybean intercropping system, which is valuable for reducing N use in agricultural ecological systems.

2. Materials and Methods

2.1. Experimental Design

A field experiment was conducted at the experimental station of Northeast Agricultural University, Acheng District, Heilongjiang Province, China. The station is in the temperate continental monsoon climate, with an annual precipitation of 530 mm and annual accumulated temperature of 2800 degrees Celsius. The soil type was black soil, which is mollisol according to the USDA soil classification. The basic physical and chemical properties of the soil were as follows: soil OM (organic matter): 29.6 g/kg; TN (total N): 1.45 g/kg; (AP) available P: (Olsen P) 28.9 mg/kg; AK (available K): 122.52 mg/kg; AN (available nitrogen): 126 mg/kg; pH: 6.1 (1:2.5 w/v water).

The experiment was conducted in a two-factor split-plot design. The main factor was nitrogen application, with the three nitrogen application rates (N0, N1, and N2) in monoculture soybean and maize/soybean intercropping systems, respectively. The N0, N1, and N2 application rates were 0 kg N/ha, 40 kg N/ha, and 80 kg N/ha, respectively. The planting patterns were the secondary factor w, namely soybean monocropping (SS) and maize/soybean intercropping (IS). In the monoculture plots, soybean was grown throughout the eight rows. Soybean plant spacing was 8.5 cm, and planting density was 196,000/ha. Maize row spacing was 60 cm, and plant spacing was 17 cm. Urea was used as a nitrogen fertilizer for each treatment, with the same application amount in intercropping and monoculture. The basic fertilizer was applied with soybean per the standard treatment guidelines, and the application amount was 0 kg N/ha, 40 kg N/ha, and 80 kg N/ha, respectively. Superphosphate and potassium sulfate was used as the phosphorous potassium fertilizer 120 kg P_2O_5/ha was supplied as triple superphosphate, and 100 kg K_2O/ha was supplied as potassium sulfate. The experiment had 6 treatments with 3 replications and a total of 18 plots. Each plot's dimensions were 4.8 m (width) × 10 m (length) and included eight rows with a row spacing of 0.6 m. The maize variety "Xianyu 335" and soybean variety "Dongnong 252" were planted. The sowing time was 2 May 2019, and the harvest time was 1 October 2019.

2.2. Plant Sampling and Analysis

Five soybean plants were taken at random in the sampling area at the mature stages of the crop. After the harvest, plants were taken back to the laboratory, and the stalks and grains were separated, dried in an oven at 105 °C for 30 min, and then dried at 80 °C to a constant weight. The samples were ground thoroughly for the ^{15}N abundance determination and N concentration analysis.

The samples were digested with H_2SO_4-H_2O_2, and the N concentration was measured by the Kjeldahl method. The ^{15}N abundance was measured with a stable isotope mass spectrometer (ISOPRIME).

2.3. Analysis of Soil Physicochemical Characteristics

The Kjeldahl distillation method [29] was used for the total nitrogen (TN) determination. The available P (AP) was extracted with a 0.5 mol/L $NaHCO_3$ solution and then measured using the molybdenum antimony-D-isoascorbic acid colorimetry (MADAC) method [30]. The available potassium (AK) was determined by flame photometry using a 1 mol/L NH_4OAc neutral extraction. The available nitrogen (AN) was determined by the diffusion absorption method [31]. The soil pH was measured in a 1:2.5 soil–water suspension using a glass electrode. The soil organic matter (OM) was determined by the method of soil digestion with hot acid dichromate [32].

2.4. Measurement of Soybean Nodules and Nodules Dry Weight

Soybean root nodules were collected at the pod setting stage. Five soybean plants from each treatment were randomly selected for the determination. The roots of soybean plants were excavated from the field (20 cm(long) × 20 cm(wide) × 25 cm(deep)), then placed in sealed bags and taken then back to the laboratory. Roots were shaken off the excess loose soil on a kraft paper. The soil was collected and sieved to wash off the soil and pick out the nodules. The soil attached to the roots was brushed off and passed through a 2 mm sieve. The soil samples were then placed into a centrifuge tube and stored in a −80 °C refrigerator for the analysis of nitrogen-fixing bacteria diversity. The nodules attached to the roots and the sieves were counted. Then the nodules were dried at 80 °C to a constant weight, and their dry weight was taken.

2.5. Determination of Nitrogen-Fixing Bacteria Diversity

2.5.1. DNA Extraction and PCR Amplification

Total microbial genomic DNA was extracted from rhizosphere soil samples using the E.Z.N.A.® soil DNA Kit (Omega Bio-tek, Norcross, GA, USA) according to the manufacturer's instructions. The quality and concentration of DNA were determined by 1.0% agarose gel electrophoresis and a NanoDrop® ND-2000 spectrophotometer (Thermo Scientific Inc., Waltham, MA, USA) and were kept at −80 °C prior to further use. The *nifH-F_nifH-R* variable region was amplified by PCR with *nifH-F* (AAAGGYGGWATCG-GYAARTCCACCAC) and *nifH-R* (TTGTTSGCSGCRTACATSGCCATCAT) [33,34] primers and sequenced using the Illumina Miseq platform. The PCR reaction mixture including 4 µL 5 × Fast Pfu buffer, 2 µL 2.5 mM dNTPs, 0.8 µL Forward Primer (5 µM), 0.8 µL Reverse Primer (5 µM), 0.4 µL Fast Pfu polymerase, 0.2 µlBSA,10 ng of template DNA, and ddH_2O to a final volume of 20 µL. PCR amplification cycling conditions were as follows: initial denaturation at 95 °C for 3 min, followed by 35 cycles of denaturing at 95 °C for 30 s, annealing at 55 °C for 30 s and extension at 72 °C for 45 s, and single extension at 72 °C for 10 min, and end at 10 °C. All samples were amplified in triplicate. The PCR product was extracted from 2% agarose gel and purified using the AxyPrep DNA Gel Extraction Kit (Axygen Biosciences, Union City, CA, USA) according to the manufacturer's instructions and was quantified using Quantus™ Fluorometer (Promega, Madison, WI, USA).

2.5.2. Illumina MiSeq Sequencing

Purified amplicons were pooled in equimolar amounts and paired-end sequenced on an Illumina MiSeq PE300 platform (Illumina, San Diego, CA, USA) according to the standard protocols by Majorbio Bio-Pharm Technology Co. Ltd. (Shanghai, China). The raw sequencing reads were deposited into the NCBI Sequence Read Archive (SRA) database (Accession Number: PRJNA945238).

2.5.3. Data Processing

Raw FASTQ files were de-multiplexed using an in-house perl script and then quality-filtered by fastp version 0.19.6 and merged by FLASH version 1.2.7 with the following criteria:

(i) the 300 bp reads were truncated at any site receiving an average quality score of <20 over a 50 bp sliding window, and the truncated reads shorter than 50 bp were discarded; reads containing ambiguous characters were also discarded; (ii) only overlapping sequences longer than 10 bp were assembled according to their overlapped sequence. The maximum mismatch ratio of the overlap region is 0.2. Reads that could not be assembled were discarded; (iii) samples were distinguished according to the barcode and primers, and the sequence direction was adjusted, exact barcode matching, and 2 nucleotide mismatches in primer matching. The bacterial composition and differences between groups were analyzed and compared on the Majorbio I-Sanger cloud platform and in the *nifH* database. The RDP classifier Bayes Algorithm was used to perform taxonomic analysis on 97% similar OTU representative sequences, and statistics at each classification level in each sample community were performed (http://sourceforge.net/projects/rdp-classifier/ accessed on 12 December 2019). This part of the sequencing process was completed by Shanghai Majorbio Bio-pharm Technology Co., Ltd., (Shanghai, China).

2.6. Determination of nifH Gene

ABI 7500 fluorescence quantitative PCR instrument (USA) and SYBRGreen Real-Time PCR kit (ABI Power SybrGreen qPCR Master Mix(2X)) were used. The copy number of bacterial nifH gene was quantitatively analyzed by the SYBRGreen I method. The primers were *nifH-F* (AAAGGYGGWATCGGYAARTCCACCAC) and *nifH-R* (TTGTTSGCSGCRTACATSGCCATCAT) [33,34].

The conversion formula of gene copy number is (copies/μL) = concentration (ng/μL) $\times 10^{-9} \times 6.02 \times 10^{23}$/(molecular weight × 660).

The copy number was calibrated using CT = −k lg X0 + b. The number of qpcr was repeated 3 times, which was normalized by internal reference genes.

Preparation of the standard curve: ten times gradient dilution of each constructed plasmid; 45 μL diluent + 5 μL plasmid; generally conduct 4–6 points, through the preliminary experiment, respectively, and select 10^{-3}~10^{-8} diluent of standard for the preparation of the standard curve.

2.7. Method of Calculating Soybean Nitrogen Fixation Efficiency

After the samples were harvested, $\delta^{15}N$ data were obtained by the elemental analyser-isotope ratio mass spectrometer, and then were calculated:

$$\delta^{15}N = [\text{atom\%}\ ^{15}N_{(sample)} - \text{atom\%}\ ^{15}N_{(standard)}]/\text{atom\%}\ ^{15}N_{(standard)} \times 1000$$

$\delta^{15}N$ is the difference between ^{15}N of the sample and ^{15}N of the atmosphere, atom% $^{15}N_{(sample)}$ is the ^{15}N atomic abundance of the sample, and atom% $^{15}N_{(standard)}$ is the atmospheric standard ^{15}N atomic abundance (0.3663%).

Since grain and straw were treated separately, the above formula $\delta^{15}N$ was calculated by the weighted average of the two ^{15}N.

$$\delta^{15}N_{maize} = [N_{maize\ grain} \times \delta^{15}N_{maize\ grain} + N_{maize\ stover} \times \delta^{15}N_{maize\ stover}]/(N_{maize\ grain} + N_{maize\ stover})$$

$$\delta^{15}N_{soybean} = [N_{soybean\ grain} \times \delta^{15}N_{soybean\ grain} + N_{soybean\ stover} \times \delta^{15}N_{soybean\ stover}]/(N_{soybean\ grain} + N_{soybean\ stover})$$

In the above formula, $\delta^{15}N_{maize\ grain}$ represents $\delta^{15}N$ of maize grain, $\delta^{15}N_{maize\ stover}$ represents $\delta^{15}N$ of maize stover, $\delta^{15}N_{soybean\ grain}$ represents $\delta^{15}N$ of soybean grain, and $\delta^{15}N\ s_{oybean\ stover}$ represents $\delta^{15}N$ of soybean stalk.

$$\%Ndfa = (\delta^{15}N_{maize} - \delta^{15}N_{soybean})/(\delta^{15}N_{maize} - B) \times 100$$

The B value corresponds to the $\delta^{15}N$ value of the nitrogen-fixing soybean plants with no nitrogen supplied but with all other nutrient requirements supplemented. Thus, it considers legume plants that use atmospheric nitrogen as the only source of nitrogen.

Ndfa = %Ndfa × (Nsoybean grain + Nsoybean straw) corresponds to the nitrogen fixing of nitrogen-fixing bacteria in soybean from the air.

Nitrogen uptake = (Nsoybean grain + Nsoybean straw) − Ndfa [35].

2.8. Statistical Analysis

The statistical analysis was carried out by R language (R, V4.2.1); that is, the difference of different treatments was analyzed by one-way variance analysis (ANOVA), and the significance of the difference by Duncan multiple repetition range test ($p < 0.05$); the difference of two-ways ANOVA variance analyzed the differences among different factors. The R language "fBasics package" checks the normal distribution; "ggplot2 package" was used for data visualization and analysis; "vegan package" for RDA analysis; "ggcor package" for Mantel Test analysis; "cancor package" for Person correlation analysis.

3. Results

3.1. Soybean Nitrogen Fixation Efficiency, Biological N Fixation and N Uptake from the Soil

The nitrogen-fixing efficiency of nitrogen-fixing bacteria in soybean was significantly affected by nitrogen application rates ($p < 0.01$) and cultivation patterns ($p < 0.01$). The soybean nitrogen fixation efficiency decreased as nitrogen application rates increased, with rates of 69%, 59%, and 42% measured in the N0, N1, and N2 treatments, respectively. In contrast, intercropped soybean showed significantly higher nitrogen fixation efficiency compared to monocropped soybean, with increases of 20%, 21%, and 10% observed compared to monoculture, respectively (Figure 1a).

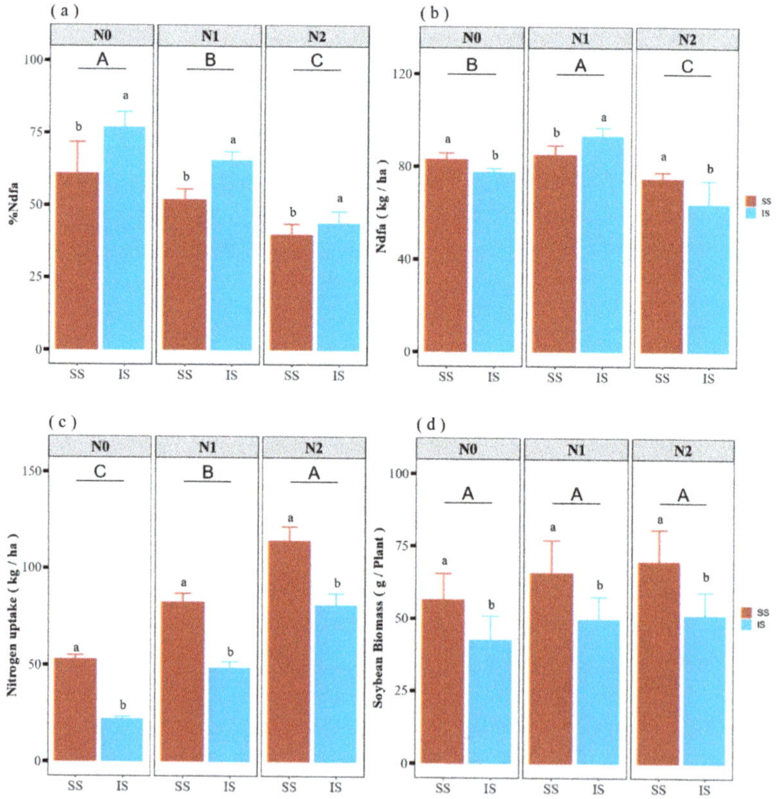

Figure 1. Nitrogen fixation efficiency (**a**) and biological nitrogen fixation (**b**) and nitrogen uptake in the soil (**c**) and biomass (**d**) of soybeans under different nitrogen application levels. Different lowercase letters indicate significant differences between single intercropping at the same nitrogen supply level ($p < 0.05$), and different capital letters indicate significant differences between the same nitrogen supply level ($p < 0.05$). Data are presented as a Mean ± SD ($n = 5$).

The biological nitrogen fixing of nitrogen-fixing bacteria in soybean was determined by the ^{15}N method. The soybean biological nitrogen fixation was significantly influenced by nitrogen levels ($p < 0.01$) and cultivation patterns ($p < 0.01$) and was 80 kg/ha, 91 kg/ha, and 70 kg/ha (average) in the N0, N1, and N2 treatments, respectively. However, the biological nitrogen fixation in the N1 treatment was significantly higher compared to other treatments and was increased by 4% compared to monocropping. On the other hand, biological nitrogen fixation in monocropping was 8% and 16% higher compared to intercropping in the N0 and N2 treatments (Figure 1b). In contrast, soil nitrogen uptake by soybean showed an opposite response to that of nitrogen fixation efficiency with regard to intercropping (Figure 1c). Nitrogen uptake from the soil increased with nitrogen application rates ($p < 0.01$). However, in the soybean–maize intercropping system, nitrogen uptake from the soil decreased by 57%, 41%, and 32% in N0, N1, and N2 treatments, respectively, compared to the monoculture ($p < 0.01$).

The soybean biomass showed an increasing trend with the increase of nitrogen application rate, but the effect of nitrogen application rate on the soybean biomass did not reach a significant level (Figure 1d). On the other hand, the soybean biomass was significantly influenced by different cultivation methods, and it was significantly higher under monoculture than intercropping. Specifically, the soybean biomass under monoculture increased by 32.35%, 32.20%, and 35.98% compared with intercropping, respectively, at N0, N1, and N2 levels.

3.2. Soybean Nodules Numbers and Nodules Dry Weight

Nodule formation is a crucial factor for symbiotic nitrogen fixation and a key indicator for studying biological nitrogen fixation. The number of nodules decreased significantly with increasing nitrogen application rates. However, when soybean was intercropped with maize, the number of nodules increased by 10%, 22%, and 21% in the N0, N1, and N2 treatments compared to the monoculture (Table 1, $p < 0.01$). No significant differences were observed in nodule dry weight among different treatments. On average, the soybean nodules' dry weight in the N2 treatment was lower than in the N0 and N1 treatments.

Table 1. Soybean nodules numbers and nodules dry weight with different nitrogen application rates.

Treatments	No.of Nodule Per Plant (no/Plant)			DM of Nodule Per Plant (g/Plant)		
	SS	IS	AVE	SS	IS	AVE
N0	196 ± 6.02 b	218 ± 4.35 a	207 ± 17.91 A	0.87 ± 0.15 a	0.99 ± 0.16 a	0.93 ± 0.15 A
N1	164 ± 9.45 b	209 ± 3.51 a	187 ± 13.47 B	0.80 ± 0.09 a	0.88 ± 0.02 a	0.84 ± 0.07 A
N2	119 ± 11.2 b	151 ± 14.4 a	135 ± 21.16 C	0.51 ± 0.01 a	0.57 ± 0.11 a	0.54 ± 0.08 B
ANOVE		*p*			*p*	
N		0.00			0.00	
C		0.00			0.11	
N×C		0.03			0.93	

NOTE: N indicates nitrogen application rates, C indicates planting patterns; N×C shows the interaction between nitrogen application rates and planting patterns. Different lowercase letters indicate significant differences between single intercropping at the same nitrogen supply level, and different capital letters indicate significant differences between different nitrogen supply levels. The following table is the same. Data are presented as a Mean ± SD ($n = 5$).

3.3. Alpha Diversity and Composition of Nitrogen-Fixing Bacteria of Soybean Rhizosphere Soil

A total of 115396 reads were obtained from sequencing and 66.22% of them could be assigned. The Shannon index (Figure 2a) and Ace index (Figure 2b) of the nitrogen-fixing bacteria community in soybean rhizosphere soil were significantly influenced by the nitrogen application rates ($p < 0.01$). The Shannon index decreased when the nitrogen application rates increased from N0 to N1, while no significant difference was observed between N1 and N2 treatments. Intercropping maize significantly increased the Shannon index in the soybean rhizosphere compared to monocropping in N0 (0 kg N/ha) and N1 (40 kg N/ha) treatments. However, the Shannon index of intercropped soybean was significantly decreased by 1.05 times in N2 (80 kg N/ha) treatment when compared to the corresponding monocropping cultivation. Similar results were found for the ACE index in soybean rhizosphere soil. Nitrogen application significantly influenced the ACE index

($p < 0.05$), while planting patterns did not have a significant effect. However, a significant interaction was observed between nitrogen application and planting patterns on the ACE index ($p < 0.01$).

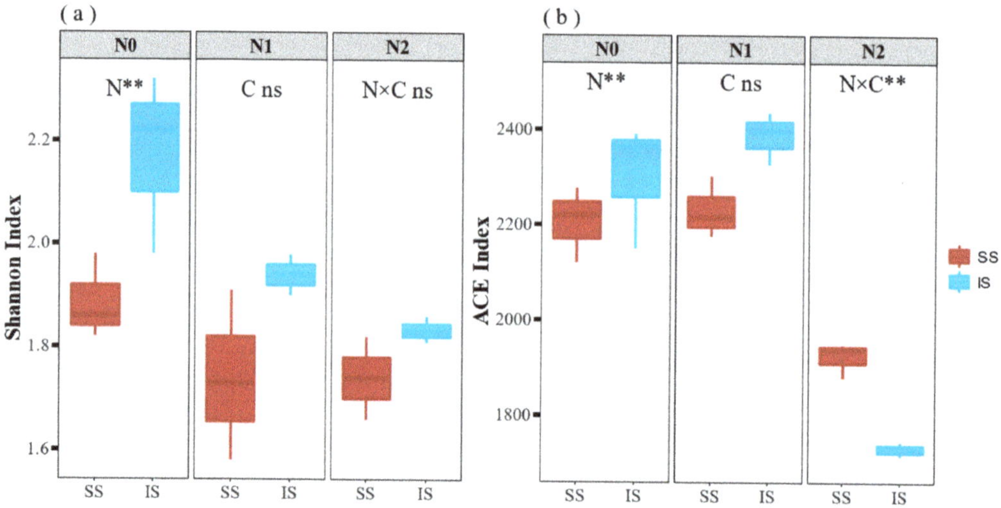

Figure 2. The Shannon index (**a**) and ACE index (**b**) of nitrogen-fixing bacteria were measured in soybean under different nitrogen levels. N indicates nitrogen application rates; C indicates planting patterns; N×C shows the interaction between nitrogen application rates and planting patterns by two ways ANOVA. ns and **, respectively, indicate that the difference is not significant, and difference is significant at levels of $p < 0.01$. Data are presented as a Mean ± SD ($n = 3$).

The top five dominant genera in soybean rhizosphere soil with more than 95% of total abundance were *Bradyrhizobium*, *Skermanella*, *g_unclassified_p_Proteobacteria*, *g_unclassified_k__norank_d__Bacteria*, and Azohydromonas (Figure 3a). Among these, *Bradyrhizobium* was the most abundant genus accounting for more than 70% of the total abundance. The lowest *g_unclassified_k__norank_d__Bacteria* abundance is only about 3% of the total abundance. The abundance of Azohydromonas was significantly different under different nitrogen levels when single cropping was not considered. It was found that the abundance of Azohydromonas under the N0 treatment was significantly increased by 2.62 times compared with the N2 treatment.

Different nitrogen application rates and intercropping did not significantly affect *Bradyrhizobium*, *Skermanella*, *g_unclassified_p_Proteobacteria*, and *g_unclassified_k__norank_d__Bacteria* in soybean rhizosphere, but *Azohydromonas* was significantly affected by nitrogen application rates ($p < 0.01$). *g_unclassified_k__norank_d__Bacteria* was significantly affected by the planting pattern ($p < 0.05$) and the interaction between nitrogen application and planting pattern ($p < 0.01$). The highest abundance of *Azohydromonas* was found in the N0 treatment (0 kg N/ha) and the lowest in N2 (80 kg N/ha). The rhizosphere *Azohydromonas* abundance of intercropping soybean was significantly increased by 35%, 7%, and 42% compared with that of monoculture under N0, N1, and N2 nitrogen levels. When nitrogen application rates were 0 kg/ha or 40 kg/ha, *g_unclassified_k__norank_d__Bacteria* abundance was higher compared to monocropping. However, in the 80 kg N/ha application rate, it was lower than intercropping.

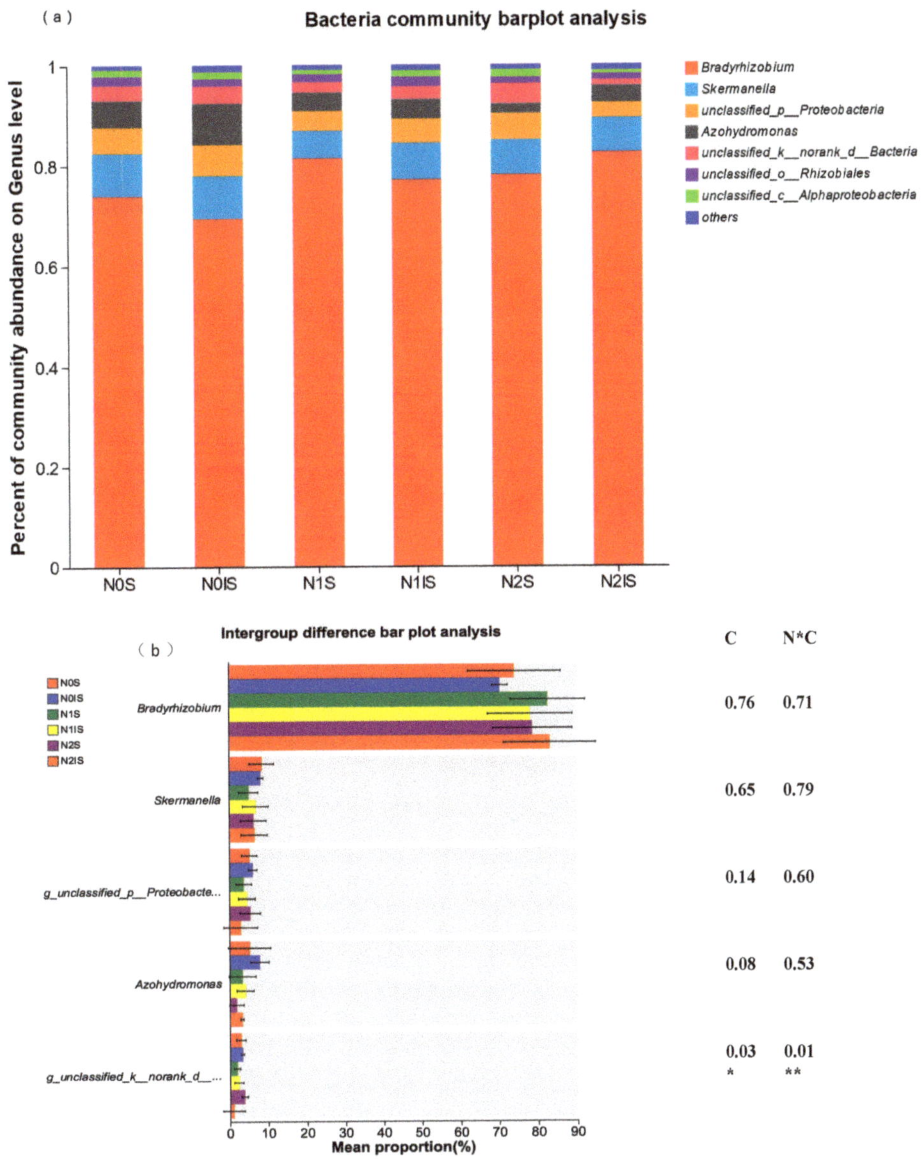

Figure 3. The soybean nitrogen-fixing bacteria community composition (**a**), differences between groups (**b**) under different nitrogen application levels. N indicates nitrogen application rates; C indicates planting patterns; N×C shows the interaction between nitrogen application rates and planting patterns by two ways ANOVA. ns and * and **, respectively, indicate that the difference is not significant, and difference is significant at levels of $p < 0.05$ and $p < 0.01$. Data are presented as a Mean ± SD ($n = 3$).

3.4. Relationship between the Nitrogen-Fixing Bacteria Community and Soil Physicochemical Characteristics

The soybean rhizosphere soil physicochemical characteristics were affected by nitrogen application rates and planting patterns (Table 2). Specifically, nitrogen content (AN), available phosphorus (AP), and pH were significantly influenced by nitrogen application rates ($p < 0.01$), with an increase observed as nitrogen application rates increased. The available phosphorus (AP) was also significantly affected by the cultivation pattern ($p < 0.05$), with the soil AP in the monocropping system being significantly higher compared to intercropping. However, no significant difference was found in available potassium (AK) and organic matter (OM) among different nitrogen treatments and cultivation patterns.

Table 2. Soybean soil physicochemical characteristics of different treatments with different nitrogen application rates.

Treatments	AN (mg/kg)			AP (mg/kg)			AK (mg/kg)			pH			OM (g/kg)		
	SS	IS	AVE	SS	IS	AVE	SS	IS	AVE	SS	IS	AVE	SS	IS	AVE
N0	132 ± 1.52 a	131 ± 4.35 a	131 ± 3.01 B	71 ± 0.55 b	70 ± 0.26 a	70 ± 0.48 B	141 ± 1.07 a	143 ± 2.50 a	142 ± 1.85 A	6.66 ± 0.08 a	6.60 ± 0.03 a	6.63 ± 0.06 B	30 ± 0.65 a	30 ± 1.03 a	30 ± 0.77 A
N1	132 ± 3.51 a	132 ± 1.01 a	132 ± 2.31 B	71 ± 0.14 b	70 ± 0.39 a	71 ± 0.32 A	141 ± 1.55 a	143 ± 1.12 a	142 ± 1.39 A	6.67 ± 0.08 a	6.65 ± 0.08 a	6.66 ± 0.07 AB	31 ± 0.65 a	31 ± 1.79 a	31 ± 0.68 A
N2	155 ± 2.02 a	154 ± 4.93 a	155 ± 3.37 A	72 ± 0.56 b	71 ± 0.05 a	71 ± 0.57 A	142 ± 0.57 a	142 ± 2.95 a	142 ± 1.90 A	6.75 ± 0.02 a	6.72 ± 0.02 a	6.73 ± 0.02 A	31 ± 0.28 a	31 ± 0.97 a	31 ± 0.70 A
ANOVE		p			p			p			p			p	
N		0.00			0.01			0.95			0.03			0.29	
C		0.67			0.03			0.40			0.25			0.35	
N×C		0.95			0.59			0.71			0.84			0.89	

NOTE: N indicates nitrogen application rates, C indicates planting patterns; N×C shows the interaction between nitrogen application rates and planting patterns. Different lowercase letters indicate significant differences between single intercropping at the same nitrogen supply level, and different capital letters indicate significant differences between different nitrogen supply levels. The following table is the same. Data are presented as a Mean ± SD (n = 3).

RDA analysis by "vegan package "was conducted to investigate the relationship between soil environmental factors and nitrogen-fixing bacterial communities in the soybean rhizosphere. The first two RDA axes explained 16.86% of the variance in bacterial community composition (Figure 4a). In the RDA analysis, the eigenvalues of DCA1 and DCA2 were 0.0473 and 0.0131, respectively. Based on the RDA, AP, AN, and pH were found to have a higher contribution rate to RDA1, while OM and AK had a higher contribution to RDA2. However, pH, AN, AP, OM, and AK did not significantly affect the nitrogen-fixing bacterial communities (Table 3). To further examine the correlation between soil physicochemical characteristics and nitrogen-fixing bacterial communities, we conducted the Mantel test using the "ggcor" package. The results indicated no significant relationship between pH, AN, AP, OM, AK, and the nitrogen-fixing bacterial communities in the soybean rhizosphere soil (Figure 4b).

Table 3. RDA analysis results of nitrogen-fixing bacteria communities in soybean rhizosphere soil with different nitrogen application rates.

Soil Characteristic	RDA1	RDA2	r^2	p
AN	0.8445	0.5356	0.2044	0.176
OM	0.3247	0.9458	0.0527	0.683
AP	0.8044	0.594	0.2444	0.114
AK	−0.5725	−0.8199	0.0502	0.681
pH	0.9566	0.2915	0.0672	0.577

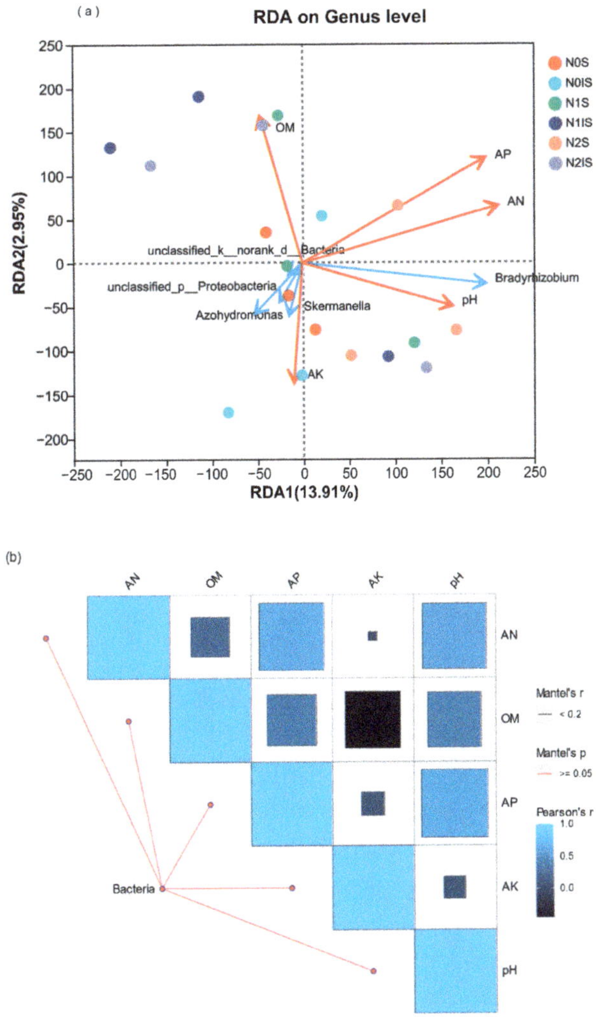

Figure 4. RDA (**a**) and Mantel test (**b**) analysis of soil characteristics of soybean under different nitrogen application levels. Color gradient and size represent Pearson correlation coefficients of physicochemical factors. The edge width represents the Mantel R statistic of the corresponding correlation coefficient, and the color indicates the Mantel p statistic. AN, soil available nitrogen; OM, soil organic matter; TN, soil total nitrogen content; AP; Soil available phosphorus content; AK, soil available potassium.

3.5. *nifH* Gene Copies of Nitrogen-Fixing Bacteria in Soybean and Maize Rhizosphere Soil

The *nifH* gene abundance was significantly affected by nitrogen application ($p < 0.01$) but not significantly affected by planting patterns ($p > 0.05$) (Figure 5a). The highest *nifH* gene copy number (6.25×10^7 copies/g soil) was observed in the N1 treatment and the lowest in the N2 treatment (3.54×10^7 copies/g soil). Intercropping with maize increased the *nifH* gene copies in the N0 and N1 treatments by 11% and 13%, respectively, but it resulted in a decrease of 23% in the N2 treatment compared to monocropping.

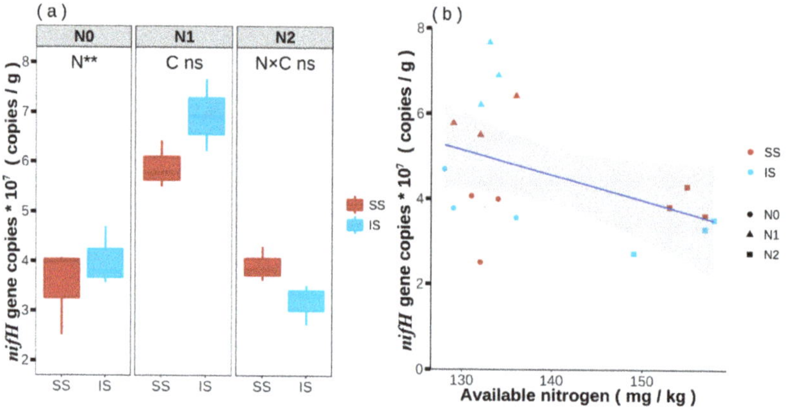

Figure 5. The *nifH* gene copies in soybean rhizosphere soil with different nitrogen levels ($\times 10^7$) (**a**) and their correlation with soil available nitrogen (**b**) with different nitrogen application rates. N indicates nitrogen application rates; C indicates planting patterns; N×C shows the interaction between nitrogen application rates and planting patterns by two ways ANOVA. ns and **, respectively, indicate that the difference is not significant, and difference is significant at levels of $p < 0.01$. Data are presented as a Mean ± SD ($n = 3$).

3.6. Correlation nifH Gene Copies and Soil Available and Biological Quantification of Nitrogen Fixation

To further elucidate the relationship between soil available nitrogen, *nifH* gene copies, and nitrogen fixation, we conducted Pearson correlation analyses. The results showed a significant negative correlation ($p < 0.05$) between soil available nitrogen and *nifH* gene copies in soybean rhizosphere soil, with a linear relationship expressed with the equation $y = -0.0606x + 13.02$ and an $r^2 = 0.2243$ (Figure 5b).

4. Discussion

4.1. Nitrogen Fixation of Intercropped Soybean Affected by Nitrogen Application Rates

Several studies have shown that intercropping systems can improve the nitrogen fixation efficiency of legumes, consistent with the results of our research (Figure 1a) [36]. When soybean and maize are intercropped, the roots of maize can grow into the area where soybean roots grow, leading to competition for available soil nitrogen [8]. As a result, soybean roots are in a low nitrogen environment, which can stimulate nitrogen fixation from the air and thus improve the soybean nitrogen fixation efficiency. In addition, proper nitrogen application could also improve legume growth [22]. We obtained similar results, showing that soybean nitrogen fixation was the highest in the N1 treatment (40 kg N/ha) (Figure 1b). The increase in nitrogen fertilizer in intercropped soybean significantly negatively affected the nodule number and dry weight (Table 1). Nodules not only serve as sites for atmospheric nitrogen (N2) fixation but also provide an energy source for rhizobia [37]. This decrease in nodule formation ultimately impacted soybeans' ability to fix nitrogen. Our results suggest that maize and soybean intercropping did not completely enhance soybean nitrogen fixation. In fact, the nitrogen fixation rates in the N0 and N2 intercropping treatments were still 1.08 and 1.19 times lower, respectively, compared to monocropping. However, in the N1 intercropped soybean treatment, nitrogen fixation was 1.04 times higher than monocropping. The lower nitrogen fixation in the intercropping system compared to monoculture was attributed to the competition from intercropped maize, which resulted in a decrease in soybean biomass in the intercropping system [38]. Our study also yielded similar results, as the soybean biomass under monocropping was significantly higher than that under intercropping.

The nitrogen fixation efficiency is closely related to the soybean nodule number, size, and dry weight. In our study, nodules were crucial in improving nitrogen fixation efficiency. After intercropping soybean with maize and applying the appropriate nitrogen fertilizers, the soybean's nitrogen fixation was enhanced. Our results indicated that nitrogen fixation efficiency in the N0 treatment was the highest, but nitrogen fixation in the N1 treatment was found to be the highest among all the treatments. Therefore, we suggest using 40 kg N/ha nitrogen fertilizer to maximize the soybean–maize intercropping system productivity.

4.2. Diversity of Nitrogen-Fixing Bacteria in the Rhizosphere Soil of Intercropped Soybean Affected by N Application

Nitrogen fixation and utilization in soil ecological environments are closely related to the nitrogen-fixing bacterial community structure and diversity. Nitrogen-fixing bacteria play an essential role in the nitrogen cycle by converting atmospheric nitrogen into a form available to plants [39]. The microbial diversity index is a crucial indicator of soil microbial diversity and can also reflect the quality of farmland ecosystems [40]. Intercropping practices can alter the diversity of nitrogen-fixing microorganisms [41]. In our study, we observed that intercropping of soybean resulted in a significantly higher Shannon index than monocropping (Figure 2a,b).

Previous studies have shown that intercropping can have varying effects on the nitrogen-fixing bacteria communities' diversity depending on the specific crop combinations and soil conditions. For example, cassava/peanut intercropping increased microbial diversity compared to peanut monoculture [42], while legume/oat intercropping increased the diversity of oat nitrogen-fixing bacteria communities [43]. However, intercropping may not significantly affect the Shannon index in some cases, such as in a maize/soybean/cotton intercropping system [44]. These discrepancies may be due to differences in soil environments and the effects of varying nitrogen application levels [45]. Therefore, it is important to consider the specific crop combinations and soil conditions when evaluating the impact of intercropping on nitrogen-fixing bacteria diversity.

In addition, different nitrogen application rates significantly impacted the nitrogen-fixing bacteria diversity in soybean. An appropriate nitrogen fertilization regime can increase soil bacterial diversity and richness [46]. However, previous studies have shown that a high nitrogen application level (100 kg/ha) reduced bacterial community diversity and richness in soybean rhizosphere [47], consistent with the results of our study. Therefore, high nitrogen application can result in a decrease in the rhizosphere microbial diversity in legumes.

Based on the species taxonomic analysis, the nitrogen-fixing bacteria community composition in soybean rhizosphere soil under different nitrogen application levels was investigated. A total of seven bacterial genera were detected in the soybean rhizosphere soil, with *Bradyrhizobium* being the dominant genus (Figure 3a). The nitrogen-fixing bacteria abundance in soybean was affected by both intercropping and nitrogen application.

The abundance of *g_unclassified_k__norank_d__Bacteria* in the soybean rhizosphere was significantly influenced by the planting patterns ($p < 0.05$), as shown in Figure 3b. Intercropping has been shown to change the composition of nitrogen-fixing microorganisms and increase the nitrogen-fixing bacteria species diversity [18], in agreement with our results. As the dominant genus, *g_unclassified_k__norank_d__Bacteria* was significantly more abundant in intercropping compared to monoculture under the N0 and N1 treatments. This suggests that intercropping can improve the nitrogen-fixing bacteria community composition in the appropriate nitrogen application.

Furthermore, the nitrogen application rate significantly impacted the community structure of nitrogen-fixing bacteria in the soybean/maize intercropping system. For instance, the abundance of Azohydromonas in the soybean rhizosphere decreased with the increase in nitrogen application rate. This finding is consistent with previous studies indicating that mycorrhizal fungi have a weaker ability to infect plant roots in high-nitrogen environments, which can ultimately lead to a reduction in soil microbial activity [48,49].

Thus, it is crucial to apply the appropriate nitrogen fertilization rates to maintain the stability and diversity of the soil microbial community.

Our study showed that nitrogen application rates and planting patterns significantly affected soil physicochemical characteristics, with soil AN, AP, and pH significantly affected by nitrogen application rates. Similar research has shown that farmland management can significantly affect the soil's physicochemical properties and biodiversity [50], with AN being the crucial factor affecting microbial diversity [51] and pH significantly affecting nitrogen-fixing microorganisms [52]. Changes in soil physicochemical properties due to different nitrogen application rates and planting patterns can alter the nitrogen-fixing bacteria community structure. However, in our study, the RDA results indicated that the soil pH and AN content had no significant effects on the soybean rhizosphere soil.

4.3. NifH Gene Copies in the Rhizosphere Soil of Intercropped Soybean

In different soil ecological environments, the *nifH* gene diversity was significantly different [53]. Planting methods have also been shown to affect the *nifH* gene copy number, as demonstrated by [52], who studied forest soil under different cultivation modes. Moreover, the soybean–maize intercropping increased *nifH* gene copies by 11% and 13% in N0 and N1 treatments compared with monoculture, respectively (Figure 4a). This suggests that intercropping can increase *nifH* gene copies in appropriate nitrogen application. It has been reported that the *nifH* gene expression, responsible for nitrogen fixation, is affected by soil nitrogen concentration [54,55]. Correlation analysis indicated that the *nifH* gene copy number was significantly negatively correlated with soil-available nitrogen (Figure 5b). The highest *nifH* gene copies were observed in the N1 intercropping treatment of soybean. This is likely because maize requires a significant amount of nitrogen during the tasseling period, intensifying competition for soil nitrogen. It stimulated soybean to regulate root exudates, altering the nitrogen-fixing bacteria community to enhance its nitrogen-fixing function to meet its own needs. As shown from previous studies, reducing nitrogen levels can improve the nitrogen fixation intensity and nitrogen-fixing bacteria number in maize and soybean intercropping systems [18].

Excessive nitrogen can have a negative impact on the activity of nitrogen-fixing microorganisms [56]. Our study found that the *nifH* gene copy numbers, an indicator of nitrogen fixation, were significantly affected by different nitrogen application rates in soybean rhizosphere soil. The highest *nifH* gene copy numbers were observed in the N1 treatment in soybean, indicating that appropriate nitrogen application can increase gene copies. However, excessive nitrogen application can inhibit the activity of nitrogen-fixing bacteria and decrease the abundance of the nifH gene, which is consistent with previous research results [54,57]. Moreover, excessive nitrogen fertilizer can lead to soil acidification, which is not conducive to the growth of microorganisms [58].

5. Conclusions

Nitrogen-fixing bacteria diversity of intercropped soybean rhizosphere was significantly affected by nitrogen application rates and planting patterns. *Bradyrhizobium* was the dominant genus. The nitrogen-fixing efficiency, quantity of *nifH* gene copies, and nodules number of intercropped soybean were increased under different nitrogen application rates due to the decrease of AN in the rhizosphere of soybean. The Pearson correlation analysis indicated that the available nitrogen content in the soil significantly negatively impacted the abundance of *nifH* genes in the soybean rhizosphere soil. These results provide an important idea for improving the nitrogen-fixing efficiency and reducing the use of chemical nitrogen fertilizer in legume/Gramineae intercropping systems.

Author Contributions: Z.C.: Experiment design, Writing—original draft preparation, sampling and analysis. L.M.: investigation, funding acquisition. T.Y.: Writing—reviewing. Y.L.: Writing—original draft. Y.Z.: Data analysis. S.L.: Editing, project administration, supervision, funding acquisition. All authors have read and agreed to the published version of the manuscript.

Funding: This research was funded by National Natural Science Foundation of China (32171547).

Data Availability Statement: Most of the collected data are contained in the tables and figures in the manuscript.

Conflicts of Interest: The authors declare no conflict of interest.

References

1. da Silva, J.N.; Neto, F.B.; de Lima, J.S.S.; dos Santos, E.C.; Nunes, R.L.C.; Chaves, A.P. Production and benefits in carrot and vegetable cowpea associations under green manuring and spatial arrangements. *Rev. Cienc. Agron.* **2020**, *51*, e20197067. [CrossRef]
2. Anas, M.; Liao, F.; Verma, K.K.; Sarwar, M.A.; Mahmood, A.; Chen, Z.L.; Li, Q.; Zeng, X.P.; Liu, Y.; Li, Y.R. Fate of nitrogen in agriculture and environment: Agronomic, eco-physiological and molecular approaches to improve nitrogen use efficiency. *Biol. Res.* **2020**, *53*, 47. [CrossRef] [PubMed]
3. Nelson, K.A.; Massey, R.E.; Burdick, B.A. Harvest Aid Application Timing Affects Wheat and Relay Intercropped Soybean Yield. *Agron. J.* **2011**, *103*, 851–855. [CrossRef]
4. Saudy, H.S. Maize–cowpea intercropping as an ecological approach for nitrogen-use rationalization and weed suppression. *Arch. Agron. Soil Sci.* **2014**, *61*, 1–14. [CrossRef]
5. Coelho, M.R.R.; Marriel, I.E.; Jenkins, S.N.; Lanyon, C.V.; Seldin, L.; O'Donnell, A.G. Molecular detection and quantification of nifH gene sequences in the rhizosphere of sorghum (Sorghum bicolor) sown with two levels of nitrogen fertilizer. *Appl. Soil Ecol.* **2009**, *42*, 48–53. [CrossRef]
6. Fu, Z.-d.; Zhou, L.; Chen, P.; Du, Q.; Pang, T.; Song, C.; Wang, X.-c.; Liu, W.-g.; Yang, W.-y.; Yong, T.-w. Effects of maize-soybean relay intercropping on crop nutrient uptake and soil bacterial community. *J. Integr. Agric.* **2019**, *18*, 2006–2018. [CrossRef]
7. Chen, P.; Du, Q.; Liu, X.; Zhou, L.; Hussain, S.; Lei, L.; Song, C.; Wang, X.; Liu, W.; Yang, F.; et al. Effects of reduced nitrogen inputs on crop yield and nitrogen use efficiency in a long-term maize-soybean relay strip intercropping system. *PLoS ONE* **2017**, *12*, e0184503. [CrossRef]
8. Shao, Z.; Wang, X.; Gao, Q.; Zhang, H.; Yu, H.; Wang, Y.; Zhang, J.; Nasar, J.; Gao, Y. Root Contact between Maize and Alfalfa Facilitates Nitrogen Transfer and Uptake Using Techniques of Foliar 15N-Labeling. *Agronomy* **2020**, *10*, 360. [CrossRef]
9. Zhang, H.; Zeng, F.; Zou, Z.; Zhang, Z.; Li, Y. Nitrogen uptake and transfer in a soybean/maize intercropping system in the karst region of southwest China. *Ecol. Evol.* **2017**, *7*, 8419–8426. [CrossRef]
10. Li, L.; Zhang, F.; Li, X.; Christie, P.; Sun, J.; Yang, S.; Tang, C. Interspecific facilitation of nutrient uptake by intercropped maize and faba bean. *Nutr. Cycl. Agroecosystems* **2003**, *65*, 61–71. [CrossRef]
11. Xiao, Y.; Li, L.; Zhang, F. The interspecific nitrogen facilitation and the subsequent nitrogen transfer between the intercropped wheat and fababean. *Sci. Agric. Sin.* **2005**, *38*, 965–973.
12. Li, S.; van der Werf, W.; Zhu, J.; Guo, Y.; Li, B.; Ma, Y.; Evers, J.B. Estimating the contribution of plant traits to light partitioning in simultaneous maize/soybean intercropping. *J. Exp. Bot.* **2021**, *72*, 3630–3646. [CrossRef]
13. Fan, F.; Zhang, F.; Song, Y.; Sun, J.; Bao, X.; Guo, T.; Li, L. Nitrogen Fixation of Faba Bean (Vicia faba L.) Interacting with a Non-legume in Two Contrasting Intercropping Systems. *Plant Soil* **2006**, *283*, 275–286. [CrossRef]
14. Homulle, Z.; George, T.S.; Karley, A.J. Root traits with team benefits: Understanding belowground interactions in intercropping systems. *Plant Soil* **2021**, *471*, 1–26. [CrossRef]
15. Zhang, W.P.; Liu, G.C.; Sun, J.H.; Fornara, D.; Zhang, L.Z.; Zhang, F.F.; Li, L.; Niels, A. Temporal dynamics of nutrient uptake by neighbouring plant species: Evidence from intercropping. *Funct. Ecol.* **2016**, *31*, 469–479. [CrossRef]
16. Solanki, M.K.; Wang, Z.; Wang, F.Y.; Li, C.N.; Gupta, C.L.; Singh, R.K.; Malviya, M.K.; Singh, P.; Yang, L.T.; Li, Y.R. Assessment of Diazotrophic Proteobacteria in Sugarcane Rhizosphere When Intercropped With Legumes (Peanut and Soybean) in the Field. *Front. Microbiol.* **2020**, *11*, 1814. [CrossRef] [PubMed]
17. Malviya, M.K.; Solanki, M.K.; Li, C.-N.; Wang, Z.; Zeng, Y.; Verma, K.K.; Singh, R.K.; Singh, P.; Huang, H.-R.; Yang, L.-T.; et al. Sugarcane-Legume Intercropping Can Enrich the Soil Microbiome and Plant Growth. *Front. Sustain. Food Syst.* **2021**, *5*, 606595. [CrossRef]
18. Lai, H.; Gao, F.; Su, H.; Zheng, P.; Li, Y.; Yao, H. Nitrogen Distribution and Soil Microbial Community Characteristics in a Legume–Cereal Intercropping System: A Review. *Agronomy* **2022**, *12*, 1900. [CrossRef]
19. Hayden, H.L.; Drake, J.; Imhof, M.; Oxley, A.P.A.; Norng, S.; Mele, P.M. The abundance of nitrogen cycle genes amoA and nifH depends on land-uses and soil types in South-Eastern Australia. *Soil Biol. Biochem.* **2010**, *42*, 1774–1783. [CrossRef]
20. Nicol, G.W.; Leininger, S.; Schleper, C.; Prosser, J.I. The influence of soil pH on the diversity, abundance and transcriptional activity of ammonia oxidizing archaea and bacteria. *Environ. Microbiol.* **2008**, *10*, 2966–2978. [CrossRef]
21. Kasa, P.; Modugapalem, H.; Battini, K. Isolation, screening, and molecular characterization of plant growth promoting rhizobacteria isolates of Azotobacter and Trichoderma and their beneficial activities. *J. Nat. Sci. Biol. Med.* **2015**, *6*, 360–363. [CrossRef] [PubMed]
22. Fan, K.; Delgado-Baquerizo, M.; Guo, X.; Wang, D.; Wu, Y.; Zhu, M.; Yu, W.; Yao, H.; Zhu, Y.G.; Chu, H. Suppressed N fixation and diazotrophs after four decades of fertilization. *Microbiome* **2019**, *7*, 143. [CrossRef] [PubMed]

23. Zhang, R.; Mu, Y.; Li, X.; Li, S.; Sang, P.; Wang, X.; Wu, H.; Xu, N. Response of the arbuscular mycorrhizal fungi diversity and community in maize and soybean rhizosphere soil and roots to intercropping systems with different nitrogen application rates. *Sci. Total Environ.* **2020**, *740*, 139810. [CrossRef] [PubMed]
24. Simonsen, A.K.; Han, S.; Rekret, P.; Rentschler, C.S.; Heath, K.D.; Stinchcombe, J.R. Short-term fertilizer application alters phenotypic traits of symbiotic nitrogen fixing bacteria. *PeerJ* **2015**, *3*, e1291. [CrossRef]
25. Hui, T. Effects of long-term fertilization on nifH gene diversity in agricultural black soil. *Afr. J. Microbiol. Res.* **2012**, *6*, 2659–2666. [CrossRef]
26. Calderoli, P.A.; Collavino, M.M.; Behrends Kraemer, F.; Morras, H.J.M.; Aguilar, O.M. Analysis of nifH-RNA reveals phylotypes related to Geobacter and Cyanobacteria as important functional components of the N(2) -fixing community depending on depth and agricultural use of soil. *Microbiologyopen* **2017**, *6*, e00502. [CrossRef]
27. Nguyen, H.P.; Miwa, H.; Obirih-Opareh, J.; Suzaki, T.; Yasuda, M.; Okazaki, S. Novel rhizobia exhibit superior nodulation and biological nitrogen fixation even under high nitrate concentrations. *FEMS Microbiol. Ecol.* **2020**, *96*, fiz184. [CrossRef]
28. Martiny, J.B.H.; Bohannan, B.J.M.; Brown, J.H.; Colwell, R.K.; Fuhrman, J.A.; Green, J.L.; Horner-Devine, M.C.; Kane, M.; Krumins, J.A.; Kuske, C.R.; et al. Microbial biogeography: Putting microorganisms on the map. *Nat. Rev. Microbiol.* **2006**, *4*, 102–112. [CrossRef]
29. Bremner, J.M. Nitrogen-Total. In *Methods of Soil Analysis*; Soil Science Society of America: Madison, WI, USA, 1996; pp. 1085–1121.
30. Bray, R.H.; Kurtz, L.T. Determination of Total, Organic, and Available Forms of Phosphorus in Soils. *Soil Sci.* **1945**, *59*, 39–46. [CrossRef]
31. Mulvaney, R.L.; Khan, S.A. Diffusion Methods to Determine Different Forms of Nitrogen in Soil Hydrolysates. *Soil Sci. Soc. Am. J.* **2001**, *65*, 1284–1292. [CrossRef]
32. Kalembasa, S.J.; Jenkinson, D.S. A comparative study of titrimetric and gravimetric methods for the determination of organic carbon in soil. *J. Sci. Food Agric.* **1973**, *24*, 1085–1090. [CrossRef]
33. Cowan, D.A.; Sohm, J.A.; Makhalanyane, T.P.; Capone, D.G.; Green, T.G.A.; Cary, S.C.; Tuffin, I.M. Hypolithic communities: Important nitrogen sources in Antarctic desert soils. *Environ. Microbiol. Rep.* **2011**, *3*, 581–586. [CrossRef] [PubMed]
34. Heller, P.; Tripp, H.J.; Turk-Kubo, K.; Zehr, J.P. ARBitrator: A software pipeline for on-demand retrieval of auto-curated nifH sequences from GenBank. *Bioinformatics* **2014**, *30*, 2883–2890. [CrossRef]
35. Shearer, G.B.; Kohl, D.H. N2-Fixation in Field Settings: Estimations Based on Natural 15N Abundance. *Aust. J. Plant Physiol.* **1986**, *13*, 699–756.
36. Chamkhi, I.; Cheto, S.; Geistlinger, J.; Zeroual, Y.; Kouisni, L.; Bargaz, A.; Ghoulam, C. Legume-based intercropping systems promote beneficial rhizobacterial community and crop yield under stressing conditions. *Ind. Crops Prod.* **2022**, *183*, 114958. [CrossRef]
37. Reid, D.E.; Ferguson, B.J.; Hayashi, S.; Lin, Y.H.; Gresshoff, P.M. Molecular mechanisms controlling legume autoregulation of nodulation. *Ann. Bot.* **2011**, *108*, 789–795. [CrossRef] [PubMed]
38. Te, X.; Din, A.M.U.; Cui, K.; Raza, M.A.; Fraz Ali, M.; Xiao, J. Inter-specific root interactions and water use efficiency of maize/soybean relay strip intercropping. *Field Crops Res.* **2023**, *291*, 108793. [CrossRef]
39. Wang, H.; Li, H.; Zhang, M.; Song, Y.; Huang, J.; Huang, H.; Shao, M.; Liu, Y.; Kang, Z. Carbon Dots Enhance the Nitrogen Fixation Activity of Azotobacter Chroococcum. *ACS Appl. Mater. Interfaces* **2018**, *10*, 16308–16314. [CrossRef]
40. Lankau, R.A.; George, I.; Miao, M. Crop performance is predicted by soil microbial diversity across phylogenetic scales. *Ecosphere* **2022**, *13*, e4029. [CrossRef]
41. Venieraki, A.; Dimou, M.; Pergalis, P.; Kefalogianni, I.; Chatzipavlidis, I.; Katinakis, P. The Genetic Diversity of Culturable Nitrogen-Fixing Bacteria in the Rhizosphere of Wheat. *Microb. Ecol.* **2011**, *61*, 277–285. [CrossRef]
42. Tang, X.; Zhong, R.; Jiang, J.; He, L.; Huang, Z.; Shi, G.; Wu, H.; Liu, J.; Xiong, F.; Han, Z.; et al. Cassava/peanut intercropping improves soil quality via rhizospheric microbes increased available nitrogen contents. *BMC Biotechnol.* **2020**, *20*, 13. [CrossRef] [PubMed]
43. Yang, Y.D.; Feng, X.M.; Hu, Y.G.; Ren, C.Z.; Zeng, Z.H. Effects of legume-oat intercropping on abundance and community structure of soil N_2-fixing bacteria. *J. Appl. Ecol.* **2017**, *28*, 957–965. [CrossRef]
44. Nourbakhsh, F.; Koocheki, A.; Mahallati, M.N. Investigation of Biodiversity and Some of the Ecosystem Services in the Intercropping of Corn, Soybean and Marshmallow. *Int. J. Plant Prod.* **2019**, *13*, 35–46. [CrossRef]
45. Koocheki, A.; Solouki, H.; Karbor, S. Study of ecological aspects of Sesame (*Sesamum indicum* L.) and Mung Bean (*Vigna radiata* L.) intercropping in weed control. *Iran. J. Pulses Res.* **2016**, *7*, 27–44. [CrossRef]
46. Ren, N.; Wang, Y.; Ye, Y.; Zhao, Y.; Huang, Y.; Fu, W.; Chu, X. Effects of Continuous Nitrogen Fertilizer Application on the Diversity and Composition of Rhizosphere Soil Bacteria. *Front. Microbiol.* **2020**, *11*, 1948. [CrossRef] [PubMed]
47. Zhou, J.; Jiang, X.; Wei, D.; Zhao, B.; Ma, M.; Chen, S.; Cao, F.; Shen, D.; Guan, D.; Li, J. Consistent effects of nitrogen fertilization on soil bacterial communities in black soils for two crop seasons in China. *Sci. Rep.* **2017**, *7*, 3267. [CrossRef] [PubMed]
48. Antoninka, A.; Reich, P.B.; Johnson, N.C. Seven years of carbon dioxide enrichment, nitrogen fertilization and plant diversity influence arbuscular mycorrhizal fungi in a grassland ecosystem. *New Phytol.* **2011**, *192*, 200–214. [CrossRef]
49. Liu, Y.; Shi, G.; Mao, L.; Cheng, G.; Jiang, S.; Ma, X.; An, L.; Du, G.; Collins Johnson, N.; Feng, H. Direct and indirect influences of 8 yr of nitrogen and phosphorus fertilization on Glomeromycota in an alpine meadow ecosystem. *New Phytol.* **2012**, *194*, 523–535. [CrossRef]

50. Sun, L.; Song, F.; Liu, S.; Cao, Q.; Liu, F.; Zhu, X. Integrated agricultural management practice improves soil quality in Northeast China. *Arch. Agron. Soil Sci.* **2018**, *64*, 1932–1943. [CrossRef]
51. Jia, Y.; Liao, Z.; Chew, H.; Wang, L.; Lin, B.; Chen, C.; Lu, G.; Lin, Z. Effect of Pennisetum giganteum z.x.lin mixed nitrogen-fixing bacterial fertilizer on the growth, quality, soil fertility and bacterial community of pakchoi (*Brassica chinensis* L.). *PLoS ONE* **2020**, *15*, e0228709. [CrossRef]
52. Sepp, S.K.; Vasar, M.; Davison, J.; Oja, J.; Anslan, S.; Al-Quraishy, S.; Bahram, M.; Bueno, C.G.; Cantero, J.J.; Fabiano, E.C.; et al. Global diversity and distribution of nitrogen-fixing bacteria in the soil. *Front. Plant Sci.* **2023**, *14*, 1100235. [CrossRef] [PubMed]
53. Zou, Y.; Zhang, J.; Yang, D.; Chen, X.; Zhao, J.; Xiu, W.; Lai, X.; Li, G. Effects of different land use patterns on nifH genetic diversity of soil nitrogen-fixing microbial communities in Leymus Chinensis steppe. *Acta Ecol. Sin.* **2011**, *31*, 150–156. [CrossRef]
54. Li, Y.; Pan, F.; Yao, H. Response of symbiotic and asymbiotic nitrogen-fixing microorganisms to nitrogen fertilizer application. *J. Soils Sediments* **2019**, *19*, 1948–1958. [CrossRef]
55. Kubota, M.; Matsushita, N.; Nakamura, T.; Fukuda, K. Nitrogen fixation and nifH gene diversity in cyanobacteria living on feather mosses in a subalpine forest of Mt. Fuji. *Oecologia* **2023**, *201*, 749–760. [CrossRef]
56. Din, I.; Khan, H.; Ahmad Khan, N.; Khil, A. Inoculation of nitrogen fixing bacteria in conjugation with integrated nitrogen sources induced changes in phenology, growth, nitrogen assimilation and productivity of wheat crop. *J. Saudi Soc. Agric. Sci.* **2021**, *20*, 459–466. [CrossRef]
57. Wang, H.; Gu, C.; Liu, X.; Yang, C.; Li, W.; Wang, S. Impact of Soybean Nodulation Phenotypes and Nitrogen Fertilizer Levels on the Rhizosphere Bacterial Community. *Front. Microbiol.* **2020**, *11*, 750. [CrossRef]
58. Gu, Y.; Wang, J.; Cai, W.; Li, G.; Mei, Y.; Yang, S. Different Amounts of Nitrogen Fertilizer Applications Alter the Bacterial Diversity and Community Structure in the Rhizosphere Soil of Sugarcane. *Front. Microbiol.* **2021**, *12*, 721441. [CrossRef]

agronomy

MDPI

Article

Rhizosphere Microbiomes of *Amaranthus* spp. Grown in Soils with Anthropogenic Polyelemental Anomalies

Anna Muratova [1,*], Svetlana Gorelova [2], Sergey Golubev [1], Dilyara Kamaldinova [3] and Murat Gins [4,5]

1 Institute of Biochemistry and Physiology of Plants and Microorganisms, Saratov Scientific Centre of the Russian Academy of Sciences (IBPPM RAS), Saratov 410049, Russia
2 Natural Science Institute, Tula State University, Tula 300012, Russia
3 Institute of Fundamental Medicine and Biology, Kazan (Volga Region) Federal University, Kazan 420021, Russia
4 Agrarian and Technological Institute, Peoples' Friendship University of Russia, Moscow 117198, Russia
5 Federal State Budgetary Scientific Institution "Federal Scientific Vegetable Center", Odintsovo 143080, Russia
* Correspondence: muratova_a@ibppm.ru; Tel.: +7-(8452)-970403

Abstract: Study of rhizospheric microbial communities of plants growing under different environmental conditions is important for understanding the habitat-dependent formation of rhizosphere microbiomes. The rhizosphere bacterial communities of four amaranth cultivars were investigated in a laboratory pot experiment. *Amaranthus tricolor* cv. Valentina, *A. cruentus* cv. Dyuimovochka, and *A. caudatus* cvs. Bulava and Zelenaya Sosulka were grown for six months in three soils with different anthropogenic polyelemental anomalies and in a background control soil. After the plant cultivation, the rhizosphere soils were sampled and subjected to metagenomic analysis for the 16S rRNA gene. The results showed that the taxonomic structure of the amaranth rhizosphere microbiomes was represented by the dominant bacterial phyla Actinobacteriota and Proteobacteria. A feature of the taxonomic profile of the rhizobiomes of *A. tricolor* cv. Valentina and *A. cruentus* cv. Dyuimovochka was a large abundance of sequences related to Cyanobacteria. The formation of the amaranth rhizosphere microbiomes was largely unaffected by soils, but cultivar differences in the formation of the amaranth rhizosphere microbial structure were revealed. Bacterial taxa were identified that are possibly selected by amaranths and that may be important for plant adaptation to various habitat conditions. The targeted enrichment of the amaranth rhizosphere with members of these taxa could be useful for improving the efficacy of amaranth use for agricultural and remediation purposes.

Keywords: *Amaranthus* spp.; rhizosphere microbial communities; rhizosphere bacterial diversity; technogenically contaminated soils; phytoremediation

check for updates

Citation: Muratova, A.; Gorelova, S.; Golubev, S.; Kamaldinova, D.; Gins, M. Rhizosphere Microbiomes of *Amaranthus* spp. Grown in Soils with Anthropogenic Polyelemental Anomalies. *Agronomy* **2023**, *13*, 759. https://doi.org/10.3390/agronomy13030759

Academic Editors: Yong-Xin Liu and Peng Yu

Received: 7 February 2023
Revised: 27 February 2023
Accepted: 3 March 2023
Published: 6 March 2023

1. Introduction

In the past decade, research in rhizosphere biology has enjoyed an increased interest [1–4]. This interest is because the plant root zone is a unique niche that is saturated with physical, chemical, and biological interactions between macro- and microorganisms and between organisms and their environment. Studies on the functions of rhizosphere microorganisms have led researchers to understand their important role in plant life. Specifically, microbes improve plant nutrition through atmospheric nitrogen fixation, mobilization of hard-to-reach phosphorus, increased availability of trace elements, and siderophore production. Furthermore, microbes participate in plant growth regulation through the production of phytohormones and other phytoactive substances, and they increase plant adaptability through stimulation of the antioxidant defense, induction of systemic resistance, and protection from pathogens and organic/inorganic toxicants. Rhizosphere microbial communities are studied by both culture-based [2,5] and culture-independent methods [3,6–8], which helps to protect plants against diseases, improve yields, and increase the efficacy of plant use to restore disturbed soils (phytoremediation). Modern molecular technologies make it possible to develop tools for the artificial modeling of rhizosphere microbiomes, which is of great importance for improving agricultural biotechnologies and predicting their results [9].

The main factors affecting the formation of plant-microbial complexes are the plant species [10] and the soil type [11]. Plant-root-associated microbiomes are considered an important extension of the plant genome itself [12]. The endosphere is the most plant-species-dependent and conserved compartment of rhizobacteria, which determines the so-called plant core microbiome [13]. Components of the core microbiome can be present in and can largely determine the composition of the microbiomes of other root zones, including the rhizoplane and the rhizosphere [10,14]. By contrast, the root-free edaphosphere (bulk soil) is the most plant-independent niche, whose microbiome is characterized by greater variability and is determined by the soil type. The rhizosphere microbiome is crucial for connecting plant and soil microbiomes [12,15]. Plants control the composition of their rhizosphere microbiome via root exudates and can modify it by selecting beneficial microorganisms, thereby contributing largely to the effectiveness of agricultural biotechnologies [1,14,16].

The organisms present in the rhizosphere include bacteria, fungi, oomycetes, nematodes, protozoa, algae, viruses, archaea, and arthropods [15], with bacteria being the most numerous. Bacterial diversity in the rhizosphere can be heavily influenced by abiotic and environmental conditions and can differ depending on the soil type and the plant genotype [7]. Studies on the metagenomes of plant rhizosphere communities are extremely relevant in modern science, because they carry specific information on the rhizobiomes of economically important plants and lead researchers to develop methods for their modification so that the efficiency of a particular agricultural biotechnology can be increased [7,15,17,18]. The rhizosphere microbiome composition has already been characterized for many plant species (e.g., [15,17,19,20]).

Amaranth is a widespread plant genus, many species of which are of great economic importance. *A. caudatus*, *A. cruentus*, and some other species are ancient grain and oilseed crops, which are grown in many countries. *A. tricolor* is also grown as a vegetable plant rich in essential amino acids and biologically active substances [21]. Many cultivars of *A. caudatus*, *A. hypochondriacus*, *A. tricolor*, and other species have richly colored leaves and hanging inflorescences and are used as ornamental crops. A number of *Amaranthus* plants are able to accumulate heavy metals [22–26] and radionuclides [27] and are, therefore, regarded as promising remediators. Soil phytoremediation from both organic and inorganic pollutants is known based on plant-microbial interactions [28], and the presence in the plant rhizosphere of microorganisms resistant to pollutants and able to promote plant growth is critical to the phytoremediation of polluted soils. In this context, the study of rhizosphere microbial communities may contribute to the characterization of the remediation potential of amaranth plants. Information on amaranth-associated rhizosphere microorganisms is extremely limited, but interesting. It was reported that rhizosphere microbiomes of a number of amaranth species are characterized by a pronounced abundance of representatives of Cyanobacteria [29].

The purpose of the study was to characterize the microbial communities of four *Amaranthus* cultivars grown on soils with anthropogenic polyelemental anomalies. Comparison of the rhizosphere microbiomes of plants growing under different environmental conditions will make it possible to better understand the habitat-dependent and species-(or cultivar)-dependent formation of microbiomes of these plants.

2. Materials and Methods

2.1. Soil Sampling

The soil samples used in the pot experiment were those of urban soils based on gray forest soils (WRB, 2006: Greyic Phaeozems), which were characterized by polyelemental anomalies [30]. The soils were collected in the sanitary protection zones near Kosaya Gora Iron Works (KGIW; Tula, Russia) and Tulachermet Co. (Tula, Russia) and on Tula's central avenue, Lenin Avenue. The background soil was collected near the Yasnaya Polyana museum estate of Leo Tolstoy (Tula Region, Russia). The content of toxic substances in the background soil did not exceed the maximum permissible concentration (MPC) and approximate permissible concentration (APC); however, the iron content was quite high as compared with the world values, but minimal as compared with that of the urban soils used. Soil was sampled and prepared for the determination of toxic elements in accordance with the Russian State Standard [31]. Soil samples (~2 kg) from each location were taken from a depth of 0–25 cm from several sampling points using the envelope method. In total, at least 15 samples taken from each location were mixed and used for the pot experiment and analyses.

2.2. Soil Characterization

Soil used in our experiment was characterized on several parameters, such as type (on the basis of particle size distribution); pH; total carbon, water-soluble carbon, humus, $N\text{-}NO_3$, $N\text{-}NH_4$, and P_2O_5 contents; oil products; and metal(loid)s content. In the potassium chloride extract of soil, the potentiometric determination of pH using a glass electrode and a Mettler Toledo Delta 320 pH meter (Mettler-Toledo Instruments Shanghai Ltd., Shanghai, China) was carried out in accordance with the Russian State Standard [32]. Total organic carbon was determined according to [33]. The technique included the dichromate-wet combustion of soil organic matter by concentrated sulfuric acid and the quantitative colorimetric determination (at 590 nm) of the amount of Cr^{2+} generated by dichromate oxidation of soil organic matter. To determine the content of water-soluble carbon, aqueous (distilled water) extracts from the soil samples were obtained, and after drying, were subjected to the same procedure as the dried soil samples. Based on the data obtained, the total carbon, water-soluble carbon, and humus content were calculated. The content of mobile (available) phosphorus (P_2O_5) in mg/100 g was measured photocolorimetrically according to [34]. The method is based on the extraction of mobile compounds of phosphorus from the soil with a solution of ammonium carbonate and the subsequent photocolorimetric determination (at 710 nm) of phosphorus in the form of a blue phosphorus–molybdenum complex. Nitrates and water-soluble ammonium were measured by standard photocolorimetric methods [35,36]. The determination of nitrates included the extraction of nitrates from soil with a potassium chloride solution, reduction of nitrates to nitrites with hydrazine in the presence of copper as a catalyst, and photometric measurement (at 545 nm) of the colored diazo compound formed. The determination of exchangeable ammonium included the extraction of exchangeable ammonium from soil with a potassium chloride solution, generation of a colored indophenol compound formed by the interaction of ammonium with hypochlorite and sodium salicylate in an alkaline medium, and photometry of the colored solution (at 655 nm). All photocolorimetric measurements were carried out using an Evolution 60 UV-Vis Spectrophotometer (Thermo Scientific, Madison, WI USA). Oil products were measured gravimetrically [37]. The method is based on extraction of oil products with chloroform from air-dried soil, separation from polar compounds by liquid chromatography after replacing the solvent with hexane, and quantitative determination by gravimetric analysis. The determination of the elemental composition of soils was carried out using X-ray fluorescence analysis in the certified Laboratory of Chemical-analytical Research of GIN RAS. Soil preparation for analysis and quality control was carried out following certified methods and recommendations [30,38]. The concentrations of Mn, Fe, V, Cr, Ni, Cu, Zn, Pb, and As were determined using a serial wave XRF spectrometer "S4 Pioneer" (Bruker AXS GmbH, Karlsruhe, Germany) with a rhodium tube (capacity 4 kW). The obtained data were processed using the S4 Spectra Plus program using coefficients for correction of the routine samples matrix effects. Standard samples of the composition IAEA Soil-7, SChT-1.2 (soil), GBW-07404, 07405 (soil) were used as reference samples for soil analysis. The concentration of potentially toxic trace elements in the soil was compared with the MPCs and APCs of the metals by hygienic standards that meet international standards [39,40]. All chemical analyses of soil were performed in at least triplicate.

2.3. Experimental Design

We used four *Amaranthus* cultivars bred by the All-Russian Research Institute of Vegetable Breeding and Seed Production of the Federal Scientific Vegetable Center (VNIISSOK; Odintsovo District, Moscow Region, Russia). The specific cultivars were Valentina (*A. tricolor* L.), Dyuimovochka (*A. cruentus* L.), Bulava (*A. caudatus* L.), and Zelenaya Sosulka (*A. caudatus* L.). The description of the *Amaranthus* cultivars used is given in Table S1.

Plant seeds (~50 mg) were sown in soil-filled 2 L plastic pots. After one month, the seedlings were thinned out, leaving 10 plants per pot. Plants were cultivated under controlled conditions (day/night light cycle, 14/10 h; temperature, 22/24 °C) for 6 months. There were three replicates for each combination of *Amaranthus* cultivar and soil. At the end of growth, the plants were removed from the pots, and the soil was vigorously shaken off the roots. The rhizosphere soil adhering to young roots from 5 plants from each pot were combined from three replicates of one variant, and mixed samples were used for metagenomic analysis. The remaining soil adhering to the root surface as rhizosphere soil was carefully scraped off with sterilized tweezers.

2.4. A 16S rRNA Gene-Based Metagenomic Analysis of Rhizosphere Soil

Extraction and purification of soil DNA for metagenomic analysis was carried out using the Fast DNA®SPIN Kit for Soil (MP Biomedicals, Santa Ana, CA, USA) and a Fast Prep®24 homogenizer (MP Biomedicals, Santa Ana, CA, USA), according to the manufacturer's instructions.

A 16S rRNA sequencing library was constructed, according to the 16S metagenomics sequencing library preparation protocol (Illumina, San Diego, CA, USA), targeting the V3 and V4 hypervariable regions of the 16S rRNA gene. The initial PCR was performed with template DNA using region-specific adapters shown to have compatibility with the Illumina index and sequencing primers (forward primer: 5′-TCGTCGGCAGCGTCAGATGTGTATAAGAGACAGCCTACGGGNGGCWGCAG-3′; reverse primer: 5′-GTCTCGTGGGCTCGGAGATGTGTATAAGAGACAGGACTACHVGG GTATCTAATCC-3′) [41]. Amplification was performed using the Veriti™ Thermal Cycle, 96-Well (Applied Biosystems™, Foster City, CA, USA), according to the Illumina protocol. After the first round of amplification, PCR products were visualized using gel electrophoresis. Then, the PCR products were purified with AMPure XT magnetic beads, and the second PCR was performed using primers from a Nextera XT Index Kit (Illumina). Subsequently, purified PCR products were quantified with a Qubit dsDNA HS Assay Kit (Thermo Scientific) on a Qubit 2.0 fluorometer. The sample pool (4 nM) was denatured with 0.2 N NaOH, diluted further to 10 pM, and combined with 20% (v/v) denatured 4 pM PhiX, prepared following Illumina guidelines. Sequencing of 16S rRNA gene V3–V4 variable regions was performed on the Illumina MiSeq platform in 2 × 300 bp mode.

2.5. Bioinformatics and Statistical Analysis

Reads were analyzed using QIIME2 software, version 2022.8 (http://qiime2.org/, accessed on 25 February 2023) [42]. Before filtering, there were 118,339 read pairs per sample on average. Raw reads were processed using the DADA2 algorithm implemented in QIIME [43]. After quality filtering, chimera and phiX sequences removal, we analyzed 17,870 joined read pairs per sample on average. The taxonomy was assigned to the sequences using the Naive Bayes classifier pre-trained on the latest SILVA 138 database 99% OTUs [44]. The number of observed features varied from 601 to 1030. To characterize the richness and evenness of the bacterial community, alpha diversity indices were calculated using Chao1, Shannon, and Simpson metrics. Similarities between microbial compositions of the samples were evaluated using the beta diversity characteristics, which were estimated using weighted and unweighted Unifrac measures with further non-metric multidimensional scaling (nMDS) visualization.

Venn diagrams were constructed with Creately software (https://creately.com/lp/venn-diagram-maker/, accessed on 5 February 2023; Cinergix Pty Ltd., Melbourne, VIC,

Australia). Spearman's rank correlation coefficients were calculated with Statistica software 13 (TIBCO Software Inc. 2017, Statsoft Russia, Moscow, Russia). Other calculations were done with Excel 2019 software (Microsoft, Redmond, WA, USA).

3. Results

3.1. Soil Characteristics

For all soils used in the experiment, the main characteristics were determined: type; pH; and the total content of (including water-soluble) carbon, biogenic forms of nitrogen (NH_4 and NO_3) and phosphorus (P_2O_5), heavy metals and metalloids (Fe, Mn, V, Ni, Cu, Zn, Pb, and As), and hydrocarbons. The results are given in Table 1.

Table 1. Characteristics of the soils used in the experiment.

Properties	Background Soil (Yasnaya Polyana)	KGIW	Tulachermet	Lenin Ave.
Soil type	Clay loam	Clay loam	Sandy loam	Clay loam
pH	6.20	7.26	7.35	7.29
Total carbon (% of air-dried soil)	4.87 ± 0.05	4.41 ± 0.11	4.84 ± 0.33	3.83 ± 0.06
Water-soluble carbon (% of total carbon)	0.14 ± 0.05	0.13 ± 0.02	0.14 ± 0.06	0.12 ± 0.03
Humus (% of air-dried soil)	8.12 ± 0.16	7.14 ± 0.14	7.62 ± 0.15	6.16 ± 0.12
Humus carbon (%)	4.72 ± 0.09	4.15 ± 0.08	4.43 ± 0.09	3.58 ± 0.07
N-NO_3 (mg/kg dw)	20.5 ± 1.2	28.2 ± 1.6	38.6 ± 3.5	41.1 ± 2.4
N-NH_4 (mg/kg dw)	12.9 ±1.2	6.9 ± 0.7	9.5 ± 0.3	2.3 ± 0.2
P_2O_5 (mg/kg dw)	51.0 ± 4.8	159.5 ± 3.6	98.4 ± 7.6	218.0 ± 6.7
Oil products (g/kg dw)	1.5 ± 0.6	**2.6 ± 0.4**	**4.1 ± 1.0**	**2.5 ± 0.7**
Metal(loid)s (mg/kg dw):				
Fe	15,600 ± 1860	**78,100 ± 1280**	**120,600 ± 5830**	**37,400 ± 2160**
Mn	1300 ± 67	**5700 ± 180**	1100 ± 87	**1600 ± 58**
V	57 ± 3.0	41 ± 4.1	**136 ± 7.2**	61 ± 2.9
Ni	25 ± 3.1	31 ± 4.2	**55 ± 3.3**	35 ± 2.8
Cu	29 ± 2.3	52 ± 4.1	**75 ± 0.8**	**378 ± 0.9**
Zn	47 ± 2.0	**310 ± 9.4**	**161 ± 3.6**	186 ± 5.3
Pb	18 ± 2.1	72 ± 6.0	26 ± 0.7	59 ± 1.1
As	5.1 ± 0.4	5.9 ± 0.5	**6.4 ± 0.2**	7.3 ± 0.3

Note: Data expressed as mean ± standard deviation ($n \geq 3$). Bold type means the permissible concentrations [29,30] were exceeded for: Fe (MPC_{total} 1500 mg/kg), Mn (MPC_{total}, 1500 mg/kg), V (MPC_{total}, 100 mg/kg), Ni (APC_{total}, sandy loam, 20 mg/kg; APC_{total}, clay loam, 80 mg/kg), Cu (MPC_{total} 55 mg/kg; APC_{total}, sandy loam, 33 mg/kg), Zn (APC_{total}, clay loam, 220 mg/kg; APC_{total}, sandy loam, 55 mg/kg), As (APC_{total}, sandy loam, 2 mg/kg; APC_{total}, clay loam, 10 mg/kg), and oil products (1.0 g/kg). The oil product content in soil is not regulated at the regional level; the recommended value is 1.0 g/kg.

All soil samples were represented by gray forest soils, had neutral or close to slightly acidic pH values, and were also characterized by a high content of iron. The soils of the sanitary protection zones near Kosaya Gora Iron Works (KGIW; Tula, Russia) and Tulachermet Co. (Tula, Russia) and the urban soil of Tula's central avenue, Lenin Avenue, all have an excessive content of heavy metals and oil products, indicating human-caused pollution. The KGIW soil had a high content of Mn, Zn, and petroleum products, which exceeded MPCs by 380, 140, and 260%, respectively. The Tulachermet soil had a high content of V, Ni, Cu, Zn, As, and oil products, exceeding MPCs and APCs by 36, 275, 127, 290, 300, and 410%, respectively. The Lenin Avenue soil had a high content of Mn, Cu, and oil products, exceeding the permissible concentrations by 690, 107, and 250%, respectively. The background soil was that collected near the Yasnaya Polyana museum estate of Leo Tolstoy (Tula Region). In it, the content of environmentally regulated elements did not exceed the MPCs or APCs.

3.2. *Metagenomic Analysis of Rhizosphere Microbial Communities of Amaranthus spp.*

3.2.1. Diversity of Rhizosphere Communities

Sequencing of the 16s rRNA gene from 16 rhizosphere samples resulted in 1,915,453 raw reads. After data denoising and chimera screening, 17,870 joined read pairs per sample on average were used for further identification. Rarefaction curves obtained with the normalized OTU number almost reached saturation levels for all samples, indicating that the bacterial communities were covered well by the sequence analysis (Figure S1). The sequences with >97% similarity were combined into operational taxonomic units (OTUs). The OTUs were assigned to 38, 124, 288, 449, and 804 taxa at the phylum, class, order, family, and genus levels, respectively.

To characterize the bacterial diversity and the richness of the microbial communities, we calculated the α- and β-diversity (Table 2 and Figure 1).

Table 2. The α-diversity indices for the rhizospheric microbial communities of amaranths grown on different soils.

Soil	Plant	Observed Features	Chao1	Shannon Index	Simpson Index	Faith PD
Background	*A. tricolor* cv. Valentina	743	748.000	8.6560	0.9964	81.03
	A. cruentus cv. Dyuimovochka	716	718.111	8.6309	0.9957	73.42
	A. caudatus cv. Bulava	890	898.347	8.5701	0.9930	86.78
	A. caudatus cv. Zelenaya Sosulka	910	916.949	8.7466	0.9955	97.51
Tulachermet	*A. tricolor* cv. Valentina	741	743.258	8.6019	0.9956	87.83
	A. cruentus cv. Dyuimovochka	823	829.343	8.8715	0.9970	87.65
	A. caudatus cv. Bulava	1010	1014.614	9.2749	0.9978	92.29
	A. caudatus cv. Zelenaya Sosulka	819	831.470	8.8402	0.9967	81.70
KGIW	*A. tricolor* cv. Valentina	807	807.786	8.9055	0.9970	82.02
	A. cruentus cv. Dyuimovochka	806	806.992	8.6081	0.9940	88.70
	A. caudatus cv. Bulava	990	994.844	9.1206	0.9970	97.06
	A. caudatus cv. Zelenaya Sosulka	964	974.350	9.1507	0.9974	88.53
Lenin Ave.	*A. tricolor* cv. Valentina	832	836.833	8.9232	0.9969	85.43
	A. cruentus cv. Dyuimovochka	601	602.600	8.4401	0.9959	63.57
	A. caudatus cv. Bulava	758	766.928	8.3750	0.9926	81.42
	A. caudatus cv. Zelenaya Sosulka	707	716.784	8.5888	0.9957	74.67

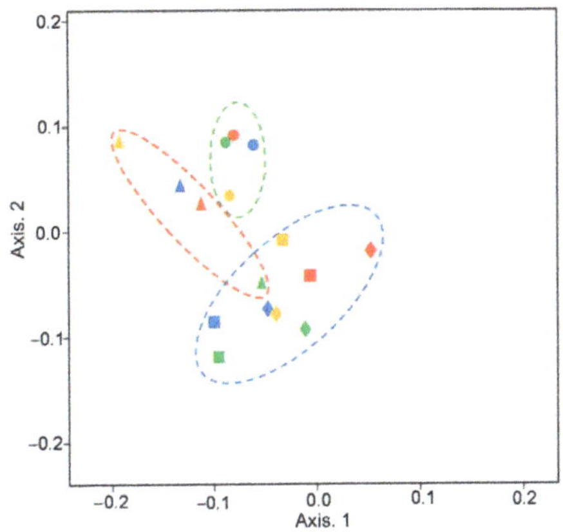

Figure 1. Principal coordinate analysis for the rhizosphere microbial communities of the *Amaranthus* cvs.: triangles (▲), *A. tricolor* cv. Valentina; circles (•), *A. cruentus* cv. Dyuimovochka; squares (■), *A. caudatus* cv. Bulava; diamonds (♦), *A. caudatus* cv. Zelenaya Sosulka; blue, background soil; yellow, Tulachermet soil; red, KGIW soil; green, Lenin Ave. soil.

The α-diversity was measured by using the species richness indices (Chao1, Shannon, and Simpson indices; Faith's phylogenetic diversity [PD], Table 2). Traditional (Shannon's and Simpson's) and phylogenetic (Faith's PD) indices of bacterial alpha-diversity in the rhizosphere communities of amaranths yielded similar conclusions: all communities were quite diverse and differed between plant species (cultivars) grown on the same soil.

A comparison of the rhizosphere microbiomes of different samples showed that the influence of soils on the formation of rhizosphere communities was not pronounced: the samples were not grouped according to soils. Yet, there were pronounced differences in rhizospheric samples according to the plant species studied (Figure 1). *A. caudatus* cv. Bulava and *A. caudatus* cv. Zelenaya Sosulka formed one cluster, which distinctly distanced itself from *A. tricolor* cv. Valentina, whereas the cluster of *A. cruentus* cv. Dyuimovochka had intersections with both *A. caudatus* cultivars.

3.2.2. Taxonomic Structure of Rhizosphere Communities

MiSeq sequencing showed that the amaranth rhizosphere communities included 804 genera of bacteria belonging to 449 families of 38 phyla.

Figure 2 illustrates the relative abundances of OTUs associated at the phylum level in the rhizosphere of the amaranth cultivars studied. In different soils in the rhizosphere communities of cv. Valentina, most OTUs were assigned to Actinobacteriota (29–39%), Proteobacteria (17–29%), Chloroflexi (8–18%), Cyanobacteria (4–18%), and Acidobacteriota (4–6%). In the rhizosphere cv. Dyuimovochka, the dominant phyla were also Actinobacteriota (25–42%), Proteobacteria (19–31%), Chloroflexi (6–12%), Cyanobacteria (3–19%), and Acidobacteriota (4–9%). In the rhizosphere cv. Bulava, the dominant phyla were Actinobacteriota (31–40%), Proteobacteria (19–25%), Chloroflexi (10–15%), and Acidobacteriota (7–12%). In the rhizosphere cv. Zelenaya Sosulka, the dominant phyla were Actinobacteriota (31–36%), Proteobacteria (21–27%), Chloroflexi (9–14%), and Acidobacteriota (8–11%). The percentages of OTUs assigned to other phyla were much smaller.

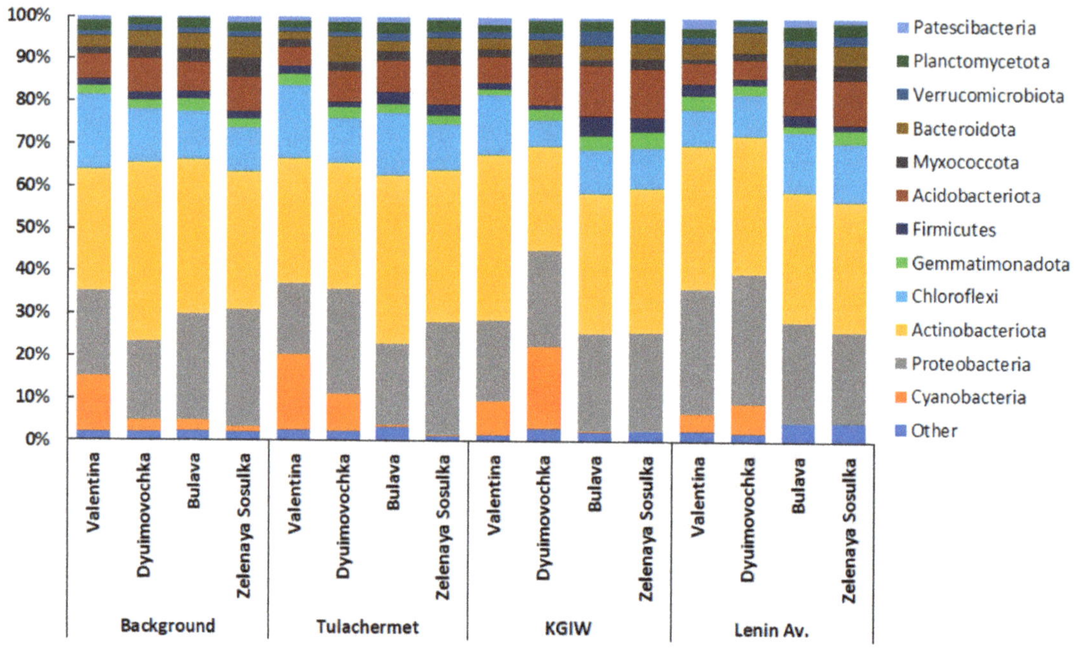

Figure 2. Relative abundances of OTUs associated at the phylum level of the microbial communities in the rhizospheres of four amaranth cultivars grown on different soils.

The list of taxa at the family level, for which the share of amaranths in the rhizosphere was ≥1%, is given in Table S2. Of the dominant Actinobacteriota phylum, 51 families were identified, of which 13 were most abundant (Table S2, Figure 3), making up 80 to 86% of all detected actinobacterial families in the amaranth rhizosphere. The family Gaiellaceae had the maximum share in the rhizospheric population of actinobacteria. Its abundance reached 1.8–3.0% in the rhizosphere of cv. Valentina, 1.5–4.6% in the rhizosphere of cv. Dyuimovochka, 2.1–3.7% in the rhizosphere of cv. Bulava, and 1.9–2.9% in the rhizosphere of cv. Zelenaya Sosulka. Other notable actinobacteria were members of the families Nocardioidaceae (1.9–4.8% of all actinobacteria) and Micromonosporaceae (0.8–5.4%) and members of the orders Solirubrobacterales (67-14 family) and MB-A2-108 (MB-A2-108 family).

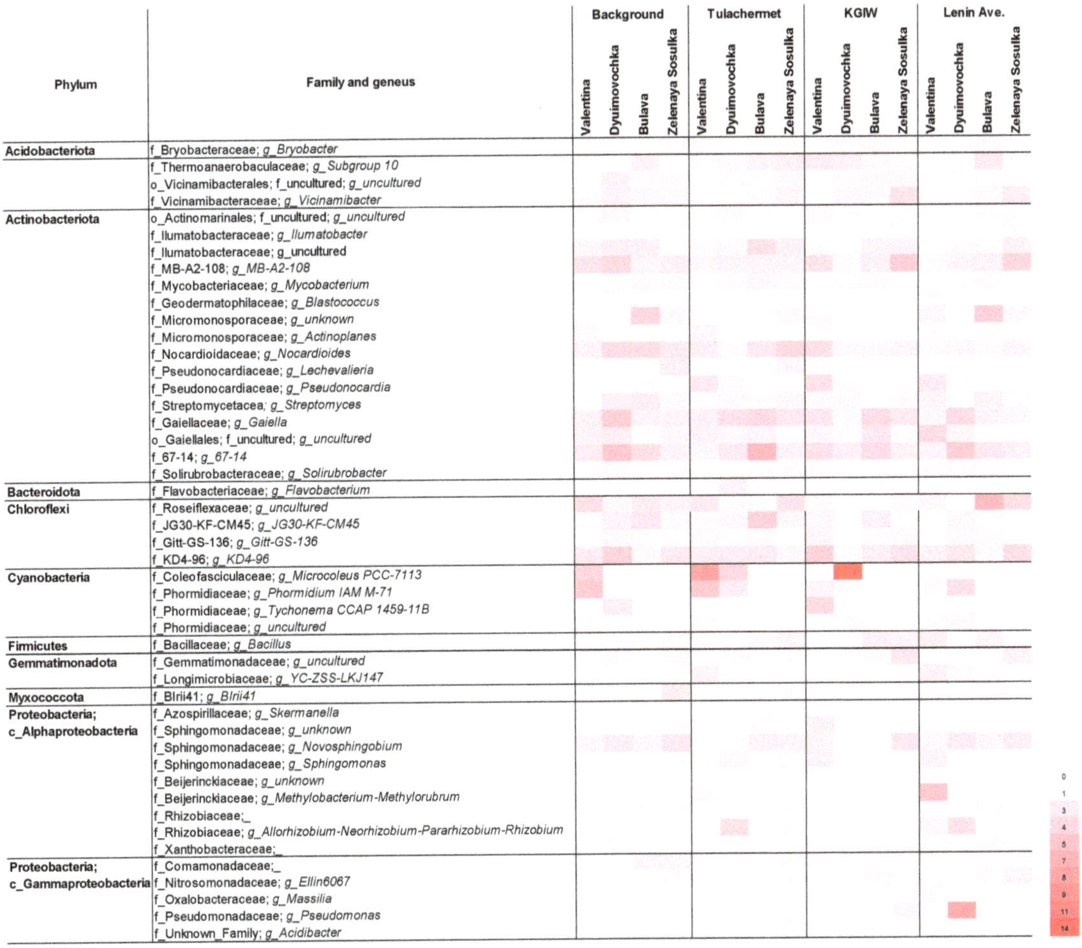

Figure 3. Heat map illustrating the relative abundances of OTUs associated at the genus level of the microbial communities in the rhizospheres of four amaranth cultivars grown on different soils.

In another dominant phylum, Proteobacteria, 93 bacterial families were identified. In total, 10 of them (Table S2, Figure 3) made the largest contribution to the structure of the amaranth rhizosphere microbiomes (54–76% of all detected families). Alphaproteobacteria was the major class of the phylum, accounting for 64–84% of all OTUs assigned to the

Proteobacteria. Among the Alphaproteobacteria, the dominant position was occupied by the families Sphingomonadaceae (3–7%) and Beijerinckiaceae (1–6%); a large proportion of OTUs also belonged to the Rhizobiaceae family (1–5%). The Gammaproteobacteria class accounted for 17 to 45% of all OTUs assigned to the Proteobacteria phylum, and the Nitrosomonadaceae and Comamonadaceae families were its predominant representatives. The contribution of the Betaproteobacteria and Deltaproteobacteria classes to the overall taxonomic structure of the amaranth microbiomes was not noticeable.

The rhizosphere microbiomes of two species of *Amaranthus* (*A. tricolor* cv. Valentina and *A. cruentus* cv. Dyuimovochka) were clearly enriched with members of the Cyanobacteria phylum, among which the Coleofasciculaceae and Phormidiaceae families dominated on all soils (Table S2, Figure 3), but were poorly represented in the rhizosphere of *A. caudatus* cv. Bulava and cv. Zelenaya Sosulka. The Phormidiaceae family was also abundant in the rhizosphere of *A. tricolor* cv. Valentina and *A. cruentus* cv. Dyuimovochka. This family was poorly represented in the rhizosphere of *A. caudatus* cv. Bulava and was not in the rhizosphere of cv. Zelenaya Sosulka. This may indicate that the maintenance of these bacterial taxa by host plants is species specific.

The species-specific changes in the taxonomic profile of the amaranth rhizosphere communities, as induced by the soil characteristics, can be seen in Figures 2 and 3. Spearman's rank correlation and principal component analysis did not reveal significant correlations between the kinds of soil used and the dominant taxa in the amaranth rhizosphere microbiomes. However, a close correlation was established between the cultivars ($r_s = 0.64$, $p < 0.05$) and the abundance of OTUs assigned to the Cyanobacteria phylum. In addition, moderate correlations were found between the cultivars and the abundance of the Acidobacteriota ($r_s = 0.55$, $p < 0.05$), Bacteroidota ($r_s = 0.52$, $p < 0.05$), and Planctomycetota ($r_s = 0.52$, $p < 0.05$) phyla in the rhizosphere communities.

3.2.3. Shared and Unique Taxa among Rhizosphere Microbial Communities

To determine which OTUs were shared by or were specific to the rhizosphere of each cultivar on the four soils, we did several comparative analyses (Figure 4).

In the background soil, the largest number of OTUs at the bacterial species level was found in the rhizosphere of *A. caudatus* cv. Zelenaya Sosulka, followed by *A. tricolor* cv. Valentina, *A. caudatus* cv. Bulava, and *A. cruentus* cv. Dyuimovochka. A total of 127 species were shared and accounted for from 30 to 39% of the rhizosphere communities of the 4 plant cultivars. The percentage of unique taxa ranged from 24 to 30%. The maximal number of unique species (125) was found for the rhizosphere of *A. caudatus* cv. Zelenaya Sosulka.

In the Tulachermet soil, the largest number of OTUs at the bacterial species level was found in the rhizosphere of *A. caudatus* cv. Bulava, followed by *A. cruentus* cv. Dyuimovochka, *A. tricolor* cv. Valentina, and *A. caudatus* cv. Zelenaya Sosulka. A total of 124 species were shared and accounted for from 28 to 34% of the rhizosphere communities of all plant cultivars. The percentage of unique taxa ranged from 23 to 29%. The maximal number of unique species (125) was found for the rhizosphere of *A. caudatus* cv. Bulava.

In the KGIW soil, the largest number of OTUs at the bacterial species level was found in the rhizosphere of *A. caudatus* cv. Bulava, followed by *A. cruentus* cv. Dyuimovochka, *A. caudatus* cv. Zelenaya Sosulka, and *A. tricolor* cv. Valentina. A total of 121 species were shared and accounted for from 28 to 33% of the rhizosphere communities of all plant cultivars. The percentage of unique taxa ranged from 25 to 30%. The maximal number of unique species (128) was found for the rhizosphere of *A. caudatus* cv. Bulava.

In the Lenin Ave. soil, the largest number of OTUs at the bacterial species level was found in the rhizospheres of *A. tricolor* cv. Valentina, followed by *A. caudatus* cv. Zelenaya Sosulka, *A. caudatus* cv. Bulava, and *A. cruentus* cv. Dyuimovochka. Only 88 species were shared and accounted for from 23 to 25% of the rhizosphere communities of the 4 plant cultivars. The percentage of unique taxa ranged from 24 to 29%. The maximal number of unique species (133) was found for the rhizosphere of *A. tricolor* cv. Valentina.

Among the unique taxa identified in the rhizosphere microbiome of each amaranth variety, those taxa that occur in at least two different soils were determined (Tables S3–S6). Overall, 53, 39, 50, and 52 such taxa were revealed in the rhizosphere of *A. tricolor* cv. Valentina, *A. cruentus* cv. Dyuimovochka, and *A. caudatus* cvs. Bulava and Zelenaya Sosulka, respectively. In addition, seven, two, eight, and six such unique taxa were revealed in the rhizosphere of cvs. Valentina, Dyuimovochka, Bulava, and Zelenaya Sosulka, respectively, grown in three different soils studied (Table 3). Most of those taxa were represented by uncultured bacteria.

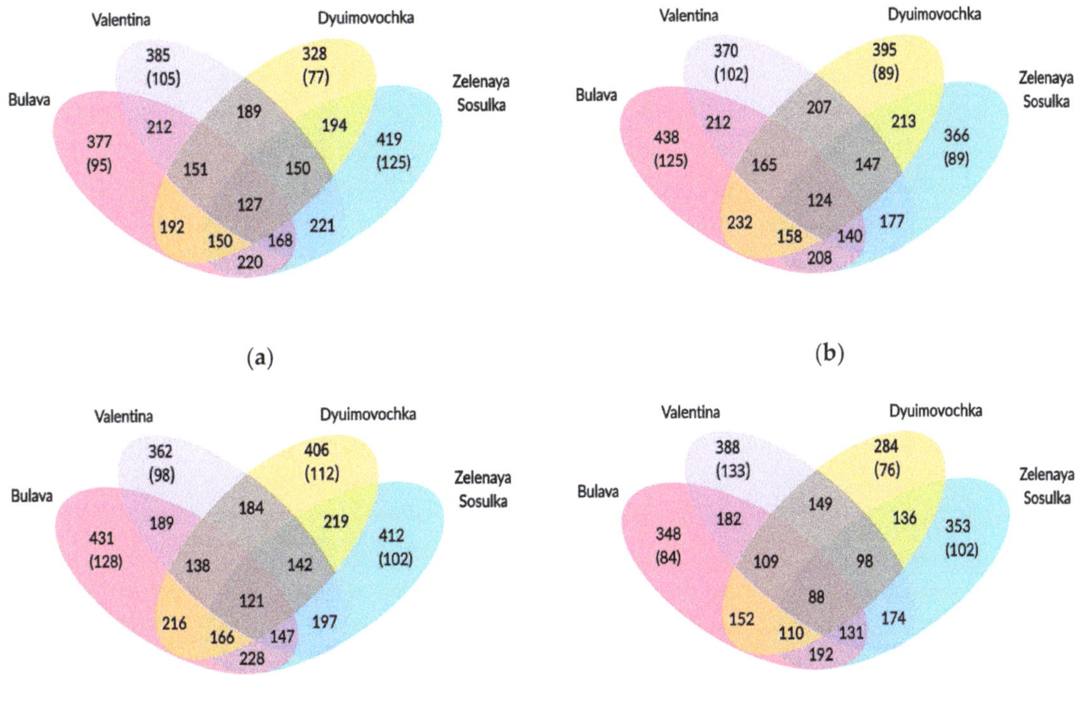

Figure 4. Venn diagram depicting shared and unique OTUs among the rhizosphere microbial communities of the amaranths grown on the background and polluted soils. The number of unique taxa is given in brackets. (**a**) Background; (**b**) Tulachermet; (**c**) KGIW; (**d**) Lenin Ave.

Table 3. Unique bacterial taxa revealed in the rhizosphere of *Amaranthus* plants and occurred in three different soils.

Plant Cultivars	Bacterial Taxa	
A. tricolor cv. Valentina	*g_Chloronema; s_Scytonema tolypothrichoides*	Background; Tulachermet; KGIW
	g_Leptolyngbya_VRUC_135;s_uncultured bacterium	Background; Tulachermet; Lenin Ave.
	g_Alsobacter; s_Alsobacter_metallidurans	Tulachermet; KGIW; Lenin Ave.
	g_C0119; s_uncultured_bacterium	Background; Tulachermet; Lenin Ave.
	g_Rhodocytophaga; s_uncultured Bacteroidetes	Background; Tulachermet; KGIW
	f_Rhodanobacteraceae; *g_uncultured; s_uncultured bacterium*	Background; Tulachermet; KGIW
	g__DS-100; s_unknown	Tulachermet; KGIW; Lenin Ave.

Table 3. *Cont.*

Plant Cultivars	Bacterial Taxa	
A. cruentus cv. Dyuimovochka	*g_Methylophilus; s_uncultured bacterium*	Tulachermet; KGIW; Lenin Ave.
	g_CENA518; s_uncultured_bacterium	Tulachermet; KGIW; Lenin Ave.
A. caudatus cv. Bulava	*g_Nakamurella; g_unknown; s_unknown*	Background; Tulachermet; KGIW
	g_Pir4_lineage; s_uncultured Pirellula	Background; Tulachermet; Lenin Ave.
	f_Moraxellaceae; *g_uncultured; s_uncultured gamma*	Background; KGIW; Lenin Ave.
	f_0319-7L14; *g_0319-7L14; s_uncultured bacterium*	Background; Tulachermet; Lenin Ave.
	o_Planctomycetales; f_uncultured; *g_uncultured*	Background; Tulachermet; KGIW
	f_Phycisphaeraceae; *g_uncultured*	Tulachermet; KGIW; Lenin Ave.
	f_Kapabacteriales; *g_Kapabacteriales*	Tulachermet; KGIW; Lenin Ave.
	f_Sericytochromatia; *g_Sericytochromatia*	Background; Tulachermet; KGIW
A. caudatus cv. Zelenaya Sosulka	f_Vicinamibacteraceae; *g_Vicinamibacteraceae;s_uncultured Geothrix*	Tulachermet; KGIW; Lenin Ave.
	f_Xanthomonadaceae; *g_unknown; s_unknown*	Background; Tulachermet; KGIW
	f_Rhodanobacteraceae; *g_Ahniella; s_uncultured bacterium*	Background; Tulachermet; Lenin Ave.
	f_Latescibacterota; *g_Latescibacterota; s_uncultured soil*	Background; Tulachermet; Lenin Ave.
	f_Blastocatellaceae; *g_Aridibacter; s_uncultured bacterium*	Background; KGIW; Lenin Ave.
	f_Subgroup 22; *g_Subgroup 22*	Tulachermet; KGIW; Lenin Ave.

The bacterial taxa listed in Table 3 are unique to each of the four plant cultivars studied and can be considered as plant-specific.

4. Discussion

Studies characterizing the rhizosphere microbiomes of *Amaranthus* plants are extremely scarce. Nambisan et al. [29] found that the Cyanobacteria phylum is distinctly enriched in the roots and the rhizosphere soil of the three grain amaranths—*A. hupochondriacus*, *A. cruentus*, and *A. caudatus*. Our study confirmed that the peculiarity of the rhizosphere microbial communities of two of the three *Amaranthus* species (*A. tricolor* cv. Valentina and *A. cruentus* cv. Dyuimovochka) is the distinct presence of cyanobacteria along with the dominant bacterial phyla, such as Proteobacteria, Actinobacteriota, and Chloroflexi. In addition, we obtained the first data on the rhizosphere microbiome of another amaranth species, *A. tricolor*. In Nambisan et al.'s research [29], the negative controls were other plant species (*Beta vulgaris, Cicer arietinum*, and *Solanum lycopersicum*), and in their rhizobiomes, no predominance of cyanobacteria was observed. In our study, besides the amaranth rhizosphere, we also analyzed the rhizosphere of other plants under the same experimental conditions—*Sorghum bicolor* cv. Sucro and *Sorghum bicolor* cv. Biomass. The rhizosphere of these plants showed no predominance of cyanobacteria either [20]. An analysis of the taxonomic profile of the cyanobacteria found in the amaranth rhizobiomes made it possible to identify members of the dominant *Microcoleus* and *Phormidium* genera. The presence of these genera in the rhizosphere of various plants was also noted earlier [4,45]. The accumulated data indicate that the amaranths may have specifically selected cyanobacteria from the surrounding soil microflora. In turn, cyanobacteria favor the growth of the amaranth plants, possibly through their plant-beneficial characteristics. It is known that cyanobacteria can produce plant-growth-promoting substances, including auxins, gibberellins, cytokinins, abscisic acids, vitamins, and amino acids [4,46–48]. Cyanobacteria can also add organic matter, synthesize and liberate amino acids and vitamins, reduce the content of soil oxidizable matter, provide oxygen to the submerged rhizosphere, ameliorate salinity, buffer the pH, solubilize phosphates, and increase the efficiency of fertilizer use in crop plants [46]. Cyanobacteria such as *Nostoc* and *Microcoleus* can form associations with cycads [49] and *Gunnera* [50] and can fix nitrogen [45], either as free-living organisms or in association with host plants, in which they reside in specific tissues. Cyanobacteria such as *Calothrix* and *Anabena* can be used as biofertilizers [51]. Rhizospheric cyanobacteria are still insufficiently studied, although interest in their use for soil fertilization and plant-growth promotion is increasing steadily [4]. The established abundance of cyanobacteria in the amaranth rhizosphere requires further and deeper research on the part they play in plant vital activity.

To determine the contribution of different plant species (or cultivars) to the formation of their rhizosphere community in a particular soil, we compared their taxonomic composition, identifying shared and unique taxa (Figure 4). It was revealed that only about a third of taxa were shared between all four amaranths studied, and also about a third of taxa were unique. The contribution of *A. tricolor* cv. Valentina and *A. cruentus* cv. Dyuimovochka to the selection of cyanobacteria in their rhizosphere was confirmed. In addition, we revealed that two cultivars of *A. caudatus* selected different unique taxa in their rhizosphere, which suggests the cultivar-specific formation of rhizosphere microbiomes by amaranths (Tables S2–S5). No debated that plant root exudates are the principal connecting link between plant and plant-root-associated microbial communities [1,14,16]. In turn, the composition of root exudates is determined by plant physiology, which is different among plant species and even cultivars.

Current views hold that the relative content of individual taxa in the microbial community is a biological indicator of the soil status—for example, heavy metal pollution [52]. The urban soils used in our experiment were polluted by heavy metals. Many taxa identified in the amaranth rhizosphere microbiomes had previously been described as resistant to heavy metals or as oil-degraders, and they may be important for the resistance of amaranth to the human-caused pollution of soil. These taxa include (1) members of Proteobacteria, such as the genera *Pseudomonas* [18,53,54], *Novosphingobium* [55,56], *Sphingomonas* [57], *Rhizobium* [58,59], and *Massilia* [18,60]; (2) members of Actinobacteriota, such as *Mycobacterium*, *Nocardiodes*, *Streptomyces* [61], and the family Gaiellaceae [62]; (3) members of the Cyanobacteria phylum, such as the genus *Microcoleus* [63]; and (4) members of Firmicutes, such as the genus *Bacillus* [18]. In the rhizosphere of the amaranths used in this study, some of these taxa were present in large numbers (the genera *Gaiella*, *Nocardiodes*, *Microcoleus*, *Novosphingobium*, and *Pseudomonas*), whereas others were found in minor or single samples. Of note, increased numbers of members of Gaiellaceae had previously been found by Sun et al. [62] in the metal-polluted rhizosphere of crops. Those authors found a significant correlation between the increased abundance of Gaiellaceae-related bacteria and various metals and metalloids, and they concluded that these bacteria play a potentially active ecological part in the interaction with soil metals.

Rhizospheric microbial communities exert a great effect on the phytoremediation of metal-contaminated soils, not only by changing the bioavailability of metals [64], but also by promoting plant growth under pollutant stress through the fixation of nitrogen; production of phytohormones (indole-3-acetic acid, cytokinins, and gibberellins), siderophores, and enzymes (1-aminocyclopropane-1-carboxylate deaminase); and transformation of nutrients [65–67]. However, only 2–5% of rhizosphere microorganisms contribute to plant growth, and plants naturally select these beneficial microorganisms, which help them to grow and survive, especially under unfavorable conditions [3]. In the rhizospheric microbial communities of the amaranths used in this study, alongside cyanobacteria, we identified other groups of rhizobacteria with plant-growth-promoting potential (Figure 3). These included *Bacillus* (Bacillaceae, [68,69]), *Sphingomonas* (Sphingomonadaceae, [70]), *Streptomyces* (Streptomycetaceae, [69,71]), *Pseudomonas* (Pseudomonadaceae, [72]), and *Rhizobium* (Rhizobiaceae, [73]). Although the percentages of these taxa varied between cultivars and soils, their mere presence indicates that the rhizosphere microbiomes have the potential to promote amaranth growth on all urban soils tested.

We were unable to reveal any significant influence of the soils on the taxonomic structure of the amaranth rhizospheric microbiomes. Only about one third of the taxa identified in the cultivar rhizosphere microbiomes were common to all soil samples (Figure 4), which may indicate the maintenance of a specific plant microbiome, regardless of the kind of soil. We supposed that peculiarities and differences in the plant root exudate composition among plant cultivars had stronger differences than among soils, which resulted in the formation of different rhizosphere microbiomes depending more on the plants.

It is known that the composition of rhizosphere microbial communities is determined not only (and possibly not so much) by the soil type, but also by the plants. In our case, all soils studied were Greyic Phaeozems with slight differences, and all plants tested were Amaranthus belonging to different species. The compositions of the studied rhizosphere communities were quite similar, but the analysis of the obtained results revealed a greater influence of plants than soils.

For the final characterization of the amaranth core rhizobiomes, additional studies of the endosphere microbiome of each plant are required.

5. Conclusions

Limited data are available on the structure of the rhizosphere microbial communities of *Amaranthus* spp. We conducted a comparative study on the rhizosphere microbiomes of four *Amaranthus* cultivars (*A. tricolor* cv. Valentina, *A. cruentus* cv. Dyuimovochka, *A. caudatus* cv. Bulava, and *A. caudatus* cv. Zelenaya Sosulka). The cultivars were grown in an unpolluted (background) soil and in three polluted soils with polyelemental anomalies. The *A. tricolor* rhizosphere microbiome was characterized for the first time. The taxonomic structure of the amaranth rhizosphere microbiomes was represented by the dominant bacterial phyla Actinobacteriota, Proteobacteria, and Chloroflexi and by the phyla Cyanobacteria, Acidobacteriota, Planctomycetota, and Bacteroidota. In the taxonomic profile of the rhizobiomes of two *Amaranthus* species (*A. tricolor* cv. Valentina and *A. cruentus* cv. Dyuimovochka), there was a significant abundance of OTUs associated with the Cyanobacteria phylum. Bacterial taxa were identified that are possibly selected by amaranth plants during their coexistence and that may be important for plant adaptation to various habitat conditions, including polluted soils. The targeted enrichment of the amaranth rhizosphere microbiomes with members of these taxa could be useful for improving the efficacy of amaranth use for agricultural and remediation purposes.

Supplementary Materials: The following supporting information can be downloaded at: https://www.mdpi.com/article/10.3390/agronomy13030759/s1, Table S1: Characteristics of Amaranthus cultivars used in the research; Table S2: Taxonomic structure of the rhizosphere microbial communities of the amaranths studied; Tables S3–S6: Unique rhizosphere taxa occurred in at least two different soils planted with amaranths studied; Figure S1: Dependence of the number of detected taxa (OTUs) on the number of sequences obtained for amaranth rhizosphere soil samples.

Author Contributions: Conceptualization, S.G. (Svetlana Gorelova) and A.M.; methodology, A.M. and D.K.; software, A.M.; validation, S.G. (Sergey Golubev), S.G. (Svetlana Gorelova), and M.G.; formal analysis, D.K. and A.M.; investigation, A.M. and S.G. (Sergey Golubev); resources, S.G. (Svetlana Gorelova); data curation, A.M.; writing—original draft preparation, A.M.; writing—review and editing, S.G. (Svetlana Gorelova) and S.G. (Sergey Golubev); visualization, A.M.; supervision, A.M. and M.G.; project administration, S.G. (Svetlana Gorelova) and M.G.; funding acquisition, S.G. (Svetlana Gorelova). All authors have read and agreed to the published version of the manuscript.

Funding: This research was funded by the Russian Foundation for Basic Research, project no. 19-29-05257.

Institutional Review Board Statement: Not applicable.

Informed Consent Statement: Not applicable.

Data Availability Statement: Raw reads are deposited in the NCBI under BioProject ID PRJNA917498 in the fastq format (https://www.ncbi.nlm.nih.gov/bioproject/PRJNA917498, accessed on 1 March 2023).

Acknowledgments: We are grateful to Anatoliy V. Gorbunov, Senior Researcher of the Laboratory of Chemical-analytical Research of GIN RAS, for X-ray fluorescence analysis of soils. We are also grateful to Dmitry N. Tychinin for his assistance in the preparation of the English text of this paper.

Conflicts of Interest: The authors declare no conflict of interest. The funders had no role in the design of the study; in the collection, analyses, or interpretation of data; in the writing of the manuscript; or in the decision to publish the results.

References

1. Bais, H.P.; Broeckling, C.D.; Vivanco, J.M. Root exudates modulate plant–microbe interactions in the rhizosphere. In *Secondary Metabolites in Soil Ecology. Soil Biology*; Karlovsky, P., Ed.; Springer: Berlin/Heidelberg, Germany, 2008; Volume 14, pp. 241–252.
2. Fan, D.; Schwinghamer, T.; Smith, D.L. Isolation and diversity of culturable rhizobacteria associated with economically important crops and uncultivated plants in Québec, Canada. *Syst. Appl. Microbiol.* **2018**, *41*, 629–640. [CrossRef] [PubMed]
3. Mohanram, S.; Kumar, P. Rhizosphere microbiome: Revisiting the synergy of plant-microbe interactions. *Ann. Microbiol.* **2019**, *69*, 307–320. [CrossRef]
4. Srivastava, R.; Kanda, T.; Yadav, S.; Mishra, R.; Atri, N. Cyanobacteria in rhizosphere: Dynamics, diversity, and symbiosis. In *Plant, Soil and Microbes in Tropical Ecosystems. Rhizosphere Biology*; Dubey, S.K., Verma, S.K., Eds.; Springer: Singapore, 2021; pp. 51–69.
5. Moreno-Espíndola, I.P.; Ferrara-Guerrero, M.J.; de León-González, F.; Rivera-Becerril, F.; Mayorga-Reyes, L.; Pérez, N.O. Enzymatic activity and culturable bacteria diversity in rhizosphere of amaranth, as indicators of crop phenological changes. *Bot. Sci.* **2018**, *96*, 640–649. [CrossRef]
6. Kumar, A.; Dubey, A. Rhizosphere microbiome: Engineering bacterial competitiveness for enhancing crop production. *J. Adv. Res.* **2020**, *24*, 337–352. [CrossRef] [PubMed]
7. Castellano-Hinojosa, A.; Strauss, S.L. Insights into the taxonomic and functional characterization of agricultural crop core rhizobiomes and their potential microbial drivers. *Sci. Rep.* **2021**, *11*, 10068. [CrossRef]
8. Moorthy, A.; Balasundaram, U. Rhizosphere metagenomics: Methods and challenges. In *Omics Science for Rhizosphere Biology, Rhizosphere Biology*; Pudake, R.N., Sahu, B.B., Kumari, M., Sharma, A.K., Eds.; Springer Nature Singapore Pte Ltd.: Singapore, 2021; pp. 1–20.
9. Ke, J.; Wang, B.; Yoshikuni, Y. Microbiome engineering: Synthetic biology of plant-associated microbiomes in sustainable agriculture. *Trends Biotechnol.* **2021**, *39*, 244–261. [CrossRef]
10. Ofek-Lalzar, M.; Sela, N.; Goldman-Voronov, M.; Green, S.J.; Hadar, Y.; Minz, D. Niche and host-associated functional signatures of the root surface microbiome. *Nat. Commun.* **2014**, *5*, 4950. [CrossRef]
11. Schreiter, S.; Ding, G.-C.; Heuer, H.; Neumann, G.; Sandmann, M.; Grosch, R.; Kropf, S.; Smalla, K. Effect of the soil type on the microbiome in the rhizosphere of field-grown lettuce. *Front. Microbiol.* **2014**, *5*, 144. [CrossRef]
12. Sun, X.; Song, B.; Xu, R.; Zhang, M.; Gao, P.; Lin, H.; Sun, W. Root-associated (rhizosphere and endosphere) microbiomes of the Miscanthus sinensis and their response to the heavy metal contamination. *J. Environ. Sci.* **2021**, *104*, 387–398. [CrossRef]
13. Neu, A.T.; Allen, E.E.; Roy, K. Defining and quantifying the core microbiome: Challenges and prospects. *Proc. Natl. Acad. Sci. USA* **2021**, *118*, e2104429118. [CrossRef]
14. Chen, Y.; Ding, Q.; Chao, Y.; Wei, X.; Wang, S.; Qiu, R. Structural development and assembly patterns of the root-associated microbiomes during phytoremediation. *Sci. Tot. Environ.* **2018**, *644*, 1591–1601. [CrossRef]
15. Mendes, R.; Garbeva, P.; Raaijmakers, J.M. The rhizosphere microbiome: Significance of plant beneficial, plant pathogenic, and human pathogenic microorganisms. *FEMS Microbiol. Rev.* **2013**, *37*, 634–663. [CrossRef] [PubMed]
16. Bell, T.H.; Cloutier-Hurteau, B.; Al-Otaibi, F.; Turmel, M.-C.; Yergeau, E.; Courchesne, F.; St-Arnaud, M. Early rhizosphere microbiome composition is related to the growth and Zn uptake of willows introduced to a former landfill. *Environ. Microbiol.* **2015**, *17*, 3025–3038. [CrossRef] [PubMed]
17. Kour, D.; Rana, K.L.; Yadav, N.; Yadav, A.N.; Kumar, A.; Meena, V.S.; Singh, B.; Chauhan, V.S.; Dhaliwal, H.S.; Saxena, A.K. Rhizospheric microbiomes: Biodiversity, mechanisms of plant growth promotion, and biotechnological applications for sustainable agriculture. In *Plant Growth Promoting Rhizobacteria for Agricultural Sustainability (From Theory to Practices)*; Kumar, A., Meena, V.S., Eds.; Springer Nature Singapore Pte Ltd.: Singapore, 2019; pp. 19–65.
18. Barra Caracciolo, A.; Terenzi, V. Rhizosphere microbial communities and heavy metals. *Microorganisms* **2021**, *9*, 1462. [CrossRef]
19. Lundberg, D.S.; Lebeis, S.L.; Paredes, S.H.; Yourstone, S.; Gehring, J.; Malfatti, S.; Tremblay, J.; Engelbrektson, A.; Kunin, V.; del Rio, T.G.; et al. Defining the core *Arabidopsis thaliana* root microbiome. *Nature* **2012**, *488*, 86–90. [CrossRef]
20. Muratova, A.Y.; Gorelova, S.V.; Sungurtseva, I.Y.; Zelenova, N.A. Rhizospheric microbiomes of *Sorghum bicolor* grown on soils with anthropogenic polyelement anomalies. In *BIO Web of Conferences PLAMIC2020*; EDP Sciences: Les Ulis, France, 2020; Volume 23, p. 03008.
21. Gins, M.S. *Biologicheski Aktivnyye Veshchestva Amaranta Amarantin: Svoystva, Mekhanizmy Deystviya i Prakticheskoye Ispol'zovaniye (Biologically Active Substances of Amaranth Amaranthine: Properties, Mechanisms of Action and Practical Use)*; RUDN: Moscow, Russia, 2002; 183p. (In Russian)
22. Shevyakova, N.I.; Cheremisina, A.I.; Kuznetsov, V.V. Phytoremediation potential of *Amaranthus* hybrids: Antagonism between nickel and iron and chelating role of polyamines. *Russ. J. Plant Physiol.* **2011**, *58*, 634–642. [CrossRef]
23. Chinmayee, M.D.; Mahesh, B.; Pradesh, S.; Mini, I.; Swapna, T.S. The assessment of phytoremediation potential of invasive weed *Amaranthus spinosus* L. *Appl. Biochem. Biotechnol.* **2012**, *167*, 1550–1559. [CrossRef]
24. Ziarati, P.; Alaedini, S. The phytoremediation technique for cleaning up contaminated soil by *Amaranthus* sp. *J. Environ. Anal. Toxicol.* **2014**, *4*, 1000208.
25. Odiyi, B.; Ologundudu, F.A.; Adegbite, T. Phytoremediation potential of *Amaranthus hybridus* L. (Caryophyllales: Amaranthaceae) on soil amended with brewery effluent. *Braz. J. Biol. Sci.* **2019**, *6*, 401–411. [CrossRef]

26. Gorelova, S.V.; Gins, M.S.; Frontasyeva, M.V. Phytoextraction of toxic elements by *Amaranthus tricolor* grown on technogenically polluted soils in open ground conditions. *Chim. Techno Acta* **2022**, *9*, 202292S8. [CrossRef]
27. Song, N.; Zhang, X.; Wang, F.; Zhang, C.; Tang, S. Elevated CO_2 increases Cs uptake and alters microbial communities and biomass in the rhizosphere of *Phytolacca americana* Linn (pokeweed) and *Amaranthus cruentus* L. (purple amaranth) grown on soils spiked with various levels of Cs. *J. Environ. Radioact.* **2012**, *112*, 29–37. [CrossRef] [PubMed]
28. Mishra, J.; Singh, R.; Arora, N.K. Alleviation of heavy metal stress in plants and remediation of soil by rhizosphere microorganisms. *Front. Microbiol.* **2017**, *8*, 1706. [CrossRef]
29. Nambisan, S.; Sunil, M.; Choudhary, B.; Srinivasan, S. Cyanobacteria is uniquely enriched in the roots of grain amaranths. *bioRxiv* **2019**, 540484. [CrossRef]
30. Gorelova, S.V.; Gorbunov, A.V.; Frontasyeva, M.V.; Sylina, A.K. Toxic elements in the soils of urban ecosystems and technogenic sources of pollution. *WSEAS Trans. Environ. Dev.* **2020**, *16*, 608–618. [CrossRef]
31. *GOST 53123-2008*; Soil Quality. Sampling. Part 5. Guidance on the Procedure for the Investigation of Urban and Industrial Sites with Regard to Soil Contamination. Russian Gost: Moscow, Russian, 2008. Available online: https://docs.cntd.ru/document/1200074384 (accessed on 1 October 2022).
32. *GOST 26483-85*; Soils. Preparation of Salt Extract and Determination of Its pH by CINAO Method. Russian Gost: Moscow, Russian, 1985. Available online: https://docs.cntd.ru/document/1200023490 (accessed on 21 February 2023).
33. Malahov, S.G. Interim Guidelines on Soil Pollution Control. 1984. Available online: https://files.stroyinf.ru/Data2/1/4294847/4294847412.htm (accessed on 21 February 2023).
34. *GOST 26205-91*; Soils. Determination of Mobile Compounds of Phosphorus and Potassium by Machigin Method Modified by CINAO. Russian Gost: Moscow, Russian, 1991. Available online: https://docs.cntd.ru/document/1200023449 (accessed on 21 February 2023).
35. *GOST 26488-85*; Soils. Determination of Nitrates by CINAO Method. Russian Gost: Moscow, Russian, 1985. Available online: https://docs.cntd.ru/document/1200023495 (accessed on 21 February 2023).
36. *GOST 26489-85*; Soils. Determination of Exchangeable Ammonium by CINAO Method. Russian Gost: Moscow, Russian, 1986. Available online: https://docs.cntd.ru/document/1200023496 (accessed on 21 February 2023).
37. *PND F 16.1:2:2.2:2.3:3.64-10*; Quantitative Chemical Analysis of Soils. Method for Measuring the Mass Fraction of Oil Products in Samples of Soils, Soils, Bottom Sediments, Silts, Sewage Sludge, Production and Consumption Wastes by the Gravimetric Method. Federal Service for Environmental, Technological and Nuclear Supervision: Moscow, Russia, 2010. Available online: https://files.stroyinf.ru/Data2/1/4293807/4293807051.htm (accessed on 21 February 2023).
38. *PND F 16.1.42-04*; Measurement Method for Measuring the Mass Fraction of Metals and Metal Oxides in Soil Powder Samples by X-Ray Fluorescence Analysis. Federal Service for Environmental, Technological and Nuclear Supervision: Moscow, Russia, 2004. (In Russian)
39. *GN 2.1.7.2041-06*; Predel'no Dopustimye Koncentracii (PDK) Himicheskih Veshchestv v Pochve: Gigienicheskie Normativy.—M.: Federal'nyj Centr Gigieny i Epidemiologii Rospotrebnadzora. Maximum Permissible Concentrations (MPCs) of Chemicals in the Soil: Moscow, Russia, 2006; 15p. Available online: https://files.stroyinf.ru/Data2/1/4293850/4293850511.pdf (accessed on 21 February 2023). (In Russian).
40. *GN 2.1.7.2511-09*; Orientirovochno Dopustimye Koncentracii (ODK) Himicheskih Veshchestv v Pochve: Gigienicheskie Normativy.—M.: Federal'nyj Centr Gigieny i Epidemiologii Rospotrebnadzora. Approximate allowable concentrations (AAC) of Chemical Substances in Soils: Moscow, Russia, 2009; 10p. Available online: https://files.stroyinf.ru/Data2/1/4293828/4293828439.pdf (accessed on 21 February 2023). (In Russian)
41. Klindworth, A.; Pruesse, E.; Schweer, T.; Peplies, J.; Quast, C.; Horn, M.; Glöckner, F.O. Evaluation of general 16S ribosomal RNA gene PCR primers for classical and next-generation sequencing-based diversity studies. *Nucleic Acids Res.* **2013**, *41*, e1. [CrossRef] [PubMed]
42. Bolyen, E.; Rideout, J.R.; Dillon, M.R.; Bokulich, N.A.; Abnet, C.C.; Al-Ghalith, G.A.; Alexander, H.; Alm, E.J.; Arumugam, M.; Asnicar, F.; et al. Reproducible, interactive, scalable and extensible microbiome data science using QIIME 2. *Nat. Biotechnol.* **2019**, *37*, 852–857. [CrossRef]
43. Callahan, B.J.; McMurdie, P.J.; Rosen, M.J.; Han, A.W.; Johnson, A.J.A.; Holmes, S.P. DADA2: High-resolution sample inference from Illumina amplicon data. *Nat. Methods* **2016**, *13*, 581–583. [CrossRef]
44. Yilmaz, P.; Parfrey, L.W.; Yarza, P.; Gerken, J.; Pruesse, E.; Quast, C.; Schweer, T.; Peplies, J.; Ludwig, W.; Glockner, F.O. The SILVA and "All-species Living Tree Project (LTP)" taxonomic frameworks. *Nucleic Acids Res.* **2014**, *42*, D643–D648. [CrossRef]
45. Jing, H.; Xia, X.; Liu, H.; Zhou, Z.; Wu, C.; Nagarajan, S. Anthropogenic impact on diazotrophic diversity in the mangrove rhizosphere revealed by nifH pyrosequencing. *Front. Microbiol.* **2015**, *6*, 1172. [CrossRef]
46. Prasanna, R.; Nain, L.; Ancha, R.; Srikrishna, J.; Joshi, M.; Kaushik, B.D. Rhizosphere dynamics of inoculated cyanobacteria and their growth-promoting role in rice crop. *Egypt. J. Biol.* **2009**, *11*, 26–36.
47. Suresh, A.; Soundararajan, S.; Elavarasi, S.; Oscara, F.L.; Thajuddin, N. Evaluation and characterization of the plant growth promoting potentials of two heterocystous cyanobacteria for improving food grains growth. *Biocatal. Agric. Biotechnol.* **2019**, *17*, 647–652. [CrossRef]
48. Toribio, A.J.; Suárez-Estrella, F.; Jurado, M.M.; López, M.J.; López-González, J.A.; Moreno, J. Prospection of cyanobacteria producing bioactive substances and their application as potential phytostimulating agents. *Biotechnol. Rep.* **2020**, *26*, e00449. [CrossRef] [PubMed]

49. Lindblad, P. Cyanobacteria in symbiosis with cycads. In *Prokaryotic Symbionts in Plants. Microbiology Monographs*; Pawlowski, K., Ed.; Springer: Heidelberg, Germany, 2008; Volume 8, pp. 225–233.
50. Bergman, B.; Johansson, C.; Söderbäck, E. The *Nostoc-Gunnera* symbiosis. *New Phytol.* **1992**, *122*, 379–400. [CrossRef]
51. Chittora, D.; Meena, M.; Barupal, T.; Swapnil, P.; Sharma, K. Cyanobacteria as a source of biofertilizers for sustainable agriculture. *Biochem. Biophys. Rep.* **2020**, *22*, 100737. [CrossRef]
52. Hermans, S.M.; Buckley, H.L.; Case, B.S.; Curran-Cournane, F.; Taylor, M.; Lear, G. Bacteria as emerging indicators of soil condition. *Appl. Environ. Microbiol.* **2017**, *83*, e02826-16. [CrossRef] [PubMed]
53. Dasgupta, D.; Ghosh, R.; Sengupta, T.K. Biofilm-mediated enhanced crude oil degradation by newly isolated *Pseudomonas* species. *Int. Sch. Res. Notices* **2013**, *2013*, 250749. [CrossRef]
54. Ueshima, M.; Ginn, B.R.; Haack, E.A.; Szymanowski, J.E.S.; Fein, J.B. Cd adsorption onto *Pseudomonas putida* in the presence and absence of extracellular polymeric substances. *Geochim. Cosmochim. Acta* **2008**, *72*, 5885–5895. [CrossRef]
55. Chettri, B.; Singh, A.K. Kinetics of hydrocarbon degradation by a newly isolated heavy metal tolerant bacterium *Novosphingobium panipatense* P5:ABC. *Bioresour. Technol.* **2019**, *294*, 122190. [CrossRef]
56. Lee, S.Y.; Lee, Y.Y.; Cho, K.S. Effect of *Novosphingobium* sp. CuT1 inoculation on the rhizoremediation of heavy metal- and diesel-contaminated soil planted with tall fescue. *Environ. Sci. Pollut. Res.* **2022**, *30*, 16612–16625. [CrossRef]
57. Asaf, S.; Numan, M.; Khan, A.L.; Al-Harrasi, A. Sphingomonas: From diversity and genomics to functional role in environmental remediation and plant growth. *Crit. Rev. Biotechnol.* **2020**, *40*, 138–152. [CrossRef]
58. Nocelli, N.; Bogino, P.C.; Banchio, E.; Giordano, W. Roles of extracellular polysaccharides and biofilm formation in heavy metal resistance of rhizobia. *Materials* **2011**, *9*, 418. [CrossRef] [PubMed]
59. Gupta, P.; Diwan, B. Bacterial exopolysaccharide mediated heavy metal removal: A review on biosynthesis, mechanism and remediation strategies. *Biotechnol. Rep.* **2017**, *13*, 58–71. [CrossRef] [PubMed]
60. Feng, G.D.; Yang, S.Z.; Li, H.P.; Zhu, H.H. *Massilia putida* sp. nov., a dimethyl disulfide-producing bacterium isolated from wolfram mine tailing. *Int. J. Syst. Evol. Microbiol.* **2016**, *66*, 50–55. [CrossRef]
61. Chikere, C.B.; Okpokwasili, G.C.; Chikere, B.O. Monitoring of microbial hydrocarbon remediation in the soil. *3 Biotech* **2011**, *1*, 117–138. [CrossRef] [PubMed]
62. Sun, W.; Xiao, E.; Krumins, V.; Häggblom, M.M.; Dong, Y.; Pu, Z.; Li, B.; Wang, Q.; Xiao, T.; Li, F. Rhizosphere microbial response to multiple metal(loid)s in different contaminated arable soils indicates crop-specific metal-microbe interactions. *Appl. Environ. Microbiol.* **2018**, *84*, e00701–e00718. [CrossRef] [PubMed]
63. Diestra, E.; Esteve, I.; Burnat, M.; Maldonado, J.; Solé, A. Isolation and characterization of a heterotrophic bacterium able to grow in different environmental stress conditions, including crude oil and heavy metals. *Commun. Curr. Res. Educ. Top. Trends Appl. Microbiol.* **2007**, *1*, 90–99.
64. Jing, Y.; He, Z.; Yang, X. Role of soil rhizobacteria in phytoremediation of heavy metal contaminated soils. *J. Zhejiang Univ. Sci. B (Biomed. Biotechnol.)* **2007**, *8*, 192–207. [CrossRef]
65. Santoyo, G.; Moreno-Hagelsieb, G.; Orozco-Mosqueda, M.C.; Glick, B.R. Plant growth-promoting bacterial endophytes. *Microbiol. Res.* **2016**, *183*, 92–99. [CrossRef]
66. Ojuederie, O.; Babalola, O. Microbial and plant-assisted bioremediation of heavy metal polluted environments: A review. *Int. J. Environ. Res. Public Health* **2017**, *14*, 1504. [CrossRef]
67. Hussain, S.S.; Mehnaz, S.; Siddique, K.H.M. Harnessing the plant microbiome for improved abiotic stress tolerance. In *Plant Microbiome: Stress Response*; Egamberdieva, D., Ahmad, P., Eds.; Springer Nature Singapore Pte Ltd: Singapore, 2018; pp. 21–43.
68. Babu, A.G.; Kim, J.-D.; Oh, B.-T. Enhancement of heavy metal phytoremediation by *Alnus firma* with endophytic *Bacillus thuringiensis* GDB-1. *J. Hazard. Mater.* **2013**, *250–251*, 477–483. [CrossRef]
69. Chen, Z.-J.; Tian, W.; Li, Y.-J.; Sun, L.-N.; Chen, Y.; Zhang, H.; Li, Y.Y.; Han, H. Responses of rhizosphere bacterial communities, their functions and their network interactions to Cd stress under phytostabilization by *Miscanthus* spp. *Environ. Pollut.* **2021**, *287*, 117663. [CrossRef] [PubMed]
70. Chen, B.; Zhang, Y.; Rafiq, M.T.; Khan, K.Y.; Pan, F.; Yang, X.; Feng, Y. Improvement of cadmium uptake and accumulation in Sedum alfredii by endophytic bacteria *Sphingomonas* SaMR12: Effects on plant growth and root exudates. *Chemosphere* **2014**, *17*, 367–373. [CrossRef] [PubMed]
71. Ali, A.; Guo, D.; Mahar, A.; Wang, Z.; Muhammad, D.; Li, R.; Wang, P.; Shen, F.; Xue, Q.; Zhang, Z. Role of *Streptomyces pactum* in phytoremediation of trace elements by *Brassica juncea* in mine polluted soils. *Ecotoxicol. Environ. Saf.* **2017**, *144*, 387–395. [CrossRef] [PubMed]
72. Preston, G.M. Plant perceptions of plant growth-promoting *Pseudomonas*. *Philos. Trans. R. Soc. Lond. B Biol. Sci.* **2004**, *359*, 907–918. [CrossRef] [PubMed]
73. Deshwal, V.K.; Singh, S.B.; Kumar, P.; Chubey, A. Rhizobia unique plant growth promoting rhizobacteria: A review. *Int. J. Life Sci.* **2013**, *2*, 74–86.

agronomy

MDPI

Article

Can Sugarcane Yield and Health Be Altered with Fully Mechanized Management?

Jian Xiao [1,2,3], Tian Liang [2], Shangdong Yang [1,2,*] and Hongwei Tan [2,*]

1 Guangxi Key Laboratory of Agro-Environment and Agro-Products Safety, National Demonstration Center for Experimental Plant Science Education, Agricultural College, Guangxi University, Nanning 530004, China
2 Guangxi Key Laboratory of Sugarcane Genetic Improvement, Guangxi Academy of Agricultural Sciences, Nanning 530007, China
3 Longping Branch, College of Biology, Hunan University, Changsha 410125, China
* Correspondence: ysd706@gxu.edu.cn (S.Y.); hongwei_tan@163.com (H.T.)

Abstract: At present, fully mechanized cultivation (FMC) has begun to be utilized in commercial sugarcane production in China. To provide new insights into whether cane yield and health are altered by fully mechanized cultivations, the cane yield and endophytic microbial community structure in stems of sugarcane that underwent fully mechanized cultivation (FMC) and conventional artificial cultivation (CAC) were compared. The results showed that the diversity and richness of endophytic microorganisms, except for the bacterial richness in the stems of sugarcane, could be significantly increased by using FMC. Meanwhile, in comparison with CAC, the relative abundance of Proteobacteria and Ascomycota increased under FMC. Moreover, some dominant endophytic bacterial genera, such as *Acidovorax*, *Microbacterium*, and *Paenibacillus*, and some dominant endophytic fungal genera, such as *Scleroramularia*, *Tetraplosphaeria*, and *Dinemasporium*, were found to be significantly enriched in cane stems under FMC treatments. Additionally, the endophytic microbial functions in sugarcane stems were not significantly altered by FMC treatments. Our results suggest that cane growth, yield, and health are not significantly altered by FMC. The results also indicate that fully mechanized management can be developed as a sustainable method in sugarcane production.

Keywords: sugarcane; fully mechanized cultivation; endophytic microbial community structure

check for updates

Citation: Xiao, J.; Liang, T.; Yang, S.; Tan, H. Can Sugarcane Yield and Health Be Altered with Fully Mechanized Management? *Agronomy* **2023**, *13*, 153. https://doi.org/10.3390/agronomy13010153

Academic Editors: Yong-Xin Liu and Peng Yu

Received: 24 November 2022
Revised: 29 December 2022
Accepted: 30 December 2022
Published: 3 January 2023

1. Introduction

Sugarcane (*Saccharum officinarum* L.), an important economic tropical crop cultivated worldwide, provides 80% of the world's sugar production and is also a crucial source of biofuel for ethanol production [1–3]. China is the third-most prominent sugar-producing country in the world. Recently, 75% and 90% of total sugar production, globally and in China, respectively, came from sugarcane [4]. In China, approximately 90% of the sugarcane crops are planted in the southern and southwestern regions, including Guangxi, Guangdong, and Yunnan provinces. In particular, Guangxi Province is the top sugarcane production area, accounting for more than 65% of sugar production in China since 1993 [5]. However, the steadily rising cost of labor in sugarcane cultivation has greatly increased the costs of the sugar industry. Savings in time, energy, and costs are advantages of agricultural mechanization [6]. Moreover, the productivity of agricultural land and processing efficiency can also be substantially increased by mechanization [6]. To develop a sustainable sugarcane industry, upgrading the agricultural mechanization and equipment used for sugarcane production is necessary [7,8].

Plant endophytic microbiomes have been shown to occur in both the cooperative and competitive interactions of plants. Although the cooperating endophytic microbiome performs beneficial functions for the plant, the competing microbiome has negative consequences [9]. Endophytic bacteria often display well-controlled multiplication inside plant niches, which is modulated by the plant's defense system [9,10]. The effects of the endophytic microorganisms on the host can be described as alleviating the host's abiotic stress; protecting the host against biotic stress (pathogens and herbivores); and providing nutritional support to the host by boosting nitrogen, phosphate, iron, and other nutrients [11]. Endophytes have been isolated from a variety of plant species, with the majority of them occurring in the host plant's rhizosphere, where they enter host plants through the roots and colonize the roots' intercellular spaces [12].

However, previous studies on the fully mechanized management of sugarcane have mainly explored its effects on soil (health and fertility) and sugarcane yield. Endophytic microbial communities have rarely been used for evaluation. In particular, the collective response of endophytic microbial (bacterial and fungal) systems in stems of sugarcane, including the microbial community structures, functions, and symbiotic network patterns, to the successive mechanized management of sugarcane fields has not been reported.

The aim of this study is to answer the following questions: (1) Can the endophytic microbial communities in the stems of sugarcane be altered by fully mechanized management? (2) What kinds of endophytic microorganisms are specially or significantly enriched in the stems of sugarcane under fully mechanized management? (3) How does fully mechanized cultivation affect the symbiotic network patterns of endophytic bacteria and endophytic fungi? We anticipate that our findings will provide new insights into the effects of mechanized cultivation on modern agricultural production.

2. Materials and Methods

2.1. Study Site

The experiment was conducted at the Experimental Base of the Sugarcane Research Institute, Guangxi Academy of Agricultural Sciences, which is located in Longan County (107°598″ E and 23°637″ N), Guangxi, China. The experimental site is located in the subtropical monsoon climate zone, which is rich in sunshine and rainfall. The mean summer and winter temperatures are 32 °C and 24 °C, respectively. The average annual temperature is approximately 21.7 °C, and the annual precipitation is approximately 1227–1691 mm, with rainfall mostly concentrated from June to September. The soil type is dominated by Quaternary red soil.

2.2. Experimental Design and Implementation

Firstly, the experiment was carried out beginning in the spring of 2019. The sugarcane cultivar Guitang 44 was used in this study. Two treatments were implemented as follows: (1) the fully mechanized cultivation of sugarcane (FMC, i.e., land preparation, sowing, and harvesting are all carried out by using different machines), and (2) the conventional artificial cultivation of sugarcane (CAC, i.e., land preparation, sowing, and harvesting are all carried out by hand labor operations only) as a control. Meanwhile, there were three replicate plots per treatment, and the size of each plot was 667 m^2.

The processes of FMC are described as follows: (1) For land preparations, weeding and deep plowing (mean depth of plowing is about 40 cm) were carried out first using tillage machinery (1LHT-440, Kaifeng, China); second, the rocks were cleaned up, and the roots or leaves of the sugarcane were broken up using a disc harrow (1GKN-300, Lianyungang, China); and third, furrowing was conducted using a furrowing machine (1LK-3D, Nanning, China). (2) Fertilizing, sowing, and mulching were performed simultaneously using a combined planter (2CZY-2, Beijing, China). (3) Land leveling was conducted after fertilizing, sowing, and mulching by using land-leveling equipment (3ZPF-1.36, Nanning, China). (4) During sugarcane growth, the actions of cultivation, fertilization, and banking were conducted several times by using a cultivator-hiller (3ZFS-2, Xuzhou, China). (5) Cane harvesting was conducted using a cane harvester (Austoft-4000, London, OH, USA).

The same treatments were performed identically for the CAC of sugarcane. However, all the above processes were performed by hand. The sowing density was approximately 90,000 buds per ha. All plots were fertilized with 300 kg ha^{-2} of urea, 75 kg ha^{-2} of K_2O, and 300 kg ha^{-2} of calcium superphosphate per season [4]. At the seedling and elongation stages of sugarcane, top dressings with 30% and 70% of the total fertilizer usage, respectively, were applied.

2.3. Plant Sampling

Plant samples were collected after a 3-year experimental setup in the early harvesting stage (December 2021), and six plant samples were obtained from each plot using a S-sampling technique and mixed as biological replicates. Each treatment was replicated three times. Plant samples were collected randomly according to the method described by Yang et al. [5] and Xiao et al. [13]. First, these samples were placed in sealed sterile bags and labeled for return to the lab. Second, the stem samples were rinsed and wiped for 2 min with sterile water by using a soft brush to remove impurities from the cane surfaces, and then were washed with 75% ethanol for 1 min, following a wash with a 1% NaClO solution for 3 min. Finally, all the stems were washed with sterile water for 0.5 min, and then sterile paper was used to remove surface water [14]. To determine the success of the sterilization of the cane surface, 100 μL of water from each washing step was placed on a Luria–Bertani (LB) agar plate (g/L) (NaCl-10, tryptone-5, yeast extract-5, and agar-20) and incubated at 25 °C for 7 d. No colonies developed on the plates, confirming that they were thoroughly sterilized. The sterilization of the stem surface samples was completed before detection and analysis of the endophytic microorganisms [15]. The stems were placed in sterile bags and stored at −80 °C for pending DNA extraction.

2.4. Analysis of Microbial Diversity

Microbial community genomic DNA was extracted from stem samples using a E.Z.N.A.® DNA Kit (Omega Bio-Tek, Norcross, GA, USA) according to the manufacturer's instructions. The DNA extract was analyzed on a 1% agarose gel, and the DNA's concentration and purity was determined using a NanoDrop 2000 UV–vis spectrophotometer (Thermo Scientific, Wilmington, NC, USA). PCR amplification and sequencing of the total DNA extracted from the plant samples were performed by Shanghai Majorbio Bio-Pharm Technology Co., Ltd., Shanghai, China. The endophytic bacterial primers 799F (5′-AACMGGATTAGATACCCKG-3′) and 1192R (5′-ACGGGCGGTGTGTRC-3′) from the V5-V7 region (endophytic bacterial 16S rRNA gene) were amplified first, and the primers 799F (5′-AACMG GATTAGATACCCKG-3′) and 1193R (5′-ACGTCATCCCCACCTTCC-3′) from the V5-V7 region were amplified second; meanwhile, ITS1F (5′-CTTGGTCATTTAGAGGAAGTAA-3′) and ITS2R (5′-GCTGCGTTCTTC ATCGATGC-3′) primers were employed to amplify the fungal ITS1 region via a ABI GeneAmp® 9700 PCR thermocycler (ABI, Vernon, CA, USA) using standard PCR protocols and conditions. The PCR products were recovered using 2% agar-gel electrophoresis, purified using an AxyPrep DNA Gel Extraction Kit (Axygen, New York, NY, USA), and quantified using a Quantus fluorometer (Promega, Madison, WI, USA). The purified amplicons were pooled in equimolar quantities and paired-end sequenced (2 × 300) on an Illumina MiSeq platform (Illumina, San Diego, CA, USA) according to the standard protocols of Majorbio Bio-Pharm Technology Co., Ltd. (Shanghai, China). Raw reads were deposited in the NCBI Sequence Read Archive (SRA) database (accession number: SRP371574).

2.5. Statistical Analyses

The experimental data were analyzed using Excel 2019 and SPSS Statistics 21.0 (IBM Corp., Armonk, New York, NY, USA). A T-test and Wilcoxon rank-sum test were used to analyze the significant differences in the statistical analyses ($p < 0.05$).

Quantitative insights into microbial ecology (QIIME) (version 1.17) was used to truncate the 300 bp reads (average quality score <20 over a 50 bp sliding window). Operational taxonomic units (OTUs) with a 97% similarity cut-off were clustered using UPARSE (version 7.1, http://drive5.com/uparse/, accessed on 9 April 2022), and chimeric sequences were identified and removed [16]. The taxonomy of each OTU representative sequence was analyzed via the RDP Classifier (http://rdp.cme.msu.edu/, accessed on 9 April 2022) against the 16S and ITS rRNA databases, using a confidence threshold of 0.7 [5].

Alpha diversities of the bacterial and fungal communities were calculated using Mothur (version v.1.30.2, https://mothur.org/wiki/calculators/, accessed on 9 April 2022). The Shannon and Ace indices were used to represent the diversity and richness of the endophytic microbial (bacterial and fungal) community, respectively. Meanwhile, a Wilcoxon rank-sum test was also performed to evaluate the diversity and richness of the microbial communities under the FMC and CAC treatments ($p < 0.05$). A principal component analysis (PCA) based on the unweighted UniFrac and a partial least squares discriminant analysis (PLS-DA) was performed to evaluate the extent of the similarity of the endophytic microbial communities, and the R language (version 3.3.1) tool was used for statistical analysis and graphing [13]. OTU tables with a 97% similarity level were selected for microbial community composition and Venn diagram analysis, and the R language (version 3.3.1) tool was used for statistics and graphing. The vegan package of the R language (version 3.3.1) tool was used and graphed for microbial community heatmap analysis. A linear discriminant analysis (LDA) was performed using LEfSe (http://huttenhower.sph.harvard.edu/galaxy/root?tool_id=lefse_upload, accessed on 9 April 2022) on samples according to different grouping conditions that were based on taxonomic composition to identify clusters that had a significant differential impact on sample delineation [3]. A correlation network analysis was performed by using NetworkX on the plant samples. BugBase was used for the phenotypic prediction of the microbiome. BugBase (https://bugbase.cs.umn.edu/index.html, accessed on 9 April 2022) was used to identify the high-level phenotypes present in microbiome samples and enabled the use of phenotype prediction as a microbiome analysis tool. PICRUSt was used to estimate the functional components of bacterial communities using the Kyoto Encyclopedia of Genes and Genomes (KEGG) dataset [13]. Functional predictions of the fungal communities were performed with the Fungi Functional Guild (FUN Guild) tool [16]. An online data analysis was conducted using the free online platform Majorbio Cloud Platform (www.majorbio.com) from the Majorbio Bio-Pharm Technology Co., Ltd. (Shanghai, China). The data were visualized by ImageGP (https://onlinelibrary.wiley.com/doi/10.1002/imt2.5, accessed on 9 April 2022).

3. Results

3.1. Sugarcane Yields

In comparison with the CAC treatment, the cane yields increased by 13.84%, 16.88%, and 15.57% under the FMC treatment in 2019, 2020, and 2021, respectively (Table 1). The difference between the FMC and CAC treatments was not statistically significant ($p > 0.05$). However, the results still indicate that the cane yields could be improved by the FMC treatment.

Table 1. Cane yields between FMC and CAC treatments (t ha^{-1}).

Treatments	2019	2020	2021
FMC	100.25 ± 2.01 a	98.58 ± 0.89 a	101.94 ± 0.80 a
CAC	88.06 ± 0.64 a	84.34 ± 0.67 a	88.21 ± 0.71 a

All data are presented as the mean ± standard deviation (SD). A T-test was performed ($p < 0.05$). Same letters within a column indicate no significant differences among treatments at $p > 0.05$. FMC—fully mechanized cultivation; CAC—conventional artificial cultivation.

3.2. Diversity of Endophytic Bacteria and Fungi in Sugarcane Stems

The results show that the endophytic microbial diversity (Shannon) and richness (Ace) indices of the sugarcane stems under the FMC treatment were not significantly different from stems under the CAC treatment (Figure 1a–d). The results suggest that the diversity and richness of the endophytic microorganisms in sugarcane stems were not significantly changed by the FMC treatment as compared with the CAC treatment.

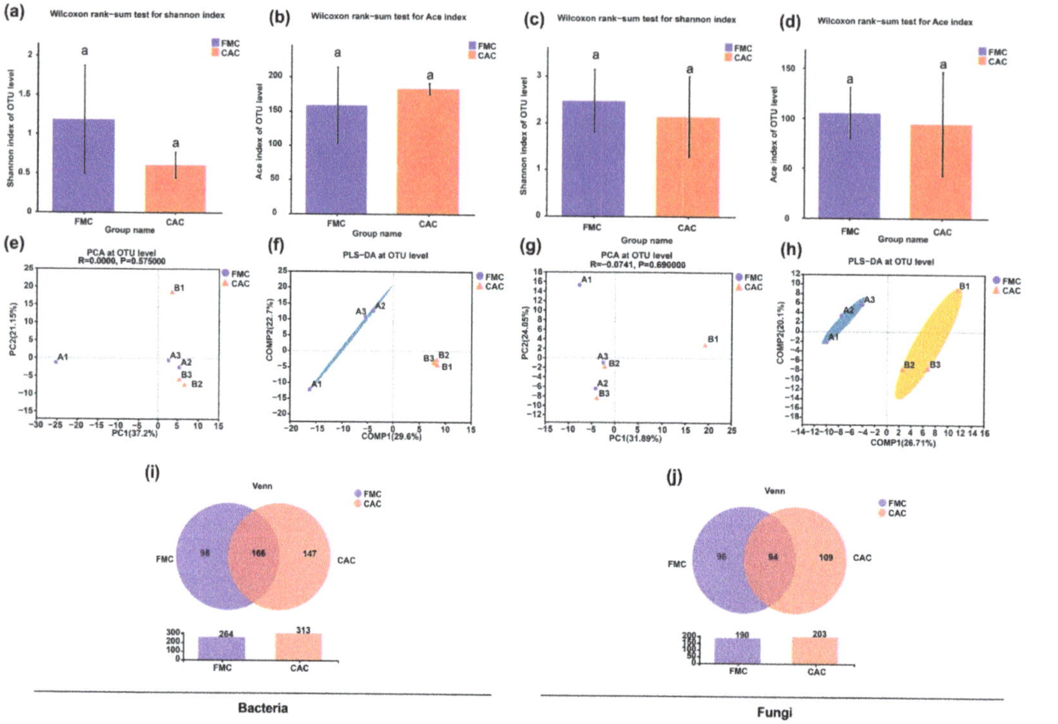

Figure 1. Comparison of endophytic microbiota structures in stems of sugarcane at a similarity level of 97% between FMC and CAC treatments (OTU level). (**a**) The Shannon index indicates endophytic bacterial diversity. (**b**) The Ace index indicates endophytic bacterial richness. (**c**) The Shannon index indicates endophytic fungal diversity. (**d**) The Ace index indicates endophytic fungal richness. (**e**) PCA of endophytic bacteria communities. (**f**) PLS-DA score plot of endophytic bacteria communities. (**g**) PCA of endophytic fungi communities. (**h**) PLS-DA score plot of endophytic fungi communities. (**i**) Venn diagram analyses of endophytic bacteria. (**j**) Venn diagram analyses of endophytic fungi. FMC—fully mechanized cultivation; CAC—conventional artificial cultivation. Same letters on bars within a figure indicate no significant differences in mean ranks among treatments at $p > 0.05$.

The unweighted UniFrac principal component analysis (PCA) and partial least squares discriminant analysis (PLS-DA) were also performed to evaluate the extent of the similarity of the endophytic microbial communities at the operational taxonomic unit (OTU) level. The results showed that the microbial communities of FMC and CAC were clustered separately, but there were also similarities in the microbial compositions between FMC and CAC treatments (Figure 1e–h). In addition, both the total and the unique numbers of microorganisms in sugarcane stems under the FMC treatment were lower than those under the CAC treatment at the OTU level (Figure 1i,j).

3.3. Composition of Endophytic Bacteria and Fungi in Sugarcane Stems at Different Levels

The dominant endophytic microorganisms (bacteria and fungi) were referred to as those with relative abundance percentages greater than 1% (Figures 2 and 3). A Wilcoxon rank-sum test was also performed for the endophytic microorganisms with relative abundance percentages at a phylum and genus level ($p < 0.05$). However, the results showed that there were no significant differences in this measure between the FMC and CAC treatments.

At the phylum level, the proportions of dominant endophytic bacterial phyla in sugarcane under the CAC treatment, from high to low, were Proteobacteria at 94.11%,

Actinobacteriota at 4.81%, and others at 1.09%. In contrast, the proportions of dominant endophytic bacterial phyla in sugarcane under the FMC treatment were Proteobacteria at 96.49%, Actinobacteriota at 2.56%, and others at 0.95%. The relative abundance of Proteobacteria increased in canes under the FMC treatment as compared with those under the CAC treatment. Meanwhile, the relative abundances of Actinobacteriota and other bacteria in sugarcane under the FMC treatment were all lower than those under the CAC treatment (Figure 2a).

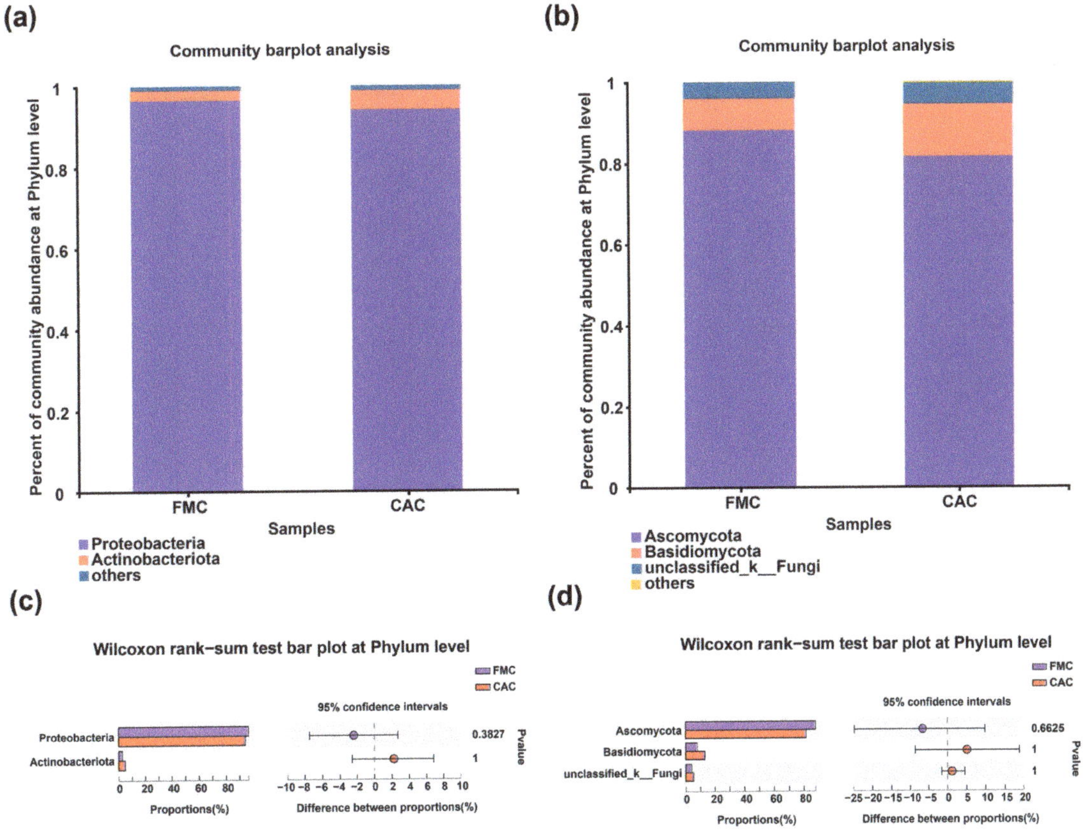

Figure 2. Compositions of endophytic microorganisms in the stems of sugarcane under FMC and CAC treatments at phylum level. (**a**) The proportions of the dominant endophytic bacteria. (**b**) The proportions of the dominant endophytic fungi. (**c**) Test for significant difference in the bacterial abundance between groups. (**d**) Test for significant difference in the fungal abundance between groups. FMC—fully mechanized cultivation; CAC—conventional artificial cultivation.

In addition, the proportions of endophytic dominant fungal phyla in CAC sugarcane, from high to low, were Ascomycota at 81.56%, Basidiomycota at 12.98%, and unclassified_k__Fungi at 5.14%. By contrast, the proportions of dominant endophytic fungal phyla in FMC sugarcane were Ascomycota at 88.15%, Basidiomycota at 7.88%, and unclassified_k__Fungi at 3.95%. The relative abundance of Ascomycota increased in FMC-treated sugarcane compared with that under the CAC treatment. Meanwhile, the relative abundances of Basidiomycota and unclassified_k__Fungi were lower under the FMC treatment compared with those under the CAC treatment (Figure 2b).

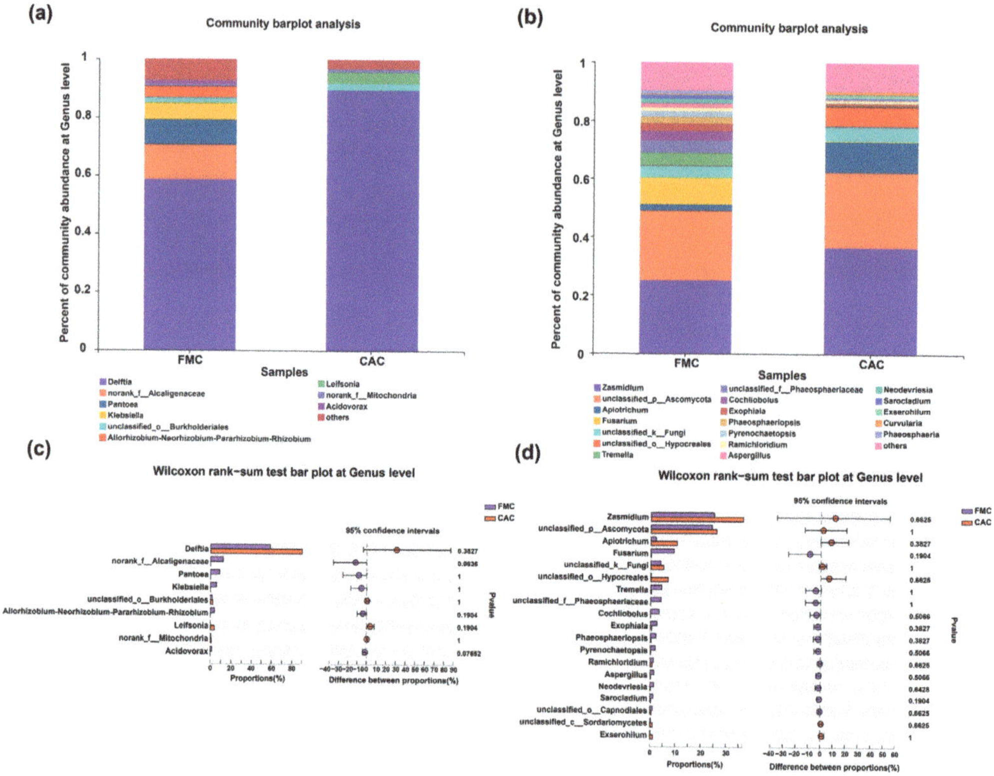

Figure 3. Compositions of endophytic microorganisms in the stems of sugarcane under FMC and CAC treatments at genus level. (a) The proportions of the dominant endophytic bacteria. (b) The proportions of the dominant endophytic fungi. (c) Test for significant difference in the bacterial abundance between groups. (d) Test for significant difference in the fungal abundance between groups. FMC—fully mechanized cultivation; CAC—conventional artificial cultivation.

At the genus level, the relative abundances of *Delftia* and unclassified_o__*Burkholderiales* were lower under the FMC treatment as compared with the CAC treatment. However, norank_f__*Alcaligenaceae*, *Pantoea*, *Klebsiella*, *Allorhizobium-Neorhizobium-Pararhizobium-Rhizobium*, and *Acidovorax* were the special, dominant endophytic bacterial genera in FMC plots. *Leifsonia* and norank_f__*Mitochondria* were the unique, dominant endophytic bacterial genera in CAC plots (Figure 3a).

Additionally, the proportions of *Zasmidium*, unclassified_p__*Ascomycota*, *Apiotrichum*, unclassified_k__Fungi, and others were also lower under the FMC treatment as compared with the CAC treatment. Moreover, *Fusarium*, *Tremella*, unclassified_f__*Phaeosphaeriaceae*, *Cochliobolus*, *Exophiala*, *Phaeosphaeriopsis*, *Pyrenochaetopsis*, *Ramichloridium*, *Aspergillus*, *Neodevriesia*, *Sarocladium*, and *Phaeosphaeria* were the special, dominant endophytic fungal genera under the FMC treatment. In contrast, unclassified_o__*Hypocreales*, *Exserohilum*, and *Curvularia* were the only unique, dominant endophytic fungal genera under the CAC treatment (Figure 3b).

An LEfSe analysis was also conducted to identify endophytic microbes in sugarcane stems under the FMC treatment. A total of 39 bacterial and 12 fungal clades (from phylum to genus) exhibited significant differences in their cladogram structure (LDA > 2.0).

As seen in Figure 4, there was no significant enrichment of dominant endophytic bacteria and fungi detected between the FMC and CAC treatments at the phylum level.

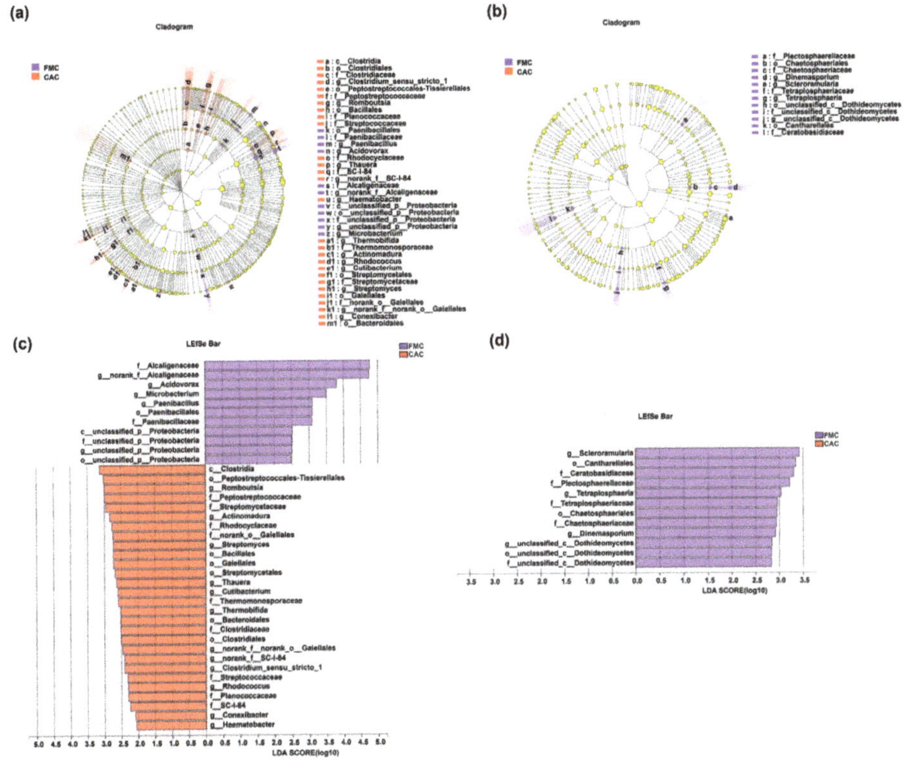

Figure 4. Cladogram showing the phylogenetic distribution of the bacterial (**a**) and fungal (**b**) lineages associated with stems of sugarcane under FMC and CAC treatments. Indicator bacteria (**c**) and fungi (**d**) with LDA scores of 2.0 or greater in microbial communities associated with stems of sugarcane under FMC and CAC treatments (LEfSe). Circles indicate phylogenetic levels from phylum to genus. The diameter of each circle is proportional to the abundance of the group. Different prefixes indicate different levels (p—phylum; c—class; o—order; f—family; g—genus). FMC—fully mechanized cultivation; CAC—conventional artificial cultivation.

Meanwhile, the endophytic bacteria, such as *Romboutsia*, *Actinomadura*, *Streptomyces*, *Thauera*, *Cutibacterium*, *Thermobifida*, norank_f__norank_o__*Gaiellales*, norank_f__SC-I-84, *Clostridium*_sensu_stricto_1, *Rhodococcus*, *Conexibacter*, and *Haematobacter*, were significantly enriched in canes under the CAC treatment at the genus level; in contrast, norank_f__*Alcaligenaceae*, *Acidovorax*, *Microbacterium*, *Paenibacillus*, and unclassified_p__*Proteobacteria* were significantly enriched under the FMC treatment at the genus level.

Moreover, in comparison with the FMC treatment, there was no significant enrichment of dominant endophytic fungi under the CAC treatment at the genus level; however, *Scleroramularia*, *Tetraplosphaeria*, *Dinemasporium*, and unclassified_c__*Dothideomycetes* were significantly enriched under the FMC treatment at the genus level.

The correlation between the significant enrichment of endophytic bacterial and fungal genera is important. Therefore, an interaction network was constructed to further elucidate the relationship between the significant enrichment of the bacterial and fungal genera. Based on the abundance of the endophytic bacterial and fungal genera, Spearman's rank correlation coefficients were calculated to reflect the correlations between them (Figure 5). The results showed that the enrichment of the endophytic fungal genera (*Scleroramularia*, *Tetraplosphaeria*, and *Dinemasporium*) was positively correlated with the enrichment of the endophytic bacterial genus *Acidovorax*.

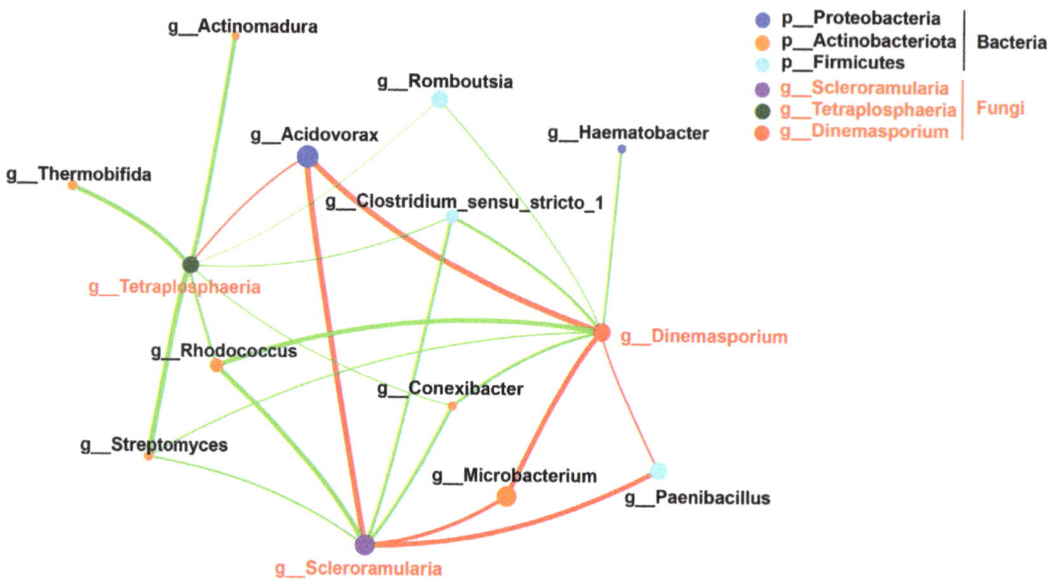

Figure 5. Correlation network analysis of significantly enriched bacterial and fungal genera. The Spearman coefficients that showed the significant enrichment of bacterial and fungal genera were also calculated to reflect the correlation between species, where the absolute value of the correlation coefficient ≥ 0.5, with $p < 0.05$. The sizes of the nodes in Figure 5 indicate the abundances of species, and different colors indicate different species; the color of the connecting lines indicates a positive and negative correlation, where red indicates a positive correlation, green indicates a negative correlation, and the thickness of the lines indicates the magnitude of the correlation coefficient. The thicker the line, the higher the correlation between species; the more lines, the closer the connection between the nodes (p—phylum; g—genus).

Based on the BugBase analysis, it was found that the endophytic bacterial phenotypes in sugarcane stems under FMC and CAC treatments were mainly classified into nine groups (Figure 6). Meanwhile, a Wilcoxon rank-sum test was also performed for the nine bacterial phenotype groups under FMC and CAC treatments ($p < 0.05$). The results showed that the abundances of these nine bacterial phenotypes were not significantly different for FMC and CAC treatments. However, the abundant percentages of Stress_Tolerant, Forms_ Biofilms, and Contains_ Mobile_ Elements in the bacterial community increased in the stems under the FMC treatment as compared with those under the CAC treatment. This result indicates that the stress resistance of sugarcane could be improved by FMC treatment.

PICRUSt2 and FUNGuild were carried out to predict bacterial and fungal functions, respectively. Meanwhile, a Wilcoxon rank-sum test was also performed to evaluate the functions of the bacterial and fungal communities under FMC and CAC treatments ($p < 0.05$). The results showed that the functions of endophytic bacteria (Figure 7a) and fungi (Figure 7b) in sugarcane under the FMC treatment were not significantly different from those under the CAC treatment.

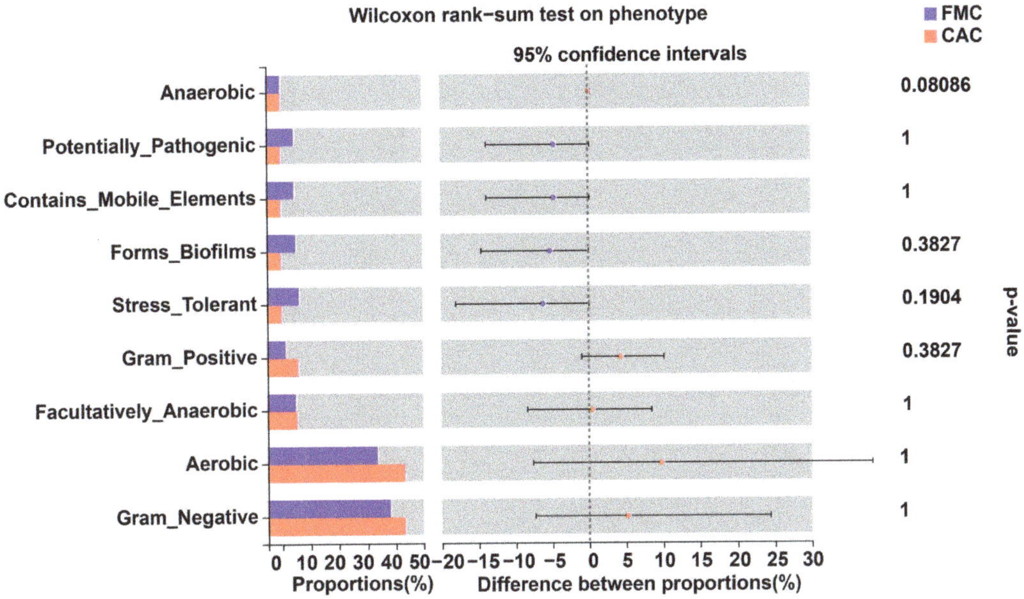

Figure 6. Endophytic bacterial community phenotypes from FMC and CAC treatments identified by BugBase-predicted analysis. FMC—fully mechanized cultivation; CAC—conventional artificial cultivation.

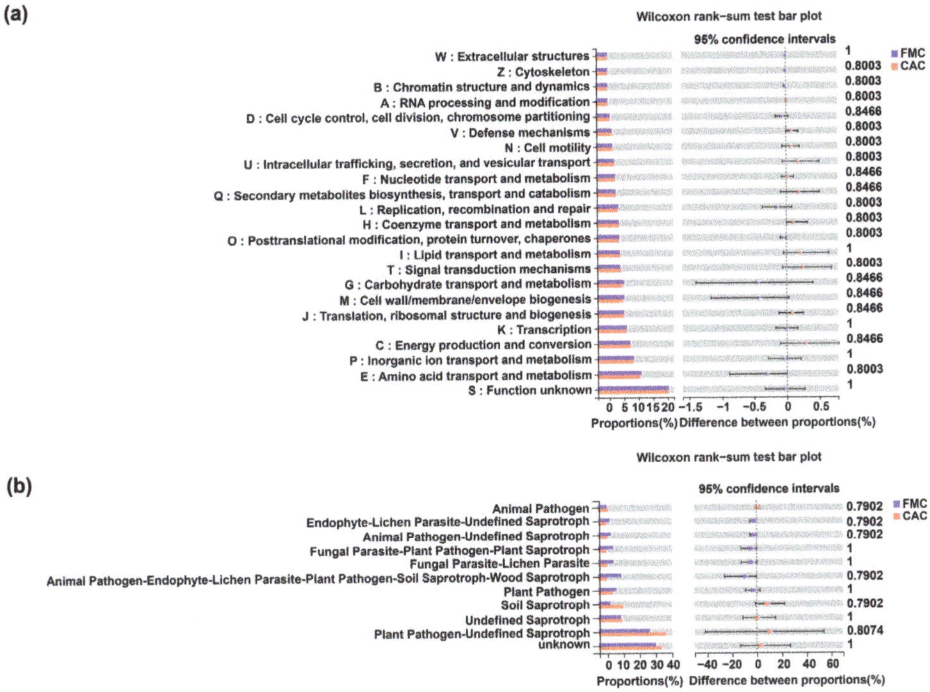

Figure 7. Functional predictions of the bacterial (**a**) and fungal (**b**) communities in stems of sugarcanes under FMC and CAC treatments. FMC—fully mechanized cultivation; CAC—conventional artificial cultivation.

4. Discussion

Hartman et al. [17] found that management type and tillage intensity were the main causes of bacteria and fungi in roots. Similarly, in comparison with the CAC treatment, we found that for the endophytic microbial compositions, the endophytic bacteria as well as the endophytic fungal compositions in sugarcane were significantly changed by the FMC treatment. For example, in comparison with the CAC treatment, the relative abundance of Proteobacteria in stems under the FMC treatment was increased. Endophytic Proteobacteria have been demonstrated to be a Plant Growth-Promoting Rhizobacteria (PGPR) [18].

At the genus level, *Pantoea*, *Klebsiella*, *Allorhizobium-Neorhizobium-Pararhizobium-Rhizobium*, *Acidovorax*, *Microbacterium*, and *Paenibacillus* were the special or significantly enriched dominant endophytic bacterial genera under the FMC treatment. Previous studies reported that *Pantoea* is a type of polysaccharide-producing, IAA-producing, iron carrier-producing, phosphate-solubilizing, and antagonistic-to-pathogenic fungi, and is part of the functional fungal genus. *Pantoea* also has the function of promoting plant growth and development [19,20]. Meanwhile, *Klebsiella*, as one of the nitrogen-fixing bacteria in sugarcane, can have a pro-growth effect [21]. Bigott et al. [22] also suggested that *Allorhizobium-Neorhizobium-Pararhizobium-Rhizobium* might be helpful in promoting plant growth and improving resistance against abiotic stress. Moreover, *Acidovorax*, known as one of the commensal species or plant-helpful bacteria, can produce secondary metabolites and hormones to promote plant development while simultaneously being antagonistic to plant pathogens [23,24]. Furthermore, *Microbacterium*, a producer of secondary metabolites that belongs to the Gram-positive group of bacteria and is not acid-fast, can also produce carotenoids with antioxidant and coloring properties [25]. *Paenibacillus*, characterized by nitrogen fixation, phosphate solubilization, phytohormone indole-3-acetic acid (IAA) synthesis, and siderophore release, can improve crop growth. The enrichment of the dominant endophytic bacterial genera under the FMC treatment not only can help to defend against insect herbivores such as nematodes but can also resist plant pathogens [26].

Additionally, the enrichment of endophytic fungi, such as *Scleroramularia*, *Tetraplosphaeria*, and *Dinemasporium*, was also found under the FMC treatment. *Sclerоramularia*, a new, potentially species-rich genus of epiphytic fungus, is often found on the fruit surfaces of several hosts. This implies that it may have many untapped niches to be investigated [27]. Meanwhile, *Dinemasporium* can produce bioactive metabolites with antibacterial, antifungal, and antialgal activities [28].

5. Conclusions

In comparison with the CAC treatment, even though the diversity and richness of endophytic microorganisms in sugarcane stems under the FMC treatment were not significantly different, the relative abundances of Proteobacteria and Ascomycota increased under the FMC treatment. Additionally, some dominant endophytic bacterial genera, such as *Acidovorax*, *Microbacterium*, and *Paenibacillus*, and some dominant endophytic fungal genera, such as *Scleroramularia*, *Tetraplosphaeria*, and *Dinemasporium*, which belong to beneficial microbes, were all significantly enriched under the FMC treatments. All of our results suggest that cane growth and health are not only positively impacted by the FMC treatment, but that this treatment could be considered a sustainable method for future sugarcane production.

Author Contributions: Conceptualization, S.Y. and H.T.; data curation, J.X.; formal analysis, S.Y.; funding acquisition, H.T.; investigation, T.L.; methodology, J.X. and T.L.; project administration, J.X. and T.L.; resources, T.L.; software, J.X.; visualization, J.X.; writing—original draft, J.X.; writing—review and editing, J.X. and S.Y. All authors have read and agreed to the published version of the manuscript.

Funding: This research was funded by the National Key R&D Program of China (2020YFD1000600), the FAO project (FAO/CPR/3804), and the Guangxi Major Science and Technology Project (GuikeAA22117006).

Institutional Review Board Statement: Not applicable.

Informed Consent Statement: Not applicable.

Data Availability Statement: Raw reads were deposited in the NCBI Sequence Read Archive (SRA) database (accession number: SRP371574).

Conflicts of Interest: The authors declare no conflict of interest.

References

1. Singh, R.K.; Singh, P.; Li, H.B.; Song, Q.Q.; Guo, D.J.; Solanki, M.K.; Verma, K.K.; Malviya, M.K.; Song, X.P.; Lakshmanan, P.; et al. Diversity of nitrogen-fixing rhizobacteria associated with sugarcane: A comprehensive study of plant-microbe interactions for growth enhancement in *Saccharum* spp. *BMC Plant Biol.* **2020**, *20*, 220. [CrossRef] [PubMed]
2. Pang, Z.; Dong, F.; Liu, Q.; Lin, W.; Hu, C.; Yuan, Z. Soil metagenomics reveals effects of continuous sugarcane cropping on the structure and functional pathway of rhizospheric microbial community. *Front. Microbiol.* **2021**, *12*, 369. [CrossRef]
3. Yang, S.D.; Xiao, J.; Liang, T.; Tan, H.W. Response of bacterial compositions to the use of slow-release fertilizers with long-acting agents and synergists. *Appl. Soil Ecol.* **2023**, *182*, 104699. [CrossRef]
4. Tayyab, M.; Yang, Z.; Zhang, C.; Islam, W.; Lin, W.; Zhang, H. Sugarcane monoculture drives microbial community composition, activity and abundance of agricultural-related microorganisms. *Environ. Sci. Pollut. Res.* **2021**, *28*, 48080–48096. [CrossRef] [PubMed]
5. Yang, S.D.; Xiao, J.; Liang, T.; He, W.Z.; Tan, H.W. Response of soil biological properties and bacterial diversity to different levels of nitrogen application in sugarcane fields. *AMB Express* **2021**, *11*, 172. [CrossRef] [PubMed]
6. Fatah, G.S.A.; Verona, S.L.; Nugraheni, S.D. The analysis of labor efficiency on sugarcane cultivation through mechanization application. *IOP Conf. Ser. Earth Environ. Sci.* **2022**, *974*, 012127. [CrossRef]
7. Tayyab, M.; Fallah, N.; Zhang, C.; Pang, Z.; Islam, W.; Lin, S.; Lin, W.; Zhang, H. Sugarcane cultivar-dependent changes in assemblage of soil rhizosphere fungal communities in subtropical ecosystem. *Environ. Sci. Pollut. Res.* **2022**, *29*, 20795–20807. [CrossRef]
8. Tayyab, M.; Islam, W.; Noman, A.; Pang, Z.; Li, S.; Lin, S.; Lin, W.; Zhang, H. Sugarcane cultivars manipulate rhizosphere bacterial communities' structure and composition of agriculturally important keystone taxa. *3 Biotech* **2022**, *12*, 32. [CrossRef]
9. Kumar, M.; Kumar, A.; Sahu, K.P.; Patel, A.; Reddy, B.; Sheoran, N.; Charishma, K.; Rajashekara, H.; Bhagat, S.; Rathour, R. Deciphering core-microbiome of rice leaf endosphere: Revelation by metagenomic and microbiological analysis of aromatic and non-aromatic genotypes grown in three geographical zones. *Microbiol. Res.* **2021**, *246*, 126704. [CrossRef]
10. Ashajyothi, M.; Kumar, A.; Sheoran, N.; Ganesan, P.; Gogoi, R.; Subbaiyan, G.K.; Bhattacharya, R. Black pepper (*Piper nigrum* L.) associated endophytic Pseudomonas putida BP25 alters root phenotype and induces defense in rice (*Oryza sativa* L.) against blast disease incited by *Magnaporthe oryzae. Biol. Control* **2020**, *143*, 104181. [CrossRef]
11. Teheran-Sierra, L.G.; Funnicelli, M.I.G.; de Carvalho, L.A.L.; Ferro, M.I.T.; Soares, M.A.; Pinheiro, D.G. Bacterial communities associated with sugarcane under different agricultural management exhibit a diversity of plant growth-promoting traits and evidence of synergistic effect. *Microbiol. Res.* **2021**, *247*, 126729. [CrossRef] [PubMed]
12. Vandana, U.K.; Rajkumari, J.; Singha, L.P.; Satish, L.; Alavilli, H.; Sudheer, P.D.V.N.; Chauhan, S.; Ratnala, R.; Satturu, V.; Mazumder, P.B.; et al. The Endophytic Microbiome as a Hotspot of Synergistic Interactions, with Prospects of Plant Growth Promotion. *Biology* **2021**, *10*, 101. [CrossRef] [PubMed]
13. Xiao, J.; Chen, S.Y.; Sun, Y.; Yang, S.D.; He, Y. Differences of rhizospheric and endophytic bacteria are recruited by different watermelon phenotypes relating to rind colors formation. *Sci. Rep.* **2022**, *12*, 6360. [CrossRef] [PubMed]
14. Ren, F.; Dong, W.; Yan, D.H. Organs, cultivars, soil, and fruit properties affect structure of endophytic mycobiota of Pinggu peach trees. *Microorganisms* **2019**, *7*, 322. [CrossRef] [PubMed]
15. Di, Y.N.; Kui, L.; Singh, P.; Liu, L.F.; Xie, L.Y.; He, L.L.; Li, F.S. Identification and characterization of bacillus subtilis B9: A diazotrophic plant growth-promoting endophytic bacterium isolated from sugarcane root. *J. Plant Growth Regul.* **2022**. [CrossRef]
16. Xiao, J.; Chen, S.; Sun, Y.; Wu, S.; Liang, W.; Yang, S. Effects of mechanical weeding on soil fertility and microbial community structure in star anise (*Illicium verum* Hook.f.) plantations. *PLoS ONE* **2022**, *17*, e0266949. [CrossRef] [PubMed]
17. Hartman, K.; van der Heijden, M.G.A.; Wittwer, R.A.; Banerjee, S.; Walser, J.C.; Schlaeppi, K. Cropping practices manipulate abundance patterns of root and soil microbiome members paving the way to smart farming. *Microbiome* **2018**, *6*, 14. [CrossRef]
18. Bruto, M.; Prigent-Combaret, C.; Muller, D.; Moënne-Loccoz, Y. Analysis of genes contributing to plant-beneficial functions in plant growth-promoting rhizobacteria and related Proteobacteria. *Sci. Rep.* **2014**, *4*, 6261. [CrossRef]
19. Ansari, F.A.; Ahmad, I. Isolation, functional characterization and efficacy of biofilm-forming rhizobacteria under abiotic stress conditions. *Antonie Van Leeuw.* **2019**, *112*, 1827–1839. [CrossRef]
20. Xiao, J.; Chen, S.Y.; Sun, Y.; Yang, S.D.; Tan, H.W. Effect of intercropping with different legume crops on endophytic bacterial diversity of sugarcanes. *Chin. J. Trop. Crops* **2021**, *42*, 3188–3198. (In Chinese) [CrossRef]
21. Mehnaz, S. Bacteria in Agrobiology: Crop Ecosystems. Plant Growth-Promoting Bacteria Associated with Sugarcane. In *Bacteria in Agrobiology: Crop Ecosystems*; Maheshwari, D.K., Ed.; Springer: Berlin/Heidelberg, Germany, 2011; pp. 165–187.
22. Bigott, Y.; Gallego, S.; Montemurro, N.; Breuil, M.C.; Pérez, S.; Michas, A.; Martin-Laurent, F.; Schröder, P. Fate and impact of wastewater-borne micropollutants in lettuce and the root- associated bacteria. *Sci. Total Environ.* **2022**, *831*, 154674. [CrossRef] [PubMed]

23. Siani, R.; Stabl, G.; Gutjahr, C.; Schloter, M.; Radl, V. *Acidovorax* pan-genome reveals specific functional traits for plant beneficial and pathogenic plant-associations. *Microb. Genom.* **2021**, *7*, 000666. [CrossRef] [PubMed]
24. Giordano, P.R.; Chaves, A.M.; Mitkowski, N.A.; Vargas, J.M. Identification, characterization, and distribution of *Acidovorax avenae* subsp. *avenae* associated with creeping bentgrass etiolation and decline. *Plant Dis.* **2012**, *96*, 1736–1742. [CrossRef] [PubMed]
25. Xie, F.; Niu, S.; Lin, X.; Pei, S.; Jiang, L.; Tian, Y.; Zhang, G. Description of *Microbacterium luteum* sp. nov.; *Microbacterium cremeum* sp. nov.; and *Microbacterium atlanticum* sp. nov.; three novel C50 carotenoid producing bacteria. *J. Microbiol.* **2021**, *59*, 886–897. [CrossRef] [PubMed]
26. Grady, E.N.; MacDonald, J.; Liu, L.; Richman, A.; Yuan, Z.C. Current knowledge and perspectives of *Paenibacillus*: A review. *Microb. Cell Fact.* **2016**, *15*, 203. [CrossRef]
27. Li, H.; Sun, G.; Batzer, J.C.; Crous, P.W.; Groenewald, J.Z.; Karakaya, A.; Gleason, M.L. *Scleroramularia* gen. nov. associated with sooty blotch and flyspeck of apple and pawpaw from the Northern Hemisphere. *Fungal Divers.* **2011**, *46*, 53–66. [CrossRef]
28. Krohn, K.; Sohrab, M.H.; van Ree, T.; Draeger, S.; Schulz, B.; Antus, S.; Kurtán, T. Dinemasones A, B and C—New bioactive metabolites from the endophytic gungus *Dinemasporium strigosum*. *Eur. J. Org. Chem.* **2008**, *2008*, 5638–5646. [CrossRef]

MDPI
St. Alban-Anlage 66
4052 Basel
Switzerland
www.mdpi.com

Agronomy Editorial Office
E-mail: agronomy@mdpi.com
www.mdpi.com/journal/agronomy

www.ingramcontent.com/pod-product-compliance
Lightning Source LLC
Chambersburg PA
CBHW061818210926
51850CB00006B/739